PRAISE FOR *THE PERFECT BET*

"An elegant and amusing account. . . . Anyone planning to enter a casino or place an online bet would be advised to keep this book handy."

—*Wall Street Journal*

"This book is full of magic. It's brimming with clever people and clever ideas. . . . Even if gambling is not your thing, you should give *The Perfect Bet* a chance. Its subject matter may seem a little racy to apply to your everyday stick-or-twist decisions, but there's wisdom here, too."

—*New Scientist*

"[*The Perfect Bet* is] terrific: beautifully written, solidly researched, and full of surprises. It's also practical. Even if you don't spend time [in] Las Vegas you'll be surprised by the clarity it will bring to your day."

—*New York Times* Numberplay blog

"With anecdotes of mathematicians, physicists, and programmers outwitting 'the house,' it's both a paean to human ingenuity and a Robin Hood tale of wealth redistribution."

—*Telegraph* (UK)

"Kucharski's book is an interesting account of the interactions between gambling, science, and mathematics, but what makes it so readable is the way the stories are told—always with a human angle that makes the narrative easier to follow and principles behind it easier to understand."

—*Prospect (UK)*

"Brilliant new book."

—CardsChat

"It's an interesting spin on the gambling industry, and might serve as yet another qualified reason to visit a casino this weekend."

—Casino.Org News

"Kucharski delivers a fascinating read." —*Publishers Weekly*

"A lucid yet sophisticated look at the mathematics of probability as it's played out on gaming tables, arenas, and fields. . . . [G]amblers and math buffs alike will enjoy it for its smart approach to real-world problems."

—*Kirkus Reviews*

"Kucharski does a remarkable job of telling the story of how gambling has influenced science and science has influenced gambling. He manages to make it a good read while providing a scholarly underpinning."

—J. Doyne Farmer, director of the complexity economics program, Institute for New Economic Thinking, Oxford Martin School

"With an entertaining writing style, Adam Kucharski guides us through the history and state of the art of the perfect bet, showing us how mathematics and computers are used to come up with optimal ways to gamble, play games, bluff, and invest our money. Extremely well-written and carefully researched. I highly recommend it."

—Arthur Benjamin, author of *The Magic of Math: Solving for X and Figuring Out Why*

"This book contains so many great stories of how smart people have used maths, statistics, and science to try and beat the odds—legally. It almost, but not quite, makes me want to take up gambling."

—David Spiegelhalter, Winton Professor for the Public Understanding of Risk, University of Cambridge

"In *The Perfect Bet*, Adam Kucharski takes us on a wild ride through the history, psychology, mathematics, and technology of gaming—a remarkable look behind the curtain of what most people think is intuitive, but isn't."

—Paul Offit, author of *Bad Faith* and *Autism's False Prophets*

THE PERFECT

How Science and Math Are Taking the Luck Out of Gambling

Adam Kucharski

BASIC BOOKS

NEW YORK

Basic Books
Hachette Book Group
1290 Avenue of the Americas, New York, NY 10104
www.basicbooks.com/

Printed in the United States of America

Originally published by Basic Books in 2016
First Trade Paperback Edition: August 2017

Published by Basic Books, an imprint of Perseus Books, LLC,
a subsidiary of Hachette Book Group, Inc.

Print book interior design by Linda Mark

The Library of Congress has catalogued the hardcover as follows:
Kucharski, Adam (Mathematician)
The perfect bet : how science and math are taking the luck out of
gambling / Adam Kucharski.
pages cm
Includes bibliographical references and index.
ISBN 978-0-465-05595-1 (hardcover)—ISBN 978-0-465-09859-0 (ebook)
1. Games of chance (Mathematics) 2. Gambling. 3. Gambling systems.
4. Probabilities. I. Title.
QA271.K83 2015
519.2'7—dc23
2015034255
10 9 8 7 6 5 4 3 2 1

ISBN: 978-1-5416-9723-2 (paperback)

LSC-C

10 9 8 7 6 5 4 3 2 1

For my parents

Luck is probability taken personally.

—Chip Denman

CONTENTS

INTRODUCTION

IN JUNE 2009, A BRITISH NEWSPAPER TOLD THE STORY OF ELLIOTT Short, a former financial trader who'd made over £20 million betting on horse races. He had a chauffeur-driven Mercedes, kept an office in the exclusive Knightsbridge district of London, and regularly ran up huge bar tabs in the city's best clubs. According to the article, Short's winning strategy was simple: always bet against the favorite. Because the highest-rated horse doesn't always win, it was possible to make a fortune using this approach. Thanks to his system, Short had made huge profits on some of Britain's best-known races, from £1.5 million at Cheltenham Festival to £3 million at Royal Ascot.

There was just one problem: the story wasn't entirely true. The profitable bets that Short claimed to have made at Cheltenham and Ascot had never been placed. Having persuaded investors to pour hundreds of thousands of pounds into his betting system, he'd spent much of the money on holidays and nights out. Eventually, his investors started asking questions, and Short was arrested. When the case went to trial in April 2013, Short was found guilty of nine counts of fraud and was sentenced to five years in prison.

It might seem surprising that so many people were taken in. But there is something seductive about the idea of a perfect betting system. Stories of successful gambling go against the notion that casinos and bookmakers are unbeatable. They imply that there are flaws in games of chance, and that these can be exploited by anyone sharp enough to spot them. Randomness can be reasoned with, and fortune controlled by formulae. The idea is so appealing that, for as long as many games have existed, people have tried to find ways to beat them. Yet the search for the perfect bet has not only influenced gamblers. Throughout history, wagers have transformed our entire understanding of luck.

WHEN THE FIRST ROULETTE wheels appeared in Parisian casinos in the eighteenth century, it did not take long for players to conjure up new betting systems. Most of the strategies came with attractive names, and atrocious success rates. One was called "the martingale." The system had evolved from a tactic used in bar games and was rumored to be foolproof. As its reputation spread, it became incredibly popular among local players.

The martingale involved placing bets on black or red. The color didn't matter; it was the stake that was important. Rather than betting the same amount each time, a player would double up after a loss. When players eventually picked the right color, they would therefore win back all the money lost on earlier bets plus a profit equal to their initial stake.

At first glance, the system seemed flawless. But it had one major drawback: sometimes the required bet size would increase far beyond what the gambler, or even casino, could afford. Following the martingale might earn a player a small profit initially, but in the long run solvency would always get in the way of strategy. Although the martingale might have been popular, it was a tactic that no one could afford to carry out successfully. "The martingale is as elusive as the soul," as writer Alexandre Dumas put it.

One of the reasons the strategy lured in so many players—and continues to do so—is that mathematically it appears perfect. Write down the amount you've bet and the amount you could win, and you'll always come out on top. The calculations have a flaw only when they meet reality. On paper, the martingale seems to work fine; in practical terms, it's hopeless.

When it comes to gambling, understanding the theory behind a game can make all the difference. But what if that theory hasn't been invented yet? During the Renaissance, Gerolamo Cardano was an avid gambler. Having frittered away his inheritance, he decided to make his fortune by betting. For Cardano, this meant measuring how likely random events were.

Probability as we know it did not exist in Cardano's era. There were no laws about chance events, no rules about how likely something was. If someone rolled two sixes while playing dice, it was simply good luck. For many games, nobody knew precisely what a "fair" wager should be.

Cardano was one of the first to spot that such games could be analyzed mathematically. He realized that navigating the world of chance meant understanding where its boundaries lay. He would therefore look at the collection of all possible outcomes, and then home in on the ones that were of interest. Although two dice could land in thirty-six different arrangements, there was only one way to get two sixes. He also worked out how to deal with multiple random events, deriving "Cardano's formula" to calculate the correct odds for repeated games.

Cardano's intellect was not his only weapon in card games. He also carried a long knife, known as a poniard, and was not opposed to using it. In 1525, he was playing cards in Venice and realized his opponent was cheating. "When I observed that the cards were marked, I impetuously slashed his face with my poniard," Cardano said, "though not deeply."

In the decades that followed, other researchers chipped away at the mysteries of probability, too. At the request of a group of Italian nobles, Galileo investigated why some combinations of dice faces appeared more often than others. Astronomer Johannes Kepler also took time off from studying planetary motion to write a short piece on the theory of dice and gambling.

The science of chance blossomed in 1654 as the result of a gambling question posed by a French writer named Antoine Gombaud. He had been puzzled by the following dice problem. Which is more likely: throwing a single six in four rolls of a single die, or throwing double sixes in twenty-four rolls of two dice? Gombaud believed the two events would occur equally often but could not prove it. He wrote to his mathematician friend Blaise Pascal, asking if this was indeed the case.

To tackle the dice problem, Pascal enlisted the help of Pierre de Fermat, a wealthy lawyer and fellow mathematician. Together, they built on Cardano's earlier work on randomness, gradually pinning down the basic laws of probability. Many of the new concepts would become central to mathematical theory. Among other things, Pascal and Fermat defined the "expected value" of a game, which measured how profitable it would be on average if played repeatedly. Their research showed that Gombaud had been wrong: he was more likely to get a six in four rolls of one die than double sixes in twenty-four rolls of two dice. Still, thanks to Gombaud's gambling puzzle, mathematics had gained an entirely new set of ideas. According to mathematician Richard Epstein, "Gamblers can rightly claim to be the godfathers of probability theory."

As well as helping researchers understand how much a bet is worth in purely mathematical terms, wagers have also revealed how we value decisions in real life. During the eighteenth century, Daniel Bernoulli wondered why people would often prefer low-risk bets to ones that were, in theory, more profitable. If expected profit was not driving their financial choices, what was?

Bernoulli solved the wager problem by thinking in terms of "expected utility" rather than expected payoff. He suggested that the same amount of money is worth more—or less—depending on how much a person already has. For example, a single coin is more valuable to a poor person than it is to a rich one. As fellow researcher Gabriel Cramer said, "The mathematicians estimate money in proportion to its quantity, and men of good sense in proportion to the usage that they may make of it."

Such insights have proved to be very powerful. Indeed, the concept of utility underpins the entire insurance industry. Most people prefer to make regular, predictable payments than to pay nothing and risk getting hit with a massive bill, even if it means paying more on average. Whether we buy an insurance policy or not depends on its utility. If something is relatively cheap to replace, we are less likely to insure it.

Over the following chapters, we will find out how gambling has continued to influence scientific thinking, from game theory and statistics to chaos theory and artificial intelligence. Perhaps it shouldn't be surprising that science and gambling are so intertwined. After all, wagers are windows into the world of chance. They show us how to balance risk against reward and why we value things differently as our circumstances change. They help us to unravel how we make decisions and what we can do to control the influence of luck. Encompassing mathematics, psychology, economics, and physics, gambling is a natural focus for researchers interested in random—or seemingly random—events.

The relationship between science and betting is not only benefiting researchers. Gamblers are increasingly using scientific ideas to develop successful betting strategies. In many cases, the concepts are traveling full circle: methods that originally emerged from academic curiosity about wagers are now feeding back into real-life attempts to beat the house.

THE FIRST TIME PHYSICIST Richard Feynman visited Las Vegas in the late 1940s, he went from game to game, working out how much he could expect to win (or, more likely, lose). He decided that although craps was a bad deal, it wasn't that bad: for every dollar he bet, he could expect to lose 1.4 cents on average. Of course, that was the expected loss over a large number of attempts. When Feynman tried the game, he was particularly unlucky, losing five dollars right away. It was enough to put him off casino gambling for good.

Nevertheless, Feynman made several trips to Vegas over the years. He was particularly fond of chatting with the showgirls. During one trip, he had lunch with a performer named Marilyn. As they were eating, she pointed out a man strolling across the grass. He was a well-known professional gambler named Nick Dandolos, or "Nick the Greek." Feynman found the notion puzzling. Having calculated the odds for each casino game, he couldn't work out how Nick the Greek could consistently make money.

Marilyn called Nick the Greek over to their table, and Feynman asked how it was possible to make a living gambling. "I only bet when the odds are in my favor," Nick replied. Feynman didn't understand what he meant. How could the odds ever be in someone's favor?

Nick the Greek told Feynman the real secret behind his success. "I don't bet on the table," he said. "Instead, I bet with people around the table who have prejudices—superstitious ideas about lucky numbers." Nick knew the casino had the edge, so he made wagers with naive fellow gamblers instead. Unlike the Parisian gamblers who used the martingale strategy, he understood the games, and understood the people playing them. He had looked beyond the obvious strategies—which would lose him money—and found a way to tip the odds in his favor. Working out the numbers hadn't been the tricky part; the real skill was turning that knowledge into an effective strategy.

Although brilliance is generally less common than bravado, stories of other successful gambling strategies have emerged over the years. There are tales of syndicates that have successfully exploited lottery loopholes and teams that have profited from flawed roulette tables. Then there are the students—often of the mathematical variety—who have made small fortunes by counting cards.

Yet in recent years these techniques have been surpassed by more sophisticated ideas. From the statisticians forecasting sports scores to the inventors of the intelligent algorithms that beat human poker players, people are finding new ways to take on casinos and bookmakers. But who are the people turning hard science into hard cash? And—perhaps more importantly—where did their strategies come from?

Coverage of winning exploits often focuses on who the gamblers were or how much they won. Scientific betting methods are presented as mathematical magic tricks. The critical ideas are left unreported; the theories remain buried. But we should be interested in how these tricks are done. Wagers have a long history of inspiring new areas of science and generating insights into luck and decision making. The methods have also permeated wider society, from technology to finance. If we can uncover the inner workings of modern betting strategies, we can find out how scientific approaches are continuing to challenge our notions of chance.

From the simple to the intricate, from the audacious to the absurd, gambling is a production line for surprising ideas. Around the globe, gamblers are dealing with the limits of predictability and the boundary between order and chaos. Some are examining the subtleties of decision making and competition; others are looking at quirks of human behavior and exploring the nature of intelligence. By dissecting successful betting strategies, we can find out how gambling is still influencing our understanding of luck—and how that luck can be tamed.

1

THE THREE DEGREES OF IGNORANCE

BENEATH LONDON'S RITZ HOTEL LIES A HIGH-STAKES CASINO. It's called the Ritz Club, and it prides itself on luxury. Croupiers dressed in black oversee the ornate tables. Renaissance paintings line the walls. Scattered lamps illuminate the gold-trimmed decor. Unfortunately for the casual gambler, the Ritz Club also prides itself on exclusivity. To bet inside, you need to have a membership or a hotel key. And, of course, a healthy bankroll.

One evening in March 2004, a blonde woman walked into the Ritz Club, chaperoned by two men in elegant suits. They were there to play roulette. The group weren't like the other high rollers; they turned down many of the free perks usually doled out to big-money players. Still, their focus paid off, and over the course of the night, they won £100,000. It wasn't exactly a small sum, but it was by no means unusual by Ritz standards. The following night the group returned to the casino and again perched beside a roulette table. This time their winnings were much larger. When they eventually cashed in their chips, they took away £1.2 million.

Casino staff became suspicious. After the gamblers left, security looked at the closed-circuit television footage. What they saw was enough to make them contact the police, and the trio were soon arrested at a hotel not far from the Ritz. The woman, who turned out to be from Hungary, and her accomplices, a pair of Serbians, were accused of obtaining money by deception. According to early media reports, they had used a laser scanner to analyze the roulette table. The measurements were fed into a tiny hidden computer, which converted them into predictions about where the ball would finally land. With a cocktail of gadgetry and glamour, it certainly made for a good story. But a crucial detail was missing from all the accounts. Nobody had explained precisely how it was possible to record the motion of a roulette ball and convert it into a successful prediction. After all, isn't roulette supposed to be random?

THERE ARE TWO WAYS to deal with randomness in roulette, and Henri Poincaré was interested in both of them. It was one of his many interests: in the early twentieth century, pretty much anything that involved mathematics had at some point benefited from Poincaré's attention. He was the last true "Universalist"; no mathematician since has been able to skip through every part of the field, spotting crucial connections along the way, like he did.

As Poincaré saw it, events like roulette appear random because we are ignorant of what causes them. He suggested we could classify problems according to our level of ignorance. If we know an object's exact initial state—such as its position and speed—and what physical laws it follows, we have a textbook physics problem to solve. Poincaré called this the first degree of ignorance: we have all the necessary information; we just need to do a few simple calculations.

The second degree of ignorance is when we know the physical laws but don't know the exact initial state of the object, or cannot measure it accurately. In this case we must either improve our measurements or limit our predictions to what will happen to the object in the very near future. Finally, we have the third, and most extensive, degree of ignorance. This is when we don't know the initial state of the object or the physical laws. We can also fall into the third level of ignorance if the laws are too intricate to fully unravel. For example, suppose we drop a can of paint into a swimming pool. It might be easy to predict the reaction of the swimmers, but predicting the behavior of the individual paint and water molecules will be far more difficult.

We could take another approach, however. We could try to understand the effect of the molecules bouncing into each other without studying the minutiae of the interactions between them. If we look at all the particles together, we will be able to see them mix together until—after a certain period of time—the paint spreads evenly throughout pool. Without knowing anything about the cause, which is too complex to grasp, we can still comment on the eventual effect.

The same can be said for roulette. The trajectory of the ball depends on a number of factors, which we might not be able to grasp simply by glancing at a spinning roulette wheel. Much like for the individual water molecules, we cannot make predictions about a single spin if we do not understand the complex causes behind the ball's trajectory. But, as Poincaré suggested, we don't necessarily have to know what causes the ball to land where it does. Instead, we can simply watch a large number of spins and see what happens.

That is exactly what Albert Hibbs and Roy Walford did in 1947. Hibbs was studying for a math degree at the time, and his friend Walford was a medical student. Taking time off from their studies at the University of Chicago, the pair went to Reno to see whether roulette tables were really as random as casinos thought.

Most roulette tables have kept with the original French design of thirty-eight pockets, with numbers 1 to 36, alternately colored black and red, plus 0 and 00, colored green. The zeros tip the game in the casinos' favor. If we placed a series of one-dollar bets on our favorite number, we could expect to win on average once in every thirty-eight attempts, in which case the casino would pay thirty-six dollars. Over the course of thirty-eight spins, we would therefore put down thirty-eight dollars but would only make thirty-six dollars on average. That translates into a loss of two dollars, or about five cents per spin, over the thirty-eight spins.

The house edge relies on there being an equal chance of the roulette wheel producing each number. But, like any machine, a roulette table can have imperfections or can gradually wear down with use. Hibbs and Walford were on the hunt for such tables, which might not have produced an even distribution of numbers. If one number came up more often than the others, it could work to their advantage. They watched spin after spin, hoping to spot something odd. Which raises the question: What do we actually mean by "odd"?

WHILE POINCARÉ WAS IN France thinking about the origins of randomness, on the other side of the English Channel Karl Pearson was spending his summer holiday flipping coins. By the time the vacation was over, the mathematician had flipped a shilling twenty-five thousand times, diligently recording the results of each throw. Most of the work was done outside, which Pearson said "gave me, I have little doubt, a bad reputation in the neighbourhood where I was staying." As well as experimenting with shillings, Pearson got a colleague to flip a penny more than eight thousand times and repeatedly pull raffle tickets from a bag.

To understand randomness, Pearson believed it was important to collect as much data as possible. As he put it, we have "no ab-

solute knowledge of natural phenomena," just "k
sensations." And Pearson didn't stop at coin tosses
In search of more data, he turned his attention to t
of Monte Carlo.

Like Poincaré, Pearson was something of a polymath. In addition to his interest in chance, he wrote plays and poetry and studied physics and philosophy. English by birth, Pearson had traveled widely. He was particularly keen on German culture: when University of Heidelberg admin staff accidently recorded his name as Karl instead of Carl, he kept the new spelling.

Unfortunately, his planned trip to Monte Carlo did not look promising. He knew it would be near impossible to obtain funding for a "research visit" to the casinos of the French Riviera. But perhaps he didn't need to watch the tables. It turned out that the newspaper *Le Monaco* published a record of roulette outcomes every week. Pearson decided to focus on results from a four-week period during the summer of 1892. First he looked at the proportions of red and black outcomes. If a roulette wheel were spun an infinite number of times—and the zeros were ignored—he would have expected the overall ratio of red to black to approach 50/50.

Out of the sixteen thousand or so spins published by *Le Monaco*, 50.15 percent came up red. To work out whether the difference was down to chance, Pearson calculated the amount the observed spins deviated from 50 percent. Then he compared this with the variation that would be expected if the wheels were random. He found that a 0.15 percent difference wasn't particularly unusual, and it certainly didn't give him a reason to doubt the randomness of the wheels.

Red and black might have come up a similar number of times, but Pearson wanted to test other things, too. Next, he looked at how often the same color came up several times in a row. Gamblers can become obsessed with such runs of luck. Take the night of August

18, 1913, when a roulette ball in one of Monte Carlo's casinos landed on black over a dozen times in a row. Gamblers crowded around the table to see what would happen next. Surely another black couldn't appear? As the table spun, people piled their money onto red. The ball landed on black again. More money went on red. Another black appeared. And another. And another. In total, the ball bounced into a black pocket twenty-six times in a row. If the wheel had been random, each spin would have been completely unrelated to the others. A sequence of blacks wouldn't have made a red more likely. Yet the gamblers that evening believed that it would. This psychological bias has since been known as the "Monte Carlo fallacy."

When Pearson compared the length of runs of different colors with the frequencies that he'd expect if the wheels were random, something looked wrong. Runs of two or three of the same color were scarcer than they should have been. And runs of a single color—say, a black sandwiched between two reds—were far too common. Pearson calculated the probability of observing an outcome at least as extreme as this one, assuming that the roulette wheel was truly random. This probability, which he dubbed the *p* value, was tiny. So small, in fact, that Pearson said that even if he'd been watching the Monte Carlo tables since the start of Earth's history, he would not have expected to see a result that extreme. He believed it was conclusive evidence that roulette was not a game of chance.

The discovery infuriated him. He'd hoped that roulette wheels would be a good source of random data and was angry that his giant casino-shaped laboratory was generating unreliable results. "The man of science may proudly predict the results of tossing halfpence," he said, "but the Monte Carlo roulette confounds his theories and mocks at his laws." With the roulette wheels clearly of little use to his research, Pearson suggested that the casinos be closed down and their assets donated to science. However, it later emerged that

Pearson's odd results weren't really due to faulty wheels. Although *Le Monaco* paid reporters to watch the roulette tables and record the outcomes, the reporters had decided it was easier just to make up the numbers.

Unlike the idle journalists, Hibbs and Walford actually watched the roulette wheels when they visited Reno. They discovered that one in four wheels had a bias of some sort. One wheel was especially skewed, so betting on it caused the pair's initial one-hundred-dollar stake to grow rapidly. Reports of their final profits differ, but whatever they made, it was enough to buy a yacht and sail it around the Caribbean for a year.

There are plenty of stories about gamblers who've succeeded using a similar approach. Many have told the tale of the Victorian engineer Joseph Jagger, who made a fortune exploiting a biased wheel in Monte Carlo, and of the Argentine syndicate that cleaned up in government-owned casinos in the early 1950s. We might think that, thanks to Pearson's test, spotting a vulnerable wheel is fairly straightforward. But finding a biased roulette wheel isn't the same as finding a profitable one.

In 1948, a statistician named Allan Wilson recorded the spins of a roulette wheel for twenty-four hours a day over four weeks. When he used Pearson's test to find out whether each number had the same chance of appearing, it was clear the wheel was biased. Yet it wasn't clear how he should bet. When Wilson published his data, he issued a challenge to his gambling-inclined readers. "On what statistical basis," he asked, "should you decide to play a given roulette number?"

It took thirty-five years for a solution to emerge. Mathematician Stewart Ethier eventually realized that the trick wasn't to test for a nonrandom wheel but to test for one that would be favorable when betting. Even if we were to look at a huge number of spins and find substantial evidence that one of the thirty-eight numbers came up more often than others, it might not be enough to make a profit. The number would have to appear on average at least once every thirty-six spins; otherwise, we would still expect to lose out to the casino.

The most common number in Wilson's roulette data was nineteen, but Ethier's test found no evidence that betting on it would be profitable over time. Although it was clear the wheel wasn't random, there didn't seem to be any favorable numbers. Ethier was aware that his method had probably arrived too late for most gamblers: in the years since Hibbs and Walford had won big in Reno, biased wheels had gradually faded into extinction. But roulette did not remain unbeatable for long.

WHEN WE ARE AT our deepest level of ignorance, with causes that are too complex to understand, the only thing we can do is look at a large number of events together and see whether any patterns emerge. As we've seen, this statistical approach can be successful if a roulette wheel is biased. Without knowing anything about the physics of a roulette spin, we can make predictions about what might come up.

But what if there's no bias or insufficient time to collect lots of data? The trio that won at the Ritz didn't watch loads of spins, hoping to identify a biased table. They looked at the trajectory of the roulette ball as it traveled around the wheel. This meant escaping not just Poincaré's third level of ignorance but his second one as well.

This is no small feat. Even if we pick apart the physical processes that cause a roulette ball to follow the path it does, we cannot necessarily predict where it will land. Unlike paint molecules crashing into water, the causes are not too complex to grasp. Instead, the cause can be too small to spot: a tiny difference in the initial speed of the ball makes a big difference to where it finally settles. Poincaré argued that a difference in the starting state of a roulette ball—one so tiny it escapes our attention—can lead to an effect so large we cannot miss it, and then we say that the effect is down to chance.

The problem, which is known as "sensitive dependence on initial conditions," means that even if we collect detailed measurements about a process—whether a roulette spin or a tropical storm—a small oversight could have dramatic consequences. Seventy years before mathematician Edward Lorenz gave a talk asking "Does the flap of a butterfly's wings in Brazil set off a tornado in Texas?" Poincaré had outlined the "butterfly effect."

Lorenz's work, which grew into chaos theory, focused chiefly on prediction. He was motivated by a desire to make better forecasts about the weather and to find a way to see further into the future. Poincaré was interested in the opposite problem: How long does it take for a process to become random? In fact, does the path of a roulette ball ever become truly random?

Poincaré was inspired by roulette, but he made his breakthrough by studying a much grander set of trajectories. During the nineteenth century, astronomers had sketched out the asteroids that lay scattered along the Zodiac. They'd found that these asteroids were pretty much uniformly distributed across the night sky. And Poincaré wanted to work out why this was the case.

He knew that the asteroids must follow Kepler's laws of motion and that it was impossible to know their initial speed. As Poincaré put it, "The Zodiac may be regarded as an immense roulette board on which the Creator has thrown a very great number of small balls." To understand the pattern of the asteroids, Poincaré therefore decided to compare the total distance a hypothetical object travels with the number of times it rotates around a point.

Imagine you unroll an incredibly long, and incredibly smooth, sheet of wallpaper. Laying the sheet flat, you take a marble and set it rolling along the paper. Then you set another going, followed by several more. Some marbles you set rolling quickly, others slowly. Because the wallpaper is smooth, the quick ones soon roll far into the distance, while the slow ones make their way along the sheet much more gradually.

The marbles roll on and on, and after a while you take a snapshot of their current positions. To mark their locations, you make a little cut in the edge of the paper next to each one. Then you remove the marbles and roll the sheet back up. If you look at the edge of the roll, each cut will be equally likely to appear at any position around the circumference. This happens because the length of the sheet—and hence the distance the marbles can travel—is much longer than the diameter of the roll. A small change in the marbles' overall distance has a big effect on where the cuts appear on the circumference. If you wait long enough, this sensitivity to initial conditions will mean that the locations of the cuts will appear random. Poincaré showed the same thing happens with asteroid orbits. Over time, they will end up evenly spread along the Zodiac.

To Poincaré, the Zodiac and the roulette table were merely two illustrations of the same idea. He suggested that after a large number of turns, a roulette ball's finishing position would also be completely random. He pointed out that certain betting options would tumble into the realm of randomness sooner than others. Because roulette slots are alternately colored red and black, predicting which of the two appears meant calculating exactly where the ball will land. This would become extremely difficult after even a spin or two. Other options, such as predicting which half of the table the ball lands in, were less sensitive to initial conditions. It would therefore take a lot of spins before the result becomes as good as random.

Fortunately for gamblers, a roulette ball does not spin for an extremely long period of time (although there is an oft-repeated myth that mathematician Blaise Pascal invented roulette while trying to build a perpetual motion machine). As a result, gamblers can—in theory—avoid falling into Poincaré's second degree of ignorance by measuring the initial path of the roulette ball. They just need to work out what measurements to take.

THE RITZ WASN'T THE first time a story of roulette-tracking technology emerged. Eight years after Hibbs and Walford had exploited that biased wheel in Reno, Edward Thorp sat in a common room at the University of California, Los Angeles, discussing get-rich-quick schemes with his fellow students. It was a glorious Sunday afternoon, and the group was debating how to beat roulette. When one of the others said that casino wheels were generally flawless, something clicked in Thorp's mind. Thorp had just started a PhD in physics, and it occurred to him that beating a robust, well-maintained wheel wasn't really a question of statistics. It was a physics problem. As Thorp put it, "The orbiting roulette ball suddenly seemed like a planet in its stately, precise and predictable path."

In 1955, Thorp got hold of a half-size roulette table and set to work analyzing the spins with a camera and stopwatch. He soon noticed that his particular wheel had so many flaws that it made prediction hopeless. But he persevered and studied the physics of the problem in any way he could. On one occasion, Thorp failed to come to the door when his in-laws arrived for dinner. They eventually found him inside rolling marbles along the kitchen floor in the midst of an experiment to find out how far each would travel.

After completing his PhD, Thorp headed east to work at the Massachusetts Institute of Technology. There he met Claude Shannon, one of the university's academic giants. Over the previous decade, Shannon had pioneered the field of "information theory," which revolutionized how data are stored and communicated; the work would later help pave the way for space missions, mobile phones, and the Internet.

Thorp told Shannon about the roulette predictions, and the professor suggested they continue the work at his house a few miles outside the city. When Thorp entered Shannon's basement, it became clear quite how much Shannon liked gadgets. The room was an inventor's playground. Shannon must have had a $100,000 worth of motors, pulleys, switches, and gears down there. He even had a pair of huge polystyrene "shoes" that allowed him to take strolls on the water of a nearby lake, much to his neighbors' alarm. Before long, Thorp and Shannon had added a $1,500 industry-standard roulette table to the gadget collection.

MOST ROULETTE WHEELS ARE operated in a way that allows gamblers to collect information on the ball's trajectory before they bet. After setting the center of the roulette wheel spinning counterclockwise, the croupier launches the ball in a clockwise direction, sending it circling around the wheel's upper edge. Once the ball has looped around a few times, the croupier calls "no more bets" or—if casinos like their patter to have a hint of Gallic charm—"*rien ne va plus*." Eventually, the ball hits one of the deflectors scattered around the edge of the wheel and drops into a pocket. Unfortunately for gamblers, the ball's trajectory is what mathematicians call "nonlinear": the input (its speed) is not directly proportional to the output (where it lands). In other words, Thorp and Shannon had ended up back in Poincaré's third level of ignorance.

Rather than trying to dig themselves out by deriving equations for the ball's motion, they instead decided to rely on past observations. They ran experiments to see how long a ball traveling at a certain speed would remain on the track and used this information to make predictions. During a spin, they would time how long it took for a ball to travel once around the table and then compared the time to their previous results to estimate when it would hit a deflector.

The calculations needed to be done at the roulette table, so at the end of 1960, Thorp and Shannon built the world's first wearable computer and took it to Vegas. They tested it only once, as the wires were unreliable, needing frequent repairs. Even so, it seemed like the computer could be a successful tool. Because the system handed gamblers an advantage, Shannon thought casinos might abandon roulette once word of the research got out. Secrecy was therefore of the utmost importance. As Thorp recalled, "He mentioned that social network theorists studying the spread of rumors claimed that two people chosen at random in, say, the United States are usually linked by three or fewer acquaintances, or 'three degrees of separation.'" The idea of "six degrees of separation" would eventually creep into popular culture, thanks to a highly publicized 1967 experiment by sociologist Stanley Milgram. In the study, participants were asked to help a letter get to a target recipient by sending it to whichever of their acquaintances they thought were most likely to know the target. On average, the letter passed through the hands of six people before eventually reaching its destination, and the six degrees phenomenon was born. Yet subsequent research has shown that Shannon's suggestion of three degrees of separation was probably closer to the mark. In 2012, researchers analyzing Facebook connections—which are a fairly good proxy for real-life acquaintances—found that there are an average of 3.74 degrees of separation between any two people. Evidently, Shannon's fears were well founded.

TOWARD THE END OF 1977, the New York Academy of Sciences hosted the first major conference on chaos theory. They invited a diverse mix of researchers, including James Yorke, the mathematician who first coined the term "chaotic" to describe ordered yet unpredictable phenomena like roulette and weather, and Robert May, an ecologist studying population dynamics at Princeton University.

Another attendee was a young physicist from the University of California, Santa Cruz. For his PhD, Robert Shaw was studying the motion of running water. But that wasn't the only project he was working on. Along with some fellow students, he'd also been developing a way to take on the casinos of Nevada. They called themselves the "Eudaemons"—a nod to the ancient Greek philosophical notion of happiness—and the group's attempts to beat the house at roulette have since become part of gambling legend.

The project started in late 1975 when Doyne Farmer and Norman Packard, two graduate students at UC Santa Cruz, bought a refurbished roulette wheel. The pair had spent the previous summer toying with betting systems for a variety of games before eventually settling on roulette. Despite Shannon's warnings, Thorp had made a cryptic reference to roulette being beatable in one of his books; this throwaway comment, tucked away toward the end of the text, was enough to persuade Farmer and Packard that roulette was worth further study. Working at night in the university physics lab, they gradually unraveled the physics of a roulette spin. By taking measurements as the ball circled the wheel, they discovered they would be able to glean enough information to make profitable bets.

One of the Eudaemons, Thomas Bass, later documented the group's exploits in his book *The Eudaemonic Pie*. He described how, after honing their calculations, the group hid a computer inside a shoe and used it to predict the ball's path in a number of casinos. But there was one piece of information Bass didn't include: the equations underpinning the Eudaemons' prediction method.

MOST MATHEMATICIANS WITH AN interest in gambling will have heard the story of the Eudaemons. Some will also have wondered whether such prediction is feasible. When a new paper on roulette appeared in the journal *Chaos* in 2012, however, it revealed that someone had finally put the method to the test.

Michael Small had first come across *The Eudaemonic Pie* while working for a South African investment bank. He wasn't a gambler and didn't like casinos. Still, he was curious about the shoe computer. For his PhD, he'd analyzed systems with nonlinear dynamics, a category that roulette fell very nicely into. Ten years passed, and Small moved to Asia to take a job at Hong Kong Polytechnic University. Along with Chi Kong Tse, a fellow researcher in the engineering department, Small decided that building a roulette computer could be a good project for undergraduates.

It might seem strange that it took so long for researchers to publicly test such a well-known roulette strategy. However, it isn't easy to get access to a roulette wheel. Casino games aren't generally on university procurement lists, so there are limited opportunities to study roulette. Pearson relied on dodgy newspaper reports because he couldn't persuade anyone to fund a trip to Monte Carlo, and without Shannon's patronage, Thorp would have struggled to carry out his roulette experiments.

The mathematical nuts and bolts of roulette have also hindered research into the problem. Not because the math behind roulette is too complex but because it's too simple. Journal editors can be picky about the types of scientific papers they publish, and trying to beat roulette with basic physics isn't a topic they usually go for. There has been the occasional article about roulette, such as the paper Thorp published that described his method. But though Thorp gave enough away to persuade readers—including the Eudaemons—that computer-based prediction could be successful, he omitted the details. The crucial calculations were notably absent.

Once Small and Tse had convinced the university to buy a wheel, they got to work trying to reproduce the Eudaemons' prediction method. They started by dividing the trajectory of the ball

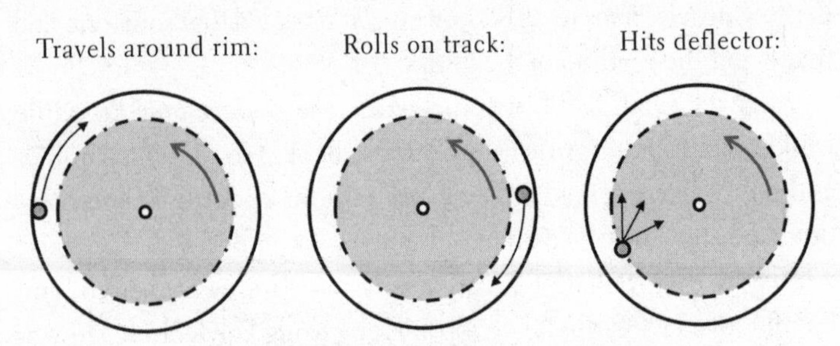

FIGURE 1.1. The three stages of a roulette spin.

into three separate phases. When a croupier sets a roulette wheel in motion, the ball initially rotates around the upper rim while the center of the wheel spins in the opposite direction. During this time, two competing forces act on the ball: centripetal force keeping it on the rim, and gravity pulling it down toward the center of the wheel.

The pair assumed that as the ball rolls, friction slows it down. Eventually, the ball's angular momentum decreases so much that gravity becomes the dominant force. At this point, the ball moves into its second phase. It leaves the rim and rolls freely on the track between the rim and the deflectors. It moves closer to the center of the wheel until it hits one of the deflectors scattered around the circumference.

Until this point, the ball's trajectory can be calculated using textbook physics. But once it hits a deflector, it scatters, potentially landing in one of several pockets. From a betting point of view, the ball leaves a cozy predictable world and moves into a phase that is truly chaotic.

Small and Tse could have used a statistical approach to deal with this uncertainty. However, for the sake of simplicity, they decided to define their prediction as the number the ball was next to when it hit a deflector. To predict the point at which the ball would clip one of the deflectors, Small and Tse needed six pieces of information:

the position, velocity, and acceleration of the ball, and the same for the wheel. Fortunately, these six measurements could be reduced to three if they considered the trajectories from a different standpoint. To an onlooker watching a roulette table, the ball appears to move in one direction and the wheel in the other. But it is also possible to do the calculations from a "ball's-eye view," in which case it's only necessary to measure how the ball moves relative to the wheel. Small and Tse did this by using a stopwatch to clock the times at which the ball passed a specific point.

One afternoon, Small ran an initial series of experiments to test the method. Having written a computer program on his laptop to do the calculations, he set the ball spinning, taking the necessary measurements by hand, as the Eudaemons would have done. As the ball traveled around the rim a dozen or so times, he gathered enough information to make predictions about where it would land. He only had time to run the experiment twenty-two times before he had to leave the office. Out of these attempts, he predicted the correct number three times. Had he just been making random guesses, the probability he would have got at least this many right (the p value) was less than 2 percent. This persuaded him that the Eudaemons' strategy worked. It seemed that roulette really could be beaten with physics.

Having tested the method by hand, Small and Tse set up a high-speed camera to collect more precise measurements about the ball's position. The camera took photos of the wheel at a rate of about ninety frames per second. This made it possible to explore what happened after the ball hit a deflector. With the help of two engineering students, Small and Tse spun the wheel seven hundred times, recording the difference between their prediction and the final outcome. Collecting this information together, they calculated the probability of the ball landing a specified distance away from the predicted pocket. For most of the pockets, this probability wasn't particularly large or small; it was pretty much what they'd have expected if picking pockets at random. Some patterns did emerge, however. The ball landed in the predicted pocket far more often than it would have if the process were down to chance. Moreover, it rarely landed on the numbers that lay on the wheel directly before the predicted pocket. This made sense because the ball would have to bounce backward to get to these pockets.

The camera showed what happened in the ideal situation—when there was very good information about the trajectory of the ball—but most gamblers would struggle to sneak a high-speed camera into a casino. Instead, they would have had to rely on measurements taken by hand. Small and Tse found this wasn't such a disadvantage: they suggested that predictions made with a stopwatch could still provide gamblers with an expected profit of 18 percent.

After announcing his results, Small received messages from gamblers who were using the method in real casinos. "One guy sent me detailed descriptions of his work," he said, "including fabulous photos of a 'clicker' device made from a modified computer mouse strapped to his toe." The work also came to the attention of Doyne Farmer. He was sailing in Florida when heard about Small and Tse's paper. Farmer had kept his method under wraps for over thirty years because—much like Small—he disliked casinos. The trips he made to Nevada during his time with the Eudaemons were enough to convince him that gambling addicts were being exploited by the industry. If people wanted to use computers to beat roulette, he didn't want to say anything that would hand the advantage back to the casinos. However, when Small and Tse's paper was published, Farmer decided it was time to finally break his silence. Especially because there was an important difference between the Eudaemons' approach and the one the Hong Kong researchers had suggested.

Small and Tse had assumed that friction was the main force slowing the ball down, but Farmer disagreed. He'd found that air resistance—not friction—was the main reason for the ball slowing down. Indeed, Farmer pointed out that if we placed a roulette table in a room with no air (and hence no air resistance), the ball would spin around the table thousands of times before settling on a number.

Like Small and Tse's approach, Farmer's method required that certain values be estimated while at the roulette table. During their casino trips, the Eudaemons had three things to pin down: the amount of air resistance, the velocity of the ball when it dropped off the rim of the wheel, and the rate at which the wheel was decelerating. One of the biggest challenges was estimating air resistance and drop velocity. Both influenced the prediction in a similar way: assuming a smaller resistance was much like having an increased velocity.

It was also important to know what was happening around the roulette ball. External factors can have a big effect on a physical process. Take a game of billiards. If you have a perfectly smooth table, a shot will cause the balls to ricochet in a cobweb of collisions. To predict where the cue ball will go after a few seconds, you'd need to know precisely how it was struck. But if you want to make longer-term predictions, Farmer and his colleagues have pointed out it's not enough to merely know about the shot. You also need to take into account forces such as gravity—and not just that of the earth. To predict exactly where the cue ball will travel after one minute, you have to include the gravitational pull of particles at the edge of the galaxy in your calculations.

When making roulette predictions, obtaining correct information about the state of the table is crucial. Even a change in the weather can affect results. The Eudaemons found that if they calibrated their calculations when the weather was sunny in Santa Cruz, the arrival of fog would cause the ball to leave the track half a rotation earlier than they had expected. Other disruptions were closer to home. During one casino visit, Farmer had to abandon betting because an overweight man was resting against the table, tilting the wheel and messing up the predictions.

The biggest hindrance for the group, though, was their technical equipment. They implemented the betting strategy by having one person record the spins and another place the bets, so as not to raise the suspicions of casino security. The idea was that a wireless signal would transmit messages telling the player with the chips which number to bet on. But the system often failed: the signal would disappear, taking the betting instructions with it. Although the group had a 20 percent edge over the casino in theory, these technical problems meant it was never converted into a grand fortune.

As computers have improved, a handful of people have managed to come up with better roulette devices. Most rarely make it into the news, with the exception of the trio who won at the Ritz in 2004. On that occasion, newspapers were particularly quick to latch on to the story of a laser scanner. Yet when journalist Ben Beasley-Murray talked to industry insiders a few months after the incident, they dismissed suggestions that lasers were involved. Instead, it was likely the Ritz gamblers used mobile phones to time the spinning wheel. The basic method would have been similar to the one the Eudaemons used, but advances in technology meant it could be implemented much more effectively. According to ex-Eudaemon Norman Packard, the whole thing would have been pretty easy to set up.

It was also perfectly legal. Although the Ritz group were accused of obtaining money by deception—a form of theft—they hadn't actually tampered with the game. Nobody had interfered with the ball or switched chips. Nine months after the group's initial arrest, police therefore closed the case and returned the £1.3 million haul. In many ways, the trio had the UK's wonderfully archaic gambling laws to thank for their prize. The Gaming Act, which was signed in 1845, had not been updated to cope with the new methods available to gamblers.

Unfortunately, the law does not hand an advantage only to gamblers. The unwritten agreement you have with a casino—pick the correct number and be rewarded with money—is not legally binding in the UK. You can't take a casino to court if you win and it doesn't pay up. And although casinos love gamblers with a losing system, they are less keen on those with winning strategies. Regardless of which strategy you use, you'll have to escape house countermeasures. When Hibbs and Walford passed $5,000 in winnings by hunting for biased tables in Reno, the casino shuffled the roulette tables around to foil them. Even though the Eudaemons didn't need to watch the table for long periods of time, they still had to beat a hasty retreat from casinos on occasion.

AS WELL AS DRAWING the attention of casino security, successful roulette strategies have something else in common: all rely on the fact that casinos believe the wheels are unpredictable. When they aren't, people who have watched the table for long enough can exploit the bias. When the wheel is perfect, and churns out numbers that are uniformly distributed, it can be vulnerable if gamblers collect enough information about the ball's trajectory.

The evolution of successful roulette strategies reflects how the science of chance has developed during the past century. Early efforts to beat roulette involved escaping Poincaré's third level of ignorance, where nothing about the physical process is known. Pearson's work on roulette was purely statistical, aiming to find patterns in data. Later attempts to profit from the game, including the exploits at the Ritz, took a different approach. These strategies tried to overcome Poincaré's second level of ignorance and solve the problem of roulette's outcome being incredibly sensitive to the initial state of the wheel and ball.

For Poincaré, roulette was a way to illustrate his idea that simple physical processes could descend into what seems like randomness. This idea formed a crucial part of chaos theory, which emerged as a new academic field in the 1970s. During this period, roulette was always lurking in the background. In fact, many of the Eudaemons would go on to publish papers on chaotic systems. One of Robert Shaw's projects demonstrated that the steady rhythm of droplets from a dripping tap turns into an unpredictable beat as the tap is unscrewed further. This was one of the first real-life examples of a "chaotic transition" whereby a process switches from a regular pattern to one that is as good as random. Interest in chaos theory and roulette does not appear to have dampened over the years. The topics can still capture the public imagination, as shown by the extensive media attention given to Small and Tse's paper in 2012.

Roulette might be a seductive intellectual challenge, but it isn't the easiest—or most reliable—way to make money. To start with, there is the problem of casino table limits. The Eudaemons played for small stakes, which helped them keep a low profile but also put a cap on potential winnings. Playing at high-stakes tables might bring in more money, but it will also bring additional scrutiny from casino security. Then there are the legal issues. Roulette computers are banned in many countries, and even if they aren't, casinos are understandably hostile toward anyone who uses one. This makes it tricky to earn good profits.

For these reasons, roulette is really only a small part of the scientific betting story. Since the shoe-computer exploits of the Eudaemons, gamblers have been busy tackling other games. Like roulette, many of these games have a long-standing reputation for being unbeatable. And like roulette, people are using scientific approaches to show just how wrong that reputation can be.

2

A BRUTE FORCE BUSINESS

OF THE COLLEGES OF THE UNIVERSITY OF CAMBRIDGE, GONVILLE and Caius is the fourth oldest, the third richest, and the second biggest producer of Nobel Prize winners. It's also one of the few colleges that serves three-course formal dinners every night, which means that most students end up well acquainted with the college's neo-Gothic dining hall and its unique stained glass windows.

One window depicts a spiraling DNA helix, a nod to former college fellow Francis Crick. Another shows a trio of overlapping circles in tribute to John Venn. There is also a checkerboard situated in the glass, each square colored in a seemingly random way. It's there to commemorate one of the founders of modern statistics, Ronald Fisher.

After winning a scholarship at Gonville and Caius, Fisher spent three years studying at Cambridge, specializing in evolutionary biology. He graduated on the eve of the First World War and tried to join the British Army. Although he completed the medical exams several times, he failed on each occasion because of poor eyesight. As a result, he spent the war teaching mathematics at a number of prominent English private schools, publishing a handful of academic papers in his spare time.

As the conflict drew to a close, Fisher began to search for a new job. One option was to join Karl Pearson's laboratory, where he had been offered the role of chief statistician. Fisher wasn't particularly keen on this option: the previous year, Pearson had published an article criticizing some of his research. Still reeling from the attack, Fisher declined the job.

Instead, Fisher took a job at the Rothamsted Experimental Station, where he turned his attention to agricultural research. Rather than just being interested in the results of experiments, Fisher wanted to make sure that experiments were designed to be as useful as possible. "To consult the statistician after an experiment is finished is often merely to ask him to conduct a post mortem examination," he said. "He can perhaps say what the experiment died of."

Considering the work at hand, Fisher was puzzled about how to scatter different crop treatments across a plot of land during an experiment. The same problem appears when conducting medical trials across a large geographic area. If we are comparing several different treatments, we want to make sure they are scattered across a wide region. But if we distribute them by picking locations at random, there is a chance that we will repeatedly pick similar locations. In which case, a treatment ends up concentrated only in one area, and we have a pretty lousy experiment.

Suppose we want to test four treatments across sixteen trial sites, arranged in a four-by-four grid. How can we scatter the treatments across the area without risking all of them ending up in the same place? In his landmark book *The Design of Experiments*, Fisher suggested that

C	D	B	A
B	A	D	C
D	C	A	B
A	B	C	D

Latin square

A	A	B	D
A	B	D	B
A	C	D	C
D	C	C	B

Lousy square

FIGURE 2.1.

the four treatments be distributed so that they appear in each row and column only once. If the field had good soil at one end and poor land at the other, all treatments therefore would be exposed to both conditions. As it happened, the pattern Fisher proposed had already found popularity elsewhere. It was common in classical architecture, where it was known as a Latin square, as shown in Figure 2.1.

The stained glass window at Gonville and Caius College shows a larger version of a Latin square, with the letters—one for each type of treatment—replaced by colors. As well as earning a tribute in an ancient hall, Fisher's ideas are still used today. The problem of how to construct something that is both random and balanced arises in many industries, including agriculture and medicine. It also comes up in lottery games.

Lotteries are designed to cost players money. They originated as a palatable form of tax, often to support major building projects. The Great Wall of China was financed with profits from a lottery run by the Han dynasty; proceeds from a lottery organized in 1753 funded the British Museum; and many of the Ivy League universities were built on takings from lotteries arranged by colonial governments.

Modern lotteries are made up of several different games, with scratchcards a lucrative part of the business. In the United Kingdom, they make up a quarter of the National Lottery's revenues, and American state lotteries earn tens of billions of dollars from ticket sales. Prizes run into the millions, so lottery operators are careful to limit the supply of winning cards. They can't put random numbers below the scratch-off foil, because there is a chance that could produce more prizes than they could afford to pay out. Nor would it be wise to send batches of cards to places arbitrarily, because one town could end up with all the "lucky" tickets. Scratchcards need to include an element of chance to make sure the game is fair, but operators also need to tweak the game somehow to ensure that there aren't a huge number of winners or too many in one place. To quote statistician William Gossett, they need "controlled randomness."

FOR MOHAN SRIVASTAVA, THE idea that scratchcards follow certain rules started with a joke present. It was June 2003, and he'd been given a handful of cards, including one with a collection of tic-tac-toe games. When he scratched off the foil, he discovered three symbols in a line, which netted him three dollars. It also got him thinking about how the lottery keeps track of the different prizes.

Srivastava worked as a statistician in Toronto, and he suspected that each card contained a code that identified whether it was a winner. Code breaking was something he had always found interesting; he'd known Bill Tutte, the British mathematician who had broken the Nazi Lorenz cipher in 1942, an achievement later described as "one of the greatest intellectual feats of World War II." On the way to collect his prize from a local gas station, Srivastava started to wonder how the lottery might go about distributing the tic-tac-toe scratchcards. He had plenty of experience with such algorithms. He worked as a consultant for mining companies, which hired him to hunt down gold deposits. In high school, he'd even written a computer version of tic-tac-toe for an assignment. He noticed that each foil panel on the scratchcard had a three-by-three grid of numbers printed on it. Perhaps these numbers were the key?

Later that day, Srivastava stopped by the gas station again and bought a bundle of scratchcards. Examining the numbers, he found that some numbers appeared several times on the card, and some only once. As he sifted through a pile of cards, he spotted the fact that if a row contained three of these unique numbers, it usually signaled a winning card. It was a simple and effective method. The challenge was finding such cards.

Unfortunately, it turns out that winning cards aren't all that common. For example, during the early hours of April 16, 2013, a car smashed through the doors of a convenience store in Kentucky. A woman jumped out, grabbed a display containing 1,500 scratchcards, and drove away. By the time she was arrested a few weeks later, she'd managed to claim a mere $200 in prizes.

Even though Srivastava had a reliable—and legal—method for finding profitable scratchcards, it didn't mean he could turn it into a lucrative business. He worked out how long it would take to sort through all potential cards to find the "lucky" ones and realized that he was better off sticking with his existing job. Having decided it wouldn't be worth changing careers, Srivastava thought the lottery might like to know about his discovery. First he tried to get ahold of them by phone, but, perhaps thinking he was just another gambler with a dodgy system, they didn't return his calls. So, he divided twenty untouched scratchcards into two groups—one of winners, and one of losers—and mailed them to the lottery's security team by courier. Srivastava got a phone call from the lottery later that day. "We need to talk," they said.

The tic-tac-toe games were soon removed from stores. According to the lottery, the problem was due to a design flaw. But since

2003, Srivastava has looked at other lotteries in the United States and Canada and suspects some may still be producing scratchcards with the same problem.

In 2011, a few months after *Wired* magazine featured Srivastava's story, reports emerged of an unusually successful scratchcard player in Texas. Joan Ginther had won four jackpots in the Texas scratchcard lottery between 1993 and 2010, bringing in a total of $20.4 million. Was it just down to luck? Although Ginther has never commented on the reason for her multiple wins, some have speculated that her statistics PhD might have had something to do with it.

It's not just scratchcards that are vulnerable to scientific thinking. Traditional lotteries do not include controlled randomness, yet they are still not safe from mathematically inclined players. And when lotteries have a loophole, a winning strategy can begin with something as innocuous as a college project.

EVEN WITHIN A UNIVERSITY as famously offbeat as the Massachusetts Institute of Technology, Random Hall has a reputation for being a little quirky. According to campus legend, the students who first lived there in 1968 wanted to call the dorm "Random House" until the publisher of that name sent them a letter to object. The individual floors have names, too. One is called Destiny, a result of its cash-strapped inhabitants selling the naming rights on eBay; the winning bid was $36 from a man who wanted to name it after his daughter. The hall even has its own student-built website, which allows occupants to check whether the bathrooms or washing machines are available.

In 2005, another plan started to take shape in the corridors of Random Hall. James Harvey was nearing the end of his mathematics degree and needed a project for his final semester. While searching for a topic, he became interested in lotteries.

The Massachusetts State Lottery was set up in 1971 as a way of raising extra revenue for the government. The lottery runs several different games, but the most popular are Powerball and Mega-Millions. Harvey decided that a comparison of the two games could make for a good project. However, the project grew—as projects often do—and Harvey soon began to compare his results with other games, including one called Cash WinFall.

The Massachusetts Lottery introduced Cash WinFall in autumn 2004. Unlike games such as Powerball, which were played in other states too, Cash WinFall was unique to Massachusetts. The rules were simple. Players would choose six numbers for each two-dollar ticket. If they matched all six in the draw, they won a jackpot of at least half a million dollars. If they matched some but not all the numbers, they won a smaller sum. The lottery designed the game so that $1.20 of every $2.00 would be paid out in prizes, with the rest being spent on local good causes. In many ways, WinFall was like all the other lottery games. However, it had one important difference. Usually, when nobody wins the jackpot in a lottery, the prize rolls over to the next draw. If there's no winning ticket next time, it rolls over again and continues to do so until somebody eventually matches all the numbers. The problem with rollovers is that winners—who are good publicity for a lottery—can be rare. And if no smiling faces and giant checks appear in the newspapers for a while, people will stop playing.

Massachusetts Lottery faced precisely that difficulty in 2003, when its Mass Millions game went without a winner for an entire year. They decided that WinFall would avoid this awkward situation by limiting the jackpot. If the prize money rose to $2 million without a winner, the jackpot would "roll down" and instead be split among the players who had matched three, four, or five numbers.

Before each draw, the lottery published its estimate for the jackpot, which was based on ticket sales from previous draws. When the estimated jackpot reached $2 million, players knew that the money would roll down if nobody matched six numbers. People soon spotted that the odds of winning money were far better in a roll-down week than at other times, which meant ticket sales always surged before these draws.

As he studied the game, Harvey realized that it was easier to make money on WinFall than on other lotteries. In fact, the expected payoff was sometimes positive: when a roll down happened, there was at least $2.30 waiting in prize money for every $2.00 ticket sold.

In February 2005, Harvey formed a betting group with some of his fellow MIT students. About fifty people chipped in for the first batch of tickets—raising $1,000 in total—and tripled their money when their numbers came up. Over the next few years, playing the lottery became a full-time job for Harvey. By 2010, he and a fellow team member incorporated the business. They named it Random Strategies Investments, LLC, after their old MIT accommodation.

Other syndicates got in on the action, too. One team consisted of biomedical researchers from Boston University. Another group was led by retired shop owner (and mathematics graduate) Gerald Selbee, who had previously had success with a similar game elsewhere. In 2003, Selbee had noticed a loophole in a Michigan lottery game that also included roll downs. Gathering a thirty-two-person-strong betting group, Selbee spent two years bulk-buying tickets—and netting jackpots—before the lottery was discontinued in 2005. When Selbee's syndicate heard about WinFall, they turned their attention to Massachusetts. There was a good reason for the influx of such betting teams. Cash WinFall had become the most profitable lottery in the United States.

DURING THE SUMMER OF 2010, the WinFall jackpot again approached the roll-down limit. After a $1.59 million prize went unclaimed on August 12, the lottery estimated that the jackpot for the next draw would be around $1.68 million. With a roll down surely only two or three draws away, betting syndicates started to prepare. By the end of the month, they planned to have thousands more dollars in winnings.

But the roll down didn't arrive two draws, or even three draws, later. It came the following week, on August 16. For some reason, there had been a huge increase in ticket sales, enough to drive the total prize money past $2 million. This flood of sales triggered a premature roll down. The lottery officials were as surprised as anyone: they had never sold that many tickets when the estimated jackpot was so low. What was going on?

When WinFall was introduced, the lottery had looked into the possibility of somebody deliberately nudging the draw into a roll down by buying up a large number of tickets. Aware that ticket sales depended on the estimated jackpot—and potential roll downs—the lottery didn't want to get caught out by underestimating the prize money.

They calculated that a player who used stores' automated lottery machines, which churned out tickets with arbitrary numbers, would be able to place one hundred bets per minute. If the jackpot stood at less than $1.7 million, the player would need to buy over five hundred thousand tickets to push it above the $2 million limit. Because this would take well over eighty hours, the lottery didn't think anyone would be able to tip the total over $2 million unless the jackpot was already above $1.7 million.

The MIT group thought otherwise. When James Harvey first started looking at the lottery in 2005, he'd made a trip to the town of Braintree, where the lottery offices were based. He wanted to get ahold of a copy of the guidelines for the game, which would outline precisely how the prize money was distributed. At the time, nobody could help him. But in 2008, the lottery finally sent him the guidelines. The information was a boost for the MIT group, which until then had been relying on their own calculations.

Looking at past draws, the group found that if the jackpot failed to top $1.6 million, the estimate for the next prize was almost always below the crucial value of $2 million. Pushing the draw over the limit on August 16 had been the result of extensive planning. As well as waiting for an appropriate jackpot size—one close to but below $1.6 million—the MIT group had to fill out around 700,000 betting slips, all by hand. "It took us about a year to ramp up to it," Harvey later said. The effort paid off: it's been estimated that they made around $700,000 that week.

Unfortunately, the profits did not continue for much longer. Within a year, the *Boston Globe* had published a story about the loophole in WinFall and the betting syndicates that had profited from it. In the summer of 2011, Gregory Sullivan, Massachusetts Inspector General, compiled a detailed report on the matter. Sullivan pointed out that the actions of the MIT group and others were entirely legal, and he concluded that "no one's odds of having a winning ticket were affected by high-volume betting." Still, it was clear that some people were making a lot of money from WinFall, and the game was gradually phased out.

Even if WinFall hadn't been canceled, the Boston University syndicate told the inspector general that the game wouldn't have remained profitable for betting teams. More people were buying tickets in roll-down weeks, so the prizes were split into smaller and smaller chunks. As the risk of losing money increased, the potential rewards were shrinking. In such a competitive environment, it was crucial to obtain an edge over other teams. The MIT group did this by understanding the game better than many of their competitors: they knew the probabilities and the payoffs and exactly how much advantage they held.

Betting success is not just limited by competition, however. There is also the not-so-small matter of logistics. Gerard Selbee pointed out that if a group wanted to maximize their profits during a roll-down week, they needed to buy 312,000 betting slips, because this was the "statistical sweet spot." The process of buying so many tickets was not always straightforward. The ticket machines would jam in humid weather and run slowly when low on ink. On one occasion, a power outage got in the way of the MIT group's preparations. And some stores would refuse to serve teams altogether.

There was also the question of how to store and organize all the tickets they bought. Syndicates had to keep millions of losing tickets in boxes to show to tax auditors. Moreover, it was a headache to find the winning slips. Selbee claims to have won around $8 million since he starting tackling lotteries in 2003. But after a draw, he and his wife would have to work for ten hours a day examining their collection of tickets to identify the profitable ones.

SYNDICATES HAVE LONG USED the tactic of buying up large combinations of numbers—a method known as a "brute force attack"—to beat lotteries. One of the best-known examples is the story of Stefan Klincewicz, an accountant who hatched a plan to win the Irish National Lottery in 1990. Klincewicz had noticed that it would cost him just under £1 million to buy enough tickets to cover every potential combination, thereby guaranteeing a winning ticket when the draw was made. But the strategy would only work if the jackpot was big enough. While waiting for a large rollover to appear, Klincewicz gathered a twenty-eight-man syndicate. Over a period of six months, the group filled out thousands upon thousands of lottery tickets. When a rollover of £1.7 million was eventually announced for the bank holiday draw in May 1992, they put their plan into action. Picking lottery terminals in quieter locations, the team started placing the necessary bets.

The surge in activity caught the attention of lottery officials, who tried to stop the syndicate by shutting off the terminals they were using. As a result, the group members were able to buy up only 80 percent of the possible number combinations. It wasn't enough to guarantee a win, but it was enough to put luck on their side; when the draw was announced, the syndicate had the winning numbers in their collection. Unfortunately, there were also two other winners, so the group had to share the jackpot. They still ended up with a profit of £310,000, however.

Simple brute force approaches like these do not require many calculations to pull off. The only real obstacle is buying enough tickets. It's more a question of manpower than mathematics, and this reduces the exclusivity of the methods. Whereas roulette players have only to outwit the casino, lottery syndicates often have to compete with other teams attempting to win the same jackpot.

Despite the ongoing competition, some betting syndicates have managed to repeatedly—and legally—turn a profit. Their stories illustrate another difference from roulette betting. Rather than acting alone or in small teams below the official radar, many lottery syndicates have formed companies. They have investors, and they file tax returns. The contrast reflects a wider shift in the world of scientific gambling. What were once individual efforts have grown into an entire industry.

3

FROM LOS ALAMOS TO MONTE CARLO

BILL BENTER IS ONE OF THE WORLD'S MOST SUCCESSFUL GAMBLERS. Based in Hong Kong, his betting syndicate has bet—and won—millions of dollars on horse races over the years. But Benter's gambling career didn't begin with racing. It didn't even begin with sports.

When he was a student, Benter came across a sign in an Atlantic City casino. "Professional card counters are prohibited from playing at our tables." It wasn't a particularly effective deterrent. After reading the sign, only one thought came to mind: card counting works. It was the late 1970s, and casinos had spent the previous decade or so clamping down on a tactic they saw as cheating. Much of the blame—or perhaps credit—for the casinos' losses goes to Edward Thorp. In 1962, Thorp published *Beat the Dealer*, which described a winning strategy for blackjack.

Although Thorp has been called the father of card counting, the idea for a perfect blackjack strategy was actually born in a military barracks. Ten years before Thorp released his book, Private Roger Baldwin had been playing cards with fellow soldiers at the Aberdeen Proving Ground in Maryland. When one of the men suggested a game of blackjack, conversation turned to the rules of the game. They agreed on the basic format. Each player receives two cards, with the dealer getting one faceup and one facedown. Players then choose whether to hit, taking another card in the hope of getting a total bigger than the dealer's, or stand, sticking with their current number. If taking a card sends a player's total over twenty-one, they go "bust" and lose their stake.

Once all players have made their choice, it's the dealer's turn. One of the soldiers pointed out that in Las Vegas, the dealer must stand with a total of seventeen or higher. Baldwin was amazed. The dealer had to follow fixed rules? Whenever he'd played in private games, the dealer was free to do whatever he wanted. Baldwin, who had a master's degree in mathematics, realized this could help him in a casino. If the dealer was subject to strict constraints, it should be possible to find a strategy that would maximize his chances of success.

Like all casino games, blackjack is designed to give the house an edge. Although the dealer and player both appear to have the same aim—drawing cards to get a total near to twenty-one—the dealer has the advantage because the player always goes first. If the player asks for one card too many, and overshoots the target, the dealer wins without doing anything.

Looking at some example blackjack hands, Baldwin noticed his odds improved if he took the value of the dealer's faceup card into account when making decisions. If the dealer had a low card, there was a good chance the dealer would have to draw several cards, increasing the risk of the total going over twenty-one. With a six, for example, the dealer had a 40 percent chance of going bust. With a ten, that probability was halved. Baldwin could therefore get away with standing on a lower total if the dealer had a six, because it was likely that the rules would force the dealer to draw too many cards.

In theory, it would be simple for Baldwin to build these ideas into a perfect strategy. In practice, however, the vast number of potential blackjack hands made the task near impossible to do with pen and paper. To make things worse, a player's choices in a casino weren't just limited to hitting or standing. Players also had the option of doubling their stake, on condition that they would receive one more card to go with the two they already had. Or, if they had received a pair of cards showing the same number, they could "split" these into two separate hands.

Baldwin wouldn't be able to do all the work by hand, so he asked Wilbert Cantey, a sergeant and fellow math graduate, if he could use the base's calculator. Intrigued by Baldwin's idea, Cantey agreed to help, as did James McDermott and Herbert Maisel, two other soldiers who worked in the analytics division.

While Thorp was working on his roulette predictions in Los Angeles, the four men spent their evenings working out the best way to beat the dealer. After several months of calculations, they arrived at what they thought was the optimal strategy. But their perfect system didn't turn out to be, well, perfect. "In statistical terms, we still had a negative expectation," Maisel later said. "Unless you got lucky, you'd still lose in the long run." Even so, by the group's calculations they had managed to reduce the casino's edge to a mere 0.6 percent. In contrast, a player who simply copied the dealer's rules—always standing on seventeen or higher—could expect to lose 6 percent of the time. The four men published their findings in 1956, in a paper titled 'The Optimum Strategy in Blackjack."

It happened that Thorp had already booked a trip to Las Vegas when the paper came out. It was meant to be a relaxing holiday with his wife: a few days of dinner tables rather than blackjack tables. But just before they left, a UCLA professor told Thorp about the soldiers' research. Ever curious, Thorp wrote down the strategy and took it along on his trip.

When Thorp tried the strategy in a casino one evening, slowly reading from a crib sheet while he sat at the table, his fellow gamblers thought he was crazy. Thorp was drawing cards when he should stick and turning down cards when he should take them. He doubled his bet after receiving weak cards. He even chose to split his paltry pair of eights when the dealer had a much stronger hand. What on earth was he thinking?

Despite Thorp's apparently reckless strategy, he didn't run out of chips. One by one, the other players left the table with empty pockets, but Thorp remained. Eventually, having lost eight of his ten dollars, Thorp called it a night. But the little excursion had convinced him that the soldiers' strategy worked better than any other. It also made him wonder how it could be improved.

To simplify calculations, Baldwin had assumed that cards were dealt randomly, with each of the fifty-two cards in the deck having an equal chance of appearing. But blackjack isn't really so random. Unlike roulette, in which each spin is—or at least should be—independent of the last, blackjack has a form of memory: over time, the dealer gradually works through the deck.

Thorp was convinced that if he could record which cards had previously been dealt, it would help him anticipate what might come up next. And because he already had a strategy that in theory pretty much broke even, having information about whether the next card would be high or low was enough to tip the game in his favor. He soon found that even a tactic as simple as keeping track of the number of tens in the deck could turn a profit. By counting cards, Thorp gradually turned the research of the four Aberdeen soldiers—later dubbed the "Four Horsemen of Aberdeen"—into a winning strategy.

Although Thorp made money from blackjack, it wasn't the main reason he made all those trips to Vegas. He saw it more as an academic obligation. When he'd first mentioned the existence of a winning strategy, the response wasn't exactly positive. People ridiculed the idea, just as gamblers had done during his first attempt. After all, Thorp's research challenged the widely held assumption that blackjack couldn't be defeated. *Beat the Dealer* was Thorp's way of proving that his theory was right.

THAT SIGN IN ATLANTIC CITY always stuck with Bill Benter, so when he heard about Thorp's book during a year studying abroad at Bristol University, he headed to the local library to get a copy. He had never seen anything so remarkable. "It showed that nothing was invulnerable," he said. "Old maxims about the house always having the edge were no longer true." Upon his return to the United States, Benter decided to take some time away from study. Switching his university campus in Cleveland, Ohio, for the casinos of Las Vegas, he set to work putting Thorp's system into action. The decision was to prove extremely lucrative: in his early twenties, Benter was making about $80,000 a year from blackjack.

During this time, Benter met an Australian who was also making a tidy sum from card counting. Whereas Benter had gone straight from lecture theaters to casino floors, Alan Woods had started training as an actuary after leaving college. In 1973, his firm was commissioned by the Australian government to calculate the house edge on games in the country's first legal casino. The project introduced Woods to the idea of profitable blackjack systems, and over the next few years he spent his weekends beating casinos around the globe. By the time he met Benter, Woods was a full-time blackjack player. But things were becoming harder for successful gamblers like them.

In the years since Thorp had published his strategy, casinos had become better at spotting card counters. One of the biggest problems with counting—aside from the mental focus required—is that you have to see a lot of cards before you have enough information to make predictions about the rest of the deck. During this time, you have little choice but to use Baldwin's optimal system and bet small amounts to limit your losses. And when you eventually decide that the upcoming cards are likely to be favorable, you need to dramatically increase your stakes to make the most of your advantage. This gives a clear signal to any casino staff on the hunt for card counters. "It's easy to learn how to count cards," as one blackjack professional put it. "It's hard to learn how to get away with it."

Keeping a mental note of card values isn't illegal in Nevada (or anywhere else, for that matter), but that didn't mean Thorp and his strategy were welcome in Las Vegas. Because casinos are private property, they can ban anyone they please. To evade security, Thorp started to wear disguises on his visits. With casinos on the lookout for big changes in betting patterns, gamblers started to search for a better way to play blackjack. Rather than count cards until things looked promising, what if it were possible to predict the order of the entire deck?

MOST MATHEMATICIANS IN THE early twentieth century had read Poincaré's work on probability, but it seemed that hardly anyone truly understood it. One of a few who did was Émile Borel, another mathematician based at the University of Paris. Borel was particularly interested in an analogy Poincaré had used to describe how random interactions—like paint in water—eventually settled down to equilibrium.

Poincaré had compared the situation to the process of card shuffling. If you know the initial order of a deck of cards, randomly swapping a few cards around won't completely mess up the order. Your knowledge about the original deck will therefore still be useful. As more and more shuffles are made, however, this knowledge becomes less and less relevant. Much like paint and water mixing over time, the cards gradually become uniformly distributed, with each card being equally likely to appear at any point in the deck.

Inspired by Poincaré's work, Borel found a way to calculate how quickly the cards would converge to this uniform distribution. His research is still used today when calculating the "mixing time" of a random process, whether card shuffles or chemical interactions. The work also helped blackjack players tackle a growing problem.

To make life difficult for card counters, casinos had started using multiple decks—sometimes combining as many as six—and shuffling the cards before all of them had been dealt. Because this made it harder to keep count, casinos hoped it would help stamp out any player advantage. They didn't realize that the changes also made it harder to shuffle the cards effectively.

During the 1970s, casinos often used a "dovetail shuffle" to mix up the cards. To perform the shuffle, the pack is split in two, and then the two the halves are riffled together. If the cards are perfectly riffled, with cards from each half alternating as they fall, no information is lost: the original order can be recovered by simply looking at every other card. Even if cards fall randomly from each half, however, some information remains.

FIGURE 3.1. A dovetail shuffle. (Credit: Todd Klassy)

Suppose you have a deck of thirteen cards. If you perform a dovetail shuffle, the cards might end up swapping around as follows:

A 2 3 4 5 6 7 8 9 10 J Q K

⇓

A 2 3 4 5 6 7   8 9 10 J Q K

⇓

A 2 **8** 3 **9 10** 4 5 **J** 6 **Q K** 7

The shuffled deck is far from random. Instead, there are two clear sequences of rising numbers (shown in boldface and plain text above). Actually, several card tricks rely on this fact: if a card is placed into an ordered deck and the deck is shuffled once or twice, the extra card will usually stand out because it won't fit into a rising sequence.

For a fifty-two-card deck, mathematicians have shown that a dealer should shuffle the cards at least half a dozen times so as not to leave any detectable patterns. However, Benter found that casinos would rarely bother to be so diligent. Some dealers would shuffle the deck two or three times; others seemed to think that just once was enough.

In the early 1980s, players began to use hidden computers to keep track of the deck. They would enter information by pressing a switch, and the computer would vibrate when a favorable situation cropped up. Keeping track of the shuffles meant it didn't matter whether casinos used several decks. It also helped players avoid giving a clear signal to casino security. If the computer indicated that good cards were likely to come up on the next hand, players didn't have to increase their bets substantially to profit. Unfortunately for gamblers, the advantage no longer exists: computer-aided betting has been illegal in American casinos since 1986.

Even without the clampdown on technology, there was another problem for players like Woods and Benter. Much like Thorp, they had gradually found themselves banned from most casinos around the globe. "Once you become well known," Benter said, "it's a very small world." With casinos refusing to let them play, the pair eventually decided to abandon blackjack. Rather than leave the industry, however, they instead planned to take on a much grander game.

WEDNESDAY NIGHTS AT THE Happy Valley racecourse are busy. Seriously busy. Tucked behind the skyscrapers of Hong Kong Island on a patch of what used to be swampland, over thirty thousand spectators crowd into its stands. Cheers rise above the sound of engines and car horns from the nearby Wan Chai district. The jostling and the noise are signs of how much money is at stake. Gambling is a big part of life at Happy Valley: an average of $145 million was bet during each race day in 2012. To put that in perspective, in that same year the Kentucky Derby set a new American record for betting on a horse race. The total was $133 million.

Happy Valley is managed by the Hong Kong Jockey Club, which also runs the Saturday races at Sha Tin racecourse across the bay in Kowloon. The Jockey Club is a nonprofit organization and has a reputation for running a good operation: gamblers are confident that the races are honest.

Betting in Hong Kong operates on a so-called pari-mutuel system. Rather than gamblers placing a bet with a bookmaker at fixed odds, gamblers' money goes into a pool, with the odds depending on how much has already been wagered on each horse. As an example, suppose there are two horses racing. A total of $200 has been bet on the first, and $300 on the second. Adding these together gives the total betting pool. The race organizers begin by subtracting a fee: in Hong Kong it's 19 percent, which, if the total was $500, would leave

TABLE 3.1. An example tote board.

	Amount bet	Odds
Horse 1	$200	2.03
Horse 2	$300	1.35

$405 in the pot. Then they calculate each horse's odds—the amount you'd get back if you wagered $1 on it—by taking the total available winnings ($405) and dividing it by the amount bet on that horse, as shown in Table 3.1.

Invented by Parisian businessman Joseph Oller, who also founded the Moulin Rouge cabaret club, pari-mutuel betting requires constant calculations and recalculations to produce correct odds. Since 1913, these calculations have been easier thanks to the invention of the "automatic totalizator," commonly known as the "tote board." Its Australian inventor, George Julius, had originally wanted to build a vote-counting machine, but his government had no interest in his design. Undeterred, Julius tweaked the mechanism to calculate betting odds instead and sold it to a racetrack in New Zealand.

In a pari-mutuel system, spectators are effectively betting against each other. The race organizers take the same cut regardless of which horse wins. The odds therefore depend only on which horse the bettors think will do well. Of course, people have all sorts of different methods for picking a successful horse. They might go for one that's been putting in some impressive performances. Perhaps it's won a few races recently or looked confident in practice. It might run well in certain weather. Or have a respected jockey. Maybe it's currently a good weight or a good age.

If enough people bet, we might expect the pari-mutuel odds to settle down to a "fair" value, which reflects the horse's true chances of winning. In other words, the betting market is efficient, bringing together all the scattered bits of information about each horse until there's nothing left to give anyone an advantage. We might expect that to happen, but it doesn't.

When the tote board shows a horse has odds of 100, it suggests bettors think its chance of winning is around 1 percent. Yet it seems people are often too generous about a weaker horse's chances. Statisticians have compared the money people throw at long shots with the amount those horses actually win and have found that the probability of victory is often much lower than the odds imply. Conversely, people tend to underestimate the prospects of the horse that is the favorite to win.

The favorite-long-shot bias means top horses are often more likely to win than their odds suggest. However, betting on them isn't necessarily a good strategy. Because the track takes a cut in a pari-mutuel system, there is a hefty handicap to overcome. Whereas card counters only have to improve on the Four Horsemen's method, which almost breaks even, sports bettors need a strategy that will be profitable even when the track charges 19 percent.

The favorite-long-shot bias might be noticeable, but it's rarely that severe. Nor is it consistent: the bias is larger at some racetracks than at others. Still, it shows that the odds don't always match a horse's chances of winning. Like blackjack, the Happy Valley betting market is vulnerable to smart gamblers. And in the 1980s, it became clear that such vulnerability could be extremely profitable.

HONG KONG WASN'T WOODS'S first attempt at a betting system for horse racing. He'd spent 1982 in New Zealand with a group of professional gamblers, hoping that their collective wisdom would be enough to spot horses with incorrect odds. Unfortunately, it was a year of mixed success.

Benter had a background in physics and an interest in computers, so for the races at Happy Valley, the pair planned to employ a more scientific approach. But winning at the racetrack and winning at blackjack involved very different sets of problems. Could mathematics really help predict horse races?

A visit to the University of Nevada's library brought the answer. In a recent issue of a business journal, Benter spotted an article by Ruth Bolton and Randall Chapman, two researchers based at the University of Alberta in Canada. It was called "Searching for Positive Returns at the Track." In the opening paragraph, they hinted at what followed over the next twenty pages. "If the public makes systematic and detectable errors in establishing the betting odds," they wrote, "it may be possible to exploit such a situation with a superior wagering strategy." Previously published strategies had generally concentrated on well-known discrepancies in racing odds, like the favorite-long-shot bias. Bolton and Chapman had taken a different approach. They'd developed a way to take available information about each horse—such as percentage of races won or average speed—and convert it into an estimate of the probability that horse would win. "It was the paper that launched a multi-billion dollar industry," Benter said. So, how did it work?

TWO YEARS AFTER HIS work on the roulette wheels of Monte Carlo, Karl Pearson met a gentleman by the name of Francis Galton. A cousin of Charles Darwin, Galton shared the family passion for science, adventure, and sideburns. However, Pearson soon noticed a few differences.

When Darwin developed his theory of evolution, he'd taken time to organize the new field, introducing so much structure and direction that his fingerprints can still be seen today. Whereas Darwin was an architect, Galton was an explorer. Much like Poincaré, Galton was happy to announce a new idea and then wander off in search of another. "He never waited to see who was following him," Pearson said. "He pointed out the new land to biologist, to anthropologist, to psychologist, to meteorologist, to economist, and left them to follow or not at their leisure."

Galton also had an interest in statistics. He saw it as a way to understand the biological process of inheritance, a subject that had fascinated him for years. He'd even roped others into studying the topic. In 1875, seven of Galton's friends received sweet pea seeds, with instructions to plant them and return the seeds from their progeny. Some people received heavy seeds; some light ones. Galton wanted to see how the weights of the parent seeds were related to those of the offspring.

Comparing the different sizes of the seeds, Galton found that the offspring were larger than the parents if the parents were small, and smaller than them if the parents were large. Galton called it "regression towards mediocrity." He later noticed the same pattern when he looked at the relationship between heights of human parents and children.

Of course, a child's appearance is the result of several factors. Some of these might be known; others might be hidden. Galton realized it would be impossible to unravel the precise role of each one. But using his new regression analysis, he would be able to see whether some factors contributed more than others. For example, Galton noticed that although parental characteristics were clearly important, sometimes features seemed to skip generations, with characteristics coming from grandparents, or even great-grandparents. Galton believed that each ancestor must contribute some amount to the heritage of a child, so he was delighted when he heard that a horse breeder in Pittsburg, Massachusetts, had published a diagram illustrating the exact process he'd been trying to describe. The breeder, a man by the name of A. J. Meston, used a square to represent the child, and then divided it into smaller squares to show the contribution each ancestor made: the bigger the square, the bigger the contribution. Parents

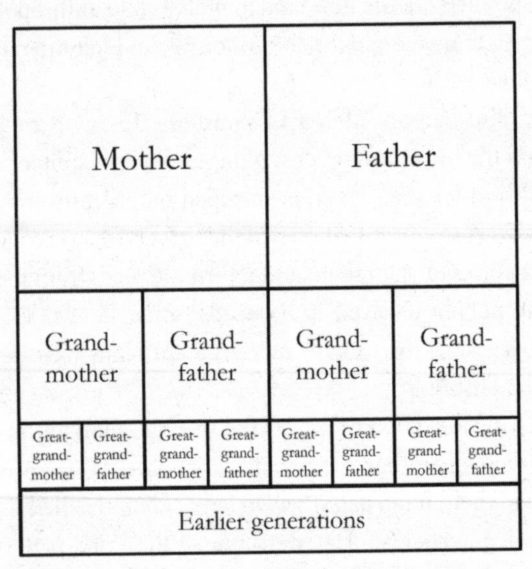

FIGURE 3.2. A. J. Meston's illustration of inheritance.

took up half the space; grandparents a quarter; great-grandparents an eighth, and so on. Galton was so impressed with the idea that he wrote a letter to the journal *Nature* in January 1898 suggesting that they reprint it.

Galton spent a good deal of time thinking about how outcomes, such as child size, were influenced by different factors, and he was meticulous about collecting data to support this research. Unfortunately, his limited mathematical background meant he couldn't take full advantage of the information. When he met Pearson, Galton didn't know how to calculate precisely how much a change in a particular factor would affect the outcome.

Galton had yet again pointed to a new land, and it was Pearson who filled it with mathematical rigor. The pair soon started to apply the ideas to questions about inheritance. Both viewed regression to the mediocre as a potential problem: they wondered how society could make sure that "superior" racial characteristics were not lost in subsequent generations. In Pearson's view, a nation could be improved by "insuring that its numbers are substantially recruited from the better stocks."

From a modern viewpoint, Pearson is a bit of a contradiction. Unlike many of his peers, he thought men and women should be treated as social and intellectual equals. Yet at the same time, he used his statistical methods to argue that certain races were superior to others; he also claimed that laws restricting child labor turned children into social and economic burdens. Today, that's all rather unsavory. Nevertheless, Pearson's work has been hugely influential. Not long after Galton's death in 1911, Pearson established the world's first statistics department at University College London. Building on the diagram Galton had sent to *Nature*, Pearson developed a method for "multiple regression": out of several potentially influential factors, he worked out a way to establish how related each was to a given outcome.

Regression would also provide the backbone for the University of Alberta researchers' racing predictions. Whereas Galton and Pearson used the technique to examine the characteristics of a child, Bolton and Chapman employed it to understand how different factors affected a horse's chances of winning. Was weight more important than percentage of recent races won? How did average speed compare with the reputation of the jockey?

Bolton's first exposure to the world of gambling had come at a young age. "When I was a toddler my Dad took me to the track," she said, "and apparently my little hand picked the winning horse." Despite her early success, it was the last time that she went to the races. Two decades later, however, she found herself picking winners once again, this time with a far more robust method.

The idea for a horseracing prediction method had taken shape in the late 1970s, while Bolton was a student at Queens University in

Canada. Bolton had wanted to learn more about an area of economics known as choice modeling, which aims to capture the benefits and costs of a certain decision. For her final-year dissertation, Bolton teamed up with Chapman, who was researching problems in that area. Chapman, who had a long-standing interest in games, had already accumulated a collection of horse racing data, and together the pair examined how the information could be used to forecast race results. The project was not just the start of an academic partnership; the researchers married in 1981.

Two years after the wedding, Bolton and Chapman submitted the horse racing research to the journal *Management Science*. At the time, prediction methods were growing in popularity, which meant the work received a lot of scrutiny. "The paper spent a long time in review," Bolton said. The research eventually went through four rounds of revisions before appearing in print in the summer of 1986.

In their paper, Bolton and Chapman assumed that a particular horse's chances of winning depended on its quality, which they calculated by bringing together several different measurements. One of these was the starting position. A lower number meant the horse was starting nearer the inside of the track, which should improve a horse's chances, because it means a shorter distance to run. The pair therefore expected regression analysis to show that an increase in starting number would lead to a decrease in quality.

Another factor was the weight of a horse. It was less clear how this would affect quality. Weight restrictions at some races penalize heavier horses, but faster horses often have a higher weight. Old-school racing pundits might try to come up with opinions about which is more important, but Bolton and Chapman didn't need to take such views: they could simply let the regression analysis do the hard work and show them how weight was related to quality.

In Bolton and Chapman's model of a horse race, the quality measurement depended on nine possible factors, including weight, average speed in recent races, and starting position. To illustrate how the different factors contribute to a horse's quality, it's tempting to use a setup similar to the diagram Galton sent *Nature*. However, real life is not as simple as such illustrations suggest. Although Galton's diagram shows how relatives might shape the characteristics of a child, the picture is incomplete because not everything is inherited. Environmental factors can also influence things, and these might not always be visible or known. Moreover, the neat boxes—for mother, father, and so on—are likely to overlap: if a child's father has a certain characteristic, the grandfather or grandmother might have it, too. So, you can't say that each contributing factor is completely independent of the others. The same is true for horse racing. As well as the nine performance-related factors, Bolton and Chapman therefore included an uncertainty factor in their prediction of horse quality. This accounted for unknown influences on horse performance as well as the inevitable quirks of a particular race.

Once the pair had measured the horses' quality, they converted the measurements into predictions about each animal's chance of victory. They did this by calculating the total amount of quality across all the horses in the race. The probability a particular horse would win depended on how much the horse contributed to this overall total.

To work out which factors would be useful for making predictions, Bolton and Chapman compared their model to data from two hundred races. Handling the information was a feat in itself, with race results stored on dozens of computer punch cards. "When I got the data, it was in a big box," Bolton said. "For years, I carried that box around." Getting the results into the computer was also a challenge: it took about an hour to enter the data for each race.

Of the nine factors Bolton and Chapman tested, the pair found that average speed was the most important in deciding where a horse would finish. In contrast, weight didn't seem to make any difference to predictions. Either it was irrelevant or any effect it did have was covered by another factor, in a similar way to how a grandfather's influence on a child's appearance might be covered by the contribution from the father.

It can be surprising which certain factors turn out to be most important. In an early version of Bill Benter's model, the number of races a horse had previously run made a big contribution to the predictions. However, there was no intuitive reason why it was so crucial. Some gamblers might try to think up an explanation, but Benter avoided speculating about specific causes. This is because he knew that different factors were likely to overlap. Rather than try to interpret why something like number of races appears to be important, he instead concentrated on putting together a model that could reproduce the observed race results. Just like the gamblers who searched for biased roulette tables, he could obtain a good prediction without pinning down the precise underlying causes.

In other industries, of course, it might be necessary to isolate how much a certain factor affects an outcome. While Galton and Pearson had been studying inheritance, the Guinness brewery had been trying to improve the life span of its stout. The task fell to William Gossett, a promising young statistician who had spent the winter of 1906 working in Pearson's lab.

Whereas betting syndicates have no control over factors like the weight of a horse, Guinness could alter the ingredients it put in its beer. In 1908, Gossett used regression to see how much hops influenced the drinkable life span of beer. Without hops, the company could expect beer to last between twelve and seventeen days; adding the right amount of hops could increase the life span by several weeks.

Betting teams aren't particularly interested in knowing why certain factors are important, but they do want to know how good their predictions are. It might seem easiest to test the predictions against the racing data the team had just analyzed. Yet this would be an unwise approach.

Before he worked on chaos theory, Edward Lorenz spent the Second World War as a forecaster for the US Air Corps in the Pacific. One autumn in 1944, his team made a series of perfect predictions about weather conditions on the flight path between Siberia and Guam. At least they were perfect according to the reports from aircraft flying that route. Lorenz soon realized what was causing the incredible success rate. The pilots, busy with other tasks, were just repeating the forecast as the observation.

The same problem appears when syndicates test betting predictions against the data used to calibrate the model. In fact, it would be easy to build a seemingly perfect model. For each racing result, they could include a factor that indicates which horse came in first. Then they could tweak these factors until they fitted perfectly with the horses that actually won each race. It would look like they've got a flawless model, when all they've really done is dress up the actual results as a prediction.

If teams want to know how well a strategy will work in the future, they need to see how good it is at predicting *new* events. When collecting information on past races, syndicates therefore put a chunk of the results to one side. They use the rest of the data to evaluate the factors in their model; once this is done, they test the predictions against the collection of yet-to-be-used results. This allows teams to check how the model might perform in real life.

Testing strategies against new data also helps ensure that models satisfy the scientific principle of Occam's razor, which states that if you have to choose between several explanations for an observed event, it is best to pick the simplest. In other words, if you want to build a model of a real-life process, you should shave away the features that you can't justify.

Comparing predictions against new data helps betting teams avoid throwing too many factors into a model, but they still need to assess how good the model actually is. One way to measure the accuracy of a prediction is to use what statisticians call the "coefficient of determination." The coefficient ranges from 0 to 1 and can be thought of as measuring the explanatory power of a model. A value of 0 means that the model doesn't help at all, and bettors might as well pick the winning horse at random; a value of 1 means the predictions line up perfectly with the actual results. Bolton and Chapman's model had a value of 0.09. It was better than randomly choosing horses, but there were still plenty of things that the model wasn't capturing.

Part of the problem was the data they had used. The two hundred races they'd analyzed came from five American racetracks. This meant there was a lot of hidden information: horses would have raced against a range of opponents, in different conditions, with a variety of jockeys. It might have been possible to overcome some of these problems with a lot of racing data, but with only two hundred races? It was doubtful. Still, the strategy could potentially work, if only the race conditions were a bit less variable.

IF YOU HAD TO put together an experiment to study horse racing, it would probably look a lot like Hong Kong. With races happening on one of two tracks, your laboratory conditions are going to be fairly consistent. The subjects of your experiment won't vary too much either: in the United States, tens of thousands of horses race all over the country; in Hong Kong, there is a closed pool of about a thousand horses. With around six hundred races a year, these horses race against each other again and again, which means you can observe similar events several times, just as Pearson always tried to. And, unlike Monte Carlo and its lazy roulette reporters, in Hong Kong there's also plenty of publicly available data on the horses and their performances.

When Benter first analyzed the Hong Kong data, he found that at least five hundred to a thousand races were needed to make good predictions. With fewer than this, there wasn't enough information to work out how much each factor contributed to performance, which meant the model wasn't particularly reliable. In contrast, including more than a thousand races didn't lead to much improvement in the predictions.

In 1994, Benter published a paper outlining his basic betting model. He included a table that showed how his predictions compared to actual race outcomes. The results looked pretty good. Apart from a few discrepancies here and there, the model was remarkably close to reality. However, Benter warned that the results hid a major flaw. If anyone had tried to bet using the predictions, it would have been catastrophic.

SUPPOSE YOU CAME INTO some money and wanted to use the windfall to buy a little bookstore somewhere. There are a couple of ways you could go about it. Having drawn up a short list of stores you might buy, you could go into each one, check the inventory, quiz the management, and examine the accounts. Or you could bypass the paperwork and simply sit outside and count how many customers go in and how many books they come out with. These contrasting strategies reflect the two main ways people approach investing. Researching a company to its core is known as "fundamental analysis," whereas watching how other people view the company over time is referred to as "technical analysis."

Bolton and Chapman's predictions used a fundamental approach. Such methods rely on having good information and sifting through it in the best way possible. The views of pundits don't feature in the analysis. It doesn't matter what other people are doing and which horses they are choosing. The model ignores the betting market. It's like making predictions in a vacuum.

Although it might be possible to make predictions about races in isolation, the same cannot be said for betting on them. If syndicates want to make money at the track, they need to outwit other gamblers. This is where purely fundamental approaches can run into problems. When Benter compared the predictions from his fundamental model with the public odds, he noticed a worrying bias. He'd used the model to find "overlays": horses that, according to the model, have a better chance of winning than their odds imply. These were the horses he would bet on if he were hoping to beat other gamblers. Yet when Benter looked at actual race results, the overlays did not win as often as the predictions suggested. It seemed that the true chances of these horses winning lay somewhere between the probability given by the model and the probability implied by the betting odds. The fundamental approach was clearly missing something.

Even if a betting team has a good model, the public's views on a horse's chances—as shown by the odds on the tote board—aren't completely irrelevant, because not every gambler picks horses based on publicly available information. Some people might know about the jockey's strategy for the race or the horse's eating and workout schedule. When they try to capitalize on this privileged information, it changes the odds on the board.

It makes sense to combine the two available sources of expertise, namely, the model and opinion of other gamblers (as shown by the odds on the tote board). This is the approach Benter advocated. His model still ignores the public odds initially. The first set of predictions is made as if there is no such thing as betting. These predictions are then merged with the public's view. The probability each horse will win is a balance between the chance of the horse winning in the model and the chance of victory according to the current odds. The scales can tip one way or the other: whichever produces the combined prediction that lines up best with actual results. Strike the right balance, and good predictions can become profitable ones.

WHEN WOODS AND BENTER arrived in Hong Kong, they did not meet with immediate success. While Benter spent the first year putting together the statistical model, Woods tried to make money exploiting the long-shot-favorite bias. They had come to Asia with a bankroll of $150,000; within two years, they'd lost it all. It didn't help that investors weren't interested in their strategy. "People had so little faith in the system that they would not have invested for 100 percent of the profits," Woods later said.

By 1986, things were looking better. After writing hundreds of thousands of lines of computer code, Benter's model was ready to go. The team had also collected enough race results to generate decent predictions. Using the model to select horses, they took home $100,000 that year.

Disagreements meant the partnership ended after that first successful season. Before long, Woods and Benter had formed rival syndicates and continued to compete against each other in Hong Kong. Although Woods later admitted that Benter's team had the better model, both groups saw their profits rise dramatically over the next few years.

Several betting syndicates in Hong Kong now use models to predict horse races. Because the track takes a cut, it's difficult to make money on simple bets such as picking the winner. Instead, syndicates chase the more complicated wagers on offer. These include the trifecta: to win, gamblers must predict the horses that will finish first, second, and third in correct order. Then there's the triple trio, which involves winning three trifectas in a row. Although the payoffs for these exotic bets can be huge, the margin for error is also much smaller.

One of the drawbacks with Bolton and Chapman's original model is that it assumes the same level of uncertainty for all the horses. This makes the calculations easier, but it means sacrificing some realism. To illustrate the problem, imagine two horses. The first is a bastion of reliability, always finishing the race in about the same time. The second is more variable, sometimes finishing much quicker than the first, but sometimes taking much longer. As a result, both horses take the same time on average to run a race.

If just these two horses are racing, they will have an equal probability of winning. It might as well be a coin toss. But what if several horses are in the race, each with a different level of uncertainty? If a betting team wants to pick the top three accurately, they need to account for these differences. For years, this was beyond the reach of even the best horse racing models. In the past decade, though, syndicates have found a way to predict races with a varying amount of uncertainty looming over each horse. It's not just recent increases in computing power that have made this possible. The predictions also rely on a much older idea, originally developed by a group of mathematicians working on the hydrogen bomb.

ONE EVENING IN JANUARY 1946, Stanislaw Ulam went to bed with a terrible headache. When he woke up the next morning, he'd lost the ability to speak. He was rushed to a Los Angeles hospital, where concerned surgeons drilled a hole in his skull. Finding his brain severely inflamed as a result of infection, they treated the exposed tissue with penicillin to halt the disease.

Born in Poland, Ulam had left Europe for the United States only weeks before his country fell to the Nazis in September 1939. He was a mathematician by training and had spent most of the Second

World War working on the atomic bomb at Los Alamos National Laboratory. After the conflict ended, Ulam joined UCLA as a professor of mathematics. It wasn't his first choice: amid rumors that Los Alamos might close after the war, Ulam had applied to several higher-profile universities that all turned him down.

By Easter 1946, Ulam had fully recovered from his operation. The stay in the hospital had given him time to consider his options, and he decided to quit his job at UCLA and return to Los Alamos. Far from shutting it down, the government was now pouring money into the laboratory. Much of the effort was going into building a hydrogen bomb, nicknamed the "Super." When Ulam arrived, several obstacles were still in the way. In particular, the researchers needed a means to predict the nuclear chain reactions involved in a detonation. This meant working out how often neutrons collide—and hence how much energy they would give off—inside a bomb. To Ulam's frustration, this couldn't be calculated using conventional mathematics.

Ulam did not enjoy grinding away at problems for hours, as many mathematicians spent their time doing. A colleague once recalled him trying to solve a quadratic equation on a blackboard. "He furrowed his brow in rapt absorption, while scribbling formulas in his tiny handwriting. When he finally got the answer, he turned around and said with relief, 'I feel I have done my work for the day.'"

Ulam preferred to focus on creating new ideas; others could fill in the technical details. It wasn't just mathematical puzzles he tackled in inventive ways. While working at the University of Wisconsin during the winter of 1943, he'd noticed that several of his colleagues were no longer showing up to work. Soon afterward, Ulam received an invitation to join a project in New Mexico. The letter didn't say what was involved. Intrigued, Ulam headed to the campus library and tried to find out all he could about New Mexico. It turned out that there was only one book about the state. Ulam looked at who'd checked it out recently. "Suddenly, I knew where all my friends had disappeared to," he said. Glancing over the others' research interests, he quickly pieced together what they were all working on out in the desert.

WITH HIS HYDROGEN BOMB calculations turning into a series of mathematical cul-de-sacs, Ulam remembered a puzzle he'd thought about during his stay in the hospital. While recovering from surgery, he had passed the time playing solitaire. During one game, he'd tried to work out the probability of a certain card arrangement appearing. Faced with having to calculate a vast set of possibilities—the sort of monotonous work he usually tried to avoid—Ulam realized it might be quicker just to lay out the cards several times and watch what happened. If he repeated the experiment enough times, he would end up with a good idea of the answer without doing a single calculation.

Wondering whether the same technique could also help with the neutron problem, Ulam took the idea to one of his closest colleagues, a mathematician by the name of John von Neumann. The two had known each other for over a decade. It was von Neumann who'd suggested Ulam leave Poland for America in the 1930s; he'd also been the one who invited Ulam to join Los Alamos in 1943. They made quite the pair, portly von Neumann in his immaculate suits—jacket always on—and Ulam with his absent-minded fashion sense and dazzling green eyes.

Von Neumann was quick-witted and logical, sometimes to the point of being blunt. He'd once grown hungry during a train journey and had asked the conductor to send the sandwich seller his way. The request fell on unsympathetic ears. "I will if I see him," the conductor said. To which von Neumann replied, "This train is linear, isn't it?"

When Ulam described his solitaire idea, von Neumann immediately spotted its potential. Enlisting the help of another colleague, a physicist named Nicholas Metropolis, they outlined a way to solve the chain reaction problem by repeatedly simulating neutron collisions. This was possible thanks to the recent construction of a programmable computer at Los Alamos. Because they worked for a government agency, however, the trio needed a code name for their new approach. With a nod to Ulam's heavy-gambling uncle, Metropolis suggested they call it the "Monte Carlo method."

Because the method involved repeated simulations of random events, the group needed access to lots of random numbers. Ulam joked that they should hire people to sit rolling dice all day. His flippancy hinted at an unfortunate truth: generating random numbers was a genuinely difficult task, and they needed a lot of them. Even if those nineteenth-century Monte Carlo journalists had been honest, Karl Pearson would have struggled to build a collection big enough for the Los Alamos men.

Von Neumann, inventive as ever, instead came up with a method for creating "pseudorandom" numbers using simple arithmetic. Despite its being easy to implement, von Neumann knew his method had shortcomings, chiefly the fact that it couldn't generate truly random numbers. "Anyone who considers arithmetical methods of producing random digits is, of course, in a state of sin," he later joked.

As computers have increased in power, and good pseudorandom numbers have become more readily available, the Monte Carlo method has become a valuable tool for scientists. Edward Thorp even used Monte Carlo simulations to produce the strategies in *Beat the Dealer*. However, things aren't so straightforward in horse racing.

In blackjack, only so many combinations of cards can come up—too many to solve the game by hand, but not by computer. Compare this with horse racing models, which can have over a hundred factors. It's possible to tweak the contribution of each one—and hence change the prediction—in a vast number of ways. By just randomly picking different contributions, it's very unlikely you would hit on the best possible model. Every time you made a new guess, it would have the same chance of being the best one, which is hardly the most efficient way of finding the ideal strategy. Ideally, you would make each guess better than the last. This means finding an approach that includes a form of memory.

DURING THE EARLY TWENTIETH century, Poincaré and Borel weren't the only researchers curious about card shuffling. Andrei Markov was a Russian mathematician with a reputation for immense talent and immense temper. When he was young, he'd even picked up the nickname "Andrei Neistovy": Andrei the angry.

In 1907, Markov published a paper about random events that incorporated memory. One example was card shuffling. Just as Thorp would notice decades later, the order of a deck after a shuffle depends on its previous arrangement. Moreover, this memory is short-lived. To predict the effect of the next shuffle, you only need to know the current order; having additional information on the cards' arrangement several shuffles ago is irrelevant. Thanks to Markov's work, this one-step memory has become known as the "Markov property." If the random event is repeated several times, it's a "Markov chain." From card shuffling to Chutes and Ladders, Markov chains are common in games of chance. They can also help when searching for hidden information.

Remember how it takes at least six dovetail shuffles to properly mix up a deck of cards? One of the mathematicians behind that result was a Stanford professor named Persi Diaconis. A few years after Diaconis published his card shuffling paper, a local prison psychologist turned up at Stanford with another mathematical riddle. The psychologist had brought a bundle of coded messages, confiscated from prisoners. Each one was a jumble of symbols made from circles, dots, and lines.

Diaconis decided to give the code to one of his students, Marc Coram, as a challenge. Coram suspected that the messages used a substitution cipher, with each symbol representing a different letter. The difficulty was working out which letter went where. One option was to tackle the problem through trial and error. Coram could have used a computer to shuffle the letters again and again and then examined the resulting text until he hit upon a message that made sense. This is the Monte Carlo method. He would have deciphered the messages eventually, but it could have taken an absurdly long time to get there.

Rather than starting with a new random guess each time, Coram instead chose to use the Markov property of shuffling to gradually improve his guesses. First, he needed a way to measure how realistic a particular guess was. He downloaded a copy of *War and Peace* to find out how often different pairs of letters appeared together. This let him work out how common each particular pairing should be in a given piece of text.

During each round of guessing, Coram randomly switched a couple of the letters in the cipher and checked whether his guess had improved. If a message contained more realistic letter pairings than the previous guess, Coram stuck with it for the next go. If the message wasn't as realistic, he would usually switch back. But occasionally he stuck with a less plausible cipher. It's a bit like solving a Rubik's Cube. Sometimes the quickest route to the solution involves a step that at first glance takes you in the wrong direction. And, like a Rubik's Cube, it might be impossible to find the perfect arrangement by only taking steps that improve things.

The idea of combining the power of the Monte Carlo method with Markov's memory property originated at Los Alamos. When Nick Metropolis first joined the team in 1943, he'd worked on the problem that had also puzzled Poincaré and Borel: how to understand the interactions between individual molecules. It meant solving the equations that described how particles collided, a frustrating task given the crude calculators around at the time.

After years of battling with the problem, Metropolis and his colleagues realized that if they linked the brute force of the Monte Carlo method with a Markov chain, they would be able to infer the properties of substances made of interacting particles. By making smarter guesses, it would be possible to gradually uncover values that couldn't be observed directly. The technique, which became known as "Markov chain Monte Carlo," is the same one Coram would later use to decipher the prison messages.

It eventually took Coram a few thousand rounds of computer-assisted guessing to crack the prison code. This was vastly quicker than a pure brute force method would have been. It turned out that one of the prisoners' messages described the unusual origins of a fight: "Boxer was making loud and loud voices so I tell him por favour can you kick back homie cause I'm playing chess."

To break the prison code, Coram had to take a set of unobserved values (the letters that corresponded to each symbol) and estimate them using letter pairings, which he could observe. In horse racing, betting teams face a similar problem. They don't know how much uncertainty surrounds each horse, or how much each factor should contribute to predictions. But—for a particular level of uncertainty and combination of factors—they can measure how well the resulting predictions match actual race outcomes. The method is classic Ulam. Rather than trying to write down and solve a set of near-impenetrable equations, they let the computer do the work instead.

In recent years, Markov chain Monte Carlo has helped syndicates come up with better race forecasts and predict lucrative exotic results like the triple trio. Yet successful gamblers don't just need to find an edge. They also need to know how to exploit it.

IF YOU WERE BETTING $1.00 on a coin toss coming up tails, a fair payout would be $1.00. Were someone to offer you $2.00 for a bet on tails, that person would be handing you an advantage. You could expect to win $2.00 half the time and suffer a $1.00 loss the other half, which would translate into an expected profit of $0.50.

How much would you bet if someone let you scale up such a biased wager? All of your money? Half of it? Bet too much, and you risk wiping out your savings on an event that still has only a 50 percent chance of success; bet too little, and you won't be fully exploiting your advantage.

After Thorp put together his winning blackjack system, he turned his attention to the problem of such bankroll management. Given a particular edge over the casino, what was the optimal amount to bet? He found the answer in a formula known as the Kelly criterion. The formula is named after John Kelly, a gunslinging Texan physicist who worked with Claude Shannon in the 1950s. Kelly argued that, in the long run, you should wager a percentage of your bankroll equal to your expected profit divided by the amount you'll receive if you win.

For the coin toss above, the Kelly criterion would be the expected payoff ($0.50) divided by the potential winnings ($2.00). This works out to 0.25, which means you should bet a quarter of your available bankroll. In theory, wagering this amount will ensure good profits while limiting the risk of chipping away at your funds. A similar calculation can be performed for horse racing. Betting teams know the probability that a horse will win according to their model. Thanks to the tote board, they can also see what chance the public thinks it has. If the public thinks a victory is less likely than the model suggests, there might be money to be made.

Despite its success in blackjack, there are some drawbacks to the Kelly criterion, especially at the racetrack. First, the calculation assumes you know the true probability of an event. Although you know the chance of a coin coming up heads, things are less clear in horse racing: a model just gives an estimate of the chance a horse will win. If a team overestimates a horse's chances, following the Kelly criterion will cause them to bet too much, increasing their risk of going bust. Consistently overestimate by twofold—for instance, by thinking a horse has a 50 percent chance of victory when in reality it has a 25 percent chance—and it will eventually lead to certain bankruptcy. For this reason, syndicates generally bet less than the Kelly criterion would encourage, often wagering only a half or third of the suggested amount. This reduces the risk of having a "rough ride" and losing a large chunk—or worse, all—of their wealth.

Wagering a smaller amount can also help teams overcome one of the quirks of the betting market in Hong Kong. If you think betting on a certain horse will have a big expected payoff, the Kelly criterion will tell you to put a lot of money on it. In the extreme case, when you are certain of a result, you should stake everything you have. Yet in pari-mutuel betting, this is not necessarily a good idea. A horse's odds depend on the amount wagered, so the more people bet on it, the less money you'll make if it wins.

Even one large bet can shift the whole market. For instance, you might compare your model's predictions and the current odds and notice that you can expect a 20 percent return if you bet on a certain horse. Put down $1.00 and it won't change the overall odds much, so you'll still expect to bag $0.20 if the horse wins. If you have deep pockets, you might choose to bet more than $1.00. The Kelly criterion will certainly be telling you to do so. But if you make a $100.00 bet, it might lower the odds a little. So, your actual profit will be only 19 percent. Still, you've made $19.00.

You might decide to go bigger and bet $1,000. This could shift the odds quite a bit. If a few thousand dollars have already been staked on that horse, it might knock your expected profit down to

10 percent, which means a payoff of $100. Eventually, there comes a point at which putting more money on a horse actually reduces your profits. If the expected return for a $2,000 bet is only 4 percent, you'd be better off wagering a lower amount.

The potential for bets to move the market isn't the only problem you'd have to deal with. All the calculations above assume that you are the last person to bet, and so know the public odds. In reality, devising an optimal strategy isn't that straightforward. At the track, there is a lag on the tote board, sometimes of up to 30 seconds, which means more bets might come in after you've picked your horse.

The total betting pool at Happy Valley might be $300,000 when a team place their bets, but it will probably grow by at least another $100,000 by the time the race starts. Syndicates need to adjust for this influx of cash when deciding how to bet; otherwise, a strategy that initially looks like it will generate a big return could end up producing a mediocre profit. They can't assume that the extra money will be placed on random horses either. In the past decade or so, scientific betting has become more popular, and there are now several syndicates operating in Hong Kong that use models to predict races. These teams are likely to be the ones behind any last-minute betting. "The late money tends to be smart money," Bill Benter said. Teams therefore have to assume the worst: others will also bet on the favorable horse, so any potential profits will have to be divided among more people.

UNTIL SYNDICATES STARTED USING scientific approaches in Hong Kong, successful racetrack betting strategies were few and far between. The techniques are now so effective—and the wins so consistent—that teams like Benter's don't celebrate when their predictions come good. Much of the reason for Benter's early success was the unique setup available to gamblers in Hong Kong. At Happy Valley, gamblers don't have to bet in person at the track. They can call in their selections by phone. This was one of the main reasons Benter and Woods chose Hong Kong. It removed an additional complication and meant they could concentrate on updating their computer predictions rather than worrying about how to place bets in time. Combined with the good availability of data and active betting market, it made Hong Kong the ideal place to implement their strategy.

Gradually, others noticed the appeal of Hong Kong, too. As a result, it is now extremely difficult for betting teams to make money at the city's racetracks. With competition increasing in Hong Kong, the ideas first introduced by Bolton and Chapman are spreading to other regions, including the United States. Over the past decade, scientific betting has become a major part of US horse racing. It's been estimated that teams using computer predictions bet around $2 billion a year at American racetracks, almost 20 percent of the total amount wagered. This sum is all the more impressive when you consider that computer teams cannot bet at several of the large racetracks.

Betting teams are also targeting events in other countries. Like Swedish harness racing, in which horses pull drivers around the track on two-wheeled carts. Imagine a modern version of a Roman chariot race, without the swords and capes. The techniques are growing in popularity at racetracks in Australia and South Africa, too. An idea that began as a piece of academic research has turned into a truly global industry.

It is worth mentioning that it's not cheap to set up a scientific betting syndicate. To gather the necessary technology and expertise—not to mention hone the prediction method and place the bets—costs most teams at least $1 million. Because betting strategies are expensive to run, teams in the United States often seek out racetracks that offer favorable gambling conditions. Several tracks have noticed the bump in profits that comes with the syndicates' huge bets and now encourage computer-based approaches. They even strike deals with betting teams, handing out rebates if the syndicates place large volumes of bets.

These difficulties mean that, although Bolton and Chapman enjoyed the problem-solving aspect of racetrack predictions, they have never really been that interested in gambling careers. Aware of the cost and logistics involved in implementing their strategy, they were happy to remain in academia. "We would joke that we could do it," Bolton said. "Every so often we'd hear how much money was being made and how large these operations had got, but it wasn't for us."

The success of scientific betting in horse racing is all the more remarkable because historically there has been a limit to how much gamblers can predict outcomes. The problem is not limited to horse racing. Whether betting on sports or politics, it has often been difficult to get ahold of the necessary information and to create reliable models. Even if gamblers did manage to come up with a decent prediction, the strategies could be tricky to implement. But at the start of the twenty-first century, that all changed.

4

PUNDITS WITH PHDS

When a new blackjack system hit Britain in 2006, word of its success traveled quietly but quickly. No disguises were required, or card counting, or even visits to casinos. Admittedly, the profit margin was on the sort of scale that would buy pints rather than penthouses, but the system worked. All it required was a computer, a good chunk of spare time, and willingness to do something dull in return for beer money. Students loved it.

The strategy emerged as a result of the new Gambling Act, passed by the government a few months earlier. It meant that UK-based companies could now provide online casino games as well as traditional sports betting. In the rush for new customers, firms started offering signup bonuses. Bet £100 and get an extra £50 free—that sort of thing. At first glance, such a bonus doesn't seem to help much with blackjack. In an online game, it's much easier for casinos to ensure that cards are dealt randomly, making card counting impossible.

If you use the Four Horsemen's optimal blackjack strategy instead, taking the dealer's card into account when making your decision, you can expect to lose money over time. But the signup bonuses tipped things back in the players' favor. People realized that the bonuses would in effect subsidize any losses. Playing the ideal strategy, players would probably lose some of the £100—but not much—and once they'd bet the required total, they would get the bonus. They would usually have to bet this, too, before it could be withdrawn; fortunately, they could simply repeat the previous approach to limit their losses.

During 2006, gamblers hopped from website to website, sitting through hundreds of blackjack hands to build a collection of bonus money. It didn't take long for betting companies to clamp down on what they called "bonus abuse" and exclude games like blackjack from their signup offers. Although there is nothing illegal about setting up a single account to obtain a bonus—indeed, that's the point of a signup bonus—some gamblers pushed the advantage too far. The first conviction for bonus abuse came in the spring of 2012, when Londoner Andrei Osipau was jailed for three years for using fraudulent passports and identity cards to open multiple betting accounts. For those who operated within the law in 2006, profits were far more modest than the £80,000 Osipau was reported to have made. Still, the fact that these bonuses could be exploited illustrates three crucial advantages that gamblers have gained in recent years.

First, the explosion of online betting has meant a far wider range of games and gambling options. In real-life casinos, new games are generally good news for gamblers. According to professional gambler Richard Munchkin, casinos rarely understand how much of an advantage they are offering when they introduce new games. The blackjack loophole that appeared in 2006 showed that the same is true in online gambling. And when the Internet is involved, news of a successful strategy travels much, much faster. The second advantage is the ease with which gamblers can implement a potentially profitable system. Rather than having to dodge casino security or visit bookmakers, they can simply place bets online. Whether through websites or instant messaging, access is quicker and easier than ever before. Finally, the Internet has made it much easier to get ahold of the vital ingredient for many successful betting recipes. From roulette to horse racing, the limited availability of data has dictated where and how people gamble. But these limitations are fading away. As a result, people are targeting a whole host of new games.

EVERY AUTUMN, RECRUITMENT TEAMS descend on the world's best mathematics departments. Most are from the usual crowd: oil firms wanting fluid-dynamics researchers or banks trying to find specialists in probability theory. But in recent years, another type of firm has started to appear at the career events hosted by British universities. Instead of discussing business or finance, they focus on sports such as soccer. Their career presentations are rather like watching a very technical prematch analysis. Formulae and data tables—which most companies hide from prospective applicants—fill the talks. The events have more in common with a lecture than a job pitch.

Many of the approaches are familiar to mathematicians. But although researchers might use the techniques to study ice sheets or epidemics, these firms have found a very different application for the methods. They are using scientific methods to take on the bookmakers. And they are winning.

Modern soccer predictions began with what would otherwise have been a throwaway exam question. During the 1990s, Stuart Coles was a lecturer at Lancaster University, a few miles from the sweeping fells of England's Lake District. Coles specialized in extreme value theory, which deals with the sort of severe, rare events that are like nothing ever seen before. Pioneered by Ronald Fisher in the 1930s, extreme value theory is used to predict the worst-of-the-worst-case scenarios, from floods and earthquakes to wildfires and insurance losses. In short, it is the science of the very unlikely.

Coles's research spanned everything from storm surges to severe pollution. At the prompting of Mark Dixon, another researcher in the department, Coles also started to think about soccer. Dixon had become interested in the topic after looking at a statistics exam given to final-year students at Lancaster. One of the questions involved predicting the results of a hypothetical soccer match, but Dixon spotted a flaw: the method was too simple to be useful in real life. It was an interesting problem, though, and if the ideas were extended—and applied to actual soccer leagues—it might lead to an effective betting strategy.

It took a couple of years for Dixon and Coles to develop the new method and get it ready for publication. The work eventually appeared in the *Journal of Applied Statistics* in 1997. With the research finished, Coles went back to his other projects. Little did he realize how important the soccer paper would turn out to be. "It was one of those things that at the time seemed inconsequential," he said, "but looking back it had a massive impact on my life."

TO PREDICT HORSE RACES in Hong Kong, scientific betting teams assess the quality of each horse and then compare these different quality measurements to work out the probable result. It's tricky to do the same in soccer. Although it might be possible to weigh up each team's qualities and calculate which team is likely to be successful over an entire season, it is much harder to work out who is likely to win in a given match. A team that plays well against one set of opposition can look sluggish against another. Or one shot might go in while another bounces off the woodwork. Then, you have the players. Sometimes a talismanic performance will lift a whole team; sometimes a team will carry along weak players. This tangle of on-pitch activity means that things are much messier from a statistical point of view. During the 1970s, a few researchers had even come to the conclusion that a single soccer match was so dominated by chance that prediction was hopeless.

By choosing to study soccer matches, Dixon and Coles were clearly walking into difficult territory. There was one thing on their side, however. In the United Kingdom, betting odds were generally fixed several days ahead of the match. Unlike the hectic last-minute betting at Hong Kong's racetracks, anyone analyzing soccer matches would have plenty of time to come up with a prediction and compare it to the bookmakers' odds. Even better, there were plenty of potential wagers available. Thanks to a well-established soccer betting market in the United Kingdom, there are all sorts of things to bet on, from half-time score to the number of corner kicks.

Dixon and Coles chose to start with the big question: Which team was going to win? Rather than trying to predict the final result directly, they decided to estimate the number of goals that would be scored before the final whistle. To keep things simple, the pair assumed that each team would score goals at some fixed rate over the course of a game and that the probability of scoring at each point in time was independent of what had already happened in the match.

Events that obey such rules are said to follow a "Poisson process." Named after physicist Siméon Poisson, the process crops up in many walks of life. Researchers have used the Poisson process to model telephone calls to a switchboard, radioactive decay, and even neuron activity. If you assume something follows a Poisson process, you are assuming that events occur at a fixed rate. The world has no memory; each time period is independent of the others. If a match is goalless at half time, it won't make a second-half goal more likely.

Having chosen to model a soccer game as a Poisson process—and hence assuming that goals are scored at a consistent rate over the course of the match—Dixon and Coles still needed to know what the scoring rate should be. The number of goals in a match would probably vary depending on who is playing. How many goals should they expect each team to score?

Early in their 1997 paper, Dixon and Coles set out the things you need to do if you want to build a model of a soccer league. First, you need to somehow measure each team's ability. One option is to use some sort of ranking system. Perhaps you could hand teams a certain number of points after each match and then add up the total points earned over a given time period. Most soccer leagues hand out three points for a win, one for a draw, and nothing for a loss, for example. Representing each team's ability with a single number might show which team is doing well, but it's not always possible to convert rankings into good predictions. A 2009 study by Christoph Leitner and colleagues at Vienna University of Economics and Business provided a good illustration of the problem; they came up with forecasts for the Euro 2008 soccer tournament using rankings published by the sport's governing body Fédération Internationale de Football Association (FIFA) and found that the bookmakers' predictions turned out to be far more accurate. To make money betting on soccer, it seems that you need more than one measurement for each team.

Dixon and Coles suggested splitting ability into two factors: attack and defense. Attacking ability reflected a team's aptitude at scoring goals; defensive weakness indicated how poor they were at stopping them. Given a home team with a certain attacking ability and an away team with a certain defensive weakness, Dixon and Coles assumed that the expected number of goals scored by the home team was the product of three factors:

Home attacking ability × Away defensive weakness × Home advantage factor

Here the "home advantage factor" accounts for the boost teams often get when playing at home. In a similar fashion, the expected number of away goals was equal to the away team's attacking ability multiplied by the home defensive weakness (the away team didn't get any extra advantage).

To estimate each team's attacking and defensive prowess, Dixon and Coles collected several years of data on English soccer games from the top four divisions, which among them contained a total of 92 teams. Because the model included an attack and defensive ability for each team, plus an extra factor that specified the home advantage, this meant estimating a total of 185 factors. If every team had played every other team the same number of times, estimation would have been relatively straightforward. However, promotions and relegations—not to mention cup games—meant some match-ups were more common than others. Much like the races at Happy Valley, there was too much hidden information for simple calculations. To estimate each of the 185 factors, it was therefore necessary to enlist help from computational methods like the ones developed by the researchers at Los Alamos.

When Dixon and Coles used their model to make predictions about games that had been played in the 1995–1996 season, they found that the forecasts lined up nicely with the actual results. But would the model have been good enough to bet with? They tested it by going through all the games and applying a simple rule: if the model said a particular result was 10 percent more likely than the bookmakers' odds implied, it was worth betting on. Despite using a basic model and betting strategy, the results suggested that the model would be capable of outperforming the bookmakers.

Not long after publishing their work, Dixon and Coles went their separate ways. Dixon set up Atass Sports, a consultancy firm that specialized in the prediction of sports results. Later, Coles would join Smartodds, a London-based company that also worked on sports models. There are now several firms working on soccer prediction, but Dixon and Coles's research remains at the heart of many models. "Those papers are still the main starting points," said David Hastie, who co-founded soccer analytics firm Onside Analysis.

As with any model, though, the research has some weaknesses. "It's not an entirely polished piece of work," Coles has pointed out. One problem is that the measurements for teams' attacking and defending abilities don't change over the course of a game. In reality, players may tire or launch more attacks at certain points in the game. Another issue is that, in real life, draws are more common than a Poisson process would predict. One explanation might be that teams that are trailing put in more effort, with the hope of leveling the score line, whereas their opponents get complacent. But, according to Andreas Heuer and Oliver Rubner, two researchers at the University of Münster, there's something else going on. They reckon the large number of draws is because teams tend to take fewer risks—and hence are less likely to score—if the score line is even in the later stages of a game. When the pair looked at matches in the German Bundesliga from 1968 to 2011, they found that the goal-scoring rate decreased when the score was a draw. This was especially noticeable when the score was 0–0, with players preferring to settle for the "coziness of a draw."

It turned out that certain points in a game created particularly draw-friendly conditions. Heuer and Rubner found that Bundesliga goals tended to follow a Poisson process during the first eighty minutes of the match, with teams finding the net at a fairly consistent rate. It was only during the last period of play that things became more erratic, especially if the away team was leading by one or two goals in the dying minutes of the match.

By adjusting for these types of quirks, sports prediction firms have built on the work of Dixon, Coles, and others and have turned soccer betting into a profitable business. In recent years, these companies have greatly expanded their operations. But though the industry has grown, and new firms have appeared, the scientific betting industry is still relatively new in the United Kingdom. Even the most established firms started post-2000. In the United States, however, sports prediction has a much richer history—sometimes quite literally.

TO PASS TIME DURING dull high school classes, Michael Kent would often read the sports section of the newspaper. Despite living in Chicago, he followed college athletics from all over the country. As he leafed through the scores, he would get to thinking about the winning margin in each game. "A team would beat another team 28–12," he recalled, "and I would say, well how good is that?"

After high school, Kent completed a degree in mathematics before joining the Westinghouse Corporation. He spent the 1970s working in the corporation's Atomic Power Laboratory in Pittsburgh, Pennsylvania, where they designed nuclear reactors for the US Navy. It was very much a research environment: a mixture of mathematicians, engineers, and computer specialists. Kent spent the next few years trying to simulate what happens to a nuclear reactor that has coolant flowing through its fuel channels. In his spare time, he also started writing computer programs to analyze US football games. In many ways, Kent's model did for college sports what Bill Benter's did for horse races. Kent gathered together lots of factors that might influence a game's result, and then used regression to work out which were important. Just as Benter would later do, Kent waited until he had his own estimate before he looked at the betting market. "You need to make your own number," Kent said. "Then—and only then—do you look at what other people have."

STATISTICS AND DATA HAVE long been an important part of American sport. They are particularly prominent in baseball. One reason is the structure of the game: it is split into lots of short intervals, which, as well as providing plenty of opportunities to grab a hot-dog, makes the game much easier to analyze. Moreover, baseball innings can be broken down into individual battles—such as pitcher versus batter—that are relatively independent, and hence statistician-friendly.

Most of the stats that baseball fans pore over today—from batting averages to runs scored—were devised in the nineteenth century by Henry Chadwick, a sports writer who'd honed his ideas watching cricket matches in England. With the growth of computers in the 1970s, it became easier to collate results, and people gradually formed organizations to encourage research into sports statistics. One such organization was the Society for American Baseball Research, founded in 1971. Because the society's acronym was SABR, the scientific analysis of baseball became known as "sabermetrics."

Sports statistics grew in popularity during the 1970s, but several other ingredients are needed to cook up an effective betting strategy. It just so happened that Michael Kent had all of them. "I was very fortunate that a whole bunch of things came together," he said. The first ingredient was data. Not far from Kent's atomic laboratory in Pittsburgh was the Carnegie Library, which had a set of anthologies containing several years' worth of college sports scores and schedules. The good news was that these provided Kent's model with information it needed to generate robust predictions; the bad news was that each result had to be input manually. Kent also had the technology to power the model, with access to the high-speed computer at Westinghouse. His university had been one of the first in the country to get a computer, so Kent already had far more programming experience than most. That wasn't all. As well as knowing how to write computer programs, Kent understood the statistical theory behind his models. At Westinghouse, he'd worked with an engineer named Carl Friedrich, who'd shown him how to create fast, reliable computer models. "He was one of the most brilliant people I ever met," Kent said. "The guy was unbelievable."

Even with the crucial components in place, Kent's gambling career didn't get off to the best start. "Very early on, I had four huge bets," he said. "I lost them all. I lost $5,000 that Saturday." Still, he realized that the misfortunes did have some benefits. "Nothing motivated me more than losing." After working on his model at night for seven years, Kent finally decided to make sports betting his full-time job in 1979. While Bill Benter was making his first forays into blackjack, Kent left Westinghouse for Las Vegas, ready for the new college football season.

Life in the city involved a lot of new challenges. One of them was the logistics of placing the actual bets. It wasn't like Hong Kong, where bettors could simply phone in their selections. In Las Vegas, gamblers had to turn up at a casino with hard currency. Naturally, this made Kent a little nervous. He came to rely on valet parking, because it stopped him having to walk through poorly lit parking lots with tens of thousands of dollars in cash.

Because it was tricky to place bets, Kent teamed up with Billy Walters, a veteran gambler who knew how Las Vegas worked and how to make it work for them. With Walters taking care of the betting, Kent could focus on the predictions. Over the next few years, other gamblers joined them to help implement the strategy. Some assisted with the computer model, while others dealt with the bookmakers. Together, they were known as the "Computer Group," a name that would become admired by bettors almost as much as it was dreaded by casinos.

Thanks to Kent's scientific approach, the Computer Group's predictions were consistently better than Las Vegas bookmakers'. The success also brought some unwanted attention. Throughout the 1980s, the FBI suspected the group was operating illegally, conducting investigations that were driven partly by bemusement at how the group was making so much money. Despite years of scrutiny, however, the investigations didn't come to anything. There were FBI raids, and several members of the Computer Group were indicted, but all were eventually acquitted.

It has been estimated that between 1980 and 1985, the Computer Group placed over $135 million worth of bets, turning a profit of almost $14 million. There wasn't a single year in which they made a loss. The group eventually disbanded in 1987, but Kent would continue to bet on sports for the next two decades. Kent said the division of labor remained much the same: he would come up with the forecasts, and Walters would implement the betting. Kent pointed out that much of the success of his predictions came from the attention he put into the computer models. "It's the model-building that's important," he said. "You have to know how to build a model. And you never stop building the model."

Kent generally worked alone on his predictions, but he did get help with one sport. An economist at a major university on the West Coast came up with football predictions each week. The man was very private about his betting research, and Kent referred to him only as "Professor number 1." Although the economist's estimates were very good, they were different from Kent's forecasts. So, between 1990 and 2005, they would often merge the two predictions.

Kent made his name—and his money—predicting college sports such as football and basketball. But not all sports have received this level of attention. Whereas Kent was coming up with profitable football models in the 1970s, it wasn't until 1998 that Dixon and Coles sketched out a viable method for soccer betting. And some sports are even harder to predict.

ONE AFTERNOON IN JANUARY 1951, Françoise Ulam came home to find her husband Stanislaw staring out of the window. His expression was peculiar, his eyes unfocused on the garden outside. "I found a way to make it work," he said. Françoise asked him what he meant. "The Super," he replied. "It is a totally different scheme, and it will change the course of history."

Ulam was referring to the hydrogen bomb they had developed at Los Alamos. Thanks to the Monte Carlo method and other technological advances, the United States possessed the most powerful weapon that ever existed. It was the early stages of the Cold War, and Russia had fallen behind in the nuclear arms race.

Yet grand nuclear ideas weren't the only innovations appearing during this period. While Ulam had been working on the Monte Carlo method in 1947, a very different kind of weapon had emerged on the other side of the Iron Curtain. It was called the "Avtomat Kalashnikova" after its designer Mikhail Kalashnikov. In subsequent years, the world would come to know it by another name: the AK-47. Along with the hydrogen bomb, the rifle would shape the course of the Cold War. From Vietnam to Afghanistan, it passed through the hands of soldiers, guerrillas, and revolutionaries. The gun is still in use today, with an estimated 75 million AK-47s having been built to date. The main reason for the weapon's popularity lies in its simplicity. It has only eight moving parts, which means it's reliable and easy to repair. It might not be that accurate, but it rarely jams and can survive decades of use.

When it comes to building machines, the fewer parts there are, the more efficient the machine is. Complexity means more friction between the different components: for example, around 10 percent of a car engine's power is wasted because of such friction. Complexity also leads to malfunctions. During the Cold War, expensive Western rifles would jam while the simple AK-47 continued to function. The same is true of many other processes. Making things more complicated often removes efficiency and increases error. Take blackjack: the more cards a dealer uses, the harder it is to shuffle properly. Complexity also makes it harder to come up with accurate forecasts about the future. The more parts that are involved, and the more interactions going on, the harder it is to predict what will happen from limited past data. And when it comes to sport, there is one activity that involves a particularly large number of interactions, which can make predictions very difficult.

US President Woodrow Wilson once described golf as "an ineffectual attempt to put an elusive ball into an obscure hole with an implement ill adapted to the purpose." As well as having to deal with ballistics, golfers must also contend with their surroundings. Golf courses are littered with obstacles, ranging from trees and ponds to sand bunkers and caddies. As a result, the shadow of luck is never far away. A player might hit a brilliant shot, sending the ball toward the hole, only to see it collide with the flagstick and ricochet into a bunker. Or a player could slice the ball into a tree and have it bounce back into a strong position. Such mishaps are so common in golf that the rulebook even has a phrase to cover them. If the ball hits a random object or otherwise goes astray by accident, it's just the "rub of the green."

Whereas horse races in Hong Kong resemble a well-designed science experiment, golf tournaments are more likely to require one of Ronald Fisher's statistical postmortems. Over the four days of a tournament, players tee off at all sorts of different times. The location of the hole also changes between rounds—and if the tournament is in the United Kingdom, so will the weather. If that isn't bad enough, the field of potential winners is huge in a golf tournament. Whereas the Rugby World Cup has twenty teams competing for the trophy, and the UK Grand National has forty horses running, ninety-five players compete for the US Masters each year, and the other three majors are even larger.

All these factors mean that golf is particularly difficult to predict accurately. Golf has therefore been a bit of an outlier in terms of sports forecasting. Some firms are rising to the challenge—Smartodds now has statisticians working on golf prediction—but in terms of betting activity, the sport still lags far behind many others.

Even among different team sports, some games are easier to predict than others. The discrepancy comes partly down to the scoring rates. Take hockey. Teams playing in the NHL score two or three goals per game on average. Compare that to basketball, where NBA teams will regularly score a hundred points in a game. If goals are rare—as they are in hockey—a successful shot will have more impact on the game. This means that a chance event, such as a deflection or lucky shot, is more likely to influence the final result. Low-scoring games also mean fewer data points to play with. When a brilliant team beats a lousy team 1–0, there is only one scoring event to analyze.

Fortunately, it's possible to squeeze extra information out of a game. One approach is to measure performance in other ways. In hockey, pundits often use stats such as the "Corsi rating"—the difference between the number of shots directed at an opponent's net and number targeted at a team's own goal—to make predictions about score lines. The reason they use such rating systems is that the number of goals scored in previous games does not say much about a team's future ability to score goals.

Scoring is far more common in games such as basketball, but the way in which the game is played can affect predictability, too. Haralabos Voulgaris has spent years betting almost exclusively on basketball and is now one of the world's top NBA bettors. At the MIT Sloan Sports Analytics Conference in 2013, he pointed out that the nature of scoring in basketball was changing, with players attempting more long-distance three-point shots. Because randomness plays a bigger role in these types of shots, it was becoming harder to predict which team would score more points. Traditional forecasting methods assume that team members work together to get the ball near the basket and score; these approaches are less accurate when individual players make speculative attempts from farther away.

Why does Voulgaris bet on basketball rather than another sport? It comes partly down to the simple fact that he likes the game. Sifting through reams of data doesn't make for a great lifestyle if it's not interesting. It also helps that Voulgaris has lots of data to sift through. Models need to process a certain amount of data before they can churn out reliable predictions. And in basketball plenty of information is available to analyze. The same cannot be said for other sports, however. In the early days of English soccer prediction, it was a struggle to dig up the necessary data. Whereas American pundits were dealing with a flood of information, in the United Kingdom there was barely a puddle. "We don't realise how easy we have it sometimes these days," Stuart Coles said.

With soccer data hard to come by in the late 1990s, gamblers had to get ahold of the information in any way they could. In some cases, people created automated programs that would scour the handful of websites that did publish results and copy the data tables straight off the webpages. Although this "screen scraping" provided a source of data, the websites being scraped did not like gamblers taking their content and clogging up their servers. Some installed countermeasures—such as blocking certain IP addresses—to stop people taking their data.

Even in the data-rich world of US sports, there is still plenty of variation in information between different leagues. One of the reasons Kent analyzed college sports was the amount of information available. "There are so many more games in college basketball, and a lot more teams," Kent said. "You get a huge database." Having access to these data helped Kent predict the outcomes of matches and place appropriate bets beforehand.

THROUGHOUT KENT'S CAREER, SPORTS betting in Las Vegas would come to a halt once a game started. By the time the referee blew the whistle to start the match, Kent's money was already down. The gambling and the action, two things that seemed to be so closely linked, were instead separated. It wasn't until 2009, when a new company arrived in the city, that casinos finally fixed this broken Venn diagram of gambling. That company was Cantor Gaming, part of the Wall Street firm Cantor Fitzgerald. In recent years, it has become the resident bookmaker at a number of major casinos. Walk into the sports section of the Venetian, the Cosmopolitan, or the Hard Rock, and you'll find dozens of big screens and betting machines, all operated by Cantor. Crammed between coverage of everything from baseball to football, there are rows of numbers and names, showing the odds for different matches. These "betting lines" rise and fall with the noise of the crowds. The room feels like a hybrid of a sports bar and a trading floor, a place where drinks and data blend together under the eternally bright casino lights.

The numbers on Cantor's screens might reflect the emotions of the spectators, but they are actually controlled by a computer program that adjusts the betting lines throughout the course of the game. Cantor calls it the "Midas algorithm." If something happens in the game, the program updates the odds on display automatically. Thanks to Midas, in-play betting has taken off in a big way in Las Vegas.

Much of the credit for the Midas software goes to an Englishman named Andrew Garrood, who joined Cantor in 2000. Before that, he'd been working as a trader for a Japanese investment bank. The leap to Las Vegas wasn't as big as it might seem: Garrood simply went from designing models that could price financial derivatives to ones that could put a value on sports results.

Cantor's biggest statement of intent came in 2008, when it bought a company called Las Vegas Sports Consultants. This company came up with odds for bookmakers across Nevada, including nearly half the casinos in Las Vegas. Cantor wasn't just interested in its predictions, however. In buying the company, Cantor had secured an extensive database of past results for a whole range of sports. The information would form a vital part of Cantor's analysis. From baseball to football, Cantor needed to know how certain events changed a game. If the San Francisco Giants get another home run, how would it affect their chances of winning? If the New England Patriots have one last attempt to get the ball down in the dying moments of a game, how likely are they to pull it off?

According to Garrood, straightforward "vanilla" events are relatively easy to predict. For instance, it's not too difficult to work out the chances a football team will score a touchdown if they start a drive from the 20-yard line. The problem is that there could be many successes and failures during a game, some of which are subtler than others. Which events are worth knowing about? Garrood has found that most plays don't affect the outcome much. It's therefore important to pin down the crucial events, the ones that do make a big difference. This is where the enormous database comes in handy. While many gamblers are relying on gut instinct, Midas can assess just how much effect that touchdown will really have.

How does Cantor make sure it gets all its predictions right? The answer is that it doesn't try to. There is a commonly held view that firms like Cantor use models to try to nail the correct result for every game. Matthew Holt, director of sports data at Cantor, has rebuffed this myth. "We don't make lines to predict the outcome of a game," he said in 2013. "We make lines in anticipation of where the action will come."

When it comes to betting, bookmakers' aims are fundamentally different from those of gamblers. Suppose two tennis players in the US Open are perfectly matched. The game is 50/50, which means that for a $1.00 bet, a fair return would be $1.00: if a gambler bet on both players, the bettor would come out even. But a bookmaker won't offer odds that return $1.00. Instead, it might offer a payoff of $0.95. Anyone who bets on both players will therefore end up $0.05 poorer.

If the same total amount is wagered on each player, the bookmaker will lock in a profit. But what if most bets go on one of the players? The bookmaker will need to adjust the odds to make sure it stands to gain the same amount regardless of who wins. The new odds might suggest one player is less likely to come out on top. Smart gamblers, who know that both players are equally good, will therefore bet on the one with longer odds. For bookmakers that have done their job properly, this isn't a concern. They don't move their betting lines to match the true chance of a result happening. They do it to balance the books.

Every day, the Midas algorithm combines computer predictions with real betting activity, tweaking the odds as wagers come in. It performs this juggling act for dozens of different sports, simultaneously updating betting lines as games progress. To make a profit, bookmakers like Cantor have to understand where gamblers' money is going. What are they betting on? How might they react to a particular event?

Just as information flows between bettors and bookmakers, in many instances gamblers will also try to work out what their rivals are doing. When word gets out that a betting syndicate has come up with a successful strategy, others inevitably want a piece of the action. Because many betting strategies have their origins in academia, it's often possible to piece together the basic models by sifting through research articles. But sports betting is a competitive industry, which means some of the most effective techniques remain shrouded in secrecy. According to sports statistician Ian McHale, "The proprietary nature of prediction models means that the published ones rarely (if ever) represent the very best models."

If gamblers don't know who has the best strategy, it can create a tense environment. In the huge Asian markets, where many of the largest soccer bets are placed, wagers are often made via instant messaging software. At the same time, information ricochets between bookmakers and gamblers, as each tries to work out what the other is thinking and how the other will bet. "The betting grapevine is huge," as one industry insider put it. "There is a lot of paranoia."

WHEN ASIAN BOOKMAKERS GET coverage in the Western media, it's not usually for a good reason. After some suspicious bowling in cricket matches between Pakistan and England in 2010, three of Pakistan's players were handed bans for agreeing to deliver bad balls. Reporters noted that bookmakers—many of whom were based in Asia—would often target such games. The scandals have since continued. During the summer of 2013, three cricketers playing in the Indian Premier League were charged with match fixing. Police claimed bookmakers had promised upward of $40,000 to the players if they let the opposition score runs at specific times. Then, in December 2013, UK police arrested six soccer players for allegedly offering to collect yellow or red cards to order.

There is certainly a huge appetite for gambling in Asia, and not all of it aboveboard. It's been estimated that the illegal betting market in China is ten times larger than the legitimate sums handled by the Hong Kong Jockey Club. Illegal betting is also common in India. When the national cricket team plays arch rival Pakistan, the total amount wagered can approach $3 billion. Yet the Asian betting market is changing. Gamblers no longer need to track down black market bookmakers in rooms behind backstreet bars. There was a time when they would need to bring cash and a code word; now they can bet by phone or online. Glossy call centers have replaced grimy betting rooms. The new industry is a step away from the illegal black market, but it remains little regulated. This is the "gray market": modern, corporate, and opaque.

When it comes to high-stakes betting on sports such as soccer, Asia is the location of choice for many Western gamblers. The reason is simple. In Europe and the United States, bookmakers rarely take large bets. As a result, gamblers based in these regions are finding it harder to stake the sort of money that will make their strategies profitable. Despite being a prolific bettor—or rather *because* he is a prolific bettor—Haralabos Voulgaris has complained that US bookmakers are reluctant to take his bets. Even when they do, the betting limits are placed at unhelpfully low levels; he might be allowed to stake only a few thousand dollars. Not all Western bookmakers shun successful bettors, however. In the past decade, one firm has gained a reputation for accepting—and even encouraging—bets from smart gamblers.

When Pinnacle Sports started in 1998, it was clear that it had some bold ambitions. The betting limits were high, with maximum stakes larger than those offered by many existing bookmakers. Pinnacle claimed it was happy to let players bet the maximum as often as they liked. Even if a player consistently made money, Pinnacle wouldn't shut the player down. Back in 2003, such ideas went completely against established bookmaking wisdom. If you want to make money, went the dogma, don't let smart bettors place huge bets. And certainly don't let them do it again and again. So, how did Pinnacle pull it off?

Whereas all bookmakers look at overall betting activity, Pinnacle also puts a lot of effort into understanding who is placing those bets. By accepting wagers from sharp bettors, Pinnacle can get an idea of what these gamblers think might happen. It's not very different from Bill Benter combining his predictions with the public odds displayed at Happy Valley. Sometimes the public knows things that a betting syndicate—or a bookmaker—might not.

Pinnacle generally posts an initial set of odds on Sunday night. It knows these numbers might not be perfect, so only take a small amount of bets at first. It has found that the first bets almost always come from talented small-stakes bettors: because the early odds are often incorrect, sharp gamblers pile in and exploit them. But Pinnacle is happy to hand an advantage to these so-called hundred-dollar geniuses if it means ending up with much better predictions about the games. In essence, Pinnacle pays smart gamblers for information.

The strategy of purchasing information has been attempted in other walks of life, too, sometimes with controversial results. In the summer of 2003, US Senators stumbled across a Department of Defense proposal for a "policy analysis market" that would allow traders to speculate on events in the Middle East. It would be possible to bet on events such as a biochemical attack, for example, or a coup d'état, or the assassination of an Arab leader. The idea was that if anyone had inside information and tried to exploit it, the Pentagon would be able to spot the change in market activity. Investors might make a profit, but they would also reveal their hand in the process. Robin Hanson, the economist behind the proposal, pointed out that intelligence agencies by definition pay people to report unsavory details. In moral terms, he didn't see a market as any better or worse than other types of transactions.

The Senators disagreed. One called the idea "grotesque"; another said it was "unbelievably stupid." According to Hillary Clinton, the policy would create "a market in death and destruction." The proposal did not survive long in the face of such fierce opposition. By the end of July, the Pentagon had scrapped the idea. The decision was arguably an ethical rather than economic one. Although critics attacked the morality of the proposal, few disputed that betting markets can reveal valuable insights about an event. Unlike participants in an opinion poll, gamblers have a financial incentive to be right.

When they make predictions about the future, they are putting their money where their mouth (or model) is.

Today, Pinnacle canvasses gamblers' opinions on a wide range of subjects. People can bet on the identity of the next president or who will take home an Academy Award. Pinnacle has so much faith in the approach that it regularly takes large bets on popular events: in the past, it has been possible to wager half a million dollars on the soccer Champions League final. Because Pinnacle's business model relies on having accurate predictions, there are some things it doesn't take bets on. In 2008, for instance, Pinnacle dropped horse racing as a betting option because it doesn't specialize in the sport.

Companies like Pinnacle, which have found a way to combine in-house statistical predictions with the opinions of smart gamblers, have challenged traditional bookmaking. By harnessing the knowledge of smart gamblers, they have more confidence in their odds, and hence are happy to take larger bets. Yet bookmakers are not the only ones changing. In some cases, gamblers are skipping the bookmaker altogether.

DURING THE PAST DECADE or so, approaches to betting have changed dramatically. As well as wagers moving online, bookmakers have faced competition from a new type of gambling market, in the form of the betting exchange. This is much like a stock exchange, except instead of buying and selling shares, gamblers can offer and accept wagers. Perhaps the best-known betting exchange is the London-based Betfair, which handles over seven million bets a day.

Betfair's creator, Andrew Black, came up with the idea for the website during the late 1990s, when he was a programmer at the British Government Communication Headquarters in Gloucestershire. Security wouldn't let him stay on site past five o'clock, so he found himself spending each evening alone in his rural farmhouse. Having so much free time was a burden, but it was also rather fruitful. "The boredom was horrendous," he later told the *Guardian*, "but mentally I became really quite productive."

While attending college, Black had developed an interest in betting. But there were some drawbacks with the traditional way of gambling, and during those evenings in Gloucestershire, Black thought about how things could be improved. Rather than going through a bookmaker, as he'd always had to do, why not let gamblers bet directly against each other? The project meant combining ideas from financial markets, gambling, and online retail. Black, who had previously spent time as a professional gambler, stock trader, and website builder, had experience in all three of these areas.

The Betfair website launched in 2000. That summer, the company arranged for a mock funeral procession to pass through the city of London, with a coffin announcing the "death of the bookmaker." Although the stunt brought plenty of media coverage, competitors were already lurking. One rival website mimicked eBay: if someone wanted to place a bet of £1,000 at certain odds, the site would try to pair the bettor up with someone happy to take such a bet. Trying to pair people up was a bit like playing a giant online game of snap. And that sometimes meant waiting a long time for a match.

Fortunately, Betfair had a way to speed things up. If there were no takers for a wager, the website would divide up the bet between several different people. Rather than trying to find someone willing to take the full £1,000, for example, the website might slice up the total and match it with, say, five people wanting to accept a £200 wager. Whereas bookmakers had traditionally made their money by tweaking the odds on display, Betfair left the odds untouched and instead took a cut from the profits of whoever won a particular bet.

Betting exchanges like Betfair have opened up a new approach to gambling. Unlike traditional bookmakers, you aren't limited to betting for a particular result. You can also "lay" the result by taking the other side of the bet; if the result doesn't happen, you win whatever was staked.

Because you can bet both ways on a betting exchange, it's possible to make money before a match ends. Suppose a betting exchange currently displays odds of 5 for a particular team. You decide to bet £10 on them, which means you'll get back £50 if they win. Then something changes. Perhaps the opposition's star player picks up an injury. Your team is more likely to win now, so the odds drop to 2. Rather than waiting until the match ends—and risking the result going against you—you can hedge your original bet by accepting someone else's £10 bet at the lower odds. If your team wins, you'll get £50 from the first bet but have to cough up £20 for the second; if your team loses, the two bets will cancel each other out, as shown in Table 4.1. The match hasn't even started and you're guaranteed £30 if your team wins, and you won't lose anything if they don't. (Many bookmakers have since introduced a "cashout" feature, which in essence reproduces these trades.)

Because you can back and lay each result, the Betfair website displays two columns for every match, showing the best available odds on each side of the bet. Such technology has made it easier for gamblers to see what others are thinking and to take advantage of odds they believe are incorrect. Yet it's not just bookmaking that is becoming more accessible.

Table 4.1. Bets Can Be Hedged by Backing and Then Laying the Same Team

		1st bet	**2nd bet**	**Total**
Result	Your team wins	£50	–£20	£30
	Your team loses	–£10	£10	£0

SCIENTIFIC BETTING STRATEGIES HAVE traditionally been the preserve of private betting syndicates like the Computer Group or, more recently, consultancy firms like Atass. This may not be the case for much longer. Just as banks offer clients access to investment funds, some companies are letting people invest in scientific gambling methods. As Bloomberg columnist Matthew Klein put it, "If I find a guy who is good at sports betting and is willing to bet with my money in exchange for a fee, he is, for all intents and purposes, a hedge fund manager." Rather than putting money into established asset classes such as shares or commodities, investors now have the option of sports betting as an alternative asset class.

Betting might seem somewhat distant from other types of investment, but that is one of its selling points. During the 2008 financial crisis, many asset prices fell sharply. Investors often try to build a diverse portfolio to protect against such shocks; for example, they might hold stocks in several different companies in a range of industries. But when markets run into trouble, this diversity is not always enough. According to Tobias Preis, a researcher in complex systems at the University of Warwick, stocks can behave in a similar way when a financial market hits a rough period. Preis and colleagues analyzed share prices in the Dow Jones Industrial Average between 1939 and 2010 and found that stocks would go down together as the market came under more stress. "The diversification effect which should protect a portfolio melts away in times of market losses," they noted, "just when it would most urgently be needed."

The problem isn't limited to stocks. In the run-up to the 2008 crisis, more and more investors began to trade "collateralized debt obligations." These financial products gathered together outstanding loans such as home mortgages, making it possible for investors to earn money by taking on some of the lenders' risk. Although there might have been a high probability that a single person would default on a loan, investors assumed it was extremely unlikely everyone would default at the same time. Unfortunately, this assumption turned out to be incorrect. When one home lost its value during the crisis, others followed.

Advocates of sports betting point out that wagers are generally unaffected by the financial world. Games will still go ahead if the stock market takes a dive; betting exchanges will still accept wagers. A hedge fund that concentrates on sports betting should therefore be an attractive investment, because it provides diversification. It was this idea that persuaded Brendan Poots to set up a sports-focused hedge fund in 2010. Based in Melbourne, Australia, Priomha Capital aimed to give the public investors access to the traditionally private world of sports prediction.

Creating good forecasts can require additional expertise, so Priomha linked up with researchers at the Royal Melbourne Institute of Technology. To some extent, the approach is a twenty-first-century version of the Computer Group's strategy. Priomha creates a model for a particular sport, runs simulations to predict the likelihood of each result, and then compares the predictions with the current odds on betting exchanges such as Betfair.

The big difference is that investors are not restricted to betting before a game starts. Which is good news, because Poots has found that odds generally settle down to a fair value in the run-up to a fixture. "Come kick off, the market's pretty efficient," he said. "But once play starts, that's where we have a huge opportunity."

When it comes to soccer prediction, "in-play" analysis was always the natural next step. After working on final score predictions in 1997, Mark Dixon turned his attention to what happens during a soccer match. Along with fellow statistician Michael Robinson, he simulated matches using a similar model to the one he'd published with Stuart Coles, but with some important new modifications. As well as accounting for each teams' attacking strength and defensive weakness, the model included factors based on the current score and time left to play. It turned out that including in-play information led to more accurate predictions than the original Dixon-Coles model.

The model also made it possible to test popular soccer "wisdom." Dixon and Robinson noted that commentators would often tell viewers that teams were more vulnerable after they scored a goal. The researchers referred to this cliché as "immediate strike back." The idea is that after a goal goes in, attackers' concentration wobbles, which can allow the opposition back into the game. But the cliché turned out to be misguided. Dixon and Robinson found that teams weren't especially vulnerable after they have scored a goal. So, why did commentators often claim that they were?

If we come across something unusual or shocking, it stands out in our mind. According to Dixon and Robinson, "People have a tendency to overestimate the frequency of surprising events." This doesn't just happen in sports. Many worry more about terrorist attacks than bathtub accidents, despite the fact that—in the United States at least—you're far more likely to die in a bathtub than at the hands of a terrorist. Unusual events are more memorable, which also explains why people think it's easier to become a millionaire with a one-dollar lottery ticket than by playing roulette repeatedly. Although both are terrible ideas, in terms of raw probability, playing roulette again and again is more likely to generate a lucky million dollars in profit.

Betting successfully during a soccer match means identifying human biases like these. Are there certain aspects of the game that gamblers consistently misjudge? Poots has found that a few things stand out. One is the effect of goals. Just as Dixon and Robinson noted, the popular view is not always the correct one: a goal doesn't always create the shock people think that it does. Gamblers also tend to overestimate the impact of red cards. That isn't to say they don't have any effect. A team playing against an opponent with ten men will probably score at a higher rate (one 2014 study reckoned the rate would be 60 percent higher on average). But the odds often move too far, suggesting gamblers mistake a difficult situation for a hopeless one.

Following a dramatic event, the odds available on a betting exchange gradually adjust to the new situation. When things have settled down, Priomha can hedge its bets by taking the opposite position. If it backed the home team to win at long odds, perhaps after a red card, it will bet against them when the odds decrease. This way, it doesn't matter how the game ends. Like a trader who buys an item from a panicked seller and later sells it back at a higher price, the team closed their position and offloaded any remaining risk.

There are plenty of opportunities to capitalize on inaccurate odds during a game. Unfortunately, there are also fewer bets on offer, which means Priomha has to be careful not to disrupt the market with large wagers. "During play, you have to drip feed your money in," Poots said. In fact, the size of the market is one of the biggest obstacles facing funds like Priomha. Because it makes money by identifying incorrect sports odds, the more money it has to invest, the more erroneous odds it has to find.

The current plan is to manage up to $20 million of investors' money. Poots pointed out that if they tried to handle a much larger amount—such as $100 million—it would be a struggle to make reasonable returns. They might be able to find enough opportunities to bring in a 5 percent annual return, but as a hedge fund, they really want to be making double figures for their investors, and they are more likely to achieve this if they constrain the size of the fund.

Although Priomha has not reached its limit yet, as the fund grows Poots is noticing a change in who is buying into the strategy. "Our investor profile used to be someone who likes sport, and liked to have a bet," he said. "It's now becoming people who have got their pension or other funds to invest."

Priomha is not the only sports betting fund to have appeared in recent years. The London-based Fidens syndicate opened its fund to investors in 2013; two years later it was managing more than £5 million. Mathematics graduate Will Wilde heads up Fidens's trading strategy. This involves betting on ten soccer leagues around the globe, placing around three thousand wagers per year.

Stock market investing has often been compared to gambling, especially when shares are held for only a short period of time. There is a certain irony then, that gambling is increasingly seen as a viable option for investors. Not all sports betting funds have been successful, however. In 2010, investment firm Centaur launched the Galileo fund, which was designed to allow investors to profit from sports betting. The plan was to attract $100 million of investment and generate an annual return of 15 to 25 percent. The finance community watched with interest, but two years later the fund folded.

Although the ambitions of funds like Priomha are currently constrained by the size of the betting market, things could be very different if sports betting were to expand in the United States. "If America was to open up," Poots said, "the whole game changes completely." The first major hints of change came soon after Priomha was founded. Following a referendum in 2011, New Jersey governor Chris Christie signed a bill legalizing sports betting in the state. For the first time, gamblers in Atlantic City would be able to bet on games like the Super Bowl. That was the theory at least. It did not take long for professional sports leagues to bring in lawyers to halt the expansion. The case has been bouncing through the court system ever since, the main obstacle being a federal law from 1992 that prohibits sports betting in all but four states. Opponents say gambling should be limited to places like Las Vegas; New Jersey claims the law is unconstitutional and that the public supports legalized betting. Indeed, many sports leagues already allow people to put money on their predictions coming true. Every year people pay to take part in fantasy sports leagues, even though betting on a specific match outcome is still illegal.

Advocates for law changes say there are two main advantages to legalized gambling. First, it would generate more tax. It's been estimated that less than 1 percent of sports bets in the United States are placed legally. The remaining 99 percent of wagers, made through unlicensed bookmakers or offshore websites, probably run into hundreds of billions of dollars. If these bets were legal, the tax revenue would be enormous. Second, legalization means regulation, and regulation means transparency. Bookmakers and betting exchanges keep records of customers, and online firms also have bank details. According to NBA commissioner Adam Silver, legalizing gambling would bring the activity into view of government scrutiny. "I believe that sports betting should be brought out of the underground and into the sunlight where it can be appropriately monitored and regulated," he wrote in the *New York Times* in 2014.

Betting syndicates would also stand to benefit from legalization. With more bookmakers taking wagers, syndicates could place bets on a much grander scale. There is also a chance that new laws will allow syndicates to bet in Las Vegas. Currently, if gamblers want to bet on sports in the city, they still need to turn up at a casino with a handful of cash, which makes it difficult to systematically place large bets. In 2015, the Nevada senate passed a bill that would allow a group of investors to back a bettor, which is essentially what Priomha already does outside of the United States. If the bill gets through the state assembly and becomes law, many more sports hedge funds could appear. Other countries are also debating new gambling laws. In Japan, sports bettors can currently put money only on horse, boat, or cycle races. A new bill, submitted in April 2015, and supported by the prime minister, proposes to change that. New opportunities will also arise in India and China, as informal betting markets become more regulated.

According to sports journalist Chad Millman, it is not just established gamblers who would be well positioned to profit from law changes. During a visit to MIT in March 2013, Millman got talking to Mike Wohl, an MBA student at the university's business school. For his study project, Wohl had considered gambling as "the missing asset class." Wohl had a background in finance, and his analysis—along with his personal experience of betting—suggested that sports wagers could produce as good a trade-off between risk and return as investing in stocks could.

Millman pointed out that there are two extremes to the gambling spectrum. At one end are professional sports bettors, the so-called sharps who regularly place successful bets. At the other are the everyday gamblers, who don't have predictive tools or reliable strategies. In between, Millman says, are a number of people like Wohl who have the necessary skills to bet successfully but haven't yet chosen to use them. They might work in finance or research; perhaps they have MBAs or PhDs. If sports betting was to expand in the United States, these small-scale bettors would be in a good position to profit. With their quantitative backgrounds, they are already familiar with the crucial methods. They also have the necessary tools, thanks to increases in computing power and data availability. All they need now is the access.

THERE ARE CERTAIN ADVANTAGES to being a betting start-up. For one, it means more flexibility. But should new syndicates follow sports betting strategies that have already been successful? Or should they exploit their flexibility and try something else?

In retrospect, Michael Kent would look at matches in far more detail. "If I was starting over right now," he said, "I would want to have play-by-play data." The additional information would make it possible to measure individual contributions. This would be a stark contrast with his previous analysis: in his models, Kent has always treated teams as a single entity. "I have no knowledge of players," he said. "I know what the team did, but I don't know the name of the quarterback."

Some modern betting syndicates go to great lengths to measure individual performances. "We do analysis on the effect of every player in every team," Will Wilde said. "Every player has a rating that goes up or down, regardless of whether they play or not." In Hong Kong, Bill Benter's syndicate even employs people to sift through videos of races. They might look at how a horse's speed changes during the race or how well it recovers after a bump. These "video variables" make up a relatively small part of the model—about 3 percent—but they all help nudge the predictions a little closer to reality.

It is not always just a matter of collecting more data. In soccer, successful defenders can be a nightmare for statisticians. During his years playing for Milan and Italy, Paolo Maldini averaged one tackle every other game. It wasn't because he was a lazy player; it was because he didn't need to make many tackles. He held back the opposition by getting into the right positions. Raw statistics such as number of tackles can therefore be misleading. If a defender makes fewer tackles, it doesn't always mean he's getting worse. It could mean he's improving.

A similar problem crops up with cornerbacks in US football. Their job is to patrol the edges of the field, defending against attacking passes by the opposition. Good cornerbacks intercept lots of passes, but great ones won't need to: the other team will be trying to avoid them. As a result, the best cornerbacks in the NFL might touch the ball only a handful of times per season.

How can we measure a player's ability if they rarely do anything that can be measured? One option is to compare the overall team performance when a player is and isn't on the field. At the simplest level, we could look at how often a team wins when a certain individual is playing. Sometimes it is clear that a player is valuable to a team. For example, when striker Thierry Henry played for Arsenal soccer club between 1999 and 2007, the team won 61 percent of matches he appeared in. On the other hand, they won only 52 percent of the games he missed.

Counting wins is simple enough, but measuring players in this way can raise some unexpected results. In some cases, it might even appear that fan favorites are not actually that important to the team. Since Steven Gerrard made his first appearance for Liverpool in 1998, they have won half the games he's played in. Yet they have also won half the games he hasn't had a role in. Brendan Poots points out that the best clubs have strong squads, so can often cope with losing a star individual. When top players go off injured, teams adjust. "In the sum of the parts," Poots said, "the effect that they have—or their absence has—is not as great as people think."

The problem with simply tallying up wins with and without a certain player, however, is that the calculation doesn't account for the importance of those games or the strength of the opposition. Teams often field more big-name players in crucial matches, for example. One way to get around these issues is to use a predictive model. Sports statisticians often assess the importance of a particular player by comparing the predicted scores for the games the player played in with the actual results of those games. If the team performs better than expected when that player is on the field, it suggests the player is especially important to the team.

Again, it's not always the best-known players who come out on top. This is because identifying the most important player is not the same as finding the best player. The most important player—as judged by the model—might be someone without an obvious replacement or a player whose style suits the team particularly well.

To interpret the results of their predictive models, firms working on sports forecasts employ analysts with a detailed knowledge of each team. These experts can suggest why a certain player appears to be so important and what that might mean for upcoming matches. Such information is not always easy to quantify, but it might have a big effect on results. The trick is to know what the model doesn't capture and to account for such features when making predictions. Sports statistician David Hastie points out that this goes against many people's idea of a scientific betting strategy. "There is a common perception that betting is all about models," he said. "People expect a magic formula."

GAMBLERS NEED TO KNOW how to get at crucial information, whether it is quantitative, as is the case with model predictions, or of a more qualitative nature, as with human insights. Although well known for his computer models, Kent knew the importance of human experts when making predictions. He received regular updates from people with in-depth knowledge of certain sports, people whose job it was to know things that the model might not capture. "We had a guy in New York City who could tell you the starting lineup for 200 college basketball teams," he said.

Making better predictions about individual players doesn't just benefit gamblers. As techniques improve, bettors and sports teams are finding more common ground, drawn together by a common desire to anticipate what will happen in the next season, or the next game, or even the next quarter. Every spring, team managers chat with statisticians and modelers at the MIT Sloan Sports Analytics Conference. Prediction methods can be particularly useful when teams go scouting for new signings. Historically, assessing a player's value has been difficult because performances are subject to chance. A player might have an impressive—and lucky—season one year, and then have a less successful time the next.

The "*Sports Illustrated* jinx" is a well-known example of this problem: often a player who appears on the cover of *Sports Illustrated* subsequently suffers a dip in form. Statisticians have pointed out that the *Sports Illustrated* jinx is not really a jinx. Players who end up on the cover often do so because they've had an unusually good season, which was down to random variation rather than a reflection of their true ability. The drop in performance that came the following year was simply a case of regression to the mean, just as Francis Galton found while studying inheritance.

When a club signs a new player, it has to make decisions based on past accomplishments. Yet what it is really paying for is future performances. How can a sports club predict a player's true ability? Ideally, it would be possible to pull past performances apart and work out how much they were influenced by ability and by chance. Statistician James Albert has attempted to do this for baseball. By trawling through lots of different statistics for pitchers, including wins and losses, strikeouts—where the batter misses the ball three times—and runs scored against them. He found that the number of strikeouts was the most accurate representation of a pitcher's true skill, whereas statistics such as home runs conceded were more subject to chance, and hence are a poor reflection of pitching ability.

Other sports are trickier to analyze. Soccer pundits generally use simple measurements, such as goals per game, to quantify how good strikers are. But what if strikers play for a good team and benefit from having other players setting them up with scoring chances? In 2014, researchers at Smartodds and the University of Salford assessed the goal-scoring ability of different soccer players. Rather than just asking how likely a striker was to score, they split goal scoring into two components: the process of generating a shot—which could be influenced by the team performance—and the process of converting that shot into a goal. Splitting up scoring in this way led to far better predictions about future goal tallies than simple goals-per-game statistics provided. The study also produced some unexpected conclusions. For instance, it appeared that the number of shots a player had bore little relation to the team's attacking ability. In other words, good players generally end up with a similar shot count regardless of whether they are playing for a great team or a weak one. Although better teams have more shots overall, a decent player ends up being a little fish in a large scoring pool; at a struggling club, that same player can make a bigger contribution to the total. The researchers also found that it was difficult to predict how often a player would convert shots into goals. Hence, they suggest that team managers looking at a potential signing should estimate how many shots that player generates rather than how many goals the player scores.

WHEN IT COMES TO scientific sports betting, the most successful gamblers are often the ones who study games others have neglected. From Michael Kent's work on college football to Mark Dixon and Stuart Coles's research in soccer, the big money generally comes from moving away from what everyone is doing.

Over time, bookmakers and gamblers have gradually latched on to the best-known strategies. As a result, it is becoming harder to profit from major sports leagues. Erroneous odds are less common, and competitors are quick to jump on any advantage. New syndicates are therefore better off focusing their attention on lesser-known sports, where scientific ideas have often been ignored. According to Haralabos Voulgaris, this is where the biggest opportunities lie. "I would start with the minor sports," he said at the MIT Sloan Sports Analytics Conference in 2013. "College basketball, golf, NASCAR, tennis."

In minority sports, additional knowledge—whether from models or experts—can prove extremely valuable. Because crucial variables are not so well known, the difference in skill between a sharp bettor and a casual gambler can be huge. As well as helping gamblers build better predictive models, improvements in technology are also changing how bets are made. The days of suitcases full of banknotes are coming to an end. Bets can be placed online, and gamblers can control hundreds of wagers at the same time. This technology has also paved the way for new types of strategy. A large part of sports betting throughout its history has been about forecasting the correct result. But scientific betting is no longer just a matter of predicting score lines. In some cases, it is becoming possible to know nothing about the result and still make money.

5

RISE OF THE ROBOTS

"WHAT HATH GOD WROUGHT!" THE MESSAGE READ. IT WAS May 24, 1844, and the world's first long-distance telegram had just arrived in Baltimore. Thanks to Samuel Morse's new telegraph machine, the biblical quote had traveled along a wire all the way from Washington, DC. Over the next few years, single-wire telegraph systems spread around the globe, creeping into the heart of all sorts of industries. Railway companies used them to send signals between stations, while police fired off telegrams to get ahead of fleeing criminals. It wasn't long before British financiers got ahold of the telegraph, too, and realized that it could be a new way to make money.

At the time, stock exchanges in the United Kingdom operated independently in each region. This meant there were occasional differences in prices. For example, it was sometimes possible to buy a stock for one price in London and sell it for a higher price in one of the provinces. Obtain such information quickly enough, and there was a profit to be made. During the 1850s, traders used telegrams to tell each other about discrepancies, cashing in on the difference before the price changed. From 1866 onward, America and Europe were linked by a transatlantic cable, which meant traders were able to spot incorrect prices even faster. The messages that traveled down the wire were to become an important part of finance (even today, traders refer to the GBP/USD exchange rate as "cable").

The invention of the telegraph meant that if prices were out of line in two locations, traders had the means to take advantage of the situation by buying at the cheaper price and selling at the higher one. In economics, the technique is known as "arbitrage." Even before the invention of the telegraph, so-called arbitrageurs had been on the hunt for mismatched prices. In the seventeenth century, English goldsmiths would melt down silver coins if the price of silver climbed past the value of the coin. Some would even trek further afield, hauling gold from London to Amsterdam to capitalize on differences in the rate of exchange.

Arbitrage can also work in gambling. Bookmakers and betting exchanges are merely different markets trading the same thing. They all have varying levels of betting activity and contrasting opinions about what might happen, which means their odds won't necessarily line up. The trick is to find a combination of bets so that whatever happens, the payoff will be positive. Suppose you're watching a tennis match between Rafael Nadal and Novak Djokovic. If one bookmaker is offering odds of 2.1 on Nadal, and another is offering 2.1 on Djokovic, betting $100 on each player will net you $210—and cost you $100—whatever the result. Whoever wins, you walk away with a profit of $10.

Unlike syndicates working on sports prediction, which are in essence betting that their forecast is closer to the truth than the odds suggest, arbitrageurs don't need to take a view on what will happen.

Whatever the result, the strategy should lead to a guaranteed profit, so long as a gambler can spot the opportunity in the first place. But how common are arbitrage situations?

In 2008, researchers at Athens University looked at bookmakers' odds on 12,420 different soccer matches in Europe and found 63 arbitrage opportunities. Most of the discrepancies occurred during competitions such as the European Championship. This was not particularly surprising, because tournament results are generally more variable than results in leagues where teams play each other often.

The following year, a group at the University of Zurich searched for potential arbitrage in odds given by betting exchanges like Betfair as well as traditional bookmakers. When they considered both types of market, there were far more stray odds. They found it would have been possible to make a guaranteed profit on almost a quarter of games. The average return wasn't huge—around 1 to 2 percent per game—but it was clear there were enough inconsistencies to make arbitrage a viable option.

Despite the allure of arbitrage betting, there are some potential pitfalls. To be successful, gamblers need to set up accounts with a large number of bookmakers. These companies usually make it easy to deposit money but hard to withdraw it. Bets also need to be placed simultaneously: if one wager lags behind another, the odds might change, thwarting any chance of a guaranteed profit. Even if gamblers can overcome these logistical issues, they have to avoid attracting the attention of the bookmakers themselves, who generally dislike having arbitrageurs cutting into their profits.

It is not just differences between bookmakers that can be exploited. Economist Milton Friedman pointed out that there is a paradox when it comes to trading. Markets need arbitrageurs to take advantage of incorrect prices and make them more efficient. Yet, by definition, an efficient market shouldn't be exploitable, and hence shouldn't attract arbitrageurs. How can we explain this contradictory situation? In reality, it turns out that markets often have short-term inefficiencies. There are periods of time when prices (or betting odds) do not reflect what is really going on. Although the information is out there, it hasn't been processed properly yet.

After a major event—such as a goal being scored—gamblers on betting exchanges need to update their opinions on what the odds should be. During this period of uncertainty, whoever reacts to the news first will be able to place bets against opponents who have not yet adjusted their odds. There is a limited window in which to do this. Over time the market will become more efficient, and the available odds will change to reflect the new information. In 2008, a group of researchers at the University of Lancaster reported that it takes less than sixty seconds for gamblers on betting exchanges to adjust to a dramatic event in a soccer match.

Not only is the betting window small, potential gains can be modest, too. To profit a gambler would need to place a large number of bets, and place them quickly. Unfortunately, this is not something that humans are particularly good at. We take time to process information. We hesitate. We struggle with multiple tasks. As a result, some gamblers are choosing to step back from the bustle of hectic betting markets. Where humans falter, the robots are rising.

THERE ARE TWO WAYS to access the Betfair betting exchange. Most people simply go to the website, which displays the latest odds as they become available. But there is another option. Gamblers can also bypass the website and link their computers directly to the exchange. This makes it possible to write computer programs that can place bets automatically. These robot gamblers have plenty of advantages over humans: they are faster, more focused, and they can bet on dozens of games at once. The speed of betting exchanges also works in their favor. Betfair is quick to pair up people who want to bet for a particular event with those intending to bet against it. Of the 4.4 million bets placed on the day of England's opening match in the 2006 soccer World Cup, all but twenty were handled in less than a second.

Automated gamblers are increasingly common in betting. According to sports analyst David Hastie, there are plenty of bots out there searching for stray odds and exploiting other gamblers' mistakes. "These algorithms mop up any mispricing," he said. The presence of artificial arbitrageurs makes it difficult for humans to cash in on such opportunities. Even if they spot an erroneous price, it's often too late to do anything about it. The bots will already be placing bets, removing these slices of profit from the market.

Arbitrage algorithms are also becoming popular in finance. As in betting, the faster the better. Companies are doing all they can to ensure they get to the action before their competitors do. It has led to many firms placing their computers directly next to stock exchange servers. When the market reacts quickly, even a slightly longer wire can lead to a critical delay in making a trade.

Some are going to even more extreme lengths. In 2011, US firm Hibernia Atlantic started work on a new $300 million transatlantic cable, which will allow data to cross the ocean faster than ever before. Unlike previous wires, it will be directly below the flight path from New York to London, the shortest possible route between the cities. It currently takes 65 milliseconds for messages to travel the Atlantic; the new cable aims to cut that down to 59. To give a sense of the scales involved, one blink of the human eye takes 300 milliseconds.

Fast trading algorithms are helping firms learn about new events first and act on them before others do. Yet not all bots are chasing arbitrage opportunities. In fact, some have the opposite aim. While arbitrage algorithms are searching for lucrative information, other bots are trying to conceal it.

WHEN SYNDICATES BET ON horse races in Hong Kong, they know that the odds will change after they've placed their bets. This is because in pari-mutuel betting the odds depend on the size of the betting pool. Teams therefore have to account for the shift when developing a betting strategy. If they put down too much, and shift the odds too far, they might end up worse off than if they'd bet less.

The problem also appears in sports betting. If you try to put down a large amount of money on a football match, it will be the bookmakers—or betting exchange users—who move the odds against you. Let's say you want to bet $500,000 on a certain outcome. One bookmaker might offer you odds that would return double the stake. But the bookmaker might only be willing to take a bet of $100,000 at those odds. After you place that initial bet, the bookmaker's odds will probably drop. Which means you've still got $400,000 you want to bet, and you've already disrupted the market. So, if you bet another $100,000 at the new odds, you won't quite double your money. You might get even lower odds for the next chunk of cash, and things will continue to get worse with each bet you make.

Traders call the problem "slippage." Although the price initially on offer might look good, it can slip to a less favorable price as the transaction is made. How can you get around the problem? Well, you could try to hunt down a bookmaker who'll take the bet in one go. At best, this could take a while; at worst, you'll never find one. Alternatively, you could place the first bet of $100,000 and then wait and hope the bookmaker's odds will rise again so you can bet the next chunk of money. Which is clearly not the most reliable strategy either.

A better approach would be to mimic the tactics employed by betting exchanges. Betfair's early success was in part the result of the way it handled each bet. Rather than attempting to find a gambler who wanted to accept a bet of the exact same size, Betfair sliced up the bet into smaller chunks. It was far easier—and quicker—to find several users happy to take on these little wagers than to hunt down a single gambler willing to take the whole bet.

The same idea makes it possible to sneak a trade into the market with limited slippage. Instead of trying to offload the whole trade at once, so-called order-routing algorithms can slice the main trade up into a series of smaller "child" orders, which can easily be completed. For the process to work effectively, algorithms need to have good knowledge of the market. As well as having information on who's happy to take the other side of each trade—and at what price—the program has to time the transactions carefully to reduce the chances of the market moving before the trade is complete. The resulting trade is known as an "iceberg order": although competitors see small amounts of trading activity, they never know what the full transaction looks like. After all, traders don't want rivals shifting prices because they know a big order is about to arrive. Nor do they want others to know what their trading strategy is.

Because such information is valuable, some competitors employ programs that can search for iceberg trades. One example is a "sniffing algorithm," which make lots of little trades to try to detect the presence of big orders. After the sniffing program submits each trade, it measures how quickly it takes to get snapped up in the market. If there's a big order lurking somewhere, the trades might go through faster. It's a bit like dropping coins into a well and listening for the splashes to work out how deep it is.

Although bots allow gamblers and banks to carry out multiple transactions quickly, they do not always act in the interests of their owners. Left unsupervised, bots can behave in unexpected ways. And sometimes they wander deep into trouble.

BY THE TIME THE 2011 Christmas Hurdle at Dublin's Leopardstown Racecourse reached the halfway mark, the race was as good as won.

It was just after two o'clock, and the horse named Voler La Vedette was already leading by a good distance. As the hooves pounded the ground on that cold December afternoon, nobody with any sense would have bet against that horse.

Yet somebody did. Even as Voler La Vedette approached the line, the Betfair online market was displaying extremely favorable odds for the horse that was almost certain to win. It appeared that someone was happy to accept bets at odds of 28: for every £1 bet, the bettor was offering to pay £28 if the horse won. Very happy, in fact. This remarkably pessimistic gambler was offering to accept £21 million worth of bets. If Voler La Vedette came first, the gambler would be on the hook for almost £600 million.

Soon after the race finished, one Betfair user posted a message on the website's forum. Having witnessed the whole bizarre situation, the user joked that someone must have been giving bettors a Christmas bonus. Others chipped in with potential explanations for the mishap. Maybe a gambler had suffered an attack of "fat fingers" and hit the wrong number on the keyboard?

It didn't take long for another user to suggest what might really have been going on. The person had noticed something odd about that offer to match £21 million of bets. To be precise, the number displayed on the exchange was just under £21.5 million. The user pointed out that computer programs often store binary data in units that contain thirty-two values, known as "bits." So, if the rogue gambler had designed a 32-bit program to bet automatically, the largest positive number the bot would be able to input on the exchange would be 2,147,483,648 pence. Which meant that if the bot had been doubling up its bets—just as misguided Parisian gamblers used to do while betting on roulette in the eighteenth century—£21.5 million is the highest it would have been able to go.

It turned out to be a superb piece of detective work. Two days later Betfair admitted that the error had indeed been caused by a faulty bot. "Due to a technical glitch within the core exchange database," they said, "one of the bets evaded the prevention system and was shown on the site." Apparently, the bot's owner had less than £1,000 in an account at the time, so as well as fixing the glitch, Betfair voided the bets that had been made.

As several Betfair users had already pointed out, such ridiculous odds should never have been available. The two hundred or so gamblers who had bet on the race would therefore have struggled to persuade a lawyer to take their case. "You cannot win—or lose—what is not there in the first place," Greg Wood, the *Guardian*'s racing correspondent, wrote at the time, "and even the most opportunistic ambulance-chaser is likely to take one look at this fact and point to the door."

Unfortunately, the damage created by bots isn't always so limited. Computer trading software is also becoming popular in finance, where the stakes can be much higher. Six months after the Voler La Vedette bot got its odds wrong, one financial company was to discover just how expensive a troublesome program could be.

THE SUMMER OF 2012 was a busy time for Knight Capital. The New Jersey–based stockbroker was getting its computer systems ready for the launch of the New York Stock Exchange's Retail Liquidity Program on August 1. The idea of the liquidity program was to make it cheaper for customers to carry out large stock trades. The trades themselves would be executed by brokers like Knight, which would provide the bridge between the customer and the market.

Knight used a piece of software called SMARS to handle customers' trades. The software was a high-speed order router: when a trade request came in from a client, SMARS would execute a series of smaller child orders until the original request had been filled. To avoid overshooting the required value, the program kept a tally of how many child orders had been completed and how much of the original request still needed to be executed.

Until 2003, a program named Power Peg had been responsible for halting trading once the order had been met. In 2005, this program was phased out. Knight disabled the Power Peg code and installed the tally counter into a different part of the SMARS software. But, according to a subsequent US government report, Knight did not check what would happen if the Power Peg program was accidently triggered again.

Toward the end of July 2012, technicians at Knight Capital started to update the software on each of the company's servers. Over a series of days, they installed the new computer code on seven of the eight servers. However, they reportedly failed to add it to the eighth server, which still contained the old Power Peg program.

Launch day arrived, and trade orders started coming in from customers and other brokers. Although Knight's seven updated servers did their job properly, the eighth was unaware of how many requests had already been completed. It therefore did its own thing, peppering the market with millions of orders and buying and selling stocks in a rapid-fire trading spree. As the erroneous orders piled up, the tangle of trades that would later have to be unraveled grew larger and larger. While technology staff worked to identify the problem, the company's portfolio grew. Over the course of forty-five minutes, Knight bought around $3.5 billion worth of stocks and sold over $3 billion. When it eventually stopped the algorithm and unwound the trades, the error would cost it over $460 million, equivalent to a loss of $170,000 per second. The incident left a massive dent in Knight's finances, and in December of that year the company was acquired by a rival trading firm.

Although Knight's losses came from the unanticipated behavior of a computer program, technical problems are not the only enemy of algorithmic strategies. Even when automated software is working as planned, companies can still be vulnerable. If their program is too well behaved—and hence too predictable—a competitor might find a way to take advantage of it.

In 2007, a trader named Svend Egil Larsen noticed that the algorithms of one US-based broker would always respond to certain trades in the same way. No matter how many stocks were bought, the broker's software would raise the price in a similar manner. Larsen, who was based in Norway, realized that he could nudge up the price by making lots of little purchases, and then sell a large amount of stock back at the higher price. He'd become the financial equivalent of Professor Pavlov, ringing his bell and watching the algorithm respond obediently. Over the course of a few months, the tactic earned Larsen over $50,000.

Not everybody appreciated the ingenuity of his strategy. In 2010, Larsen and fellow trader Peder Veiby—who'd been doing the same thing—were charged with manipulating the market. The courts seized their profits and handed the pair suspended sentences. When the verdict was announced, Veiby's lawyer argued that the nature of the opponent had biased the ruling. Had the pair profited from a stupid human trader rather than a stupid algorithm, the court would not have reached the same conclusion. Public opinion sided with Larsen and Veiby, with the press comparing their exploits to those of Robin Hood. Their support was vindicated two years later, when the Supreme Court overturned the verdict, clearing the two men of all charges.

There are several ways algorithms can wander into dangerous territory. They might be influenced by an error in the code, or they might be running on an out-of-date system. Sometimes they take a wrong turn; sometimes a competitor leads them astray. But so far we have only looked at single events. Larsen targeted a specific broker. Knight was a lone company. Just one gambler offered ridiculous odds on Voler La Vedette. Yet there are an increasing number of algorithms in betting and finance. If a single bot can take the wrong path, what happens when lots of firms use these programs?

DOYNE FARMER'S WORK ON prediction did not end with the path of a casino roulette ball. After obtaining his PhD from UC Santa Cruz, Farmer moved to Los Alamos, and then to the Santa Fe Institute in New Mexico. While there, he developed an interest in finance. Over a few short years, he went from forecasting roulette spins to anticipating the behavior of stock markets. In 1991, he founded a hedge fund with fellow ex-Eudaemon Norman Packard. It was named Prediction Company, and the plan was to apply concepts from chaos theory to the financial world. Mixing physics and finance was to prove extremely successful, and Farmer spent eight years with the company before deciding to return to academia.

Farmer is now a professor at the University of Oxford, where he looks at the effects of introducing complexity to economics. Although there is already plenty of mathematical thinking in the world of finance, Farmer has pointed out that it is generally aimed at specific transactions. People use mathematics to decide the price of their financial products or to estimate the risk involved in certain trades. But how do all these interactions fit together? If bots influence each other's decisions, what effect could it have on the economic system as a whole? And what might happen when things go wrong?

A crisis can sometimes begin with a single sentence. At lunchtime on April 23, 2013, the following message appeared on the Associated Press's Twitter feed: "Breaking: Two Explosions in the White House and Barack Obama is injured." The news was relayed to the millions of people who follow the Associated Press on Twitter, with many of them reposting the message to their own followers.

Reporters were quick to question the authenticity of the tweet, not least because the White House was hosting a press conference at the time (which had not seen any explosions). The message indeed turned out to be a hoax, posted by hackers. The tweet was soon removed, and the Associated Press Twitter account was temporarily suspended.

Unfortunately, financial markets had already reacted to the news. Or, rather, they had overreacted. Within three minutes of the fake announcement, the S&P 500 stock index had lost $136 billion in value. Although markets soon returned to their original level, the speed—and severity—of the reaction made some financial analysts wonder whether it was really caused by human traders. Would people have really spotted an errant tweet so quickly? And would they have believed it so easily?

It wasn't the first time a stock index had ended up looking like a sharp stalactite, stretching down from the realms of sanity. One of the biggest market shocks came on May 6, 2010. When the US financial markets opened that morning, already several potential clouds were on the horizon, including the upcoming British election and ongoing financial difficulties in Greece. Yet nobody foresaw the storm that was to arrive midafternoon.

Although the Dow Jones Industrial Average had dipped a little earlier in the day, at 2:32 p.m. it started to decline sharply. By 2:42 p.m. it had lost almost 4 percent in value. The decline accelerated, and five minutes later the index was down another 5 percent. In barely twenty minutes, almost $900 billion had been wiped from the market's value. The descent triggered one of the exchange's fail-safe mechanisms, which paused trading for a few moments. This allowed prices to stabilize, and the index started to clamber back toward its original level. Even so, the drop had been staggering. So, what had happened?

Severe market disruptions can often be traced to one main trigger event. In 2013, it was the hoax Twitter announcement about the White House. Bots that scour online newsfeeds, attempting to exploit information before their competitors, would have likely picked up on this and started making trades. The story gained a curious footnote in the following year, when the Associated Press introduced automated company earnings reports. Algorithms sift through the reports and produce a couple of hundred words summarizing firms' performance in the Associated Press's traditional writing style. The change means that humans are now even more absent from the financial news process. In press offices, algorithms convert reports into prose; on trading floors, their fellow robots turn these words into trading decisions.

The 2010 Dow Jones "flash crash" was thought to be the result of a different type of trigger event: a trade rather than an announcement. At 2:32 p.m., a mutual fund had used an automated program to sell seventy-five thousand futures contracts. Instead of spreading the order over a period of time, as a series of small icebergs, the program had apparently dropped the whole thing in pretty much all at once. The previous time the fund had dealt with a trade that big, it had taken five hours to sell seventy-five thousand contracts. On this occasion, it had completed the whole transaction in barely twenty minutes.

It was undoubtedly a massive order, but it was just one order, made by a single firm. Likewise, bots that analyze Twitter feeds are relatively niche applications: the majority of banks and hedge funds do not trade in this way. Yet the reaction of these Twitter-happy algorithms led to a spike that wiped billions off the stock market. How did these seemingly isolated events lead to such turbulence?

To understand the problem, we can turn to an observation made by economist John Maynard Keynes in 1936. During the 1930s, English newspapers would often run beauty contests. They would publish a collection of girls' photos and ask readers to vote for the six they thought would be most popular overall. Keynes pointed out that shrewd readers wouldn't simply choose the girls they liked best. Instead, they would select the ones they thought everyone else would pick. And, if readers were especially sharp, they would go to the next level and try to work out which girl everyone else would expect to be the most popular.

According to Keynes, the stock market often works in much the same way. When speculating on share prices, investors are in effect trying to anticipate what everyone else will do. Prices don't necessarily rise because a company is fundamentally sound; they increase because other investors think the company is valuable. The desire to know what others are thinking means lots of second-guessing. What's more, modern markets are moving further and further away from a carefully considered newspaper contest. Information arrives fast, and so does the action. And this is where algorithms can run into trouble.

Bots are often viewed as complicated, opaque creatures. Indeed, *complex* seems to be the preferred adjective of journalists writing about trading algorithms (or any algorithm, for that matter). But in high-frequency trading, it's quite the opposite: if you want to be quick, you need to keep things simple. The more instructions you have to deal with when trading financial products, the longer things take. Rather than clogging up their bots with subtlety and nuance, creators instead limit strategies to a few lines of computer code. Doyne Farmer warns that this doesn't leave much room for reason and rationality. "As soon as you limit what you can do to ten lines of code, you're non-rational," he said. "You're not even at insect-level intelligence."

When traders react to a big event—whether a Twitter post or a major sell order—it piques the attention of the high-speed algorithms monitoring market activity. If others are selling stocks, they join in. As prices plummet, the programs follow each other's trades, driving prices further downward. The market turns into an extremely fast beauty contest, with no one wanting to pick the wrong girl. The speed of the game can lead to serious problems. After all, it's hard to work out who will move first when algorithms are faster than the eye can see. "You don't have much time to think," Farmer said. "It creates a big danger of over-reaction and herding."

Some traders have reported that mini flash crashes happen frequently. These shocks are not severe enough to grab headlines, but they are still there to be found by anyone who looks hard enough. A share price might drop in a fraction of a second, or trading activity will suddenly increase a hundredfold. In fact, there might be several such crashes every day. When researchers at the University of Miami looked at stock market data between 2006 and 2011, they found thousands of "ultrafast extreme events" in which a stock crashed or spiked in value—and recovered again—in less than a second. According to Neil Johnson, who led the research, these events are a world away from the kind of situations covered by traditional financial theories. "Humans are unable to participate in real time," he said, "and instead, an ultrafast ecology of robots rises up to take control."

WHEN PEOPLE TALK ABOUT chaos theory, they often focus on the physics side of things. They might mention Edward Lorenz and his work on forecasting and the butterfly effect: the unpredictability of the weather, and the tornado caused by the flap of an insect's wings. Or they might recall the story of the Eudaemons and roulette prediction, and how the trajectory of a billiard ball can be sensitive to initial conditions. Yet chaos theory has reached beyond the physical sciences. While the Eudaemons were preparing to take their roulette strategy to Las Vegas, on the other side of the United States ecologist Robert May was working on an idea that would fundamentally change how we think about biological systems.

Princeton University is a world away from the glittering high-rises of Las Vegas. The campus is a maze of neo-Gothic halls and sun-dappled quads; squirrels dash through ivy-clad archways, while students' distinctive orange and black scarves billow in the New Jersey wind. Look carefully and there are also traces of famous past residents. There's an "Einstein Drive," which loops in front of the nearby Institute of Advanced Study. For a while there was also a "Von Neumann corner," named after all the car accidents the mathematician reportedly had there. The story goes that von Neumann had a particularly ambitious excuse for one of his collisions. "I was proceeding down the road," he said. "The trees on the right were passing me in orderly fashion at sixty miles per hour. Suddenly one of them stepped in my path."

During the 1970s, May was a professor of zoology at the university. He spent much of his time studying animal communities. He was particularly interested in how animal numbers changed over time. To examine how different factors influenced ecological systems, he constructed some simple mathematical models of population growth.

From a mathematical point of view, the simplest type of population is one that reproduces in discrete bursts. Take insects: many species in temperate regions breed once per season. Ecologists can explore the behavior of hypothetical insect populations using an equation called "the logistic map." The concept was first proposed in 1838 by statistician Pierre Verhulst, who was investigating potential limits to population. To calculate the population density in a particular year using the logistic map, we multiply three factors together: the population growth rate, the density in the previous year, and the amount of space—and hence resources—still available. Mathematically, this takes the form:

Density in next year = Growth rate × Current density × (1–Current density)

The logistic map is built on a simple set of assumptions, and when the growth rate is small it churns out a simple result. Over a few seasons, the population settles down to equilibrium, with the population density remaining the same from one year to the next.

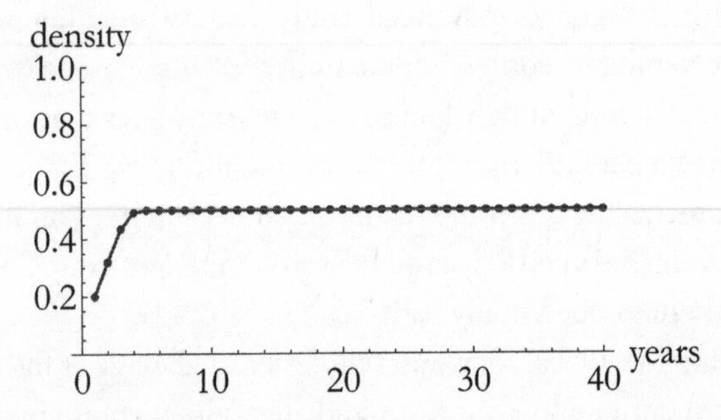

FIGURE 5.1. Results from the logistic map with a low growth rate.

The situation changes as the growth rate increases. Eventually, the population density starts to oscillate. In one year, lots of insects are hatched, which reduces available resources; next year, fewer insects survive, which makes spaces for more creatures the following year, and so on. If we sketch out how the population changes over time, we get the picture shown in Figure 5.2.

When the growth rate gets even larger, something strange happens. Rather than settle down to a fixed value, or switch between two values in a predictable way, the population density begins to vary wildly.

Remember that there is no randomness in the model, no chance events. The animal density depends on a simple one-line equation.

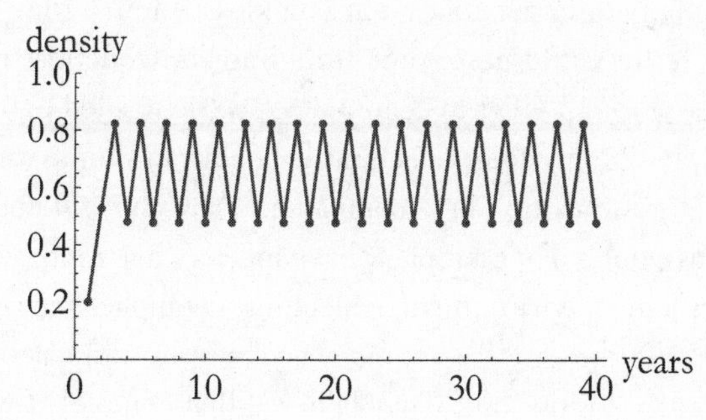

FIGURE 5.2. With a medium growth rate, the population density oscillates.

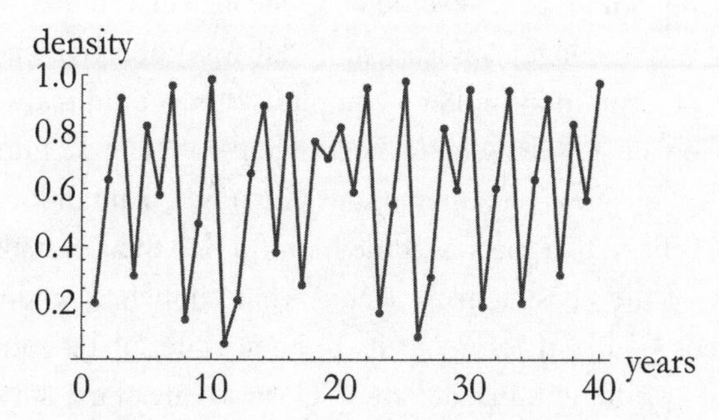

FIGURE 5.3. High growth rates lead to highly variable population dynamics.

And yet the result is a bumpy, noisy set of values, which do not appear to follow a straightforward pattern.

May found that chaos theory could explain what was going on. The fluctuations in density were the result of the population being sensitive to initial conditions. Just as Poincaré had found for roulette, a small change in the initial setup had a big effect on what happens further down the line. Despite the population following a straightforward biological process, it was not feasible to predict how it would behave far into the future.

We might expect roulette to produce unexpected outcomes, but ecologists were stunned to find that something as simple as the logistic map could generate such complex patterns. May warned that the result could have some troubling consequences in other fields, too. From politics to economics, people needed to be aware that simple systems do not necessarily behave in simple ways.

As well as studying single populations, May thought about ecosystems as wholes. For example, what happens when more and more creatures join an environment, generating a complicated web of interactions? In the early 1970s, many ecologists would have said the answer was a positive one. They believed that complexity was generally a good thing in nature; the more diversity there was in an ecosystem, the more robust it would be in the face of a sudden shock.

That was the dogma, at least, and May was not convinced it was correct. To examine whether a complex system could really be stable, he looked at a hypothetical ecosystem with a large number of interacting species. The interactions were chosen at random: some were beneficial to a species, some harmful. He then measured the stability of the ecosystem by seeing what happened when it was disrupted. Would it return to its original state, or do something completely different, like collapse? This was one of the advantages of working with a theoretical model: he could test stability without disrupting the real ecosystem.

May found that the larger the ecosystem, the less stable it would be. In fact, as the number of species grew very large, the probability of the ecosystem surviving shrank to zero. Increasing the level of complexity had a similarly harmful effect. When the ecosystem was more connected, with a higher chance of any two given species interacting with each other, it was less stable. The model suggested that the existence of large, complex ecosystems was unlikely, if not impossible.

Of course, there are plenty of examples of complex yet seemingly robust ecosystems in nature. Rainforests and coral reefs have vast numbers of different species, yet they haven't all collapsed. According to ecologist Andrew Dobson, the situation is the biological equivalent of a joke made in the early days of the European currency union. Although the euro worked in practice, observers said, it was not clear why it worked in theory.

To explain the difference between theory and reality, May suggested that nature had to resort to "devious strategies" to maintain stability. Researchers have since put forward all sorts of intricate strategies in an attempt to drag the theory closer to nature. Yet, according to Stefano Allesina and Si Tang, two ecologists at the University of Chicago, this might not be necessary. In 2013, they proposed a possible explanation for the discrepancy between May's model and real ecosystems.

Whereas May had assumed random interactions between different species—some positive, some negative—Allesina and Tang focused on three specific relationships that are common in nature. The first of these was a predator-prey interaction, with one species eating another; obviously, the predator will gain from this relationship, and the prey will lose out. As well as predation, Allesina and Tang also included cooperation, where both parties benefit from the relationship, and competition, with both species suffering negative effects.

Next, the researchers looked at whether each relationship stabilized the overall system or not. They found that excessive levels of competitive and cooperative relationships were destabilizing, whereas predator-prey relationships had a stabilizing effect on the system. In other words, a large ecosystem could be robust to disruption as long as it had a series of predator-prey interactions at its core.

So, what does all this mean for betting and financial markets? Much like ecosystems, markets are now inhabited by several different bot species. Each has a different objective and specific strengths and weaknesses. There are bots out hunting for arbitrage opportunities; they are trying to react to new information first, be it an important event or an incorrect price. Then there are the "market makers," offering to accept trades or bets on both sides and pocket the difference. These bots are essentially bookmakers, making their money by anticipating where the action will be. They buy low and sell high, with the aim of balancing their books. There are also bots trying to hide large transactions by sneaking smaller trades into the market. And there are predator bots watching for these large trades, hoping to spot a big transaction and take advantage of the subsequent shift in the market.

During the flash crash on May 6, 2010, there were over fifteen thousand different accounts trading the futures contracts involved in the crisis. In a subsequent report, the Securities and Exchange Commission (SEC) divided the trading accounts into several different categories, depending on their role and strategy. Although there has been much debate about precisely what happened that afternoon, if the crash was indeed triggered by a single event—as the SEC report suggested—the havoc that followed was not the result of one algorithm. Chances are it came from the interaction between lots of different trading programs, with each one reacting to the situation in its own way.

Some interactions had particularly damaging effects during the flash crash. In the middle of the crisis, at 2:45 p.m., there was a drought of buyers for futures contracts. High-frequency algorithms therefore traded among themselves, swapping over twenty-seven thousand futures in the space of fourteen seconds. Normality only resumed after the exchange deliberately paused the market for a few seconds, halting the runaway drop in price.

Rather than treating betting or financial markets as a set of static economic rules, it makes sense to view them as an ecosystem. Some traders are predators, feeding off weaker prey. Others are competitors, fighting over the same strategy and both losing out. Many of the ideas and warnings from ecology can therefore apply to markets. Simplicity does not mean predictability, for example. Even if algorithms follow simple rules, they won't necessarily behave in simple ways. Markets also involve webs of interactions—some strong, some brittle—which means that having lots of different bots in the same place does not necessarily help matters. Just as May showed, making an ecosystem more complex doesn't necessarily make it more stable.

Unfortunately, increased complexity is inevitable when there are lots of people looking for profitable strategies. Whether in betting or finance, ideas are less lucrative once others notice what is going on. As exploitable situations become widely known, the market gets more efficient and the advantage disappears. Strategies therefore have to evolve as existing approaches become redundant.

Doyne Farmer has pointed out that the process of evolution can be broken down into several stages. To come up with a good strategy, you first need to spot a situation that can be exploited. Next, you need to get ahold of enough data to test whether your strategy works. Just as gamblers need plenty of data to rate horses or sports teams, traders need enough information to be sure that the advantage is really there, and not a random anomaly. At Prediction Company, this process was entirely algorithm-driven. The trading strategies were what Farmer called "evolving automata," with the decision-making process mutating as the computers accumulated new experience.

The shelf life of a trading strategy depends on how easy it is to complete each evolutionary stage. Farmer has suggested that it can often take years for markets to become efficient and strategies to become useless. Of course, the bigger the inefficiency is, the easier it is to spot and exploit. Because computer-based strategies tend to be highly lucrative at first, copycats are more likely to appear. Algorithmic approaches therefore have to evolve faster than other types of strategy. "There's going to be an ongoing saga of one-upmanship," Farmer said.

RECENT YEARS HAVE SEEN a huge growth in the number of algorithms scouring financial markets and betting exchanges. It is the latest connection between two industries that have a history of shared ideas, from probability theory to arbitrage. But the distinction between finance and gambling is blurring more than ever before.

Several betting websites now allow people to bet on financial markets. As with other types of online betting, these transactions constitute gambling and hence are exempt from tax in many European countries (at least for the customer; there is still a tax burden on the bookmaker). One of the most popular types of financial wager is spread betting. In 2013, around a hundred thousand people in Britain placed bets in this way.

In a traditional bet, the stake and potential payoff are fixed. You might bet on a certain team winning or on a share price rising. If the outcome goes your way, you get the payoff. If not, you lose your stake. Spread betting is slightly different. Your profit depends not just on the outcome but also on the size of the outcome. Let's say a share is currently priced at $50, and you think it will increase in value in the next week. A spread betting company might offer you a spread bet at $1 per point over $51 (the difference between the current price and the offered number is the "spread," and how the bookmaker makes its money). For every dollar the price rises above $51, you will get $1, and for every dollar it drops below, you will lose $1. In terms of payoff, it's not that different from simply buying the share and then selling it a week later. You'll make pretty much the same amount of profit (or loss) on both the bet and the financial transaction.

But there is a crucial difference. If you make a profitable stock trade in the United Kingdom, you have to pay stamp duty and capital gains tax. If you place a spread bet, you don't. Things are different in other countries. In Australia, profits from spread betting are classed as income and are therefore subject to tax.

Deciding how to regulate transactions is a challenge in both gambling and finance. When dealing with an intricate trading ecosystem, however, it is not always clear what effects regulation will have. In 2006, the US Federal Reserve and the National Academy of Sciences brought together financiers and scientists to debate "systemic risk" in finance. The idea was to consider the stability of the financial system as a whole rather than just the behavior of individual components.

During the meeting, Vincent Reinhart, an economist at the Federal Reserve, pointed out that a single action could have multiple potential outcomes. The question, of course, is which one will prevail. The result won't depend on only what regulators do. It could also depend on how the policy is communicated and how the market reacts to news. This is where economic approaches borrowed from the physical sciences can come up short. Physicists study interactions that follow known rules; they don't generally have to deal with human behavior. "The odds on a hundred-year storm do not change because people think it has become more likely," Reinhart said.

Ecologist Simon Levin, who also attended the meeting, elaborated on the unpredictability of behavior. He noted that economic interventions—like the ones available to the Federal Reserve—aim to change individual behavior in the hope of improving the system as a whole. Although certain measures can change what individuals do, it is very difficult to stop panic spreading through a market.

Yet the spread of information is only going to get faster. News no longer has to be read and processed by humans. Bots are absorbing news automatically and handing it to programs that make trading decisions. Individual algorithms react to what others do, with decisions made on the sort of timescales that humans can never fully supervise. This can lead to dramatic, unexpected behavior. Such problems often come from the fact that high-frequency algorithms are designed to be simple and fast. The bots are rarely complex or clever: the aim is to exploit an advantage before anyone else gets there. Creating successful artificial gamblers is not always a matter of being first, however. As we shall discover, sometimes it pays to be smart.

6

LIFE CONSISTS OF BLUFFING

IN SUMMER 2010, POKER WEBSITES LAUNCHED A CRACKDOWN ON robot players. By pretending to be people, these bots had been winning tens of thousands of dollars. Naturally, their human opponents weren't too happy. In retaliation, website owners shut down any accounts that were apparently run by software. One company handed almost $60,000 back to players after discovering that bots had been winning on their tables.

It wasn't long before computer programs again surfaced in online poker games. In February 2013, Swedish police started investigating poker bots that had been operating on a state-owned poker website. It turned out that these bots had made the equivalent of over half a million dollars. It wasn't just the size of the haul that worried poker companies; it was how the money was made. Rather than taking money from weaker players in low-stakes games, the bots had been winning on high-stakes tables. Until these sophisticated computer players were discovered, few people in the industry had realized that bots were capable of playing so well.

Yet poker algorithms have not always been so successful. When bots first became popular in the early 2000s, they were easily beaten. So, what has changed in recent years? To understand why bots are getting better at poker, we must first look at how humans play games.

WHEN THE US CONGRESS put forward a bill in 1969 suggesting that cigarette advertisements be banned from television, people expected American tobacco companies to be furious. After all, this was an industry that had spent over $300 million promoting their products the previous year. With that much at stake, a clampdown would surely trigger the powerful weapons of the tobacco lobby. They would hire lawyers, challenge members of Congress, fight antismoking campaigners. The vote was scheduled to take place in December 1970, which gave the firms eighteen months to make their move. So, what did they choose to do? Pretty much nothing.

Far from hurting tobacco companies' profits, the ban actually worked in the companies' favor. For years, the firms had been trapped in an absurd game. Television advertising had little effect on whether people smoked, which in theory made it a waste of money. If the firms had all got together and stopped their promotions, profits would almost certainly have increased. However, ads did have an impact on which brand people smoked. So, if all the firms stopped their publicity, and one of them started advertising again, that company would steal customers from all the others.

Whatever their competitors did, it was always best for a firm to advertise. By doing so, it would either take market share from companies that didn't promote their products or avoid losing customers to firms that did. Although everyone would save money by cooperating, each individual firm would always benefit by advertising.

Which meant all the companies inevitably ended up in the same position, putting out advertisements to hinder the other firms. Economists refer to such a situation—where each person is making the best decision possible given the choices made by others—as a "Nash equilibrium." Spending would rise further and further until this costly game stopped. Or somebody forced it to stop.

Congress finally banned tobacco ads from television in January 1971. One year later, the total spent on cigarette advertising had fallen by over 25 percent. Yet tobacco revenues held steady. Thanks to the government, the equilibrium had been broken.

JOHN NASH PUBLISHED HIS first papers on game theory while he was a PhD student at Princeton. He'd arrived at the university in 1948, after being awarded a scholarship on the strength of his undergraduate tutor's reference, a two-sentence letter that read, "Mr. Nash is nineteen years old and is graduating from Carnegie Tech in June. He is a mathematical genius."

During the next two years, Nash worked on a version of the "prisoner's dilemma." This hypothetical problem involves two suspects caught at the scene of a crime. Each is placed in a separate cell and must choose whether to remain silent or testify against the other person. If they both keep quiet, both receive one-year sentences. If one remains silent and the other talks, the quiet prisoner gets three years and the one who blames him is released. If both talk, both are sent down for two years.

Overall, it would be best if both prisoners kept their mouths shut and took the one-year sentence. However, if you are a prisoner stuck alone in a cell, unable to tell what your accomplice is going to do, it is always better to talk: if your partner stays silent, you get off; if your partner talks, you receive two years rather than three. The Nash equilibrium for the prisoner's dilemma game therefore has both players talking. Although they will end up suffering two years in prison rather than one, neither will gain anything if one alone changes strategy. Substitute talking and silence for advertising and cutting promotions, and it is the same problem the advertising firms faced.

Nash received his PhD in 1950, for a twenty-seven-page thesis describing how his equilibrium can sometimes thwart seemingly beneficial outcomes. But Nash wasn't the first person to take a mathematical hammer to the problem of competitive games. History has given that accolade to John von Neumann. Although later known for his time at Los Alamos and Princeton, in 1926 von Neumann was a young lecturer at the University of Berlin. In fact, he was the youngest in its history. Despite his prodigious academic record, however, there were still some things he wasn't very good at. One of them was poker.

Poker might seem like the ideal game for a mathematician. At first glance, it's just a matter of probabilities: the probability you receive a good hand; the probability your opponent gets a better one. But anyone who has played poker using only probability knows that things are not so simple. "Real life consists of bluffing," von Neumann noted, "of little tactics of deception, of asking yourself what is the other man going to think I mean to do." If he was to grasp poker, he would need to find a way to account for his opponent's strategy.

Von Neumann started by looking at poker at its most basic, where it is a game between two players. To simplify matters further, he assumed that each player was dealt a single card, showing a number somewhere between 0 and 1. After both players put in a dollar to start, the first player—who we'll call Alice—has three options: fold, and therefore lose one dollar; check (equivalent to betting nothing); or bet one dollar. Her opponent then decides whether to fold, and forfeit the money, or match the bet, in which case the winner depends on whose card has the highest number.

Obviously, it is pointless for Alice to fold at the start, but should she check or bet? Von Neumann looked at all possible eventualities and worked out the expected profit from each strategy. He found that she should bet if her card shows a very low or very high number, and she should check otherwise. In other words, she should bluff only with her worst hand. This might seem counterintuitive, but it follows logic familiar to all good poker players. If her card shows an average-to-low number, Alice has two options: bluff or check. With a terrible card, Alice has no hope of winning unless her opponent folds. She should therefore bluff. Middling cards are trickier. Bluffing won't persuade someone with a decent card to fold, and it's not worth Alice betting on the off chance that her mediocre card will come out on top in a showdown. So, the best option is to check and hope for the best.

In 1944, von Neumann and economist Oskar Morgenstern published their insights in a book titled *Theory of Games and Economic Behavior.* Although their version of poker was much simpler than the real thing, the pair had cracked a problem that had long bothered players, namely, whether bluffing was really a necessary part of the game. Thanks to von Neumann and Morgenstern, there was now mathematical proof that it was.

Despite his fondness for Berlin's nightlife, von Neumann didn't use game theory when he visited casinos. He saw poker mainly as an intellectual challenge and eventually moved on to other problems. It would be several decades before players worked out how to use von Neumann's ideas to win for real.

BINION'S GAMBLING HALL IS part of the old Las Vegas. Away from the Strip's shows and fountains, it lies in the thumping downtown heart of the city. While most hotels were built with theaters and concert halls as well as casinos, Binion's was designed for gambling from the start. When it opened in 1951, betting limits were much higher than at other venues, and in the entrance a giant upturned horseshoe straddled a box displaying a million dollars in cash. Binion's was also the first casino to give free drinks to all gamblers to keep them (and their money) at the tables. So, it was only natural that when the first World Series of Poker took place in 1970, it was held at Binion's.

Over the following decades, players gathered at Binion's each year to pit their wits—and luck—against each other. Some years were especially tense. Early in the 1982 competition, Jack Straus stumbled onto a losing streak that left him with a single chip. Fighting back, he managed to win enough hands to stay in the game, eventually going on to win the whole tournament. The story goes that when Straus was later asked what a poker player needs for victory, his reply was "a chip and a chair."

On May 18, 2000, the thirty-first World Series reached its finale. Two men were left in the competition. On one side of the table was T. J. Cloutier, a poker veteran from Texas. Opposite him sat Chris Ferguson, a long-haired Californian with a penchant for cowboy hats and sunglasses. Ferguson had started the game with far more chips than Cloutier, but his lead was shrinking with each hand that was dealt.

With the players almost even, the dealer handed out yet another set of cards. They were playing Texas hold'em poker, which meant that Ferguson and Cloutier first received two personal "pocket" cards. After looking at his hand—the ninety-third of the day—Cloutier opened with a bet of almost $200,000. Sensing a chance to retake the advantage, Ferguson raised him half a million dollars. But Cloutier was confident, too. So confident, in fact, that he responded by pushing all his chips into the center of the table. Ferguson looked at his cards again. Did Cloutier really have the better hand? After pondering his options for several moments, Ferguson decided to match Cloutier's bet of almost $2.5 million.

Once two initial pocket cards have been dealt in Texas hold'em, there are up to three additional rounds of betting. The first of these is known as the "flop." Three more cards are dealt, this time placed face up on the table. If betting continues, another card—the "turn"—is revealed. Another round of betting means that the game reaches the "river," where a fifth card is shown. The winner is the player who has the best five-card hand when the two pocket cards are combined with the five communal cards.

Because Cloutier and Ferguson had both gone all in at the start, there would be no additional betting. Instead, they would have to show their pocket cards and watch as the dealer turned over each of the five additional cards. When the players showed their hands, the crowd surrounding the table knew Ferguson was in trouble. Cloutier had an ace and a queen; Ferguson had only an ace and a nine. First, the dealer turned over the flop cards: a king, a two, and a four. Cloutier still had the better hand. Next came the turn, and another king. The game would therefore be settled on the river. As the final card was revealed, Ferguson leapt from his seat. It was a nine. He'd won the game, and the tournament. "You didn't think it would be that tough to beat me, did you?" Cloutier asked Ferguson after he'd netted the $1.5 million prize money. "Yes," Ferguson replied, "I did."

UNTIL CHRIS FERGUSON'S TRIUMPHANT performance in Las Vegas, no poker player had won more than $1 million in tournament prizes. But unlike many competitors, Ferguson's extraordinary success did not rely solely on intuition or instinct. When he played in the World Series, he was using game theory.

The year before he beat Cloutier, Ferguson had completed a doctorate in computer science at UCLA. During that time, he worked as a consultant for the California State Lottery, picking apart existing games and coming up with new ones. His family members have mathematical backgrounds, too: both parents have PhDs in the subject and his father, Thomas, is a professor of mathematics at UCLA.

While studying for his doctorate, Chris Ferguson would compete for play money in some of the early Internet chat rooms. He saw poker as a challenge, and it was one he was rather good at. The chat room games didn't lead to any profit, but they did give Ferguson access to large amounts of data. Combined with improvements in computing power, this enabled him to study vast numbers of different hands, evaluating how much to bet and when to bluff.

Like von Neumann, Ferguson soon realized that poker was too complicated to study properly without making a few simplifications. Building on von Neumann's ideas, Ferguson decided to look at what happens when two players have more options. Of course, he would have more than one opponent at the start of a real poker game, but it was still useful to analyze the simple two-player scenario. Players may fold as the betting rounds progress, so by the time the endgame arrives, there are often only a couple of players left.

Yet there are still a number of things the two players might do at this point. The first player, Alice, had three simple choices in von Neumann's game—bet one dollar, check, or fold—but in a real game she might do something else, like change her bet. And the second player might not respond by matching the bet or folding. The second player might be confident like Cloutier was and raise the betting.

As more options creep into the game, picking the best one becomes more complicated. In a simple setup, von Neumann showed that players should employ "pure strategies," in which they follow fixed rules such as "if this happens, always do A" and "if that happens, always do B." But pure strategies are not always a good approach to use. Take a game of rock-paper-scissors. Picking the same option every time is admirably consistent, but the strategy is easy to beat if your opponent works out what you're doing. A better idea is to use a "mixed strategy." Rather than always going with the same approach, you should switch between one of the pure strategies—rock, paper, or scissors—with a certain probability. Ideally, you will play each of the three options in a balance that makes it impossible for your opponent to guess what you're going to do. For rock-paper-scissors, the optimal strategy against a new opponent is to choose randomly, playing each option one-third of the time.

Mixed strategies also make an appearance in poker. Analysis of the endgame suggests that you should balance the number of times you are honest and the number of times you bluff so that your opponent is indifferent to calling or folding. Like rock-paper-scissors, you don't want the other person to work out what you are likely to do. "You always want to make your opponents' decisions as difficult as possible," Ferguson said.

Sifting through the data from the chat room games, Ferguson spotted other areas for improvement. When experienced players had good hands, they would raise heavily to encourage their opponents to fold. This removed the risk of a weak hand turning into a winning hand when the communal cards were revealed. But Ferguson's research showed that the raises were too high: sometimes it was worth betting less and allowing people to remain in the game. As well as winning more money with strong hands, it meant that if a hand did lose, it wouldn't lose as much.

Through his research, Ferguson discovered that finding a successful approach to poker doesn't necessarily mean chasing profits at all costs. As he once told *The New Yorker*, the optimal strategy isn't a case of "How do I win the most?" but one of "How do I lose the least?" Novice players usually confuse the two and don't fold often enough as a result. True, it's impossible to win anything by folding, but sitting out a hand allows players to avoid costly betting rounds. Collecting together his results into detailed tables, Ferguson memorized the strategies—including when to bluff, when to bet, how much to raise—and started playing for real money. He entered his first World Series in 1995; five years later he was champion.

Ferguson has always been fond of picking up new skills. He once taught himself to throw a playing card so fast from a distance of ten feet that it could slice a carrot in two. In 2006, he decided to take on a new challenge. Starting with nothing, he would work his way up to $10,000. His aim was to show the importance of bankroll management in poker. Just as the Kelly criterion helped gamblers adjust their bet size in blackjack and sports betting, Ferguson knew it was essential to adjust his playing style to balance profit and risk.

Because he was starting with zero dollars, Ferguson's first task was to get ahold of some cash. Fortunately, some poker websites ran daily "freeroll tournaments." Hundreds of players could enter for free, with the top dozen or so receiving cash prizes. It's not often that a big-name player enters a freeroll tournament, let alone takes it seriously. When other online players found out who they were playing against, most thought it was a joke. Why was a world champion like Chris Ferguson plying his trade on the free tables?

After a few attempts, Ferguson eventually netted some all-important cash. "I remember winning my first $2 a couple of weeks into the challenge," he later wrote, "and I strategized for three days, deliberating over what game to play with it." He settled on the lowest-stakes game possible, but within one round he'd lost it all. Finding himself back to zero, he returned to the freeroll tournaments and started over again. It was clear that he would have to be extremely disciplined if he was going to reach his target.

Playing around ten hours per week, it took Ferguson nine months to get to $100 (he'd expected it to take around six). He kept going, sticking to a strict set of rules. For instance, he would only ever risk 5 percent of his bankroll in a particular game. It meant that if he lost a few rounds, he would have to go back to lower-stakes tables. Psychologically, he found it difficult to drop down a level. Ferguson was used to the excitement of high-stakes games and the profits they brought. After moving down, he would lose focus and struggle to keep to his rules. Rather than take more risks, he stepped back; it was pointless playing the game until he'd regained his concentration. The self-restraint paid off. After another nine months of careful play, Ferguson finally reached his $10,000 total.

The bankroll challenge, along with his earlier World Series victory, cemented Ferguson's reputation as a virtuoso of poker theory. Much of his success came from working on optimal strategies, but do such strategies always exist in games like poker? The question was actually one of the first that von Neumann asked when he started working on two-player games at the University of Berlin. As well as laying the foundations for the entire field, the answer would go on to cause a bitter dispute about who was the true inventor of game theory.

GAMES LIKE POKER ARE "zero-sum," with winning players' profits equal to other players' losses. When two players are involved, this means one person is always trying to minimize the opponent's payoff—a quantity the opponent will be trying to maximize. Von Neumann called it the "minimax" problem and wanted to prove that both players could find an optimal strategy in this tug-of-war. To do this, he needed to show that each player could always find a way to minimize the maximum amount that could potentially be lost, regardless of what their opponent did.

One of the most prominent examples of a zero-sum game with two players is a soccer penalty. This ends either in a goal, with the kicker winning and the goalkeeper losing, or a miss, in which case the payoffs are reversed. Keepers have very little time to react after a penalty is taken, so generally make their decision about which way to dive before the kicker strikes the ball.

Because players are either right- or left-footed, choosing the right- or left-hand side of the goal can alter their chances of scoring. When Ignacio Palacios-Heurta, an economist at Brown University, looked at all the penalties taken in European leagues between 1995 and 2000, he found that the probability of a goal varies depending on whether the kicker chooses the "natural" half of the goal. (For a right-footed player, this would be the left-hand side of the goal; for a left-footed kicker, it would be the right side.)

The penalty data showed that if the kicker picked the natural side and the keeper chose the correct direction, the kicker scored about 70 percent of the time; if the keeper got it wrong, around 90 percent of shots went in. In contrast, kickers who went for the nonnatural side scored 60 percent of shots if the keeper picked correctly and 95 percent if they didn't. These probabilities are summarized in Table 6.1.

If kickers want to minimize their maximum loss, they should therefore pick the natural side: even if the goalkeeper gets the correct direction, the player has at least a 70 percent chance of scoring. In contrast, the keeper should dive to the kicker's nonnatural side. At worst it will result in the player scoring 90 percent of the time rather than 95 percent.

If these strategies were optimal, the worse-case probabilities for kicker and keeper would be equal. This is because a penalty shootout is zero-sum: each person is trying to minimize the potential loss, which means if each plays the perfect strategy, it should minimize the maximum payoff for the opponent. Yet this is clearly not the

TABLE 6.1. The Probability of Scoring a Penalty Depends on Which Side the Kicker and Goalkeeper Choose

		Keeper	
		Natural	Non-natural
Kicker	Natural	70%	90%
	Non-natural	95%	60%

case, because the worst outcome for the player results in scoring 70 percent of the kicks, whereas the worst result for the goalkeeper leads to letting in 90 percent of shots.

The fact that the values are not equal implies that each person can adjust tactics to improve the chances of success. As in rock-paper-scissors, switching between options might be better than relying on a simple pure strategy. For example, if the kicker always chooses the natural side, the goalkeeper should occasionally pick that option, too, which would bring the 90 percent worst-case scenario down closer to 70 percent. In response, the kicker could counter this tactic by also opting for a mixed strategy.

When Palacios-Heurta calculated the best approach for the kicker and goalkeeper, he found that both should choose the natural half of the goal with 60 percent probability, and the other side the rest of the time. Like effective bluffing in poker, this would have the effect of making the other person indifferent to what is going to happen: opponents would be unable to boost their chances by changing their strategy. Both the goalkeeper and the kicker would therefore successfully limit their loss as well as minimize the other person's gain. Remarkably, the recommended 60 percent value is within a few percent of the real proportion of times players choose each side, suggesting that—whether aware of it or not—kickers and goalkeepers have already figured out the optimal strategy for penalties.

VON NEUMANN COMPLETED HIS solution to the minimax problem in 1928, publishing the work in an article titled "Theory of Parlour Games." Proving that these optimal strategies always existed was a crucial breakthrough. He later said that without the result, there would have been no point continuing his work on game theory.

The method von Neumann used to attack the minimax problem was far from simple. Lengthy and elaborate, it has been described as a mathematical "tour de force." But not everyone was impressed. Maurice Fréchet, a French mathematician, argued that the mathematics behind von Neumann's minimax work had already been in place (though von Neumann had apparently been unaware of it). By applying the techniques to game theory, he said that von Neumann had "simply entered an open door."

The approaches Fréchet was referring to were the brainchild of his colleague Émile Borel, who had developed them a few years before von Neumann. When Borel's papers were eventually published in English in the early 1950s, Fréchet wrote an introduction crediting him with the invention of game theory. Von Neumann was furious, and the pair exchanged barbed comments in the economics journal *Econometrica*.

The dispute raised two important issues about applying mathematics to real-world problems. First, it can be hard to pin down the initiator of a theory. Should credit go to the researcher who crafts the mathematical bricks or to the person who assembles them into a useful structure? Fréchet clearly thought that brick maker Borel deserved the accolades, whereas history has given the credit to von Neumann for using mathematics to construct a theory for games.

The argument also showed that major results aren't always appreciated in their original format. Despite his defense of Borel's work, Fréchet didn't think the minimax work was particularly special because mathematicians already knew about the idea, albeit in a different form. It was only when von Neumann applied the minimax concept to games that its value become apparent. As Ferguson discovered when he applied game theory to poker, sometimes an idea that seems unremarkable to scientists can prove extremely powerful when used in a different context.

While the fiery debate between von Neumann and Fréchet sparked and crackled, John Nash was busy finishing his doctorate at Princeton. By establishing the Nash equilibrium, he had managed to extend von Neumann's work, making it applicable to a wider number of situations. Whereas von Neumann had looked at zero-sum games with two players, Nash showed that optimal strategies exist even if there are multiple players and uneven payoffs. But knowing perfect strategies always exist is just the start for poker players. The next problem is working out how to find them.

MOST PEOPLE WHO HAVE a go at creating poker bots don't rummage through game theory to find optimal strategies. Instead, they often start off with rule-based approaches. For each situation that could crop up in a game, the creator puts together a series of "if this happens, do that" instructions. The behavior of a rule-based bot therefore depends on its creator's betting style and how the creator thinks a good player should act.

While earning his master's degree in 2003, computer scientist Robert Follek put together a rule-based poker program called SoarBot. He built it using a set of artificial decision-making methods known as "Soar," which had been developed by researchers at the University of Michigan. During a poker game, SoarBot acted in three phases. First, it noted the current situation, including the pocket cards it had been dealt, the values of the communal cards, and the number of players who had folded. With this information, it then ran through all its preprogrammed rules and identified all those that were relevant to the present situation.

After collecting the available options, it entered a decision phase, choosing what to do based on preferences Follek had given it. This decision-making process could be problematic. Occasionally, the set of preferences turned out to be incomplete, with SoarBot either failing to identify any suitable options or being unable to choose between two potential moves. The predetermined preferences could also be inconsistent. Because Follek had input each one individually, sometimes the program would end up containing two preferences that were contradictory. For instance, one rule might tell SoarBot to bet in a given situation while another simultaneously tried to get it to fold.

Even if more rules were added manually, the program would still hit upon an inconsistency or incompleteness from time to time. This type of problem is well known to mathematicians. One year after obtaining his PhD in 1930, Kurt Gödel published a theorem pointing out that the rules that governed arithmetic could not be both complete and consistent. His discovery shook the research community. At the time, leading mathematicians were trying to construct a robust system of rules and assumptions for the subject. They hoped this would clear up a few logical anomalies that had recently been spotted. Led by David Hilbert, who had been von Neumann's mentor in Germany, these researchers wanted to find a set of rules that was complete—so that all mathematical statements could be proved using only these rules—and consistent, with none of the rules contradicting one another. But Gödel's incompleteness theorem showed that this was impossible: whichever set of rules was specified, there would always be situations in which additional rules were needed.

Gödel's logical rigor caused problems outside academia, too. While studying for his American citizenship assessment in 1948, he told his sponsor Oskar Morgenstern that he'd spotted some inconsistencies in the US Constitution. According to Gödel, the contradictions created a legal path for someone to become a dictator. Morgenstern told him it would be unwise to bring it up in the interview.

FORTUNATELY FOR FOLLEK, THE team that originally developed the Soar technology had found a way around Gödel's problem. When a bot ran into trouble, it would teach itself an additional rule. So, if Follek's SoarBot couldn't decide what to do, it could instead pick an arbitrary option and add the choice to its set of rules. Next time the same situation popped up, it could simply search through its memory to find out what it did last time. This type of "machine learning," with the bot adding new rules as it went along, allowed it to avoid the pitfalls Gödel described.

When Follek let SoarBot compete against human and computer opponents, it became clear his program wasn't a potential champion. "It played much better than the worst human players," he said, "and much worse than the best human and software players." Actually, SoarBot played about as well as Follek did. Although he'd read up on poker strategies, his weakness as a player limited the success of his bot.

From 2004 onward, poker bots grew in popularity thanks to the arrival of cheap software that let players put together their own bot. By tweaking the settings, they could decide which rules the program should follow. With well-chosen rules, these bots could beat some opponents. But, as Follek found, the structure of rule-based bots means that they are generally only as good as their creator. And judging by bots' success rates online, most creators aren't very good at poker.

BECAUSE RULE-BASED TACTICS CAN be difficult to get right, some people have turned to game theory to improve their computer players. But it's tricky to find the optimal tactics for a game as complicated as Texas hold'em poker. Because a huge number of different possible situations could arise, it is very difficult to compute the ideal Nash equilibrium strategy. One way around the problem is to simplify things, creating an abstract version of the game. Just as stripped-down versions of poker helped von Neumann and Ferguson understand the game, making simplifications can also help find tactics that are close to the true optimal strategy.

A common approach is to collect similar poker hands into "buckets." For example, we could work out the probability a given pair of pocket cards would beat a random other hand in a showdown, and then put hands with comparable winning probabilities into the same bucket. Such approximations dramatically reduce the number of potential scenarios we have to look at.

Bucketing also crops up in other casino games. Because the aim of blackjack is to get as near to twenty-one as possible, knowing whether the next card is likely to be high or low can give a player an advantage. Card counters get ahold of this information by keeping track of the cards that have already been dealt, and hence which ones remain. But with casinos using up to six decks at once, it's impractical to memorize each individual card as it appears. Instead, counters often group cards into categories. For instance, they might split them into three buckets: high, low, and neutral. As the game progresses, they keep a count of the type of cards they have already seen. When a high card is removed from the deck, they subtract one from the count; when a low card is dealt, they add one.

In blackjack, bucketing provides only an estimate of the true count: the fewer buckets a player uses, the less accurate the count will be. Likewise, bucketing won't give poker players a perfect strategy. Rather, it leads to what are known as "near-equilibrium strategies," some of which are closer to the true optimum than others. Just as penalty takers can improve their odds by deviating from a simple pure strategy, these not-quite-perfect poker strategies can be categorized by how much players would gain by altering their tactics.

Even with the inclusion of bucketing, we still need a way to work out a near-equilibrium strategy for poker. One way to do this is to use a technique known as "regret minimization." First, we create a virtual player and give it a random initial strategy. So, it might start off by folding half of the time in a given situation, betting the other half, and never checking. Then we simulate lots and lots of games and allow the player to update its strategy based on how much it regrets its choices. For instance, if the opponent folds prematurely, the player might regret betting big. Over time, the player will work to minimize the amount of regret it has, and in the process approach the optimal strategy.

Minimizing regret means asking, "How would I have felt if I'd done things differently?" It turns out that the ability to answer this question can be critical when playing games of chance. In 2000, researchers at the University of Iowa reported that people who had injuries to parts of the brain that are related to regret—such as the orbitofrontal cortex—performed very differently in betting games compared to those without brain damage. It wasn't because the injured players failed to remember previous bad decisions. In many cases, patients with orbitofrontal damage still had a good working memory: when asked to sort a series of cards, or match up different symbols, they had few problems. The difficulties came when they had to deal with uncertainty and use their past experiences to weigh the risks involved. The researchers found that when the feeling of regret was missing from patients' decision-making process, they struggled to master games involving an element of risk. Rather than simply looking forward to try to maximize payoffs, it appears that it is sometimes necessary to look back at what might have happened and use hindsight to refine a strategy. This contrasts with much economic theory, in which the focus is often on expected gains, with people trying to maximize future payoffs.

Regret minimization is becoming a powerful tool for artificial players. By repeatedly playing games and reevaluating past decisions, bots can construct near-equilibrium strategies for poker. The resulting strategies are far more successful than simple rule-based methods. Yet such approaches still rely on making estimates, which means that against a perfect poker bot, a near-equilibrium strategy will struggle. But how easy is it to make a perfect bot for a complex game?

GAME THEORY WORKS BEST in straightforward games in which all information is known. Tic-tac-toe is a good example: after a few games, most people work out the Nash equilibrium. This is because there aren't many ways in which a game can progress: if a player gets three in a row, the game is over; players must take it in turns; and it doesn't matter which way the board is oriented. So, although there are 3^9 ways to place X's and O's on a three-by-three board, only a hundred or so of these 19,683 combinations are actually relevant.

Because tic-tac-toe is so simple, it's fairly easy to work out the perfect way to react to an opponent's move. And once both players know the ideal strategy, the game will always result in a draw. Checkers, however, is far from simple. Even the best players have failed to find the perfect strategy. But if anyone could have spotted it, it would have been Marion Tinsley.

A mathematics professor from Florida, Tinsley had a reputation for being unbeatable. He won his first world championship in 1955, holding it for four years before choosing to retire, citing a lack of decent competition. Upon his return to the championships in 1975, he immediately regained his title, thrashing all opposition. Fourteen years later, however, Tinsley's interest in the game had again started to wane. Then he heard about a piece of software being developed at the University of Alberta in Canada.

Jonathan Schaeffer is now Dean of Science at the university, but back in 1989 he was a young professor in the department of computer science. He'd become interested in checkers after spending time looking at chess programs. Like chess, checkers is played on an eight-by-eight board. Pieces move forward diagonally and capture opposition pieces by leapfrogging. Upon reaching the other side of the board, they become kings and can move backward and forward. The simplicity of its rules makes checkers interesting to game theorists because it is relatively easy to understand, and players can predict the consequences of a move in detail. Perhaps a computer could even be trained to win?

Schaeffer decided to name the fledgling checkers project "Chinook" after the warm winds that occasionally sweep over the Canadian prairies. The name was a pun, inspired by the fact that the English call the game "draughts." Helped by a team of fellow computer scientists and checkers enthusiasts, Schaeffer quickly got to work on the first challenge: how to deal with the complexity of the game. There are around 10^{20} possible positions in checkers. That's 1 followed by twenty zeros: if you collected the sand from all of the world's beaches, you'd end up with about that number of grains.

To navigate this enormous selection of possibilities, the team got Chinook to follow a minimax approach, hunting down strategies that would be least costly. At each point in the game, there were a certain number of moves Chinook could make. Each of these branched out into another set of options, depending on what its opponent did. As the game progressed, Chinook "pruned" this decision tree, removing weak branches that were likely to lose it the game and examining stronger, potentially winning, branches in detail.

Chinook also had a few tricks lined up especially for human opponents. When it spotted strategies that would eventually lead to a draw against a perfect computer opponent, it didn't necessarily ignore them. If the draw lay at the end of a long, tangled branch of options, there was a chance a human might make a mistake somewhere along the way. Unlike many game-playing programs, Chinook would often pick these antihuman strategies over an option that was actually better according to game theory.

Chinook played its first tournament in 1990, coming in second in the US National Checker Championship. This should have meant it qualified for the World Championships, but the American Checkers Federation and the English Draughts Association did not want a computer to compete. Fortunately, Tinsley didn't share their view. After a handful of unofficial games in 1990, he decided that he liked Chinook's aggressive playing style. Whereas human players would try to force a draw against him, Chinook took risks. Determined to play the computer in a tournament, Tinsley resigned his championship title. Reluctantly, the authorities decided to allow the computer match, and in 1992 Chinook played Tinsley in a "Man-Machine World Championship." Out of 39 games, Tinsley won 4 to Chinook's 2, with 33 draws.

Despite holding their own against Tinsley, Schaeffer and his team wanted to do one better. They wanted to make Chinook unbeatable. Chinook depended on detailed predictions, which made it very good but still vulnerable to chance. If they could remove this element of luck, they would have the perfect checkers player.

It might seem odd that checkers involves luck. As long as an identical series of moves are made, the game will always end with the same result. To use the mathematical term, the game is "deterministic": it is not affected by randomness like poker is. Yet when Chinook played checkers, it couldn't control the result purely through its actions, which meant that it could be beaten. In theory, it was even possible to lose to a completely incompetent opponent.

To understand why, we must look at another piece of Émile Borel's research. As well as his work on game theory, Borel was interested in very unlikely events. To illustrate how seemingly rare things will almost certainly happen if we wait long enough, he coined the infinite monkey theorem. The premise of this theorem is simple. Suppose a monkey is hammering away randomly at a typewriter (without smashing it up, as happened when a University of Plymouth team attempted this with a real monkey in 2003) and does so for an infinitely long period of time. If the monkey continues to bash away at the keys, eventually it will be almost sure to type out the complete works of Shakespeare. By sheer chance, says the theorem, the monkey will at some point hit the right letters in the order needed to reproduce all thirty-seven of the Bard's plays.

No monkey would ever live to an infinite age, let alone sit at a typewriter for that long. So, it's best to think of the monkey as a metaphor for a random letter generator, churning out an arbitrary sequence of characters. Because the letters are random, there's a chance—albeit a small one—that the first ones the monkey types are "Who's there," the opening line of *Hamlet*. The monkey might then get lucky and keep typing the correct letters until it's reproduced all of the plays. This is extremely unlikely, but it could happen. Alternatively, the monkey might type reams of utter nonsense and then finally hit good fortune with the right combination of letters. It might even type gibberish for billions of years before eventually typing the correct letters in the correct order.

Individually, each one of these events is incredibly unlikely. But because there are so many ways the monkey could end up typing out the complete works of Shakespeare—an infinite number of ways, in fact—the chances of it happening eventually are extremely high. Actually, it's almost certain to happen.

Now suppose we replaced the typewriter with a checkers board and taught our hypothetical monkey the basic rules of the game. It would therefore make a series of totally random—but valid—moves. And the infinite monkey theorem tells us that because Chinook relied on predictions, the monkey would eventually strike upon a winning combination of moves. Whereas a computer can always force a draw in tic-tac-toe, victory in checkers depended on what Chinook's opponent did. Part of the game was therefore out of its hands. In other words, winning required luck.

CHINOOK PLAYED ITS LAST competitive game in 1996. But Schaeffer and his collegues did not fully retire their champion software. Instead, they set it to work finding a checkers strategy that would never lose, no matter what the computer's opponent did. The results were finally announced in 2007, when the Alberta researchers published a paper announcing "Checkers is solved."

There are three levels of solution for a game like checkers. The most detailed, a "strong solution," describes the final outcome when perfect players pick up any game at any point, including ones where errors have already been made. This means that, whatever the starting position, we always know the optimal strategy from that point onward. Although this type of solution requires a huge amount of computation, people have found strong solutions for relatively simple games like tic-tac-toe and Connect Four.

The next type of solution is when the optimal result is known, but we only know how to reach it if we play from the start of the game. These "weak solutions" are particularly common for complicated games, where it is only feasible to look at what happens if both players make perfect moves throughout.

The most basic, an "ultraweak solution," reveals the final result when both players make a perfect sequence of moves but doesn't show what those moves are. For instance, although strong solutions have been found for Connect Four and tic-tac-toe, John Nash showed in 1949 that when any such get-so-many-in-a-row-style game is played perfectly, the player who goes second will never win. Even if we can't find the optimal strategy, we can prove this claim is true by looking at what happens if we assume that it isn't and show that our incorrect assumption leads to a logical dead-end. Mathematicians call such an approach "proof by contradiction."

To kick off our proof, let's suppose there *is* a winning sequence of moves for the second player. The first player can turn the situation to their advantage by making their opening move completely at random, waiting for the second player to reply, and then "stealing" the second player's winning strategy from that point on. In effect, the first player has turned into the second player. This "strategy-stealing" approach works because having the randomly placed extra counter on the board at the start can only improve the first player's chance of winning.

By adopting the second player's winning strategy, the first player will end up victorious. However, at the start we assumed the second player has a winning strategy. This means both players therefore win, which is clearly a contradiction. So, the only logical outcome is that the second player can never win.

Knowing that a game has an ultraweak solution is interesting but doesn't really help a player win in practice. In contrast, strong solutions, despite guaranteeing an optimal strategy, can be difficult to find when games have a lot of possible combinations of moves. Because checkers is around a million times more complicated than Connect Four, Schaeffer and colleagues focused on finding a weak solution.

When it played Marion Tinsley, Chinook made decisions in one of two ways. Early in the game, it searched through possible moves, looking ahead to see where they might lead. In the final stages, when there were fewer pieces left on the board and hence fewer possibilities to analyze, Chinook instead referred to its "endgame database" of perfect strategies. Tinsley also had a remarkable understanding of the endgame, which is partly why he was so hard to beat. This became apparent in one of his early games against Chinook in 1990. Chinook had just made its tenth move when Tinsley said, "You're going to regret that." Twenty-six moves later, Chinook resigned.

The challenge for the Alberta team was getting the two approaches to meet in the middle. In 1992, Chinook could only look ahead seventeen moves, and its endgame database only had information for situations in which there were fewer than six pieces on the board. What happened in between came down to guesswork.

Thanks to increases in computing power, by 2007 Chinook was able to search far enough into the future, and assemble a large enough endgame database, to trace out the perfect strategy from start to finish. The result, which was published in the journal *Science*, was a remarkable achievement. Yet the strategy might never have been found had it not been for the matches against Tinsley. The Alberta team later said that the Chinook project "might have died in 1990 because of a lack of human competition."

Despite it being a perfect strategy, Schaeffer wouldn't recommend using it in a game against less-skilled opponents. Chinook's early matches against humans showed that it is often beneficial to deviate from the optimal strategy if it means increasing the chances of an opponent making a mistake. This is because most players cannot see dozens of moves ahead like Chinook can. The potential for errors is even greater in games like chess and poker, where nobody knows the perfect strategy. Which raises an important question: What happens when we apply game theory to games that are too complicated to fully learn?

ALONG WITH TOBIAS GALLA, a physicist at the University of Manchester, Doyne Farmer has started to question how game theory holds up when games aren't simple. Game theory relies on the assumption that all players are rational. In other words, they are aware of the effects of the various decisions they could make, and they choose the one that benefits them most. In simple games, like tic-tac-toe or the prisoner's dilemma, it's easy to make sense of the possible options, which means players' strategies almost always end up in Nash equilibrium. But what happens when games are too complicated to fully grasp?

The complexity of chess and many forms of poker means that players, be they human or machine, haven't yet found the optimal strategy. A similar problem crops up in financial markets. Although crucial information—from share prices to bond yields—is widely available, the interactions among banks and brokers that bump and buffet the markets are too intricate to fully understand.

Poker bots try to get around the problem of complexity by "learning" a set of strategies prior to playing an actual game. But in real life, players often learn strategies *during* a game. Economists have suggested that people tend to pick strategies using "experience-weighted attraction," preferring past actions that were successful to those that were not. Galla and Farmer wondered whether this process of learning helps players find the Nash equilibrium when games are difficult. They were also curious to see what happens if the game doesn't settle down to an optimal outcome. What sort of behavior should we expect instead?

Galla and Farmer developed a game in which two computer players could each choose from fifty possible moves. Depending on which combination the two players picked, they each got a specific payoff, which had been assigned randomly before the game started. The values of these predetermined payoffs decided how competitive the game was. The payoffs varied between being zero-sum, with one player's losses equal to the other's gain, to being identical for both players. The extent of players' memory could also vary. In some games, players took account of every previous move during the learning process; in others, they put less emphasis on events further in the past.

For each degree of competitiveness and memory, the researchers looked at how players' choices changed over time as they learned to pick moves with better outcomes. When players had poor memories, the same decisions soon cropped up again and again, with players often descending into tit-for-tat behavior. But when the players both had a good memory and the game was competitive, something odd happened. Rather than settle down to equilibrium, the decisions fluctuated wildly. Like the roulette balls Farmer had attempted to track while a student, the players' choices bounced around unpredictably. The researchers found that as the number of players increased, this chaotic decision making became more common. When games are complicated, it seems that it may be impossible to anticipate players' choices.

Other patterns also emerged, including ones that had previously been spotted in real-life games. When mathematician Benoit Mandelbrot looked at financial markets in the early 1960s, he noticed that volatile periods in stock markets tended to cluster together. "Large changes tend to be followed by large changes," he noted, "and small changes tend to be followed by small changes." The appearance of "clustered volatility" has intrigued economists ever since. Galla and Farmer spotted the phenomenon in their game, too, suggesting the pattern may just be a consequence of lots of people trying to learn the complexities of the financial markets.

Of course, Galla and Farmer made several assumptions about how we learn and how games are structured. But even if real life is different, we should not ignore the results. "Even if it turns out that we are wrong," they said, "explaining why we are wrong will hopefully stimulate game theorists to think more carefully about the generic properties of real games."

ALTHOUGH GAME THEORY CAN help us identify the optimal strategy, it's not always the best approach to use when players are error-prone or have to learn. The Chinook team knew this, which is why they ensured the program picked strategies that would entice its opponents into making mistakes. Chris Ferguson was also aware of the issue. As well as employing game theory, he looked for changes in body language, adjusting his betting if players become nervous or overconfident. Players don't just need to anticipate how the perfect opponent behaves; they need to predict how *any* opponent will behave.

As we shall see in the next chapter, researchers are now delving deeper into artificial learning and intelligence. For some of them, the work has been years in the making. In 2003, an expert human player competed against one of the leading poker bots. Although the bot used game theory strategies to make decisions, it could not predict the changing behavior of its competitors. Afterward, the human player told the bot's creators, "You have a very strong program. Once you add opponent modeling to it, it will kill everyone."

7

THE MODEL OPPONENT

WHEN IT CAME TO THE GAME SHOW *JEOPARDY!* KEN JENNINGS and Brad Rutter were the best. It was 2011, and Rutter had netted the most prize money, while Jennings had gone a record seventy-four appearances without defeat. Thanks to their ability to dissect the show's famous general knowledge clues, they had won over $5 million between them.

On Valentine's Day that year, Jennings and Rutter returned for a special edition of the show. They would face a new opponent, named Watson, who had never appeared on *Jeopardy!* before. Over the course of three episodes, Jennings, Rutter, and Watson answered questions on literature, history, music, and sports. It didn't take long for the newcomer to edge into the lead. Despite struggling with the "Name the Decade" round, Watson dominated when it came to the Beatles and the history of the Olympics. Although there was a last-minute surge from Jennings, the ex-champions could not keep up. By the end of the show, Watson had racked up over $77,000, more than Jennings and Rutter combined. It was the first time Rutter had ever lost.

Watson didn't celebrate the win, but its makers did. Named after Thomas Watson, founder of IBM, the machine was the culmination of seven years of work. The idea for Watson had come during a company dinner in 2004. During the meal, an eerie silence had descended over the restaurant. Charles Lickel, IBM's research manager, realized the lack of conversation was caused by something happening on the room's television screens. Everyone was watching Ken Jennings on his phenomenal *Jeopardy!* winning streak. As Lickel looked at the screen, he realized the game could be a good test for IBM's expertise. The firm had a history of taking on human games—their Deep Blue computer had beaten chess grandmaster Garry Kasparov in 1997—but it hadn't tackled a game like *Jeopardy!* before.

To win *Jeopardy!* players need knowledge, wit, and a talent for wordplay. The show is essentially an inverted quiz. Contestants get clues about the answer and have to tell the host what the question is. So, if a clue is "5,280," the answer might be "How many feet are there in a mile?"

The finished version of Watson would use dozens of different techniques to interpret the clues and search for the correct response. It had access to the entire contents of Wikipedia and $3 million of computer processors to crunch the information.

Analyzing human language and juggling data can be useful in other less glitzy environments, too. Since Watson's victory, IBM has updated the software so that it can trawl medical databases and help with decision making in hospitals. Banks are also planning to use it to answer customer questions, while universities are hoping to employ Watson to direct student queries. By studying cookbooks, Watson is even helping chefs find new flavor combinations. In 2015, IBM collected some of the results into a "cognitive computing cookbook," which includes recipes such as a burrito with chocolate, cinnamon, and edamame beans.

Although Watson's feats on *Jeopardy!* were impressive, the show is not the ultimate test for thinking machines. There is another, arguably much bigger, challenge for artificial intelligence out there, one that predates Watson, and even Deep Blue. While Deep Blue's predecessor "Deep Thought" was gradually clambering up the chess rankings in the early 1990s, a young researcher named Darse Billings arrived at the University of Alberta. He joined the computer science department, where Jonathan Schaeffer and his team had recently developed the successful Chinook checkers program. Perhaps chess would make a good next target? Billings had other ideas. "Chess is easy," he said. "Let's try poker."

EACH SUMMER, THE WORLD'S best poker bots gather to play a tournament. In recent years, three competitors have dominated. First, there is the University of Alberta group, which currently has about a dozen researchers working on poker programs. Next, there is a team from Carnegie Mellon University in Pittsburgh, Pennsylvania, just down the road from where Michael Kent used to work while developing his sports predictions. Tuomas Sandholm, a professor in computer science, heads up the group and their work on champion bot "Tartanian." Finally, there is Eric Jackson, an independent researcher, who has created a program named "Slumbot."

The tournament consists of several different competitions, with teams tailoring their bots' personalities to each one. Some competitions are knockouts. In each round two bots go head-to-head, and the one with the smallest pile of chips at the end gets eliminated. To win these competitions, bots need strong survival instincts. They need to win only enough to get through to the next round: greed, as it were, is *not* good. In other matches, however, the winning bot is the one that gathers the most cash overall. Computer players therefore need to squeeze as much as they can out of their opponents. Bots need to go on the offensive and find ways to take advantage of the others.

Most of the bots in the competition have spent years in development, training over millions if not billions of games. Yet there are no big prize pots waiting for the winners. The creators might get bragging rights, but they won't leave with Vegas-sized rewards. So, why are these programs useful?

Whenever a computer plays poker, it is solving a problem that's familiar to all of us: how to deal with missing information. In games like chess, information is not an issue. Players can see everything. They know where the pieces are and what moves their opponent has made. Luck creeps into the game not because players can't observe events but because they are unable to process the available information. That is why there is a chance (albeit tiny) that a grandmaster could lose to a monkey picking random moves.

With a good game-playing algorithm—and a lot of computer power—it's possible to get around the information-processing problem. That's how Schaeffer and his colleagues found the perfect strategy for checkers and how a computer might one day solve chess. Such machines can beat their opponents with brute force, crunching through every possible set of moves. But poker is different. No matter how good players are, each has to cope with the fact that opponents' cards are hidden. Although the game has rules and limits, there are always unknown factors. The same problem crops up in many aspects of life. Negotiations, auctions, bargaining; they are all incomplete information games. "Poker is a perfect microcosm of many situations we encounter in the real world," Schaeffer said.

WHILE WORKING AT LOS Alamos during the Second World War, Stanislaw Ulam, Nick Metropolis, John von Neumann, and others would often play poker late into the night. The games were not particularly intense. The stakes were small, and the conversation light. Ulam said it was "a bath of refreshing foolishness from the very serious and important business that was the raison d'être of Los Alamos." During one of their games, Metropolis won ten dollars from von Neumann. He was delighted to beat a man who'd written an entire book on game theory. Metropolis used half the money to buy a copy of von Neumann's *Theory of Games and Economic Behavior* and stuck the remaining five dollars inside the cover to mark the win.

Even before von Neumann had published his book on game theory, his research into poker was well known. In 1937, von Neumann had presented his work in a lecture at Princeton University. Among the attendees, there would almost certainly have been a young British mathematician by the name of Alan Turing. At the time Turing was a graduate student visiting from the University of Cambridge. He had come to the United States to work on mathematical logic. Although he was disappointed Kurt Gödel was no longer at the university, Turing generally enjoyed his time at Princeton, even if he did find certain American habits puzzling. "Whenever you thank them for anything they say 'You're welcome,'" he told his mother in a letter. "I rather liked it at first, thinking I was welcome, but I now find it comes back like a ball against a wall, and I become positively apprehensive."

After spending a year at Princeton, Turing returned to England. Although he was based mainly in Cambridge, he also took up a part-time position with the Government Code and Cypher School at nearby Bletchley Park. When the Second World War broke out in autumn 1939, Turing found himself at the forefront of the British effort to break enemy codes. During that period, the German military encrypted radio messages using so-called Enigma machines. These typewriter-like contraptions had a series of rotors that converted keystrokes into coded text. This complexity of the encryption was a major obstacle for the code breakers at Bletchley Park. Even if Turing and his colleagues had clues about the messages—for example, certain "crib" words that were likely to appear in the text—there were still thousands of possible rotor settings to trawl through. To solve the problem, Turing designed a computer-like machine called the "bombe" to do the hard work. Once code breakers had found a crib, the bombe could identify the Enigma settings that produced the code and decipher the rest of the message.

Breaking the Enigma code was probably Turing's most famous achievement, but much like von Neumann he was also interested in games. Von Neumann's research on poker certainly grabbed Turing's attention. When Turing died in 1954, he left his friend Robin Gandy a collection of papers. Among them was a half-finished manuscript entitled "The Game of Poker," in which Turing had tried to build on von Neumann's simple analysis of the game.

Turing did not think only about the mathematical theory of games. He also wondered how games could be used to investigate artificial intelligence. According to Turing, it did not make sense to ask "can machines think?" He said the question was too vague, the range of answers too ambiguous. Rather, we should ask whether a machine is capable of behaving in a way that is indistinguishable from a (thinking) human. Can a computer trick someone into believing it is human?

To test whether an artificial being could pass for a real person, Turing proposed a game. It would need to be a fair contest, an activity that both humans and machines could succeed at. "We do not wish to penalise the machine for its inability to shine in beauty competitions," Turing said, "nor to penalise a man for losing in a race against an aeroplane."

Turing suggested the following setup. A human interviewer would talk with two unseen interviewees, one of them human and the other a machine. The interviewer would then try to guess which was which. Turing called it the "imitation game." To avoid the participants' voices or handwriting influencing things, Turing suggested that all messages be typed. While the human would be trying to help the interviewer by giving honest answers, the machine would be out to deceive its interrogator. Such a game would require a number of different skills. Players would need to process information and respond appropriately. They would have to learn about the interviewer and remember what has been said. They might be asked to perform calculations, recall facts, or tackle puzzles.

At first glance, Watson appears to fit the job description well. While playing *Jeopardy!* the machine had to decipher clues, gather knowledge, and solve problems. But there is a crucial difference. Watson did not have to play like a human to win *Jeopardy!* It played like a supercomputer, using its faster reaction times and vast databases to beat its opponents. It did not show nerves or frustration, and it didn't have to. Watson wasn't there to persuade people that it was human; it was there to win.

The same was true of Deep Blue. When it played chess against Garry Kasparov, it played in a machine-like way. It used vast amounts of computer power to search far into the future, examining potential moves and evaluating possible strategies. Kasparov pointed out that this "brute force" approach did not reveal much about the nature of intelligence. "Instead of a computer that thought and played chess like a human, with human creativity and intuition," he later said, "they got one that played like a machine." Kasparov has suggested that poker might be different. With its blend of probability and psychology and risk, the game should be less vulnerable to brute force methods. Perhaps it could even be the sort of game that chess and checkers never could, a game that needed to be learned rather than solved?

Turing saw learning as a crucial part of artificial intelligence. To win the imitation game, a machine would need to be advanced enough to pass convincingly as a human adult. Yet it did not make sense to focus only on the polished final creation. To create a working mind, it was important to understand where a mind comes from. "Instead of trying to produce a programme to simulate the adult mind," Turing said, "why not rather try to produce one which simulates the child's?" He compared the process to filling a notebook. Rather than attempting to write everything out manually, it would be easier to start with an empty notebook and let the computer work out how it should be filled.

IN 2011, A NEW type of game started appearing among the slot machines and roulette tables of Las Vegas casinos. It was an artificial version of Texas hold'em poker: the chips reduced to two dimensions, the cards dealt on a screen. In the game, players faced a single computer opponent in a two-player form of the game, commonly known as "heads-up poker."

Ever since von Neumann looked at simplified two-player games, heads-up poker has been a favorite target of researchers. This is mainly because it is much easier to analyze a game involving a pair of players than one with several opponents. The "size" of the game—measured by counting the total possible sequences of actions a player could make—is considerably smaller with only two players. This makes it much easier to develop a successful bot. In fact, when it comes to the "limit" version of heads-up poker, in which maximum bets are capped, the Vegas machines are better than most human players.

In 2013, journalist Michael Kaplan traced the origin of the machines in an article for the *New York Times*. It turned out that the poker bots owed much to a piece of software created by Norwegian computer scientist Fredrik Dahl. While studying computer science at the University of Oslo, Dahl had become interested in backgammon. To hone his skills, he created a computer program that could search for successful strategies. The program was so good that he ended up putting it on floppy disks, which he sold for $250 apiece.

Having created a skilled backgammon bot, Dahl turned his attention to the far more ambitious project of building an artificial poker player. Because poker involved incomplete information, it would be much harder for a computer to find successful tactics. To win, the machine would have to learn how to deal with uncertainty. It would have to read its opponent and weigh large numbers of options. In other words, it would need a brain.

IN A GAME LIKE poker, an action might require several decision-making steps. An artificial brain can therefore require multiple linked neurons. One neuron might evaluate the cards on display. Another might consider the amount of money on the table; a third might examine other players' bets. These neurons won't necessarily lead directly to the final decision. The results might flow into a second layer of neurons, which combine the first round of decision making in a more detailed way. The internal neurons are known as "hidden layers" because they lie between the two

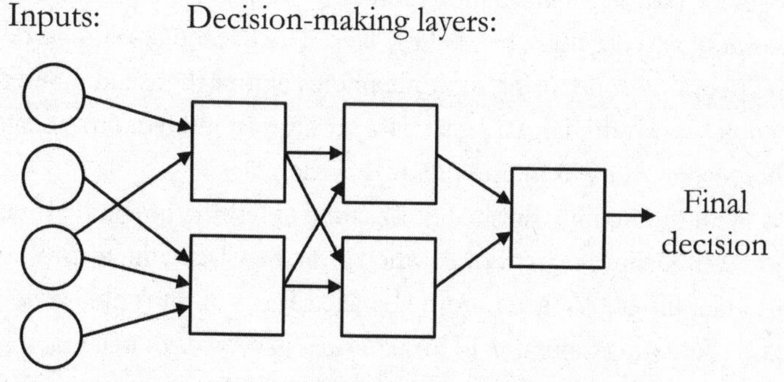

FIGURE 7.1. Illustration of a simple neural network.

visible chunks of information: what goes into the neural network and what comes out.

Neural networks are not a new idea; the basic theory for an artificial neuron was outlined in the 1940s. However, the increased availability of data and computing power means that they are now capable of some impressive feats. As well as enabling bots to learn to play games, they are helping computers to recognize patterns with remarkable accuracy.

In autumn 2013, Facebook announced an AI team that would specialize in developing intelligent algorithms. At the time, Facebook users were uploading over 350 million new photos every day. The company had previously introduced a variety of features to deal with this avalanche of information. One of them was facial recognition: the company wanted to give users the option to automatically detect—and identify—faces in their photos. In spring 2014, the Facebook AI team reported a substantial improvement in the company's facial recognition software, known as DeepFace.

The artificial brain behind DeepFace consists of nine layers of neurons. The initial layers do the groundwork, identifying where the face is in the picture and centering the image. Subsequent layers then pick out features that give a lot of clues about identity, such as the area between the eyes and eyebrows. The final neurons pull together all the separate measurements, from eye shape to mouth position, and use them to label the face. The Facebook team trained the neural network using multiple photos of four thousand different people. It was the largest facial data set ever assembled; on average, there were over a thousand pictures of each face.

With the training finished, it was time to test the program. To see how well DeepFace performed when given new faces, the team asked it to identify photos taken from "Labeled Faces in the Wild," a database containing thousands of human faces in everyday situations. The photos are a good test of facial recognition ability; the lighting isn't always the same, the camera focus varies, and the faces aren't necessarily in the same position. Even so, humans appear to be very good at spotting whether two faces are the same: in an online experiment, participants correctly matched the faces 99 percent of the time.

But DeepFace was not far behind. It had trained for so long, and had its artificial neurons rewired so many times, that it could spot whether two photos were of the same person with over 97 percent accuracy. Even when the algorithm had to analyze stills from YouTube videos—which are often smaller and blurrier—it still managed over 90 percent accuracy.

Dahl's poker program also took a long time to build experience. To train his software, Dahl set up lots of bots and got them to face off against each other in game after game. The computer programs sat through billions of hands, betting and bluffing, their artificial brains developing while they played. As the bots improved, Dahl found that they began to do some surprising things.

IN HIS LANDMARK 1952 paper "Computing Machinery and Intelligence," Turing pointed out that many people were skeptical about the possibility of artificial intelligence. One criticism, put forward by mathematician Ada Lovelace in the nineteenth century, was that machines could not create anything original. They could only do what they were told. Which meant a machine would never take us by surprise.

Turing disagreed with Lovelace, noting that "machines take me by surprise with great frequency." He generally put these surprises down to oversight. Perhaps he'd made a hurried calculation or a careless assumption while constructing a program. From rudimentary computers to high-frequency financial algorithms, this is a common problem. As we've seen, erroneous algorithms can often lead to unexpected negative outcomes.

Sometimes the error can work to a computer's advantage, however. Early in the chess match between Deep Blue and Kasparov, the machine produced a move so puzzling, so subtle, and so—well—intelligent that it threw Kasparov. Rather than grab a vulnerable pawn, Deep Blue moved its rook into a defensive position. Kasparov had no idea why it would do that. By all accounts, the move influenced the rest of the match, persuading the Russian grandmaster that he was facing an opponent far beyond anything he'd played before.

In fact, Deep Blue had no reason for choosing that particular move. Having eventually run into a situation in which it had no rules for—as predicted by Gödel's incompleteness theorem—the computer had acted randomly instead. Deep Blue's game-changing show of strategy was not an ingenious move; it was simple good luck.

Turing admitted that such surprises are still the result of human actions, with the outcomes coming from rules humans have defined (or failed to define). But Dahl's poker bots did not produce surprising actions because of human oversight. Rather, surprises were the result of the programs' learning process. During the training games, Dahl noticed that one of the bots was using a tactic known as "floating." After the three flop cards are shown, a floating player calls the opponent's bets but does not raise them. The floating player loiters, playing out the round without influencing the stakes. Once the fourth turn card is revealed, the player makes a move and raises aggressively, with the hope of scaring the opponent into folding. Dahl had not come across such a technique before, but the strategy is familiar to most good poker players. It also requires a lot of skill to pull off successfully. Not only do players need to judge the cards on display, they need to read their opponents correctly. Some are easier to scare off than others; the last thing a floating player wants is to raise aggressively and then end up in a showdown.

At first glance, such skills seem inherently human. How could a bot teach itself a strategy like this? The answer is that it is inevitable, because sometimes a play relies more on cold logic than we might think. It was just as von Neumann found with bluffing. The strategy was not a mere quirk of human psychology; it was a necessary tactic when following an optimal poker strategy.

In his *New York Times* article, Kaplan mentions that people often refer to Dahl's machine in human terms. They give it nicknames. They call it him. They even admit to talking to the metal box as if it's a real player, as if there's a person sitting behind the glass. When it comes to Texas hold'em, it appears that the bot has succeeded in making people forget that it's a computer program. If Turing's test involved a game of poker rather than a series of questions, Dahl's machine would surely pass.

Perhaps it is not particularly strange that people tend to treat poker bots as independent characters rather than viewing them as the property of the people who programmed them. After all, the best computer players are generally much better than their creators. Because the computer does all the learning, the bot doesn't need to be handed much information initially. Its human maker can therefore be relatively ignorant about game strategies, yet still end up with a strong bot. "You can do amazing things with very little knowledge," as Jonathan Schaeffer put it. In fact, despite having some of the best poker bots in the world, the Alberta poker group has limited human talent when it comes to the game. "Most of our group aren't poker players at all," researcher Michael Johanson said.

Although Dahl had created a bot that could learn to beat most players at limited-stakes poker, there was a catch. Las Vegas gaming rules stipulate that gaming machines have to behave the same against all players. They can't tailor their playing style for opponents who are skilled or inexperienced. The rules meant that Dahl's bot had to sacrifice some of its cunning before it was allowed on the casino floor. From a bot's point of view, having to follow a fixed strategy can make things more difficult. Having a rigid adult brain—rather than the flexible one of a child—prevents the machine from learning how to exploit weaknesses. This removes a big advantage, because it turns out that humans have plenty of flaws that can be exploited.

IN 2010, AN ONLINE version of rock-paper-scissors appeared on the *New York Times* website. It's still there if you want to try it. You'll be able to play against a very strong computer program. Even after a few games, most people find that the computer is pretty hard to beat; play lots of games, and the computer will generally end up in the lead.

Game theory suggests that if you follow the optimal strategy for rock-paper-scissors, and choose randomly between the three available options, you should expect to come out even. But when it comes to rock-paper-scissors, it seems that humans aren't very good at doing what's optimal. In 2014, Zhijian Wang and colleagues at Zhejiang University in China reported that people tend to follow certain behavior patterns during games of rock-paper-scissors. The researchers recruited 360 students, divided them into groups, and asked each group to play three hundred rounds of rock-paper-scissors against each other. During the games, the researchers found that many students adopted what they called a "win-stay lose-shift" strategy. Players who'd just won a round would often stick with the same action in the next round, while the losing players had a habit of switching to the option that beat them. They would swap rock for paper, for instance, or scissors for rock. Over many rounds, the players generally chose the three different options a similar number of times, but it was clear they weren't playing randomly.

The irony is that even truly random sequences can contain seemingly nonrandom patterns. Remember those lazy journalists in Monte Carlo who made up the roulette numbers? There were a lot of obstacles they'd have had to overcome to create results that appeared random. First, they would have had to make sure black and red came up similarly often in the results. The journalists actually managed to get this bit right, which meant the data passed the initial round of Karl Pearson's "Is it random?" test. However, the reporters came unstuck when it came to runs of colors, because they switched between red and black more often than a truly random sequence would.

Even if you know how randomness should look, and try to alternate between colors—or rock, paper, scissors—correctly, your ability to generate random patterns will be limited by your memory. If you had to read a list of numbers and immediately recite them, how many could you manage? Half a dozen? Ten? Twenty?

In the 1950s, cognitive psychologist George Miller noted that most young adults could learn and recite around seven numbers at a time. Try memorizing one local phone number and you'll probably be fine; attempt to remember two, and it gets tricky. This can be problematic if you're trying to generate random moves in a game; how can you ensure you use all options equally often if you can only remember the last few moves? In 1972, Dutch psychologist Willem Wagenaar observed that people's brains tend to concentrate on a moving "window" of about six to seven previous responses. Over this interval, people could alternate between options reasonably "randomly." However, they were not so good at switching between options over longer time intervals. The size of the window, around six to seven events long, could well be a consequence of Miller's earlier observation.

In the years since Miller published his work, researchers delved further into human memory capacity. It turns out that the value Miller jokingly referred to as the "magical number seven" is not so magical after all. Miller himself noted that when people had to remember only binary numbers—such as zeros and ones—they could recite a sequence of about eight digits. In fact, the size of the data "chunks" humans can remember depends on the complexity of the information. People might be able to recall seven numbers, but there is evidence they can recite only six letters or so, and five one-syllable words.

In some cases, people have learned to increase the amount of information they can recall. In memory championships, the best competitors can memorize over a thousand playing cards in an hour. They do this by changing the format of the data chunks they remember; rather than thinking in terms of raw numbers, they try to memorize images as part of a journey. Cards become celebrities or objects; the sequence becomes a series of events in which their card characters feature. This helps the competitors' brains shelve and retrieve the information more efficiently. As discussed in the previous chapter, memorizing cards also helps in blackjack, with card counters "bucketing" information to reduce the amount they have to store. Such storage problems have interested researchers looking at artificial minds as well as those working on human ones. Nick Metropolis said Stanislaw Ulam "often mused about the nature of memory and how it was implemented in the brain."

When it comes to rock-paper-scissors, machines are much better than humans at coming up with the unpredictable moves required for an optimal game theory strategy. Such a strategy is inherently defensive, of course, because it aims to limit potential losses against a perfect opponent. But the rock-paper-scissors bot on the *New York Times* website was not playing a perfect opponent. It was playing error-prone humans, who have memory issues and can't generate random numbers. The bot therefore deviated from a random strategy and started to hunt down weaknesses.

The computer had two main advantages over its human opponents. First, it could accurately remember what the human had done in previous rounds. It could recall which sequences of moves the person had played, for example, and which patterns the person was fond of. And that's when the second advantage kicked in.

The computer wasn't just using information on its current opponent. It was drawing on knowledge gained during two hundred thousand rounds of rock-paper-scissors against humans. The database came from Shawn Bayern, a law professor and an ex-computer scientist whose website runs a massive online rock-paper-scissors tournament. The competition is still going, with over half a million rounds played to date (the computer has won the majority of them). The data meant the bot could compare its current opponent with others it had played. Given a particular sequence of moves, it could work out what humans tended to do next. Rather than being interested only in randomness, the machine was instead building a picture of its opponent.

Such approaches can be particularly important in games like poker, which can have more than two players. Recall that, in game theory, optimal strategies are said to be in Nash equilibrium: no single player will gain anything by picking a different strategy. Neil Burch, one of the researchers in the University of Alberta poker group, points out that it makes sense to look for such strategies if you have a single opponent. If the game is zero-sum—with everything you lose going to your opponent, and vice versa—then a Nash equilibrium strategy will limit your losses. What's more, if your opponent deviates from an equilibrium strategy, your opponent will lose out. "In two player games that are zero-sum, there's a really good reason to say that a Nash equilibrium is the correct thing to play," Burch said. However, it isn't necessarily the best option when more players join the game. "In a three player game, that can fall apart."

Nash's theorem says players will lose out if they change their strategy unilaterally. But it doesn't say what will happen if two players swap tactics together. For instance, two of the players could decide to gang up on the third. When von Neumann and Morgenstern wrote their book on game theory, they noted that such coalitions work only when there are at least three players. "In a two-person game there are not enough players to go around," they said. "A coalition absorbs at least two players, and then nobody is left to oppose." Turing also acknowledged the potential role of coalitions in poker. "It is only etiquette, sense of fair play etc. which prevents this happening in actual games," he said.

There are two main ways to form coalitions in poker. The most blatant way to collude would be for two or more players to reveal their cards to each other. When one of them received a strong hand, they all could push the bets up gradually to squeeze more money out of their opponents. Naturally, this approach is much easier to employ while playing online. Parisa Mazrooei and colleagues at the University of Alberta suggest such collusion should really be referred to as "cheating" because players are using strategies outside the rules of the game, which stipulate that cards remain hidden.

The alternative would be for colluders to keep their cards to themselves but give signals to other players when they had a strong hand. Technically, they would be operating within the rules (if not inside the boundaries of fair play). Colluding players often follow certain betting patterns to improve their chances. If one player bets big, for example, the others follow to drive opponents out of the game. Human players would have to remember such signals, but things are easier for bots, which can have access to the exact set of programmed rules used by fellow colluders.

There are reports of unscrupulous players using both types of approach in online poker rooms. However, it can be difficult to detect collusion. If a player is matching another's bets, gradually inflating the pot, that player might be manipulating the game to help their teammate. Or the person could be just a naive beginner, trying to bluff to victory. "In any form of poker, there exist a large variety of strategy combinations that are mutually beneficial to those that apply them," Frederik Dahl has pointed out. "If they apply such strategies on purpose, we would say that they cheat by co-operating, but if it happens just by accident, we would not."

That's the problem with using game theory in poker: coalitions don't always have to be deliberate. They might just result from the strategies players choose. In many situations, there is more than one Nash equilibrium. Take driving a car. There are two equilibrium strategies: if everyone drives on the left, you will lose out if you make a unilateral decision to drive on the other side; if driving on the right is in vogue, the left is no longer the best choice.

Depending on the location of your driver's seat, one of these equilibriums will be preferable to the other. If your car is left-hand drive, for instance, you'll probably prefer it if everyone drives on the right. Obviously, having a driver's seat inconveniently positioned on the "wrong" side of the car won't be enough to make you change the side you drive on. But the situation is still a bit like having everyone else in a coalition against you (if you're feeling particularly grumpy about things). Drive on the other side of the road and you'll clearly lose out, so you just have to put up with the situation.

The same problem crops up in poker. As well as causing inconvenience, it can cost players money. Three poker players could choose Nash equilibrium strategies, and when these strategies are put together, it may turn out that two players have selected tactics that just so happen to pick on the third player. This is why three-player poker is so difficult to tackle from a game theory point of view. Not only is the game far more complicated, with more potential moves to analyze, it's not clear that hunting for the Nash equilibrium is always the best approach. "Even if you could compute one," Michael Johanson said, "it wouldn't necessarily be useful."

There are other drawbacks, too. Game theory can show you how to minimize your losses against a perfect opponent. But if your opponent has flaws—or if there are more than two players in the game—you might want to deviate from the "optimal" Nash equilibrium strategy and instead take advantage of weaknesses. One way to do this would be to start off with an equilibrium strategy, and then gradually tweak your tactics as you learn more about your opponent. Such approaches can be risky, however. Tuomas Sandholm at Carnegie Mellon University points out that players must strike a balance between exploitation and exploitability. Ideally, you want to *exploit*, taking as much as possible from weak opponents, but not be *exploitable*, and come unstuck against strong players. Defensive strategies—such as the Nash equilibrium, and the tactics employed by Dahl's poker bot—are not very exploitable. Strong players will struggle to beat them. However, this comes at the price of not exploiting weak opponents much; bad players will get off lightly. It would therefore make sense to vary strategy depending on the opponent. As the old saying goes, "Don't play the cards, play the person."

Unfortunately, learning to exploit opponents can in turn leave players vulnerable to exploitability. Sandholm calls it the "get taught and exploited problem." For example, suppose your opponent appears to play aggressively at first. When you notice this, you might adjust tactics to try to take advantage of this aggression. However, at this point the opponent could suddenly become more conservative and exploit the fact that you believe—incorrectly—that you're facing an aggressive player.

Researchers can judge the effects of such problems by measuring the exploitability of their bots. This is the maximum amount they could expect to lose if the bot had made completely the wrong assumptions about its opponent. Along with PhD student Sam Ganzfried, Sandholm has been developing "hybrid" bots, which combine defensive Nash equilibrium tactics with opponent modeling. "We would like to only attempt to exploit weak opponents," they said, "while playing the equilibrium against strong opponents."

IT'S CLEAR THAT POKER programs are getting better and better. Every year, the bots in the Annual Computer Poker Competition become smarter, and Vegas is filling up with poker machines that can beat most casino visitors. But have computers truly overtaken humans? Are the best bots really better than all people?

According to Sandholm, it's hard to say whether the crossover has happened for several reasons. To start with, you'd have to identify who the best human is. Unfortunately, it's hard to rank players definitively: poker doesn't have a clear Garry Kasparov or Marion Tinsley. "We don't really know who the best human is," Sandholm said. Games against humans are also difficult to arrange. Although there is a computer competition every year, Sandholm points out that mixed matches are far less common. "It's hard to get pros to do these man-machine matches."

There has been the occasional match-up. In 2007, professional players Phil Laak and Ali Eslami took on Polaris, a bot created by the University of Alberta group, in a series of two-player poker games. Polaris was designed to be hard to beat. Rather than attempting to exploit its opponents, it employed a strategy that was close to Nash equilibrium.

At the time, some of the poker community thought Laak and Eslami were strange choices for the match. Laak had a reputation for being hyperactive at the poker table, jumping around, rolling on the floor, doing push-ups. In contrast, Eslami was fairly unknown, having appeared in relatively few televised tournaments. But Laak and Eslami had skills that the researchers needed. Not only were they good players, they were able to say what they were thinking during a game, and they were comfortable with the unusual setup involved in a man-versus-machine match.

The venue for the match was an artificial intelligence conference in Vancouver, Canada, and the game was limit Texas hold'em: the same game Dahl's bot would later play in Vegas. Although Laak and

Eslami would play Polaris in separate games, their scores would be combined at the end of each session. It was to be humans versus the machine, with Laak and Eslami playing as a team against Polaris. To minimize the effects of luck, the card deals were mirrored: whichever cards were dealt to Polaris in one game, the human player got in the other (and vice versa). The organizers also imposed a clear margin of victory. To win, a team needed to finish with at least $250 more chips than their opponent.

On the first day, there were two sessions of play, each consisting of five hundred hands. The first session ended as a draw (Polaris had finished seventy dollars up, which was not enough to be considered a victory). In the second session, Laak was lucky enough to be dealt good cards against Polaris, which meant the computer received the same strong cards in the game against Eslami. Polaris capitalized on the advantage more than Laak did, and the bot ended the day with a clear win over the human team.

That night, Laak and Eslami met up to discuss the thousand poker hands they'd just played. The Alberta team gave them the logbook of the day's play, including all the hands that had been dealt. This helped the pair dissect the games they'd played. When they returned to the tables the next day, the humans had a much better idea of how to tackle Polaris, and they won the final two sessions. Even so, the humans were modest about their victory. "This was not a win for us," Eslami said at the time. "We survived. I played the best heads-up poker I've ever played and we just narrowly won."

The following year, there was a second man-machine competition, with a new set of human opponents. This time, seven human players would take on the University of Alberta's bot in Las Vegas. The humans were arguably some of the best around; several had career winnings totaling more than $1 million. But they would not be facing the same Polaris that lost the previous year. This was Polaris 2.0. It was more advanced and better trained. Since the games against Laak and Eslami, Polaris had played over eight billion games against itself. It was now better at exploring the vast combination of possible moves, which meant there were fewer weak links in its strategy for opponents to target.

Polaris 2.0 also put more emphasis on learning. During a match, the bot would develop a model of its opponent. It would identify which type of strategy a player was using and tailor its game to target weaknesses. Players couldn't beat Polaris by discussing tactics between games as Laak and Eslami did, because Polaris would be playing differently against each one of them. Nor could humans regain the advantage by altering their own playing style. If Polaris noticed its opponent had changed strategy, the bot would adapt to the new tactics. Michael Bowling, who headed up the Alberta team, said that many of the human players struggled against Polaris's new box of tricks; they had never seen an opponent switch strategy like that.

As before, the players paired up to take on Polaris in the limit version of Texas hold'em. There were four matches in total, spread over four days. The first two went badly for Polaris, with one a draw and the other a victory for the humans. But this time the humans did not finish strongly; Polaris won the final two matches, and with them the competition.

Whereas Polaris 2.0 drifted away from an optimal strategy to exploit its opponents, the next challenge for the Alberta team was how to make a bot that was truly unbeatable. Their existing bots could only compute an approximate Nash equilibrium, which meant there might have been a strategy out there that could beat them. Bowling and his colleagues therefore set out to find a set of flawless tactics, which in the long run would not lose money against any opponent.

Using the regret minimization approach we encountered in the previous chapter, the Alberta researchers honed the bots, and then got them to play each other again and again, at a rate of about two thousand games per second. Eventually, the bots learned how to avoid being exploited, even by a perfect player. In 2015, the team unveiled their unbeatable poker program—named Cepheus—in the journal *Science*. With a nod to the group's checkers research, the paper was titled "Heads-Up Limit Hold'em Poker Is Solved."

Some of the findings lined up with conventional wisdom. The team showed that in heads-up poker, the dealer—who goes first—holds the advantage as well as the cards. They also found that Cepheus rarely "limps" and chooses to raise or fold on its first go rather than simply calling its opponent's bet. According to Johanson, as the bot narrowed in on the optimal strategy it also started to come up with some unexpected tactics. "Every now and then we find differences between what the program chooses and what human wisdom says." For example, the final version of Cepheus chooses to play hands—such as a four and six with different suits—that many humans would fold. In 2013, the team also noticed their bot would occasionally place the minimum allowable bet rather than putting down a larger stake. Given the extent of the bot's training, this was apparently the optimal thing to do. But Burch points out that human players would see things differently. Although the computer had decided it is a smart tactic, most humans would view it as annoying. "It's almost a nuisance bet," Burch said. The polished version of Cepheus is also reluctant to bet big initially. Even when it has the best hand (a pair of aces), it will bet the maximum possible amount only 0.01 percent of the time.

Cepheus has shown that, even in complex situations, it can be possible to find an optimal strategy. The researchers point to a range of scenarios in which such algorithms could be useful, from designing coast guard patrols to medical treatments. But this was not the only reason for the research. The team ended their *Science* paper with a quote from Alan Turing: "It would be disingenuous of us to disguise the fact that the principal motive which prompted the work was the sheer fun of the thing."

Despite the breakthrough, not everyone was convinced that it represented the ultimate victory of the artificial over the biological. Michael Johanson says that many human players view limit poker as the easy option, because there is a cap on how much players can raise the betting. It means the boundaries are well defined, the possibilities constrained.

No-limit poker is seen as the bigger challenge. Players can raise by any amount and can go all in whenever they want. This creates more options, and more subtleties. The game therefore has a reputation for being more of an art than a science. And that's why Johanson would love to see a computer win. "It would attack the mystique that poker is all about psychology," he said, "and that computers couldn't possibly do it."

Sandholm says it won't be long before two-player no-limit poker has fallen to the machines. "We're working on that very actively," he said. "We may already have bots that are better than the best pros." Indeed, Carnegie Mellon's bot Tartanian put in a very strong showing in the 2014 computer poker competition. There were two types of contest for no-limit poker, and Tartanian came out on top in both. As well as winning the knockout competition, it prevailed in the total bankroll contest. Tartanian could survive when it needed to but also rack up plenty of chips against weaker opponents.

As bots get better—and beat more humans—players could end up learning poker from machines. Chess grandmasters already use computers to hone their skills in training. If they want to know how to play a particularly tricky position, the machines can show them the best way to proceed. Chess computers can come up with strategies that stretch further into the future than we humans can manage.

With computer programs cleaning up at chess, checkers, and now poker, it might be tempting to argue that humans can no longer compete at such games. Computers can analyze more data, remember more strategies, and examine more possibilities. They are able to spend longer learning and longer playing. Bots can teach themselves supposedly "human" tactics such as bluffing and even "superhuman" strategies humans haven't spotted yet. So, is there anything that computers are not so good at?

ALAN TURING ONCE NOTED that if a man tried to pretend to be a machine, "he would clearly make a very poor showing." Ask the human to perform a calculation, and he'd be much slower, not to mention more error prone, than the computer. Even so, there are still some situations that bots struggle with. When playing *Jeopardy!* Watson found the short clues the most difficult. If the host read out a single category and a name—such as "first ladies" and Ronald Reagan—Watson would take too long to search through its database to find the correct response (which is "Who is Nancy Reagan?"). Whereas Watson would beat a human contestant in a race to solve a long, complicated clue, the human would prevail if there were only a few words to go by. In quiz shows, it seems that brevity is the enemy of machines.

The same is true of poker. Bots need time to study their opponents, learning their betting style so it can be exploited. In contrast, human professionals are able to evaluate other players much more quickly. "Humans are good at making assumptions about an opponent with very little data," Schaeffer said.

In 2012, researchers at the University of London suggested that some people might be especially good at sizing up others. They designed a game, called the Deceptive Interaction Task, to test players' ability to lie and detect lies. In the task, participants were placed in groups, with one person given a cue card containing an opinion—such as "I'm in favor of reality TV"—and instructions to either lie or tell the truth. After stating the opinion, the person had to give reasons for holding that view. The others in the group had to decide whether they thought the person was lying or not.

The researchers found that people who were lying generally took longer to start to speak after receiving the cue card. Liars took an average 6.5 seconds, compared to 4.6 seconds for honest speakers. It also turned out that good liars were also effective lie detectors, much like the proverb "it takes a thief to catch a thief." Although liars appeared to be better at spotting deceit in the game, it was not clear why this was the case. The researchers suggested it might be because they were better—whether consciously or unconsciously—at picking up on others' slow responses as well as speeding up their own speech.

Unfortunately, people aren't so good at identifying the specific signs of lying. In a 2006 survey spanning fifty-eight countries, participants were asked "How can you tell when people are lying?" One answer dominated the responses, coming up in every country and topping the list in most: liars avoid eye contact. Although it's a popular lie-detection method, it doesn't appear to be a particularly good one. There's no evidence that liars avert their gaze more than truthful people. Other supposed giveaways have dubious foundations, too. It is not clear that liars are noticeably more animated or shift posture when speaking.

Behavior might not always reveal liars, but it can influence games in other ways. Psychologists at Harvard University and Caltech have shown that having certain facial expressions can lure opponents into making bad bets. In a 2010 study, they got participants to play a simplified poker game against a computer-generated player, whose face was displayed on a screen. The researchers told participants the computer would be using different styles of play but said nothing about the face on the screen. In reality, the instructions were a ruse. The computer picked moves randomly; all that changed was its face. The simulated player displayed three possible expressions, which followed stereotypes about honesty. One was seemingly trustworthy, one neutral, and one untrustworthy. The researchers found that players facing computer players with dishonest or neutral faces made relatively good choices. However, when they played "trustworthy" computer opponents, participants made significantly worse decisions, often folding when they had the stronger hand.

The researchers pointed out that the study involved a cartoon version of poker, played by beginners. Professional poker games are likely to be very different. However, the study suggests that facial expressions might not influence poker in the way we assume. "Contrary to the popular belief that the optimal poker face is neutral in appearance," the authors noted, "the face that invokes the most betting mistakes by our subjects has attributes that are correlated with trustworthiness."

Emotion can also influence overall playing style. The University of Alberta poker group has found that humans are particularly susceptible to strong-arm tactics. "In general, a lot of the knowledge that human poker pros have about how to beat other humans revolves around aggression," Michael Johanson said. "An aggressive strategy that puts a lot of pressure on opponents, making them make tough decisions, tends to be very effective." When playing humans, the bots try to mimic this behavior and push opponents into making mistakes. It seems that bots have a lot to gain by copying the behavior of humans. Sometimes, it even pays to copy their flaws.

WHEN MATT MAZUR DECIDED to build a poker bot in 2006, he knew it would have to avoid detection. Poker websites would ban anyone they suspected of running computer players. It wasn't enough to have a bot that could beat humans; Mazur would need a bot that could look human while doing it.

A computer scientist based in Colorado, Mazur worked on a variety of software projects in his spare time. In 2006, the new project was poker. Mazur's first attempt at a bot, created that autumn, was a program that played a "short stacking" strategy. This involved buying into games with very little money, and then playing very aggressively, hoping to scare off players and steal the pot. It's often seen as an irritating tactic, and Mazur discovered it wasn't a particularly successful one either. Six months in, the bot had played almost fifty thousand hands and lost over $1,000. Abandoning his flawed first draft, Mazur designed a new bot, which would play two-player poker properly. The finished bot played a tight game, choosing its moves carefully, and was aggressive in its betting. Mazur said the bot was reasonably competitive against humans in small-stakes games.

The next challenge was to avoid getting caught. Unfortunately, there wasn't much information out there to help Mazur. "Online poker sites are understandably quiet when it comes to what they look at to detect bots," he said, "so bot developers are forced to make educated guesses." While designing his poker program, Mazur therefore tried to put himself in the position of a bot hunter. "If I was trying to detect a bot, I would look at a lot of different factors, weigh them, and then manually investigate the evidence in order to make a call as far as whether a player was a bot or not."

One obvious red flag would be strange betting patterns. If a bot placed too many bets, or too quickly, it might look suspicious. Unfortunately, Mazur found his bots could sometimes behave strangely by accident. The bots worked in pairs to compete on poker websites. One of them would register for new games, and the other would play them. On one occasion, Mazur was away from his computer when the game-playing program crashed. The other bot had no idea what had happened, so it kept on registering for new games. Without the game-playing bot ready to take a seat at the table, Mazur's account skipped over twenty games in a row. Mazur later realized his bots had other quirks, too. For instance, they would often play with the same stakes for hundreds of games. Mazur points out that humans rarely behave like that: they would generally get confident (or bored) over time and move up to higher-stakes games for a while.

As well as playing sensibly, Mazur's bots also had to navigate their way around the poker websites. Mazur found that some websites had features—be they accidental or deliberate—that made automated navigation harder. Sometimes they would subtly alter what appeared on his screen, perhaps by changing the size or shape of windows or moving buttons. Such changes wouldn't cause problems for a human, but they could throw bots that had been taught to navigate a specific set of dimensions. Mazur had to get his bots to track the locations of the windows and buttons and adjust where they clicked to account for any changes.

The whole process was like a version of Turing's imitation game. To avoid detection, Mazur's bots had to convince the website they were playing like humans. Sometimes, bots even found themselves facing Turing's original test. Most poker websites include a chat feature, which lets players talk to each other. Generally, this isn't a problem; players often remain silent in poker games. But there were some conversations that Mazur decided he couldn't avoid. If someone accused his bot of being a computer program and the bot didn't reply, there was a risk that he would be reported to the website owners. Mazur therefore put together a list of terms that suspicious opponents might use. If someone mentioned words such as "bot" or "cheater" during a game, he'd get an alert and intervene. It meant he'd have to be near his computer when his bot was playing, but the alternative was potentially much worse; an unsupervised program could easily run into trouble and not know to get out of it.

It took a while for Mazur's bots to become winners: the programs didn't make money for the first eighteen months they were active. Eventually, in spring 2008, the bots started to produce modest profits. The successful run came to an abrupt end a few months later, however. On October 2, 2008, Mazur got an e-mail from the poker website informing him that his account had been suspended. So, what gave it away? "In retrospect," he said, "I think the thing that got my bot caught was that it was simply playing too many games." Mazur's bot concentrated on heads-up "Sit 'n Go" games, which commence as soon as two players join the game. "A normal player might play ten to fifteen No Limit Heads Up Sit 'n Gos in a day," Mazur said. "At its peak, my bot was playing fifty to sixty per day. That probably threw up some flags." Of course, this is only his best guess. "It's possible that it was something else entirely. I'll probably never know for sure."

Mazur wasn't actually that bothered about the loss of profit from his bot. "When my account was eventually suspended, I had not netted that much money," he said. "I would have been much better off financially if I'd actually used that time to play poker instead. But then again, I didn't build the bot to make money; I built it for the challenge."

After his account was suspended, Mazur e-mailed the poker website that had banned him and offered to explain exactly what he'd done. He knew several ways to make life even harder for bots, which he hoped might improve security for human poker players. Mazur told the company all the things they should look out for, from high volumes of games to unusual mouse movements. He even suggested countermeasures that could hinder bot development, such as varying the size and location of buttons on the screen.

Mazur also posted a detailed history of his bot's creation on his website, including screenshots and schematics. He wanted to show people that poker bots are hard to build, and there are much more useful things they could be doing with computers. "I realized that if I was going to spend that much time on a software project, I should devote that energy to more worthwhile endeavors." Looking back, however, he doesn't regret the experience. "Had I not built the poker bot, who knows where I'd be."

8

BEYOND CARD COUNTING

IF YOU EVER VISIT A LAS VEGAS CASINO, LOOK UP. HUNDREDS OF cameras cling to the ceiling like jet-black barnacles, watching the tables below. The artificial eyes are there to protect the casino's income from the quick-witted and light-fingered. Until the 1960s, casinos' definition of such cheating was fairly clear-cut. They only had to worry about things like dealers paying out on losing hands or players slipping high-value chips into their stake after the roulette ball had landed. The games themselves were fine; they were unbeatable.

Except it turned out that wasn't true. Edward Thorp found a loophole in blackjack big enough to fit a best-selling book through. Then a group of physics students tamed roulette, traditionally the epitome of chance. Beyond the casino floor, people have even scooped lottery jackpots using a mix of math and manpower.

The debate over whether winning depends on luck or skill is now spreading to other games. It may even determine the fate of the once lucrative American poker industry. In 2011, US authorities shut down a number of major poker websites, bringing an end to the "poker boom" that had gripped the country for the previous few years. The legislative muscle for the shake-up came from the Unlawful Internet Gambling Enforcement Act. Passed in 2006, it banned bank transfers related to games where the "opportunity to win is predominantly subject to chance." Although the act has helped curb the spread of poker, it doesn't cover stock trading or horseracing. So, how do we decide what makes something a game of chance?

During the summer of 2012, the answer would turn out to be worth a lot to one man. As well as taking on the big poker companies, federal authorities had also gone after people operating smaller games. That included Lawrence DiCristina, who ran a poker room on Staten Island in New York. The case went to trial in 2012, and DiCristina was convicted of operating an illegal gambling business.

DiCristina launched a motion to dismiss the conviction, and the following month he was back in court arguing his case. During the hearing, DiCristina's lawyer called economist Randal Heeb as an expert witness. Heeb's aim was to convince the judge that poker was predominantly a game of skill, and therefore didn't fall under the definition of illegal gambling. While giving evidence, Heeb presented data from millions of online poker games. He showed that, bar a few bad days, the top-ranked players won pretty consistently. In contrast, the worst players lost throughout the year. The fact that some people could make a living from poker was surely evidence that the game involved skill.

The prosecution also had an expert witness, an economist named David DeRosa. He did not share Heeb's views about poker. DeRosa had used a computer to simulate what might happen if a thousand people each tossed a coin ten thousand times. Assuming a certain outcome—such as tails—was equivalent to a win, and the number of times a particular person won the toss was totally random. And yet the results that came out were remarkably similar to those Heeb presented: a handful of people appeared to win consistently, and another group of people seemed to lose a large number of times. This wasn't evidence that a coin toss involves skill, just that—much like the infinite number of monkeys typing—unlikely events can happen if we look at a large enough group.

Another concern for DeRosa was the number of players who lost money. Based on Heeb's data, it seemed that about 95 percent of people playing online poker ended up out of pocket. "How could it be skillful playing if you're losing money?" DeRosa said. "I don't consider it skill if you lose less money than the unfortunate fellow who lost more money."

Heeb admitted that, in a particular game, only 10 to 20 percent of players were skillful enough to win consistently. He said the reason so many more people lost than won was partly down to the house fee, with poker operators taking a cut from the pot of money in each round (in DiCristina's games, the fee was 5 percent). But he did not think the apparent existence of a skilled poker elite was the result of chance. Although a small group may appear to win consistently if lots of people flip coins, good poker players generally continue to win after they've been ranked highly. The same cannot be said for the people who are fortunate with coin tosses.

According to Heeb, part of the reason good players can win is that in poker players have control over events. If bettors place a bet on a sports match or a roulette wheel, their wagers do not affect the result. But poker players can change the outcome of the game with their betting. "In poker, the wager is not in the same sense a wager on the outcome," Heeb said. "It is the strategic choice that you are making. You are trying to influence the outcome of the game."

But DeRosa argued that it doesn't make sense to look at a player's performance over several hands. The cards that are dealt are different each time, so each hand is independent of the last. If a single hand involves a lot of luck, there is no reason to think that player will have a successful round after a costly one. DeRosa compared the situation to the Monte Carlo fallacy. "If red has come up 20 times in a row in roulette," he said, "it does not mean that 'black is due.'"

Heeb conceded that a single hand involves a lot of chance, but it did not mean the game was chiefly one of luck. He used the example of a baseball pitcher. Although pitching involves skill, a single pitch is also susceptible to chance: a weak pitcher could produce a good ball, and a strong pitcher could throw a bad one. To identify the best—and worst—pitchers, we need to look at lots of throws.

The key issue, Heeb argued, is how long we must wait for the effects of skill to outweigh chance. If it takes a large number of hands (i.e., longer than most people will play), then poker should be viewed as a game of chance. Heeb's analysis of the online poker games suggested this wasn't the case. It seemed that skill overtook luck after a relatively small number of hands. After a few sessions of play, a skillful player could therefore expect to hold an advantage.

It fell to the judge, a New Yorker named Jack Weinstein, to weigh the arguments. Weinstein noted that the law used to convict DiCristina—the Illegal Gambling Business Act—listed games such as roulette and slot machines, but it did not explicitly mention poker. Weinstein said it wasn't the first time a law had failed to specify a crucial detail. In October 1926, airport operator William McBoyle helped arrange the theft of an airplane in Ottawa, Illinois. Although he was convicted under the National Motor Vehicle Theft Act, McBoyle appealed the result. His lawyers argued that the act did not explicitly cover airplanes, because it defined a vehicle as "an automobile, automobile truck, automobile wagon, motor cycle, or any other self-propelled vehicle not designed for running on rails." According to McBoyle's lawyers, this meant an airplane was not a vehicle, and so McBoyle could not be guilty of the federal crime of transporting a stolen vehicle. The US Supreme Court agreed. They noted that the wording of the law evoked the mental image of vehicles moving on land, so it shouldn't be extended to aircraft simply because it seemed that a similar rule ought to apply. The conviction was reversed.

Although poker wasn't mentioned in the Gambling Act, Judge Weinstein said this didn't automatically mean the game wasn't gambling. But the omission did mean that the role of chance in poker was up for debate. And Weinstein had found Heeb's evidence convincing. Until that summer, no court had ever ruled on whether poker was gambling under federal law. Weinstein delivered his conclusion on August 21, 2012, and ruled that poker was predominantly governed by skill rather than chance. In other words, it did not count as gambling under federal law. DiCristina's conviction was overturned.

The victory was to be short-lived, however. Although Weinstein ruled that DiCristina had not broken federal law, New York State has a stricter definition of gambling. Its laws cover any game that "depends in a material degree upon an element of chance." As a result, DiCristina's acquittal was overturned in August 2013. Weinstein's ruling on the relative role of luck and skill was not questioned. Rather, the state law meant that poker still fell under the definition of a gambling business.

The DiCristina case is part of a growing debate about how much luck comes into games like poker. Definitions like "material degree of chance" will undoubtedly raise more questions in the future. Given the close links between gambling and certain parts of finance, surely this definition would cover some financial investments, too? Where do we draw the line between flair and fluke?

IT IS TEMPTING TO sort games into separate boxes marked luck and skill. Roulette, often used as an example of pure luck, might go into one; chess, a game that many believe relies only on skill, might go in the other. But it isn't this simple. To start with, processes that we think are as good as random are usually far from it.

Despite its popular image as the pinnacle of randomness, roulette was first beaten with statistics, and then with physics. Other games have fallen to science too. Poker players have explained game theory and syndicates have turned sports betting into investments. According to Stanislaw Ulam, who worked on the hydrogen bomb at Los Alamos, the presence of skill is not always obvious in such games. "There may be such a thing as habitual luck," he said. "People who are said to be lucky at cards probably have certain hidden talents for those games in which skill plays a role." Ulam believed the same could be said of scientific research. Some scientists ran into seemingly good fortune so often that it was impossible not to suspect that there was an element of talent involved. Chemist Louis Pasteur put forward a similar philosophy in the nineteenth century. "Chance favours the prepared mind" was how he put it.

Luck is rarely embedded so deeply in a situation that it can't be altered. It might not be possible to completely remove luck, but history has shown that it can often be replaced by skill to some extent. Moreover, games that we assume rely solely on skill do not. Take chess. There is no inherent randomness in a game of chess: if two players make identical moves every time, the result will always be the same. But luck still plays a role. Because the optimal strategy is not known, there is a chance that a series of random moves could defeat even the best player.

Unfortunately, when it comes to making decisions, we sometimes take a rather one-sided view of chance. If our choices do well, we put it down to skill; if they fail, it's the result of bad luck. Our notion of skill can also be skewed by external sources. Newspapers print stories about entrepreneurs who have hit a trend and made millions or celebrities who have suddenly become household names. We hear tales of new writers who have produced instant best sellers and bands that have become famous overnight. We see success and wonder why those people were so special. But what if they are not?

In 2006, Matthew Salganik and colleagues at Columbia University published a study of an artificial "music market," in which participants could listen to, rate, and download dozens of different tracks. In total there were fourteen thousand participants, whom the researchers secretly split into nine groups. In eight of the groups, participants could see which tracks were popular with their fellow group members. The final group was the control group, in which participants had no idea what others were downloading.

The researchers found that the most popular songs in the control group—a ranking that depended purely on the merits of the songs themselves, and not on what other people were downloading—were not necessarily popular in the eight social groups. In fact, the song rankings in these eight groups varied wildly. Although the "best" songs usually racked up some downloads, mass popularity was not guaranteed. Instead, fame developed in two stages. First, randomness influenced which tracks people happened to pick early on. The popularity of these first downloaded tracks was then amplified by social behavior, with people looking at the rankings and wanting to imitate their peers. "Fame has much less to do with intrinsic quality than we believe it does," Peter Sheridan Dodds, one of the study authors, later wrote, "and much more to do with the characteristics of the people among whom fame spreads."

Mark Roulston and David Hand, statisticians at the hedge fund Winton Capital Management, point out that the randomness of popularity may also influence the ranking of investment funds. "Consider a set of funds with no skill," they wrote in 2013. "Some will produce decent returns simply by chance and these will attract investors, while the poorly performing funds will close and their results may disappear from view. Looking at the results of those surviving funds, you would think that on average they do have some skill."

The line between luck and skill—and between gambling and investing—is rarely as clear as we think. Lotteries should be textbook examples of gambling, but after several weeks of rollovers, they can produce a positive expected payoff: buy up all the combinations of numbers, and you'll make a profit. Sometimes the crossover happens the other way, with investments being more like wagers. Take Premium Bonds, a popular form of investment in the United Kingdom. Rather than receiving a fixed rate of interest as with regular bonds, investors in Premium Bonds are instead entered into a monthly prize draw. The top prize is £1 million, tax-free, and there are several smaller prizes, too. By investing in Premium Bonds, people are in effect gambling the interest they would have otherwise earned. If they instead put their savings in a regular bond, withdrew the interest, and used that money to buy rollover lottery tickets, the expected payoff would not be that different.

If we want to separate luck and skill in a given situation, we must first find a way to measure them. But sometimes an outcome is very sensitive to small changes, with seemingly innocuous decisions completely altering the result. Individual events can have dramatic effects, particularly in sports like soccer and ice hockey where goals are relatively rare. It might be an ambitious pass that sets up a winning shot or a puck that hits the post. How can we distinguish between a hockey victory that is mostly down to talent and one that benefited from lots of lucky breaks?

In 2008, hockey analyst Brian King suggested a way to measure how fortunate a particular NHL player had been. "Let's pretend there was a stat called 'blind luck,'" as he put it. To calculate his statistic, he took the proportion of total shots that a team scored while that player was on the ice and the proportion of opponents' shots that were saved, and then added these two values together. King argued that although creating shooting opportunities involves a lot of skill, there was more luck influencing whether a shot went in or not. Worryingly, when King tested out the statistic on his local NHL team, it showed that the luckiest players were getting contract extensions while the unlucky ones were being dropped.

The statistic, later dubbed "PDO" after King's online moniker, has since been used to assess the fortunes of players—and teams—in other sports, too. In the 2014 soccer World Cup, several top teams failed to make it out of the preliminary group stage. Spain, Italy, Portugal, and England all fell at the first hurdle. Was it because they were lackluster or unlucky? The England team is famously used to misfortune, from disallowed goals to missed penalties. It seems that 2014 was no different: England had the lowest PDO of any team in the tournament, with a score of 0.66.

We might think that teams with a very low PDO are just hapless. Maybe they have a particularly error-prone striker or weak keeper. But teams rarely maintain an unusually low (or high) PDO in the long run. If we analyze more games, a team's PDO will quickly settle down to numbers near the average value of one. It's what Francis Galton called "regression to mediocrity": if a team has a PDO that is noticeably above or below one after a handful of games, it is likely a symbol of luck.

Statistics like PDO can be useful for assessing how lucky teams are, but they aren't necessarily that helpful when placing bets. Gamblers are more interested in making predictions. In other words, they want to find factors that reflect ability rather than luck. But how important is it to actually understand skill?

Take horse races. Predicting events at a racetrack is a messy process. All sorts of factors could influence a horse's performance in a race, from past experience to track conditions. Some of which provide clear hints about the future, while others just muddy the predictions. To pin down which factors are useful, syndicates need to collect reliable, repeated observations about races. Hong Kong was the closest Bill Benter could find to a laboratory setup, with the same horses racing on a regular basis on the same tracks in similar conditions.

Using his statistical model, Benter identified factors that could lead to successful race predictions. He found that some came out as more important than others. In Benter's early analysis, for example, the model said the number of races a horse had previously run was a crucial factor when making predictions. In fact, it was more important than almost any other factor. Maybe the finding isn't all that surprising. We might expect horses that have run more races to be used to the terrain and less intimated by their opponents.

It's easy to think up explanations for observed results. Given a statement that seems intuitive, we can convince ourselves as to why that should be the case, and why we shouldn't be surprised at the result. This can be a problem when making predictions. By creating an explanation, we are assuming that one process has directly caused another. Horses in Hong Kong win *because* they are familiar with the terrain, and they are familiar with it *because* they have run lots of races. But just because two things are apparently related—like probability of winning and number of races run—it doesn't mean that one directly causes the other.

An oft-quoted mantra in the world of statistics is that "correlation does not imply causation." Take the wine budget of Cambridge colleges. It turns out that the amount of money each Cambridge college spent on wine in the 2012–2013 academic year was positively correlated with students' exam results during the same period. The more the colleges spent on wine, the better the results generally were. (King's College, once home to Karl Pearson and Alan Turing, topped the wine list with a spend of £338,559, or about £850 per student.)

Similar curiosities appear in other places, too. Countries that consume lots of chocolate win more Nobel prizes. When ice cream sales rise in New York City, so does the murder rate. Of course, buying ice cream doesn't make us homicidal, just as eating chocolate is unlikely to turn us into Nobel-quality researchers and drinking wine won't make us better at exams.

In each of these cases, there might be a separate underlying factor that could explain the pattern. For Cambridge colleges it could be wealth, which would influence both wine spending and exam results. Or there could be a more complicated set of reasons lurking behind the observations. This is why Bill Benter doesn't try to interpret why some factors appeared to be so important in his horseracing model. The number of races a horse has run might be related to another (hidden) factor that directly influenced performance. Alternatively, there could be an intricate trade-off between races run and other factors—like weight and jockey experience—which Benter could never hope to distill into a neat "A causes B" conclusion. But Benter is happy to sacrifice elegance and explanation if it means having good predictions. It doesn't matter if his factors are counterintuitive or hard to justify. The model is there to estimate the probability a certain horse will win, not to explain *why* that horse will win.

From hockey to horse racing, sports analysis methods have come a long way in recent years. They have enabled gamblers to study matches in more detail than ever, combining bigger models with better data. As a result, scientific betting has moved far beyond card counting.

ON THE FINAL PAGE of his blackjack book *Beat the Dealer*, Edward Thorp predicted that the following decades would see a whole host of new methods attempting to tame chance. He knew it was hopeless to try to anticipate what they might be. "Most of the possibilities are beyond the reach of our present imagination and dreams," he wrote. "It will be exciting to see them unfold."

Since Thorp made his prediction, the science of betting has indeed evolved. It has brought together new fields of research, spreading far from the felt tables and plastic chips of Las Vegas casinos. Yet the popular image of scientific wagers remains very much in the past. Stories of gambling strategies rarely stray far from the adventures of Thorp or the Eudaemons. Successful betting is viewed as a matter of card counting or watching roulette tables. Tales follow a mathematical path, with decisions reduced to basic probabilities.

But the advantage of simple equations over human ingenuity is not as clear as these stories suggest. In poker, the ability to calculate the probability of getting a particular hand is helpful but by no means a sure route to victory. Gamblers also need to account for their opponents' behavior. When John von Neumann developed game theory to tackle this problem, he found that employing deceptive tactics such as bluffing was actually the optimal thing to do. The gamblers had been right all along, even if they didn't know why.

Sometimes it's necessary to stray from mathematical perfection altogether. As researchers delve further into the science of poker, they are finding situations where game theory comes up short and where traditional gambling traits—reading opponents, exploiting weaknesses, spotting emotion—can help computer players become the best in the world. It is not enough to know just probabilities; successful bots need to combine mathematics and human psychology.

The same is true in sports. Analysts are increasingly trying to capture the individual quirks that make up a team performance.

During the early 2000s, Billy Beane famously used "sabermetrics" to identify underrated players and take the cash-strapped Oakland A's to the Major League Baseball playoffs. The techniques are now appearing in other sports. In the English Premier League, more and more soccer teams are employing statisticians to advise on team performances and potential transfers. When Manchester City won the league in 2014, they had almost a dozen analysts helping to put together tactics.

Sometimes the human element can be the dominant factor, overshadowing the statistics gleaned from available match data. After all, the probability of a goal depends both on the physics of the ball and on the psyche of the player kicking it. Roberto Martinez, manager of Everton soccer club, has suggested that mind-set is as important as performances when assessing potential signings. Managers want to know how a player will settle into a new country or whether he can cope with pressure from a hostile crowd. And, clearly, it is very hard to measure factors like this.

Measurement is often a difficult problem in sports. From the defenders who never make a tackle to the NFL cornerbacks who hardly ever touch the ball, we can't always pin down valuable information. But knowing what we are missing is crucial if we want to fully understand what is happening in a match and what might happen in the future.

When researchers develop a theoretical model of a sport, they are reducing reality to an abstraction. They are choosing to remove detail and concentrate only on key features, much like Pablo Picasso so famously did. When Picasso worked his "Bull" lithographs in the winter of 1945, he started by creating a realistic representation of the animal. "It was a superb, well-rounded bull," said an assistant watching at the time. "I thought to myself that that was that." But Picasso was not finished. After completing his first image, he moved onto a second, and then a third. As Picasso worked on each new picture, the assistant noticed the bull was changing. "It began to diminish, to lose weight," he said. "Picasso was taking away rather than adding to his composition." With each image, Picasso carved further, keeping only the crucial contours, until he reached the eleventh lithograph. Almost every detail had gone, with nothing left but a handful of lines. Yet the shape was still recognizable as a bull. In those few strokes, Picasso had captured the essence of the animal, creating an image that was abstract, but not ambiguous. As Albert Einstein once said of scientific models, it was a case of "everything should be made as simple as possible, but not simpler."

Abstraction is not limited to the worlds of art and science. It is common in other areas of life, too. Take money. Whenever we pay with a credit card, we are replacing physical cash with an abstract representation. The numbers remain the same, but superfluous details—the texture, the color, the smell—have been removed. Maps are another example of abstraction: if a detail is unnecessary, it isn't shown. Weather is abandoned when the focus is on transport and traffic; motorways vanish if we're interested in sun and showers.

Abstractions make a complex world easier to navigate. For most of us, a car accelerator is simply a device that makes the vehicle go faster. We don't care—or need to know—about the chain of events between our foot and the wheels. Likewise, we rarely look at phones as transmitters that convert sound waves to electronic signals; in daily life, they are a series of buttons that produce a conversation.

In fact, it could be argued that our entire notion of randomness is an abstraction. When we say a coin has a 50 percent chance of coming up tails, or that a roulette ball has a 1 in 37 chance of landing on a particular number, we are using an abstraction. In theory we could write down equations for the motion and solve them to predict the trajectory. But because coin flips and roulette spins are so sensitive to initial conditions, it is difficult to this do in reality. So, instead we approximate the process and assume it is unpredictable. We choose to simplify an intricate physical process for the sake of convenience.

In life, we must often choose (either consciously or subconsciously) what abstractions to use. The most extensive abstraction would not omit a single detail. As mathematician Norbert Wiener said, "The best material model of a cat is another, or preferably the same, cat." Capturing the world in such detail is rarely practical, so instead we must strip away certain features. However, the resulting abstraction is our model of reality, influenced by our beliefs and prejudices.

Sometimes abstractions have tried to deliberately influence people's perceptions. In 1947, *Time* magazine published a double-page map of Europe and Asia. Titled the "Communist Contagion," the map's perspective had been altered so the Soviet Union—colored in an ominous red hue—loomed over the rest of the world. The map's creator, a cartographer by the name of R. M. Chapin, continued the theme in subsequent issues. In 1952, a piece called "Europe from Moscow" featured the USSR rising up from the bottom of the image, its borders forming an arrow that pointed toward the West.

Even if the bias is not deliberate, models inevitably depend on their creators' aims (and resources). Recall those different horse racing models: Bolton and Chapman's model had nine factors; Bill Benter used over a hundred. Researchers have to tread a fine line when deciding on an abstraction. Simple models risk omitting crucial features, while complicated models may include unnecessary ones. The trick is to find an abstraction that is detailed enough to be useful, but simple enough to be implementable. In blackjack, for instance, card counters don't need to remember the exact value of each card; they just need enough information to tip the odds in their favor.

Of course, there is always a risk of picking the wrong abstraction, which leaves out a critical detail. Émile Borel once said that, given two gamblers, there is always one thief and one imbecile. This is not just the case when one gambler has much better information than the other. Borel pointed out that in complex situations, two people could have exactly the same information and yet come to different conclusions about the probability of an event. When the pair bet together, Borel said each one would therefore believe "that it is he who is the thief and the other the imbecile."

Poker is a good example of situation in which choice of abstraction is important. There are a huge number of possible moves in poker—far too many to compute—which means that bots have to use abstractions to simplify the game. Tuomas Sandholm has pointed out that this can cause problems. For instance, your bot might only think in terms of certain bet sizes, to avoid having to analyze every possible wager. Over time, however, the bot's view of reality will not match the true situation. "Your belief as to how much money is in the pot is no longer accurate," Sandholm said. This can leave you vulnerable to an opponent using a better abstraction, which is closer to reality.

The problem doesn't just appear in poker. The entire casino industry is built on the assumption that the games are random. Casinos treat roulette spins and blackjack shuffles as unpredictable and rely on customers sharing that view. But believing an abstraction doesn't make it correct. And when someone comes along with a better model of reality—someone like Edward Thorp or Doyne Farmer—that person can profit from the casinos' oversimplification.

Thorp and Farmer were both physics students when they began their work on casino games. In subsequent decades, other students and academics have followed their lead. Some have targeted casinos, while others have focused on sports and horse racing. Which raises the question: Why is betting so popular among scientists?

IN JANUARY 1979, A group of undergraduates at Massachusetts Institute of Technology set up an extracurricular course called "How to Gamble If You Must." It was part of the university's four-week-long independent activities period (IAP), which encouraged students to take new classes and broaden their interests. During the gambling course, participants learned about Thorp's blackjack strategy and how to count cards. Soon some of the players had decided to try out the tactics for real; first in Atlantic City, and then in Vegas.

Although the players had started with Thorp's methods, they needed a new approach if they were going to be successful. As Thorp had discovered, it was difficult to get away with solo card counting. Players have to raise their bets when the count is in their favor, which means they are likely to attract the attention of casino security. The MIT students therefore worked as a team. Some players would be spotters, whose job it was to bet the minimum stake while keeping track of the count. When the deck was sufficiently in their favor, the spotters would signal to another group—the "big players"—who would come and throw lots of money at the table. To help conceal their roles from security, the team exploited common casino stereotypes. Smart female students would put on low-cut tops and pretend to be dumb gamblers, all the while keeping count of the cards. Students with an Asian or Middle Eastern background would play the role of a rich foreigner, happy to spend their parents' money.

Although its members changed over time, the MIT team continued to take on the casinos for many years. The contrast with life in Massachusetts could not have been larger. Instead of dorm rooms and Boston rain, there were hotel suites, sunny skies, and huge profits. During the Fourth of July weekend in 1995, the team was so successful that when they met by the pool at the end of the trip, one of them was carrying a gym bag holding almost a million dollars in cash. Another time, one of the team left a paper bag containing $125,000 in a classroom at MIT. When they returned, the bag was gone. They later discovered the janitor had stored it in his locker; they only got the money back after six months of investigations by the FBI and Drug Enforcement Administration.

The MIT blackjack team has become part of gambling legend. Journalist Ben Mezrich told their story in the best-selling book *Bringing Down the House,* and the events later inspired the film *21*. Unfortunately, for modern students, however, the exploits of the MIT team have become history in more ways than one. Casinos have introduced lots more countermeasures in recent years, which means teams would struggle to reproduce the sort of success seen in the 1980s and 1990s. In fact, according to professional gambler Richard Munchkin, hardly anyone focuses exclusively on blackjack anymore. "I know very few people—people you could count on one hand—who are making a living only by counting cards," he said.

Yet the science of gambling still features at MIT. In 2012, PhD student Will Ma set up a new course as part of the independent activities period. Its official title was "15.S50," but everybody knew it as the MIT poker class. Ma, who was studying operations research, had played a lot of poker—and won a lot of money—during his undergraduate days in Canada. When he arrived at MIT, word of his success got out, and several people started asking him questions about poker. One of them was his head of department, Dimitris Bertsimas, who also had an interest in the game. Bertsimas helped Ma put together a class to teach the theory and tactics needed to win. It was a legitimate MIT class; if students passed, they could get degree credit.

The course attracted a lot of attention. In fact, so many people turned up for the first class that they had to move rooms. "It was probably one of the most popular classes during IAP," Ma said. Attendees ranged from undergraduate business students to PhD mathematicians. Ma's class also caught the eye of the online poker community. Many incorrectly believed that students were going to use their expertise to build poker software. "Through word of mouth, it somehow got twisted," Ma said. "They thought it was going to lead to a huge poker bot system with a ton of bots written by MIT students taking all the money."

As well as distancing himself from bots, Ma also had to be careful to avoid his course on poker being misinterpreted by the university. "It can be seen as gambling," he said, "and you're not supposed to teach gambling at MIT." He therefore used play money to demonstrate strategies. "I had to make sure I wasn't taking people's real money."

Ma didn't have enough time to cover every aspect of poker, so instead he tried to focus on topics that would provide the biggest benefits. "I tried to go through the steepest part of the learning curve," he said. He explained why players shouldn't be afraid to go in at the start of a round and the dangers of getting bored with folding and instead playing too many hands. Many of the lessons would be useful in other situations, too. "I tried to put it in the perspective of real life," Ma said. The poker class covered the importance of making confident moves and of not letting mistakes affect performance. Students learned how to read opponents and how to manage the image they conveyed during games. In doing so, they started to discover what luck and skill really looked like. "I think one of the things poker teaches you very well is that you can often make a good decision but not get a good result," Ma said, "or make a bad decision and get a good result."

COURSES TEACHING THE SCIENCE of gambling have cropped up in other institutions, too, from York University in Ontario to Emory University in Georgia. In these classes, students study lotteries, roulette, card shuffling, and horse races. They learn statistics and strategy, analyzing risks and weighing options. Yet, as Ma found, people can be hostile to the concept of betting in universities. Indeed, many people are against the idea of wagers in any context.

When people say they dislike betting, what they usually mean is that they dislike the betting industry. Although the two are related, they are by no means synonymous. Even if we never gambled in casinos or visited bookmakers, betting would still permeate our lives. Luck—good and bad—looms over our careers and relationships. We have to deal with hidden information and negotiate in the face of uncertainty. Risks must be balanced with rewards; optimism must be weighed against probability.

The science of gambling isn't just useful for gamblers. Studying betting is a natural way to explore the notion of luck and can therefore be a good way to hone scientific skills. Although Ruth Bolton and Randall Chapman's paper on horse racing predictions gave rise to a multi-billion-dollar betting industry, it was the only article Bolton wrote on the topic. She has spent the rest of her career working on other problems. Most of them revolve around marketing, from the effects of different pricing strategies to how businesses can manage customer relationships. Bolton admits that the horse racing paper could therefore seem like a bit of an outlier on her CV; at first glance, it doesn't really fit in with her other research. But the methods in that early racetrack study, which involved developing models and assessing potential outcomes, would go on to shape the rest of her work. "That way of thinking about the world stayed with me," she said.

Probability theory, which Bolton used to analyze horse races, is one of the most valuable analytical tools ever created. It gives us the ability to judge the likelihood of events and assess the reliability of information. As a result, it is a vital component of modern scientific research, from DNA sequencing to particle physics. Yet the science of probability emerged not in libraries or lecture theaters but among the cards and dice of bars and game rooms. For eighteenth-century mathematician Pierre Simon Laplace, it was a strange contrast. "It is remarkable that a science which began with the consideration of games of chance should have become the most important object of human knowledge."

Cards and casinos since have inspired many other scientific ideas. We have seen how roulette helped Henri Poincaré develop the early ideas of chaos theory and allowed Karl Pearson to test his new statistical techniques. We also met Stanislaw Ulam, whose card games led to the Monte Carlo method, now used in everything from 3D computer graphics to the analysis of disease outbreaks. And we have seen how game theory emerged from John von Neumann's analysis of poker.

The relationship between science and betting continues to thrive today. As ever, the ideas are flowing in both directions: gambling is inspiring new research, and scientific developments are providing new insights into betting. Researchers are using poker to study artificial intelligence, creating computers that can bluff and learn and surprise just like humans. Every year, these champion bots are coming up with new tactics that humans never knew about, or would never dare try. Meanwhile, high-speed algorithms are helping companies make bets and trades automatically, creating a complex ecosystem of interactions that has prompted new avenues of research. Sports analysts, armed with better data and faster computers, are no longer just predicting team results; they are picking apart the roles of individual players, measuring the contribution of chance and skill. From poker to betting exchanges, researchers are developing a deeper understanding of human behavior and decision making, and in turn coming up with more effective gambling strategies.

THE POPULAR IMAGE OF a scientific betting strategy is one of a mathematical magic trick. To get rich, all you need is a simple formula or a few basic rules. But, much like a magic trick, the simplicity of the performance is an illusion, concealing a mountain of preparation and practice.

As we've seen, almost any game can be beaten. But profits rarely come from lucky numbers or "foolproof" systems. Successful wagers take patience and ingenuity. They require creators who choose to ignore dogma and follow their curiosity. It might be a student like James Harvey, who wondered which lottery was the best deal and orchestrated thousands of ticket purchases to take advantage of the loophole he found. Or a physicist like Edward Thorp, rolling marbles on his kitchen floor to understand where a roulette ball would stop. It might take a business specialist like Ruth Bolton, crunching through horse racing data to find out what makes a winner. Or statisticians such as Mark Dixon and Stuart Coles, reading an undergraduate exam question about soccer prediction and wondering how the methods could be improved.

From the casinos of Monte Carlo to the racetracks of Hong Kong, the story of the perfect bet is a scientific one. Where once there were rules of thumb and old wives' tales, there are now theories guided by experiment. The reign of superstition has waned, usurped by rigor and research. Bill Benter, who made his fortune betting on blackjack and horse racing, has no doubts about who deserves the credit for the transition. "It wasn't as though streetwise Las Vegas gamblers figured out a system," he said. "Success came when an outsider, armed with academic knowledge and new techniques, came in and shone light where there had been none before."

ACKNOWLEDGMENTS

FIRST, THANKS MUST GO to my agent Peter Tallack. From proposal to publisher, his guidance over the past three years has been invaluable. I would also like to thank my editors—TJ Kelleher and Quynh Do at Basic Books, and Nick Sheerin at Profile—for taking a gamble on me, and for helping me shape the science into a story.

My parents continue to provide crucial suggestions and discussions on my writing, and for this I am eternally grateful. Thanks also to Clare Fraser, Rachel Humby, and Graham Wheeler for many useful comments on early drafts. And, of course, to Emily Conway, who has been there for me throughout with wise words and wine.

Finally, I am indebted to everyone who took the time to share their insights and experiences: Bill Benter, Ruth Bolton, Neil Burch, Stuart Coles, Rob Esteva, Doyne Farmer, David Hastie, Michael Johanson, Michael Kent, Will Ma, Matt Mazur, Richard Munchkin, Brendan Poots, Tuomas Sandholm, Jonathan Schaeffer, Michael Small, and Will Wilde. Many of these individuals have shaped entire industries with their scientific curiosity. It will be fascinating to see what comes next.

NOTES

Introduction

ix **In June 2009, a British newspaper:** Ward, Simon. "A Sacked 22-Year-Old Trainee City Trader Today Reveals How He Won a Staggering £20 Million in a Year . . . Betting on the Horses." *News of the World*, June 26, 2009.

ix **He had a chauffeur-driven Mercedes:** Duell, Mark. "'King of Betfair' Who Lived Lavish Lifestyle in Top Hotels with Chauffeur-Driven Mercedes and Clothes from Harrods after Conning Family Friends Out of £400,000 Is Jailed." *Daily Mail* Online, May 28, 2013. http://www.dailymail.co.uk/news/article-2332115/King-Betfair-stayed-hotels-splashed-chauffeur-conning-family-friends-jailed.html.

ix **The profitable bets that Short claimed:** Wood, Greg. "Short Story on Betfair System Is Pure Fiction." *Guardian Sportblog* (blog), June 29, 2009. http://www.theguardian.com/sport/blog/2009/jun/30/greg-wood-betfair-notw-story.

ix **Having persuaded investors to pour hundreds of thousands:** Duell, Mark. "Gambler, 26, Who Called Himself the 'Betfair King' Conned Friends Out of £600,000 with Betting Scam to Pay for Designer

Clothes." *Daily Mail* Online, April 23, 2013. http://www.dailymail.co.uk/news/article-2313618/Gambler-called-Betfair-king-conned friends-600–000-bogus-betting-scam.html.

x **the case went to trial:** "Criminal Sentence—Elliott Sebastian Short—Court: Southwark." TheLawPages.com, May 28, 2013. http://www.thelawpages.com/court-cases/Elliott-Sebastian-Short-11209–1.law.

x **As its reputation spread:** Ethier, Stewart. *The Doctrine of Chances: Probabilistic Aspects of Gambling* (New York: Springer, 2010), 115.

xi **"The martingale is as elusive":** Dumas, Alexandre. *One Thousand and One Ghosts* (London: Hesperus Classics, 2004).

xi **Having frittered away his inheritance:** O'Connor, J. J., and E. F. Robertson. "Girolamo Cardano." June 1998. http://www-history.mcs.st-andrews.ac.uk/Biographies/Cardan.html.

xi **nobody knew precisely what a "fair" wager:** O'Connor and Robertson, "Girolamo Cardano."

xi **deriving "Cardano's formula":** Gorroochurn, Prakash. "Some Laws and Problems of Classical Probability and How Cardano Anticipated Them." *Chance Magazine* 25, no. 4 (2012): 13–20.

xii **"When I observed that the cards were marked":** Cardan, Jerome. *Book of My Life* (New York: Dutton, 1930).

xii **At the request of a group of Italian nobles:** Ore, Oystein. "Pascal and the Invention of Probability Theory." *American Mathematical Monthly* 67, no. 5 (May 1960): 409–419.

xii **Astronomer Johannes Kepler also took time:** Epstein, Richard. *The Theory of Gambling and Statistical Logic* (Waltham, MA: Academic Press, 2013).

xii **The science of chance blossomed:** Ore, "Pascal and the Invention."

xii **he was more likely to get a six:** It's easiest to start by calculating the probability of *not* getting a six in four rolls, which is $(5/6)^4$. It therefore follows that the probability of getting at least one six is $1 - (5/6)^4 = 51.8$ percent. By the same logic, the probability of getting a double six over twenty-four throws of two dice is $1 - (35/36)^{24} = 49.1$ percent.

xii **"Gamblers can rightly claim":** Epstein, Richard. *The Theory of Gambling and Statistical Logic* (Waltham, MA: Academic Press, 2013).

xiii **Daniel Bernoulli wondered why people:** Bassett, Gilbert, Jr. "The St. Petersburg Paradox and Bounded Utility." *History of Political Economy* 19, no. 4 (1987): 517–523.

xiii **"The mathematicians estimate money":** Castelvecchi, Davide. "Economic Thinking." *Scientific American* 301, no. 82 (September 2009). doi:10.1038/scientificamerican0909–82b.

xiv **When Feynman tried the game:** Feynman, Richard. *Surely You're Joking, Mr. Feynman!* (New York: W. W. Norton, 2010).

Chapter 1

1 **It's called the Ritz Club:** Ritz Club brochure.

1 **One evening in March 2004:** Chittenden, Maurice. "Laser-Sharp Gamblers Who Stung Ritz Can Keep £1.3m." *Times* (London), December 5, 2004.

1 **The group weren't like the other high rollers:** Beasley-Murray, Ben. "Special Report: Wheels of Justice." *PokerPlayer,* January 1 2005. http://www.pokerplayer365.com/uncategorized-drafts/wheels-of-justice/.

2 **This time their winnings:** "'Laser Scam' Gamblers to Keep £1m." BBC News Online, December 5, 2004. http://news.bbc.co.uk/2/hi/uk/4069629.stm.

2 **What they saw was enough:** Chittenden, "Laser-Sharp Gamblers."

2 **It was one of his many interests:** Mazliak, Laurent. "Poincaré's Odds." *Séminaire Poincaré* XVI (2012): 999–1037.

2 **As Poincaré saw it:** Poincaré, Henri. *Science and Hypothesis* (New York: Walter Scott Publishing, 1905). (French edition published in 1902)

3 **suppose we drop a can of paint:** According to Scott Patterson, Edward Thorp once did this at a pool in Long Beach, California (with red dye rather than paint). The incident made the local paper. Source: Patterson, Scott. *The Quants* (New York: Crown, 2010).

3 **Instead, we can simply watch:** Poincaré, Henri. *Science and Method* (London: Nelson, 1914). (French edition published in 1908)

3 **Taking time off from their studies:** Ethier, Stuart. "Testing for Favorable Numbers on a Roulette Wheel." *Journal of the American Statistical Association* 77, no. 379 (September 1982): 660–665.

4 **Pearson got a colleague to flip a penny:** Pearson, K. "The Scientific Aspect of Monte Carlo Roulette." *Fortnightly Review,* February 1894.

4 **we have "no absolute knowledge of natural phenomena:** Pearson, K. *The Ethic of Freethought and Other Addresses and Essays* (London: T. Fisher Unwin, 1888).

5 **He was particularly keen on German culture:** Magnello, M. E. "Karl Pearson and the Origins of Modern Statistics: An Elastician Becomes a Statistician." *Rutherford Journal.* http://www.rutherfordjournal.org/article010107.html.

5 **the newspaper *Le Monaco*:** Pearson, "Scientific Aspect of Monte Carlo Roulette."

6 **Gamblers crowded around the table:** Huff, Darrell, and Irving Geis. *How to Take a Chance* (London: W. W. Norton, 1959), 28–29.

6 **Monte Carlo roulette confounds his theories:** Pearson, "Scientific Aspect of Monte Carlo Roulette."

7 **the reporters had decided it was easier:** MacLean, L. C., E. O. Thorp, and W. T. Ziemba, eds. *The Kelly Capital Growth Investment Criterion: Theory and Practice* (Singapore: World Scientific, 2011).

7 **Reports of their final profits differ:** Maugh, Thomas H. "Roy Walford, 79; Eccentric UCLA Scientist Touted Food Restriction." *Los Angeles Times*, May 1, 2004. http://articles.latimes.com/2004/may/01/local/me-walford1.

7 **Many have told the tale:** Ethier, "Testing for Favorable Numbers."

7 **When Wilson published his data:** Ethier, "Testing for Favorable Numbers."

9 **Poincaré had outlined the "butterfly effect:** Gleick, James. *Chaos: Making a New Science* (New York: Open Road, 2011).

9 **The Zodiac may be regarded:** Poincaré, *Science and Method.*

10 **Blaise Pascal invented roulette:** Bass, Thomas. *The Newtonian Casino* (London: Penguin, 1990).

10 **The orbiting roulette ball:** The majority of details and quotes in this section are taken from Thorp, Edward. "The Invention of the First Wearable Computer." *Proceedings of the 2nd IEEE International Symposium on Wearable Computers* (1998), 4.

13 **participants were asked to help:** Milgram, Stanley. "The Small-World Problem." *Psychology Today* 1, no. 1 (May 1967): 61–67.

13 **an average of 3.74 degrees of separation:** Backstrom, Lars, Paolo Boldi, Marco Rosa, Johan Ugander, and Sebastiano Vignal. "Four Degrees of Separation" (Cornell University Library, January 2012). http://arxiv.org/abs/1111.4570.

14 **Another attendee was a young physicist:** Gleick, "Chaos."

14 **By taking measurements:** Bass, *Newtonian Casino.*

14 **When a new paper on roulette appeared:** Small, Michael, and Chi

Kong Tse. "Predicting the Outcome of Roulette." *Chaos* 22, no. 3 (2012): 033150. doi:10.1063/1.4753920.

15 **For his PhD, he'd analyzed:** Quotes and additional details come from an interview with Michael Small in 2013.

18 **He was sailing in Florida:** Author interview with Doyne Farmer, October 2013.

19 **He'd found that air resistance:** Slezak, Michael. "Roulette Beater Spills Physics behind Victory." *New Scientist*, no. 2864 (May 12, 2012). https://www.newscientist.com/article/mg21428644-500-roulette-beater-spills-physics-behind-victory/. Additional details from author interview with Doyne Farmer, October 2013.

19 **During their casino trips:** Bass, *Newtonian Casino*.

19 **To predict exactly where the cue ball will travel:** Crutchfield, James P., J. Doyne Farmer, Norman H. Packard, and Robert S. Shaw. "Chaos." *Scientific American* 254, no. 12 (December 1986): 46–57.

20 **when journalist Ben Beasley-Murray talked:** Details about subsequent investigations are from Beasley-Murray, Ben. "Special Report: Wheels of Justice." *PokerPlayer*, January 1, 2005. http://www.pokerplayer365.com/uncategorized-drafts/wheels-of-justice/.

20 **According to ex-Eudaemon Norman Packard:** McKee, Maggie. "Alleged High-Tech Roulette Scam 'Easy to Set Up.'" *New Scientist*, March 2004.

21 **When Hibbs and Walford passed $5,000:** Ethier, "Testing for Favorable Numbers."

CHAPTER 2

23 **Of the colleges of the University of Cambridge:** Gonville and Caius. "History." http://www.cai.cam.ac.uk/history.

23 **its unique stained glass windows:** Author experience.

23 **Fisher spent three years studying at Cambridge:** O'Connor, J. J., and E. F. Robertson. "Sir Ronald Aylmer Fisher." JOC/EFR, October 2003. http://www-history.mcs.st-and.ac.uk/Mathematicians/Fisher.html.

24 **"To consult the statistician after an experiment . . .":** Fisher, Ronald. "Presidential Address to the First Indian Statistical Congress." *Sankhya* 4 (1938):14–17.

25 **The Great Wall of China was financed:** Campbell, Alex. "National

Lottery: Why Do People Still Play?" BBC News Online, October 2013. http://www.bbc.com/news/uk-24383871.

25 **proceeds from a lottery organized in 1753:** Wilson, David. "The British Museum: 250 Years On." *History Today* 52 (2002): 10.

25 **many of the Ivy League universities were built:** Lehrer, Jonah. "Cracking the Scratch Lottery Code." *Wired*, January 31, 2011. http://www.wired.com/2011/01/ff_lottery/.

26 **a quarter of the National Lottery's revenues:** Bowers, Simon. "Lottery Scratchcards Fuel Camelot Sales Boom." *Guardian*, November 18, 2011. http://www.theguardian.com/uk/2011/nov/18/national-lottery-scratchcard-sales-boom.

26 **American state lotteries earn tens of billions:** Scratchcards.org. "The Lottery Industry." http://www.scratchcards.org/featured/57121/the-lottery-industry.

26 **To quote statistician William Gossett:** Ziliak, Stephen. "Balanced Versus Randomized Field Experiments in Economics: Why W. S. Gosset aka 'Student' Matters." *Review of Behavioral Economics* 1, no. 1–2 (2014): 167–208. http://dx.doi.org/10.1561/105.00000008.

26 **how the lottery keeps track:** Lehrer, "Cracking the Scratch Lottery Code."

26 **he'd known Bill Tutte:** Yang, Jennifer. "Toronto Man Cracked the Code to Scratch-Lottery Tickets." *Toronto Star*, February 4, 2011. http://www.thestar.com/news/gta/2011/02/04/toronto_man_cracked_the_code_to_scratchlottery_tickets.html.

26 **British mathematician who had broken the Nazi Lorenz cipher:** William Tutte obituary. *Kitchener-Waterloo Record*, May 2002.

27 **a computer version of tic-tac-toe:** Yang, "Toronto Man Cracked the Code."

27 **By the time she was arrested:** George, Patrick. "Woman Crashes Car into Convenience Store to Steal 1,500 Lotto Tickets." MSN Online, May 13, 2013. http://jalopnik.com/woman-crashes-car-into-convenience-store-to-steal-1–500–504608879.

27 **"We need to talk":** Yang, "Toronto Man Cracked the Code."

28 **Joan Ginther had won four jackpots:** Rich, Nathanial. "The Luckiest Woman on Earth." *Harper's Magazine*, August 2011.

28 **wanted to call the dorm "Random House":** Roller, Dean. "Publisher's Objections Force New Dorm Name." *The Tech*, January 1968.

http://web.mit.edu/~random-hall/www/History/publisher-objections.shtml.

28 **selling the naming rights on eBay:** eBay. "eBay Item # 1700894687 Name a Floor at MITs Random Hall." http://web.mit.edu/ninadm/www/ebay.html.

28 **The hall even has its own:** Dowling, Claudia. "MIT Nerds." *Discover Magazine*, June 2005.

28 **he became interested in lotteries:** Details of the Powerball syndicate activities come from Sullivan, Gregory. "Letter to State Treasurer Steven Grossman." July 2012. http://www.mass.gov/ig/publications/reports-and-recommendations/2012/lottery-cash-winfall-letter-july-2012.pdf.

32 **"It took us about a year to ramp up to it":** Sullivan, Letter to State Treasurer Steven Grossman.

32 **The effort paid off:** Estes, Andrea. "A Game of Chance Became Anything But." *Boston Globe*, October 16, 2011. http://www.boston.com/news/local/massachusetts/articles/2011/10/16/a_game_of_chance_became_anything_but/.

32 **the *Boston Globe* had published a story:** Estes, "Game of Chance."

33 **Selbee claims to have won around $8 million:** Wile, Rob. "Retiree from Rural Michigan Tells Us the Moment He Figured Out How to Beat the State's Lottery." *Business Insider*, August 1, 2012. http://www.businessinsider.com/a-retiree-from-rural-michigan-tells-us-the-moment-he-figured-out-how-to-beat-the-states-lottery-2012–8.

Chapter 3

35 **"Professional card counters are prohibited":** Yafa, Stephen. "In the Cards." *The Rotarian*, November 2011.

35 **Thorp has been called the father of card counting:** Many sources have made this reference. One prominent example is the publisher blurb for: Thorp, Edward. *Beat the Dealer* (New York: Random House, 1962).

36 **When one of the men suggested a game of blackjack:** Kahn, Joseph P. "Legendary Blackjack Analysts Alive but Still Widely Unknown." *The Tech*, February 2008. http://tech.mit.edu/V128/N6/blackjack.html.

36 **the dealer had a 40 percent chance:** Baldwin, Roger, Wilbert E. Cantey, Herbert Maisel, and James P. McDermott. "The Optimum

Strategy in Blackjack." *Journal of the American Statistical Association* 51, no. 275 (1956): 429–439.

37 **Intrigued by Baldwin's idea:** Haney, Jeff. "They Invented Basic Strategy." *Las Vegas Sun News*, January 4, 2008.

37 **"In statistical terms":** Kahn, "Legendary Blackjack Analysts Alive."

37 **The four men published their findings:** Baldwin et al., "The Optimum Strategy in Blackjack." The four soldiers also later published a book for nonstatisticians, entitled *Playing Blackjack*. According to McDermott, it made a total of $28. (Source: Kahn, "Legendary Blackjack Analysts Alive.")

37 **It was meant to be a relaxing holiday:** Thorp, Edward. *Beat the Dealer* (New York: Random House, 1962).

38 **Thorp gradually turned the research:** Kahn, "Legendary Blackjack Analysts Alive."

38 **He saw it more as an academic obligation:** Towle, Margaret. "Interview with Edward O. Thorp." *Journal of Investment Consulting* 12, no. 1 (2011): 5–14.

39 **"It showed that nothing was invulnerable":** Author interview with Bill Benter, July 2013.

39 **Switching his university campus in Cleveland:** Yafa, Stephen. "In the Cards." *The Rotarian*, November 2011.

39 **The decision was to prove extremely lucrative:** Ibid.

39 **his firm was commissioned by the Australian government:** Dougherty, Tim. "Horse Sense." *Contingencies*, June 2009.

40 **"It's easy to learn how to count cards":** Author interview with Richard Munchkin, August 2013.

40 **To evade security:** Thorp, *Beat the Dealer*.

40 **Most mathematicians in the early twentieth century:** Mazliak, Laurent. "Poincaré's Odds." *Séminaire Poincaré* XVI (2002): 999–1037.

41 **His research is still used today:** Saloff-Coste, Laurent. "Random Walks on Finite Groups." In *Probability on Discrete Structures*, ed. Harry Kesten (New York: Springer Science & Business, 2004).

41 **To perform the shuffle:** Blood, Johnny Blood. "A Riffle Shuffle Being Performed during a Game of Poker at a Bar Near Madison, Wisconsin,. November 2005–April 2006." Source: Flickr. Image licensed under CC-BY-SA 2.0.

41 **Figure 3.1. A dovetail shuffle:** Reproduced under the CC BY-SA 2.0 license. https://www.flickr.com/photos/latitudes/66424863.

42 **For a fifty-two-card deck:** Bayer, D. B., and P. Diaconis. "Trailing the Dovetail Shuffle to Its Lair." *Annals of Applied Probability* 2, no. 2 (1992): 294–313.

42 **Benter found that casinos:** Author interview with Bill Benter, July 2013.

42 **They would enter information:** Schnell-Davis, D. W. "High-Tech Casino Advantage Play: Legislative Approaches to the Threat of Predictive Devices." *UNLV Gaming Law Journal* 3, no. 2 (2012).

42 **Unfortunately for gamblers:** Author interview with Richard Munchkin, August 2013.

43 **"Once you become well known":** Author interview with Bill Benter, July 2013.

43 **Cheers rise above the sound:** Author experience.

43 **an average of $145 million:** Lee, Simon. "Hong Kong Horse Bets Hit Record as Races Draw Young Punters." *BusinessWeek*, July 11, 2013. http://www.bloomberg.com/news/articles/2013–07–11/hong-kong-horse-bets-hit-record-as-races-draw-young-punters.

43 **the Kentucky Derby set a new American record:** "Record-Breaking Day Across-the-Board for Kentucky Derby 138." Kentucky Derby, May 6, 2012. http://www.kentuckyderby.com/news/2012/05/05/kentucky-derby-138-establishes-across-board-records.

43 **The Jockey Club is a nonprofit organization:** Rarick, Gina. "Horse Racing: Hong Kong Polishes a Good Name Worth Gold." *New York Times*, December 11, 2004. http://www.nytimes.com/2004/12/11/sports/11iht-horse_ed3_.html?_r=0.

44 **Undeterred, Julius tweaked the mechanism:** Doran, Bob. "The First Automatic Totalisator." *Rutherford Journal*. http://rutherfordjournal.org/article020109.html.

45 **sports bettors need a strategy:** Benter, William. "Computer Based Horse Race Handicapping and Wagering Systems: A Report." In *Efficiency of Racetrack Betting Markets*, ed. D. B. Hausch, V. S. Y. Lo, and W. T. Ziemba (London: Academic Press, 1994), 511–526.

45 **Unfortunately, it was a year:** Dougherty, "Horse Sense."

46 **"Searching for Positive Returns at the Track":** Bolton, R. N., and R. G. Chapman. "Searching for Positive Returns at the Track: A Multinomial Logit Model for Handicapping Horse Races." *Management Science* 32, no. 8 (1986).

46 **"It was the paper that launched":** Author interview with Bill Benter, July 2013.

46 **Karl Pearson met a gentleman:** Magnello, M. Eileen. "Karl Pearson and the Origins of Modern Statistics: An Elastician Becomes a Statistician." *Rutherford Journal.* http://www.rutherfordjournal.org/article010107.html.

46 **"He never waited to see":** Pearson, Karl. *The Life, Letters and Labours of Francis Galton* (Cambridge: Cambridge University Press, 2011).

47 **seven of Galton's friends received sweet pea seeds:** Galton, Francis. "Towards Mediocrity in Hereditary Stature." *Journal of the Anthropological Institute of Great Britain and Ireland* 15 (1986): 246–263.

48 **Galton was so impressed:** Galton, Francis. "A Diagram of Heredity." *Nature* 57 (1898): 293. http://www.esp.org/foundations/genetics/classical/fg-98.pdf.

48 **Both viewed regression to the mediocre:** Pearson, *Life, Letters and Labours.*

49 **In Pearson's view, a nation could be improved:** Pearson, Karl. *National Life from the Standpoint of Science*, 2nd ed. (Cambridge: Cambridge University Press, 1919).

49 **he also claimed that laws:** Pearson, Karl. "The Problem of Practical Eugenics" (Galton Eugenics Laboratory Lecture Series No. 5. Dulau & Co., 1909).

49 **"When I was a toddler":** Author interview with Ruth Bolton, February 2014.

52 **Some gamblers might try to think up:** Author interview with Bill Benter, July 2013.

52 **The task fell to William Gossett:** Ziliak, Stephen. "Guinnessometrics: The Economic Foundation of 'Student's' *t*." *Journal of Economic Perspectives* 22, no. 4 (2008): 199–216.

53 **The pilots, busy with other tasks:** Emanuel, Kerry. "Edward Norton Lorenz 1917–2008." *National Academy of Sciences*, 2011. ftp://texmex.mit.edu/pub/emanuel/PAPERS/Lorenz_Edward.pdf.

55 **In 1994, Benter published a paper:** Benter, "Computer Based Horse Race Handicapping."

55 **Researching a company to its core:** Investopedia.com. "Technical Analysis." http://www.investopedia.com/terms/t/technicalanalysis.asp.

57 **"People had so little faith in the system":** Dougherty, "Horse Sense."

57 **Disagreements meant the partnership ended:** Dougherty, "Horse Sense."

58 **As a result, both horses take the same time:** Author interview with Bill Benter, July 2013.

58 **Finding his brain severely inflamed:** Grimberg, Sharon (producer/writer) and Rick Groleau (producer/writer). "Race for the Superbomb," directed by Thomas Ott, aired for The American Experience series (PBS Video, 1999).

59 **It wasn't his first choice:** Rota, Gian-Carlo. "The Lost Café." *Los Alamos Science*, 1987.

59 **When he finally got the answer:** Rota, "The Lost Café."

60 **"Suddenly, I knew where":** Lounsberry, Alyse. "A-Bomb Cloaked in Mystery." *Ocala Star-Banner*, December 4, 1978, 13.

60 **"This train is linear, isn't it?":** Halmos, Paul. "The Legend of John von Neumann." *American Mathematical Monthly* 8 (1973): 382–394.

61 **"Any one who considers arithmetical methods":** Von Neumann, John. "Various Techniques Used in Connection with Random Digits." *Journal of Research of the National Bureau of Standards*, Appl. Math. Series, 1951. Quoted in Herman Heine Goldstine, *The Computer from Pascal to von Neumann* (Princeton, NJ: Princeton University Press, 2008).

62 **"Andrei Neistovy": Andrei the angry:** Mazliak, Laurent. "From Markov to Doeblin: Events in Chain" (talk given at RMR-2010, Rouen, France, June 1, 2010). http://www.proba.jussieu.fr/~mazliak/Markov_Rouen.pdf.

62 **a local prison psychologist turned up:** Described in Diaconis, P. "The Markov Chain Monte Carlo Revolution." *Bulletin of the American Mathematical Society* 46 (2009): 179–205.

64 **Metropolis and his colleagues realized:** Metropolis, Nicholas, Arianna W. Rosenbluth, Marshall N. Rosenbluth, Augusta H. Teller, and Edward Teller. "Equation of State Calculations by Fast Computing Machines." *Journal of Chemical Physics* 21 (1953): 1087. http://dx.doi.org/10.1063/1.1699114.

64 **Markov chain Monte Carlo has helped syndicates:** Author interview with Bill Benter, July 2013. Key citation: Gu, Ming Gao, and Fan Hui Kong. "A Stochastic Approximation Algorithm with Markov

Chain Monte-Carlo Method for Incomplete Data Estimation Problems." *Proceedings of the National Academy of Science USA* 95 (1998): 7270–7274.

65 **The formula is named after John Kelly:** Poundstone, W. *Fortune's Formula: The Untold Story of the Scientific Betting System That Beat the Casinos and Wall Street* (New York: Hill and Wang, 2006).

66 **Consistently overestimate by twofold:** Chapman, S. "The Kelly Criterion for Spread Bets." *IMA Journal of Applied Mathematics* 72 (2007): 43–51.

66 **This reduces the risk:** Benter, "Computer Based Horse Race Handicapping."

67 **"The late money tends to be smart money":** Author interview with Bill Benter, July 2013.

67 **The techniques are now so effective:** Dougherty, "Horse Sense."

68 **Benter and Woods chose Hong Kong:** Author interview with Bill Benter, July 2013.

68 **teams using computer predictions bet around $2 billion:** Description of current events comes from Jagow, Scott. "I, Robot: The Future of Horse Wagering?" *Paulick Report*, 2013. http://www.paulickreport.com/news/ray-s-paddock/i-robot-the-future-of-horse-race-wagering/.

68 **Like Swedish harness racing:** Author interview with Bill Benter, July 2013.

69 **"We would joke that we could do it":** Author interview with Ruth Bolton, February 2014.

CHAPTER 4

71 **When a new blackjack system hit Britain:** Author experience.

71 **The strategy emerged as a result:** House of Commons Culture, Media and Sport Committee. "The Gambling Act 2005: A Bet Worth Taking?" HC 421, 2012.

72 **The first conviction for bonus abuse:** "Man Jailed for 'Bonus Abuse.'" Metropolitan Police Online, April 2012. http://content.met.police.uk/News/Man-jailed-for-Bonus-Abuse/1400007796996/1257246745756.

72 **casinos rarely understand how much of an advantage:** Interview with Richard Munchkin, August 2013.

73 **The events have more in common:** Author experience.

74 **it is the science of the very unlikely:** De Haan, L., and A. Ferreira. Preface. In *Extreme Value Theory: An Introduction* (New York: Springer Science & Business Media, 2007).

74 **Coles's research spanned everything:** For example: Coles, Stuart, and Jonathan Tawn. "Bayesian Modelling of Extreme Surges on the UK East Coast." *Philosophical Transactions* 363, no. 1831 (2005): 1387–1406; and Coles, Stuart, and Francesca Pan. "The Analysis of Extreme Pollution Levels: A Case Study." *Journal of Applied Statistics* 23, no. 2–3 (1996): 333–348.

74 **Dixon had become interested in the topic:** Author interview with Stuart Coles, May 2013.

74 **The work eventually appeared:** Dixon, M. J., and S. G. Coles. "Modelling Association Football Scores and Inefficiencies in the Football Betting Market." *Journal of the Royal Statistical Society: Series C* 46 (1997): 2.

74 **"It was one of those things":** Author interview with Stuart Coles, May 2013.

75 **a single soccer match was so dominated by chance:** Dixon and Coles, "Modelling Association Football Scores."

75 **Researchers have used the Poisson process:** Rakocevic, G., T. Djukic, N. Filipovic, and V. Milutinovic. *Computational Medicine in Data Mining and Modeling* (New York: Springer-Verlag, 2013), 154.

76 **A 2009 study by Christoph Leitner and colleagues:** Leitner, C., A. Zeileis, and K. Hornik. "Forecasting Sports Tournaments by Ratings of (Prob)Abilities: A Comparison for the EURO 2008." *International Journal of Forecasting* 26, no. 3 (2009): 471–481.

77 **Coles would join Smartodds:** Author interview with Stuart Coles, May 2013.

78 **"Those papers are still the main starting points":** Author interview with David Hastie, March 2013.

78 **"It's not an entirely polished piece":** Author interview with Stuart Coles, May 2013.

78 **according to Andreas Heuer and Oliver Rubner:** Heuer, Andreas, and Oliver Rubner. "How Does the Past of a Soccer Match Influence Its Future? Concepts and Statistical Analysis." *PLoS ONE* 7, no. 11 (2012). doi:10.1371/journal.pone.0047678.

79 **"A team would beat another team 28–12":** Details of earlier career come from author interview with Michael Kent, October 2013.

79 **"You need to make your own number":** Author interview with Michael Kent, October 2013.

80 **Because the society's acronym was SABR:** Society for American Baseball Research. "A Guide to Sabermetric Research." http://sabr.org/sabermetrics.

81 **He came to rely on valet parking:** Thomsen, Ian. "The Gang That Beat Las Vegas." *National Sports Daily*, 1990.

82 **There were FBI raids:** Ibid.

82 **the Computer Group placed over $135 million:** Ibid.

83 **"It is a totally different scheme":** Ulam, S. M. *Adventures of a Mathematician* (Oakland: University of California Press, 1991), 311.

83 **It was called the "Avtomat Kalashnikova":** Trex, Ethan. "What Made the AK-47 So Popular?" *Mental Floss*, April 2011. http://mentalfloss.com/article/27455/what-made-ak-47-so-popular.

83 **The gun is still in use today:** Killicoat, Phillip. "Weaponomics: The Global Market for Assault Rifles" (World Bank Policy Research Working Paper 4202, Washington, DC, April 2007).

83 **Complexity means more friction:** Da Silveira, M., L. Gertz, A. Cervieri, A. Rodrigues, et al. "Analysis of the Friction Losses in an Internal Combustion Engine" (SAE Technical Paper 2012–36–0303, 2012). doi:10.4271/2012–36–0303.

84 **US President Woodrow Wilson once described golf:** "A 'Sissy Game' Was the Sport of Presidents." *Life Magazine*, July 1968, 72.

84 **If the ball hits a random object:** Mella, Mirio. "Success = Talent + Luck." Pinnacle Sports, July 15, 2015. http://www.pinnaclesports.com/en/betting-articles/golf/success-talent-luck.

85 **Teams playing in the NHL score:** ESPN. NHL page. http://espn.go.com/nhl/.

85 **where NBA teams will regularly score:** ESPN. NBA page. http://espn.go.com/nba/.

85 **stats such as the "Corsi rating":** Macdonald, Brian. "An Expected Goals Model for Evaluating NHL Teams and Players" (paper presented at MIT Sloan Sports Analytics Conference, Boston, MA, March 2–3, 2012).

85 **the nature of scoring in basketball was changing:** Predictive Sports Betting. MIT Sloan Sports Analytics Conference, Boston, MA,

March 1–2, 2013. Panel discussion with: Chad Millman, Haralabos Voulgaris, and Matthew Holt. Moderator: Jeff Ma. http://www.sloansportsconference.com/?p=9607.

86 **Sifting through reams of data:** Eden, Scott. "Meet the World's Top NBA Gambler." ESPN, February 25, 2013. http://espn.go.com/blog/playbook/dollars/post/_/id/2935/meet-the-worlds-top-nba-gambler.

86 **"We don't realise how easy we have":** Author interview with Stuart Coles, May 2013.

86 **people created automated programs:** Author interview with David Hastie, March 2013.

86 **Some installed countermeasures:** Ward, Mark. "Screen Scraping: How to Profit from Your Rival's Data." BBC News, September 30, 2013. http://www.bbc.com/news/technology-23988890.

86 **"You get a huge database":** Author interview with Michael Kent, October 2013.

87 **That company was Cantor Gaming:** Craig, Susanne. "Taking Risks, Making Odds." *New York Times*, December 24, 2010. http://dealbook.nytimes.com/2010/12/24/taking-risks-making-odds/.

87 **The room feels like a hybrid of a sports bar:** Author experience.

87 **The numbers on Cantor's screens:** Midas background from: Kaplan, Michael. "Wall Street Firm Uses Algorithms to Make Sports Betting Like Stock Trading." *Wired*, November 1, 2010. http://www.wired.com/2010/11/ff_midas/.

87 **Garrood simply went from designing models:** Eden, "Meet the World's Top NBA Gambler."

87 **Cantor wasn't just interested in its predictions:** Craig, "Taking Risks, Making Odds."

88 **Garrood has found that most plays:** Garrood comments from "Betting After the Games Are Underway." ThePostGame, January 11, 2011. http://www.thepostgame.com/blog/spread-sheet/201101/betting-after-games-are-underway.

88 **"We make lines in anticipation":** Predictive Sports Betting, MIT Sloan Sports Analytics Conference, 2013.

89 **"The proprietary nature of prediction models":** McHale, Ian. "Why Spain Will Win . . . Maybe?" *Engineering & Technology*, 5 (June 2010): 25–27.

90 **"There is a lot of paranoia":** Author interview with Rob Esteva, March 2013.

90 **After some suspicious bowling in cricket matches:** Khan, M. Ilyas. "Pakistan's Murky Cricket-Fixing Underworld." BBC News, November 3, 2011. http://www.bbc.com/news/world-asia-15576065.

90 **The scandals have since continued:** Hoult, Nick. "Indian Premier League in Crisis After Three Players Are Charged with Spot Fixing." *Telegraph*, May 16, 2013. http://www.telegraph.co.uk/sport/cricket/twenty20/ipl/10060988/Indian-Premier-League-in-crisis-after-three-players-are-charged-with-spot-fixing.html.

90 **UK police arrested six soccer players:** Hart, Simon. "DJ Campbell Arrested in Connection with Football Fixing." *Telegraph*, December 9, 2013. http://www.telegraph.co.uk/sport/football/10505343/DJ-Campbell-arrested-in-connection-with-football-fixing.html.

90 **the total amount wagered can approach $3 billion:** Wilson, Bill. "World Sport 'Must Tackle Big Business of Match Fixing.'" BBC News, November 25, 2013. http://www.bbc.com/news/business-24984787.

91 **This is the "gray market":** Hawkins, Ed. "Grey Betting Market in Asia Offers Loophole to Be Exploited." *Times* (London), November 30, 2013. http://hawkeyespy.blogspot.com/2013/11/grey-betting-market-in-asia-offers.html.

91 **Haralabos Voulgaris has complained:** Predictive Sports Betting. MIT Sloan Sports Analytics Conference.

91 **Pinnacle claimed it was happy:** Beyer, Andrew. "After Pinnacle, It's All Downhill from Here." *Washington Post*, January 17, 2007. http://www.washingtonpost.com/wp-dyn/content/article/2007/01/16/AR2007011601375.html.

91 **By accepting wagers from sharp bettors:** Noble, Simon. "Inside the Wagering Line." Pinnacle Pulse (blog), Sports Insights, February 22, 2006. https://www.sportsinsights.com/sports-betting-articles/pinnacle-pulse/the-pinnacle-pulse-2222006/.

92 **US Senators stumbled across a Department of Defense proposal:** Taylor, Elanor. "Policy Analysis Market and the Political Yuck Factor." Social Issues Research Centre, April 2004. http://www.sirc.org/articles/policy_analysis.shtml.

92 **One called the idea "grotesque":** Tran, Mark. "Pentagon Scraps Terror Betting Plans." *Guardian*, July 29, 2003. http://www.theguardian.com/world/2003/jul/29/iraq.usa1.

92 **According to Hillary Clinton:** Taylor, "Policy Analysis Market."

93 **Pinnacle has so much faith in the approach:** Wise, Gary. "Head of Sportsbook Q&A Transcript." Pinnacle Sports, August 8, 2013. http://www.pinnaclesports.com/en/betting-articles/social-media/question-answers-with-pinnacle-sports.

93 **Pinnacle dropped horse racing:** "Pinnacle Sports Halts US Horse Racing Service." Casinomeister, December 19, 2008. http://www.casinomeister.com/news/december2008/online_casino_news3/PINNACLE-SPORTS-HALTS-US-HORSE-RACING-SERVICE.php.

93 **Perhaps the best-known betting exchange:** Read, J. J., and J. Goddard. "Information Efficiency in High-Frequency Betting Markets." In *The Oxford Handbook of the Economics of Gambling*, ed. L. V. Williams and D. S. Siegel (New York: Oxford University Press, 2014).

94 **"The boredom was horrendous":** Bowers, Simon. "Odds-on Favourite." *Guardian*, June 6, 2003. http://www.theguardian.com/business/2003/jun/07/9.

94 **the company arranged for a mock funeral procession:** Clarke, Jody. "Andrew Black: Punter Who Revolutionised Gambling." *Moneyweek*, August 21, 2009. http://moneyweek.com/entrepreneurs-my-first-million-andrew-black-betfair-44933/.

94 **if someone wanted to place a bet of £1,000:** Ibid.

96 **"If I find a guy who is good at sports betting":** Klein, Matthew. "Hedge Funds Are Not Necessarily for Suckers." BloombergView, July 12, 2013. http://www.bloombergview.com/articles/2013–07–12/hedge-funds-are-not-necessarily-for-suckers.

96 **According to Tobias Preis:** Preis, Tobias, Dror Y. Kenett, H. Eugene Stanley, Dirk Helbing, and Eshel Ben-Jacob. "Quantifying the Behavior of Stock Correlations Under Market Stress." *Scientific Reports* 2 (2012). doi:10.1038/srep00752.

97 **persuaded Brendan Poots to set up:** Details and quotes from author interview with Brendan Poots, September 2013.

97 **Mark Dixon turned his attention:** Dixon, M. J., and M. E. Robinson. "A Birth Process Model for Association Football Matches." *The Statistician* 47, no. 3 (1998).

98 **you're far more likely to die in a bathtub:** Bailey, Ronald. "How Scared of Terrorism Should You Be?" *Reason Magazine*, September 6, 2011. http://reason.com/archives/2011/09/06/how-scared-of-terrorism-should.

98 **playing roulette again and again:** Spiegelhalter, David. "What's the Best Way to Win Money: Lottery or Roulette?" BBC News, October 14, 2011. http://www.bbc.com/news/uk-15309953.

98 **one 2014 study reckoned the rate:** Titman, A. C., D. A. Costain, P. G. Ridall, and K. Gregory. "Joint Modelling of Goals and Bookings in Association Football." *Journal of the Royal Statistical Society: Series A*, July 15, 2014. doi:10.1111/rssa.12075.

100 **This involves betting on ten soccer leagues:** Author interview with Will Wilde, May 2015.

100 **investment firm Centaur launched the Galileo fund:** Rovell, Darren. "Sports Betting Hedge Fund Becomes Reality." CNBC, April 7, 2010. http://www.cnbc.com/id/36218041.

101 **less than 1 percent of sports bets:** Bell, Kay. "Taxes on Gambling Winnings in Sports." Bankrate, January 2014. http://www.bankrate.com/finance/taxes/taxes-on-gambling-winnings-in-sports-1.aspx.

101 **A new bill, submitted in April 2015:** Takahashi, Maiko. "Japan Lawmakers Group Submits Legislation to Legalize Casinos." Bloomberg Business, April 28, 2015. http://www.bloomberg.com/news/articles/2015–04–28/japan-lawmakers-group-submits-legislation-to-legalize-casinos.

101 **New opportunities will also arise:** Author interview with Will Wilde, May 2015.

102 **Millman got talking to Mike Wohl:** Millman, Chad. "A New System to Bet College Football" (paper presented at MIT Sloan Sports Analytics Conference, Boston, MA, March 1–2, 2013).

102 **"I would want to have play-by-play data":** Author interview with Michael Kent, October 2013.

103 **"We do analysis on the effect":** Author interview with Will Wilde, May 2015.

103 **how a horse's speed changes:** Kaplan, Michael. "The High Tech Trifecta." *Wired*, no. 10.03 (March 2002). http://archive.wired.com/wired/archive/10.03/betting.html.

103 **These "video variables":** Dougherty, Tim. "Horse Sense: Using Applied Mathematics to Game the System." *Contingencies*, May/June 2009. http://www.contingenciesonline.com/contingenciesonline/20090506/?sub_id=qxyLfphSqUiJ#pg22.

103 **Paolo Maldini averaged one tackle:** Kuper, Simon. "How the

Spreadsheet-Wielding Geeks Are Taking over Football." *New Statesman*, June 5, 2013. http://www.newstatesman.com/culture/2013/06/how-spreadsheet-wielding-geeks-are-taking-over-football.

103 **the best cornerbacks in the NFL:** Author interview with Rob Esteva, March 2013.

104 **they won only 52 percent of the games:** Soccer statistics from: Ingle, Sean. "Why the Power of One Is Overhyped in Football." TalkingSport (blog), *Guardian*, March 24, 2013. http://www.theguardian.com/football/blog/2013/mar/24/gareth-bale-one-man-team-overhyped.

104 **"In the sum of the parts":** Author interview with Brendan Poots, September 2013.

105 **"There is a common perception that betting":** Author interview with David Hastie, March 2013.

105 **"We had a guy in New York City":** Author interview with Michael Kent, 2013.

105 **team managers chat with statisticians and modelers:** Eden, "Meet the World's Top NBA Gambler."

105 **The "*Sports Illustrated* jinx":** Wolff, Alexander. "That Old Black Magic." *Sports Illustrated*, January 21, 2002. http://www.si.com/vault/2002/01/21/317048/that-old-black-magic-millions-of-superstitious-readers—and-many-athletes—believe-that-an-appearance-on-sports-illustrateds-cover-is-the-kiss-of-death-but-is-there-really-such-a-thing-as-the-si-jinx.

106 **When a club signs a new player:** McHale, Ian, and Łukasz Szczepański. "A Mixed Effects Model for Identifying Goal Scoring Ability of Footballers." *Journal of the Royal Statistical Society: Series* A 177, no. 2 (2014): 397–417. doi:10.1111/rssa.12015.

106 **Statistician James Albert has attempted:** Albert, James. "Pitching Statistics, Talent and Luck, and the Best Strikeout Seasons of All-Time." *Journal of Quantitative Analysis in Sports* 2, no. 1 (2011).

106 **researchers at Smartodds and the University of Salford:** McHale and Szczepański, "Mixed Effects Model."

107 **Erroneous odds are less common:** Author interview with David Hastie, March 2013.

107 **"I would start with the minor sports":** Predictive Sports Betting, MIT Sloan Sports Analytics Conference.

Chapter 5

109 **"What hath God wrought!":** History of the telegram comes from: "The Birth of Electrical Communications—1837." University of Salford. http://www.cntr.salford.ac.uk/comms/ebirth.php.

110 **traders used telegrams to tell each other:** Poitras, Geoffrey. "Arbitrage: Historical Perspectives." *Encyclopedia of Quantitative Finance*, 2010. doi:10.1002/9780470061602.eqf01010.

110 **traders refer to the GBP/USD exchange rate:** Author experience.

110 **Some would even trek further afield:** Poitras, "Arbitrage: Historical Perspectives."

111 **researchers at Athens University looked at bookmakers' odds:** Vlastakis, Nikolaos, George Dotsis, and Raphael N. Markellos. "How Efficient Is the European Football Betting Market? Evidence from Arbitrage and Trading Strategies." *Journal of Forecasting* 28, no. 5 (2009): 426–444.

111 **a group at the University of Zurich searched:** Franck, Egon, Erwin Verbeek, and Stephan Nüesch. "Inter-market Arbitrage in Sports Betting" (NCER Working Paper Series no. 48, National Centre for Econometric Research, Brisbane, Queensland, Australia, October 2009). http://www.ncer.edu.au/papers/documents/WPNo48.pdf.

111 **Economist Milton Friedman pointed out:** Beinhocker, Eric. *The Origin of Wealth: Evolution, Complexity, and the Radical Remaking of Economics* (Cambridge, MA: Harvard Business Press, 2006), 396.

112 **a group of researchers at the University of Lancaster:** Buraimo, Babatunde, David Peel, and Rob Simmons. "Gone in 60 Seconds: The Absorption of News in a High-Frequency Betting Market" (working paper, from the Selected Works of Dr. Babatunde Buraimo, March 2008). http://works.bepress.com/babatunde_buraimo/17.

113 **Of the 4.4 million bets placed:** "Backing a Winner." *Computing Magazine*, January 25, 2007. http://www.computing.co.uk/ctg/analysis/1854505/backing-winnerw.

113 **"These algorithms mop up any mispricing":** Author interview with David Hastie, March 2013.

113 **It currently takes 65 milliseconds:** Williams, Christopher. "The $300m Cable That Will Save Traders Milliseconds." *Telegraph*,

September 11, 2011. http://www.telegraph.co.uk/technology/news/8753784/The-300m-cable-that-will-save-traders-milliseconds.html.

113 **one blink of the human eye:** Tucker, Andrew. "In the Blink of an Eye." Optalert, August 5, 2014. http://www.optalert.com/news/in-the-blink-of-an-eye.

114 **Traders call the problem "slippage":** Liberty, Jez. "Measuring and Avoiding Slippage." *Futures Magazine*, August 1, 2011. http://www.futuresmag.com/2011/07/31/measuring-and-avoiding-slippage.

115 **The resulting trade is known as an "iceberg order":** Almgren, Robert, and Bill Harts. "Smart Order Routing" (StreamBase White Paper, 2008). http://www.streambase.com/wp-content/uploads/downloads/StreamBase_White_Paper_Smart_Order_Routing_low.pdf.

115 **One example is a "sniffing algorithm":** Ablan, Jennifer. "Snipers, Sniffers, Guerillas: The Algo-Trading War." Reuters, May 31, 2007. http://www.reuters.com/article/2007/05/31/businesspro-usa-algorithm-strategies-dc-idUSN3040797620070531.

116 **As the hooves pounded the ground:** Details from: Rushton, Katherine. "Betfair Loses £40m on Leopardstown After 'Technical Glitch.'" *Telegraph*, December 29, 2011. http://www.telegraph.co.uk/finance/newsbysector/retailandconsumer/8983469/Betfair-loses-40m-on-Leopardstown-after-technical-glitch.html.

116 **Soon after the race finished:** Betfair forum thread: "Hope you all took advantage of betfairs xmas bonus." Geeks Toy Horseracing forum, December 28, 2011. http://www.geekstoy.com/forum/showthread.php?7065-Hope-you-all-took-advantage-of-betfairs-xmas-bonus.

117 **"Due to a technical glitch":** Webb, Peter. "£1k Account Caused £600m Betfair Error." Bet Angel Blog, December 2011. http://www.betangel.com/blog_wp/2011/12/30/1k-account-caused-600m-betfair-error/.

117 **"You cannot win—or lose":** Wood, Greg. "Betfair May Lose Out by Not Explaining How £600m Lay Bet Was Accepted." Talking Sport (blog), *Guardian*, December 30, 2011. http://www.theguardian.com/sport/blog/2011/dec/30/betfair-600m-lay-bet.

117 **The summer of 2012 was a busy time:** Details of the events come from: SEC report. "In the Matter of Knight Capital Americas LLC." File No. 3–15570. October 2013.

119 **In 2007, a trader named Svend Egil Larsen:** Details of the Larsen case come from: Stothard, Michael. "Day Traders Expose Algorithm's

Flaws." *Globe and Mail*, May 16, 2012. http://www.theglobeandmail.com/globe-investor/day-traders-expose-algorithms-flaws/article4179395/; and Stothard, Michael. "Norwegian Day Traders Cleared of Wrongdoing." *Financial Times*, May 2, 2012. http://www.ft.com/cms/s/0/e2f6d1cc-9447-11e1-bb47-00144feab49a.html#axzz3hDw6Bgnj.

120 **Farmer has pointed out:** Farmer, J. Doyne, and Duncan Foley. "The Economy Needs Agent-Based Modelling." *Nature* 460 (2009): 685–686. doi:10.1038/460685a.

120 **At lunchtime on April 23, 2013:** Foster, Peter. "'Bogus' AP Tweet About Explosion at the White House Wipes Billions off US Markets." *Telegraph*, April 23, 2013. http://www.telegraph.co.uk/finance/markets/10013768/Bogus-AP-tweet-about-explosion-at-the-White-House-wipes-billions-off-US-markets.html.

121 **One of the biggest market shocks:** Details of the flash crash come from: US Commodity Futures Trading Commission and US Securities and Exchange Commission. *Findings Regarding the Market Events of May 6, 2010.* September 30, 2010. https://www.sec.gov/news/studies/2010/marketevents-report.pdf.

122 **Algorithms sift through the reports:** Sonnad, Nikhil. "The AP's Newest Business Reporter Is an Algorithm." *Quartz*, June 30, 2014. http://qz.com/228218/the-aps-newest-business-reporter-is-an-algorithm/.

122 **To understand the problem:** Keynes, John M. *The General Theory of Employment, Interest, and Money* (London: Palgrave Macmillan, 1936).

123 **"As soon as you limit what you can do":** Quotes come from author interview with J. Doyne Farmer, October 2013.

124 **Some traders have reported:** Farrell, Maureen. "Mini Flash Crashes: A Dozen a Day." CNN Money. March 20, 2013. http://money.cnn.com/2013/03/20/investing/mini-flash-crash/.

124 **they found thousands of "ultrafast extreme events":** Johnson, Neil, Guannan Zhao, Eric Hunsader, Hong Qi, Nicholas Johnson, Jing Meng and Brian Tivnan. "Abrupt Rise of New Machine Ecology Beyond Human Response Time." *Scientific Reports* 3 (2013). doi:10.1038/srep02627.

124 **"Humans are unable to participate in real time":** Quote from: "Robots Take Over Economy: Sudden Rise of Global Ecology of Interacting Robots Trade at Speeds Too Fast for Humans" (press release, University of Miami, September 11, 2013).

125 **The campus is a maze of neo-Gothic halls:** Author experience.

125 **"The trees on the right were passing me":** Halmos, Paul. "The Legend of John von Neumann." *American Mathematical Monthly* 8 (1973): 382–394.

125 **To examine how different factors influenced ecological systems:** Details of model from: May, R. M. "Simple Mathematical Models with Very Complicated Dynamics." *Nature* 261 (1976): 459–467.

125 **This was first proposed in 1838:** Bacaër, Nicolas. "Verhulst and the Logistic Equation (1838)." *A Short History of Mathematical Population Dynamics* (2011): 35–39.

128 **May found that the larger the ecosystem:** May, Robert M. "Will a Large, Complex Ecosystem Be Stable?" *Nature* 238 (1972): 413–414. doi:10.1038/238413a0.

129 **According to ecologist Andrew Dobson:** Dobson, Andrew. "Multi-Host, Multi-Parasite Dynamics" (Infectious Disease Dynamics workshop, Isaac Newton Institute, Cambridge, UK,. August 19–23, 2013).

129 **Yet, according to Stefano Allesina and Si Tang:** Allesina, Stefano, and Si Tang. "Stability Criteria for Complex Ecosystems." *Nature* 483 (2012): 205–208. doi:10.1038/nature10832.

131 **Doyne Farmer has pointed out:** Farmer, J. Doyne. "Market Force, Ecology and Evolution." *Industrial and Corporate Change* 11, no. 5 (2002): 895–953.

132 **One of the most popular types of financial wager:** Investment Trends. *2013 UK Leveraged Trading Report*. December 23, 2013. http://www.iggroup.com/content/files/leveraged_trading_report_nov13.pdf.

133 **If you make a profitable stock trade:** HM Revenue and Customs. "General Betting Duty." 2010. https://www.gov.uk/general-betting-duty.

133 **In Australia, profits from spread betting:** Armitstead, Louise. "Treasury to Look at Spread Betting Tax Exemption After Lords Raise Concerns." *Telegraph*, November 27, 2013. http://www.telegraph.co.uk/finance/newsbysector/banksandfinance/10479460/Treasury-to-look-at-spread-betting-tax-exemption-after-Lords-raise-concerns.html.

133 **In 2006, the US Federal Reserve:** Details from: "New Directions for Understanding Systemic Risk" (report on a conference cosponsored by the Federal Reserve Bank of New York and the National Academy of Sciences, New York, NY, May 2006).

Chapter 6

135 **In summer 2010, poker websites launched:** Dance, Gabriel. "Poker Bots Invade Online Gambling." *New York Times*, March 13, 2011. http://www.nytimes.com/2011/03/14/science/14poker.html.

135 **Swedish police started investigating poker bots:** Wood, Jocelyn. "Police Investigating Coordinated Poker Bot Operation in Sweden." Pokerfuse, February 22, 2013. http://pokerfuse.com/news/poker-room-news/police-investigating-million-dollar-poker-bot-operation-sweden-21–02/.

135 **It turned out that these bots:** Jones, Nick. "Over $500,000 Repaid to Victims of Bot Ring on Svenska Spel." Pokerfuse, June 20, 2013. http://pokerfuse.com/news/poker-room-news/over-500000-repaid-to-victims-of-bot-ring-on-svenska-spel/.

135 **Until these sophisticated computer players:** Ruddock, Steve. "Alleged Poker Bot Ring Busted on Swedish Poker Site." Poker News Boy, February 24, 2013. http://pokernewsboy.com/online-poker-news/alleged-poker-bot-ring-busted-on-swedish-poker-site/13633.

136 **this was an industry that had spent over $300 million:** Surgeon General. *Preventing Tobacco Use Among Youth and Young Adults: A Report of the Surgeon General, 2012* (Washington, DC: National Center for Chronic Disease Prevention and Health Promotion Office on Smoking and Health, 2012), Table 5.3.

136 **The vote was scheduled:** McGrew, Jane. "History of Tobacco Regulation." In *Marihuana: A Signal of Misunderstanding* (report of the National Commission on Marihuana and Drug Abuse, 1972). http://www.druglibrary.org/schaffer/library/studies/nc/nc2b.htm.

136 **Far from hurting tobacco companies' profits:** McAdams, David. *Game-Changer: Game Theory and the Art of Transforming Strategic Situations* (New York: W. W. Norton, 2014), 61.

137 **Yet tobacco revenues held steady:** Hamilton, James. "The Demand for Cigarettes: Advertising, the Health Scare, and the Cigarette Advertising Ban." *Review of Economics and Statistics* 54, no. 4 (1972).

137 **"Mr. Nash is nineteen years old":** The letter was posted online by Princeton University after John Nash's death in 2015. It went viral.

138 **Despite his prodigious academic record:** Halmos, Paul. "The Legend of John von Neumann." *American Mathematical Monthly* 8 (1973): 382–394.

138 **"Real life consists of bluffing":** Harford, Tim. "A Beautiful Theory." *Forbes*, December 14, 2006. http://www.forbes.com/2006/12/10/business-game-theory-tech-cx_th_games06_1212harford.html. Original quote made in BBC show "Ascent of Man," broadcast in 1973.

138 **Von Neumann started by looking at poker:** Ferguson, Chris, and Thomas S. Ferguson. "On the Borel and von Neumann Poker Models." *Game Theory and Applications* 9 (2003): 17–32.

139 **in a book titled *Theory of Games and Economic Behavior*:** Von Neumann, John, and Oskar Morgenstern. *Theory of Games and Economic Behavior* (Princeton, NJ: Princeton University Press, 1944).

139 **Despite his fondness for Berlin's nightlife:** Dyson, Freeman. "A Walk Through Johnny von Neumann's Garden." *Notices of the AMS* 60, no. 2 (2010): 154–161.

140 **So, it was only natural:** Las Vegas: An Unconventional History. "Benny Binion (1904–1989)." PBS.org, 2005. http://www.pbs.org/wgbh/amex/lasvegas/peopleevents/p_binion.html.

140 **Early in the 1982 competition:** Monroe, Billy. "Where Are They Now—Jack Straus." Poker Works, April 11, 2008. http://pokerworks.com/poker-news/2008/04/11/where-are-they-now-jack-straus.html.

140 **the thirty-first World Series reached its finale:** Details come from video of final at: http://www.tjcloutierpoker.net/2000-world-series-of-poker-final-table-chris-ferguson-vs-tj-cloutier/. TJ Cloutier Poker. "2000 World Series of Poker Final Table—Chris Ferguson vs TJ Cloutier." October 12, 2010.

141 **"You didn't think it would be that tough":** Paulle, Mike. "If You Build It They Will Come." *ConJelCo* 31, no. 25 (May 14–18, 2000). http://www.conjelco.com/wsop2000/event27.html.

141 **no poker player had won more than $1 million:** Wilkinson, Alec. "What Would Jesus Bet?" *The New Yorker*, March 30, 2009. http://www.newyorker.com/magazine/2009/03/30/what-would-jesus-bet.

141 **consultant for the California State Lottery:** Johnson, Linda. "Chris Ferguson, 2000 World Champion." *CardPlayer Magazine* 16, no. 18 (2003).

142 **Combined with improvements in computing power:** Details from: Wilkinson, "What Would Jesus Bet?"

142 **Building on von Neumann's ideas:** Ferguson, C., and T. Ferguson. "The Endgame in Poker." In *Optimal Play: Mathematical Studies of Games and Gambling*, ed. Stewart N. Ethier and William R. Eadington

(Reno, NV: Institute for the Study of Gambling and Commercial Gaming, 2007).

143 **"You always want to make your opponents' decisions"**: Ferguson, Chris. "Sizing Up Your Opening Bet." Hendon Mob, October 7, 2007. http://www.thehendonmob.com/poker_tips/sizing_up_your_opening_bet_by_chris_ferguson.

143 **As well as winning more money:** Harford, "Beautiful Theory."

143 **"How do I win the most?"**: Wilkinson, "What Would Jesus Bet?"

144 **He once taught himself:** Johnson, "Chris Ferguson."

144 **Starting with nothing:** Details of challenge from: Ferguson, Chris. "Chris Ferguson's Bankroll Challenge." PokerPlayer, March 2009. http://www.pokerplayer365.com/poker-players/player-interviews-poker-players/read-about-chris-fergusons-bankroll-challenge-and-you-could-turn-0-into-10000/.

144 **"I remember winning my first $2"**: Ferguson. "Chris Ferguson's Bankroll Challenge."

146 **When Ignacio Palacios-Heurta:** Palacios-Heurta, Ignacio. "Professionals Play Minimax." *Review of Economic Studies* 70 (2003): 395–415.

147 **Von Neumann completed his solution:** Details of the dispute were given in: Kjedldsen, T. H. "John von Neumann's Conception of the Minimax Theorem: A Journey Through Different Mathematical Contexts." *Archive for History of Exact Science* 56 (2001).

149 **While earning his master's degree in 2003:** Follek, Robert. "SoarBot: A Rule-Based System for Playing Poker" (MSc diss., School of Computer Science and Information Systems, Pace University, 2003).

150 **Led by David Hilbert:** O'Connor, J. J., and E. F. Robertson. "Biography of John von Neumann." *JOC/EFR*, October 2003. http://www-history.mcs.st-and.ac.uk/Biographies/Von_Neumann.html.

150 **some inconsistencies in the US Constitution:** "Kurt Gödel." Institute for Advanced Study Online, 2013. https://www.ias.edu/people/godel.

151 **poker bots grew in popularity:** Kushner, David. "On the Internet, Nobody Knows You're a Bot." *Wired* 13.09 (September 2005). http://archive.wired.com/wired/archive/13.09/pokerbots.html?tw=wn_tophead_7.

151 **Just as stripped-down versions of poker:** Details of strategies given

in: Rubin, Jonathan, and Ian Watson. "Computer Poker: A Review." *Artificial Intelligence* 175 (2011): 958–987.

152 **technique known as "regret minimization":** Ibid.

153 **In 2000, researchers at the University of Iowa reported:** Bechara, A., Hanna Damasio, and Antonio R. Damasio. "Emotion, Decision Making and the Orbitofrontal Cortex." *Cerebral Cortex* 10, no. 3 (2000): 295–307. doi:10.1093/cercor/10.3.295.

153 **This contrasts with much economic theory:** Cohen, Michael D. "Learning with Regret." *Science* 319, no. 5866 (2008): 1052–1053.

154 **at the University of Alberta in Canada:** Schaeffer, Jonathan. "Marion Tinsley: Human Perfection at Checkers?" http://www.wylliedraughts.com/Tinsley.htm.

155 **The name was a pun:** Propp, James. "Chinook." *ACJ Extra*, 1999. http://faculty.uml.edu/jpropp/chinook.html.

155 **That's 10 followed by twenty zeros:** Estimate given in: Mackie, Glen. "To See the Universe in a Grain of Taranaki Sand." *North and South Magazine*, May 1999. http://astronomy.swin.edu.au/~gmackie/billions.html.

155 **Chinook "pruned" this decision tree:** Details of competition in Schaeffer, Jonathan, Robert Lake, Paul Lu, and Martin Bryant. "Chinook: The World Man-Machine Checkers Champion." *AI Magazine* 17, no. 1 (1996). doi:http://dx.doi.org/10.1609/aimag.v17i1.1208.

156 **he coined the infinite monkey theorem:** Borel, E. M. "La mécanique statique et l'irréversibilité." *Journal of Theoretical and Applied Physics*, 1913.

158 **"Checkers is solved":** Schaeffer, Jonathan, Neil Burch, Yngvi Björnsson, Akihiro Kishimoto, Martin Müller, Robert Lake, Paul Lu, and Steve Sutphen. "Checkers Is Solved." *Science* 317, no. 5844 (2007): 1518–1522. doi:10.1126/science.1144079.

158 **John Nash showed in 1949:** Demaine, Erik D., and Robert A. Hearn. "Playing Games with Algorithms: Algorithmic Combinatorial Game Theory." *Mathematical Foundations of Computer Science* (2001): 18–32. http://erikdemaine.org/papers/AlgGameTheory_GONC3/paper.pdf.

159 **Twenty-six moves later:** Schaeffer, Jonathan, and Robert Lake. "Solving the Game of Checkers." *Games of No Chance* 29 (1996): 119–133. http://library.msri.org/books/Book29/files/schaeffer.pdf.

160 **"might have died in 1990":** Schaeffer et al., "Chinook."

160 **Doyne Farmer has started to question:** Galla, Tobias, and J. Doyne Farmer. "Complex Dynamics in Learning Complicated Games." *PNAS* 110, no. 4 (2013): 1232–1236. doi:10.1073/pnas.1109672110.

162 **"Large changes tend to be followed":** Mandelbrot, Benoit. "The Variation of Certain Speculative Prices." *Journal of Business* 36, no. 4 (1963): 394–419. http://www.jstor.org/stable/2350970.

163 **"You have a very strong program":** Billings, D., N. Burch, A. Davidson, R. Holte, J. Schaeffer, T. Schauenberg, and D. Szafro. "Approximating Game-Theoretic Optimal Strategies for Full-Scale Poker." *IJCAI* (2003): 661–668. http://ijcai.org/Past%20Proceedings/IJCAI-2003/PDF/097.pdf.

Chapter 7

165 **Thanks to their ability to dissect:** Background on Watson comes from: Rashid, Fahmida. "IBM's Watson Ties for Lead on *Jeopardy* but Makes Some Doozies." EWeek, February 14, 2011. http://www.eweek.com/c/a/IT-Infrastructure/IBMs-Watson-Ties-for-Lead-on-Jeopardy-but-Makes-Some-Doozies-237890; and Best, Jo. "IBM Watson: How the *Jeopardy*-Winning Supercomputer Was Born, and What It Wants to Do Next." TechRepublic. http://www.techrepublic.com/article/ibm-watson-the-inside-story-of-how-the-jeopardy-winning-supercomputer-was-born-and-what-it-wants-to-do-next/.

166 **IBM collected some of the results:** Basulto, Dominic. "How IBM Watson Helped Me to Create a Tastier Burrito Than Chipotle." *Washington Post*, April 15, 2015. http://www.washingtonpost.com/blogs/innovations/wp/2015/04/15/how-ibm-watson-helped-me-to-create-a-tastier-burrito-than-chipotle/.

167 **"Let's try poker":** Wise, Gary. "Representing Mankind." ESPN Poker Club, August 6, 2007. http://sports.espn.go.com/espn/poker/columns/story?columnist=wise_gary&id=2959684.

167 **Finally, there is Eric Jackson:** Details and quotes from author interviews with Michael Johanson and Neil Burch, April 2014, and Tuomas Sandholm, December 2013. Additional specifics from competition online results (http://www.computerpokercompetition.org).

168 **"Poker is a perfect microcosm":** Author interview with Jonathan Schaeffer, July 2013.

169 **"a bath of refreshing foolishness":** Ulam, S. M. *Adventures of a Mathematician* (Oakland: University of California Press, 1991).

169 **young British mathematician by the name of Alan Turing:** Hodges, Andrew. *Alan Turing: The Enigma* (Princeton, NJ: Princeton University Press, 1983).

169 **"I rather liked it at first":** Turing background given in: Copeland, B. J. *The Essential Turing* (Oxford: Oxford University Press, 2004).

170 **manuscript entitled "The Game of Poker":** The game of poker. File AMT/C/18. The Papers of Alan Mathison Turing. The UK National Archives.

170 **He also wondered how games:** Details of the imitation game given in: Turing, A. M. "Computing Machinery and Intelligence." *Mind* 59 (1950): 433–460.

171 **When it played chess against Garry Kasparov:** Kasparov, Garry. "The Chess Master and the Computer." *New York Review of Books*, February 11, 2010. http://www.nybooks.com/articles/archives/2010/feb/11/the-chess-master-and-the-computer/.

172 **In 2013, journalist Michael Kaplan:** Details of Vegas bot given in: Kaplan, Michael. "The Steely, Headless King of Texas Hold 'Em." *New York Times Magazine*, September 5, 2013. http://www.nytimes.com/2013/09/08/magazine/poker-computer.html.

173 **It would have to read its opponent:** Comparison of poker and backgammon in: Dahl, Fredrik. "A Reinforcement Learning Algorithm Applied to Simplified Two-Player Texas Hold'em Poker." *EMCL '01 Proceedings of the 12th European Conference on Machine Learning* (2001): 85–96. doi:10.1007/3-540-44795-4_8.

174 **Neural networks are not a new idea:** McCulloch, Warren S., and Walter H. Pitts. "A Logical Calculus of the Ideas Immanent in Nervous Activity." *Bulletin of Mathematical Biophysics* 5 (1943): 115–133. http://www.cse.chalmers.se/~coquand/AUTOMATA/mcp.pdf.

174 **Facebook announced an AI team:** Details of AI team and DeepFace in: Simonite, Tom. "Facebook Launches Advanced AI Effort to Find Meaning in Your Posts." *MIT Technology Review*, September 20, 2013. http://www.technologyreview.com/news/519411/facebook-launches-advanced-ai-effort-to-find-meaning-in-your-posts/; and Simonite, Tom. "Facebook Creates Software That Matches

Faces Almost as Well as You Do." *MIT Technology Review*, March 17, 2014. http://www.technologyreview.com/news/525586/facebook-creates-software-that-matches-faces-almost-as-well-as-you-do/.

174 **Facebook users were uploading over 350 million:** Smith, Cooper. "Facebook Users Are Uploading 350 Million New Photos Each Day." *Business Insider*, September 18, 2013. http://www.businessinsider.com/facebook-350-million-photos-each-day-2013–9.

176 **Rather than grab a vulnerable pawn:** Description of move in: Chelminski, Rudy. "This Time It's Personal." *Wired* 9.10 (October 2001). http://archive.wired.com/wired/archive/9.10/chess.html.

176 **Deep Blue's game-changing show:** Fact that the move was random from: Silver, Nate. *The Signal and the Noise: Why So Many Predictions Fail—but Some Don't* (London: Penguin, 2012).

176 **Some are easier to scare off than others:** Bateman, Marcus. "What Does 'Floating' Mean?" Betfair Online, July 6, 2010. https://betting.betfair.com/poker/poker-strategy/what-does-floating-mean-060710.html.

177 **"Most of our group aren't poker players":** Author interview with Michael Johanson and Neil Burch, April 2014.

178 **In 2010, an online version of rock-paper-scissors:** Dance, Gabriel, and Tom Jackson. "Rock-Paper-Scissors: You vs. the Computer." *New York Times*. http://www.nytimes.com/interactive/science/rock-paper-scissors.html.

178 **In 2014, Zhijian Wang and colleagues:** Wang, Zhijian, Bin Xu, and Hai-Jun Zhou. "Social Cycling and Conditional Responses in the Rock-Paper-Scissors Game." *Scientific Reports* 4, no. 5830 (2014). doi:10.1038/srep05830.

179 **cognitive psychologist George Miller noted:** Miller, George A. "The Magical Number Seven, Plus or Minus Two: Some Limits on Our Capacity for Processing Information." *Psychological Review* 63 (1956): 81–97.

179 **Dutch psychologist Willem Wagenaar observed:** Bar-Hillel, Maya, and Willem A. Wagenaar. "The Perception of Randomness." *Advances in Applied Mathematics* 12, no. 4 (1991): 428–454. doi:10.1016/0196–8858(91)90029-I.

179 **referred to as the "magical number seven":** Jacobson, Roni. "Seven Isn't the Magic Number for Short-Term Memory." *New York Times*, September 9, 2013.

180 **the best competitors can memorize:** Lai, Angel. "World Records." http://www.world-memory-statistics.com/disciplines.php.

180 **memorizing cards also helps in blackjack:** Details about memory techniques in: Robb, Stephen. "How a Memory Champ's Brain Works." BBC News, April 7, 2009. http://news.bbc.co.uk/2/hi/uk_news/magazine/7982327.stm.

180 **"often mused about the nature of memory":** Metropolis, Nick. "The Beginning of the Monte Carlo Method." Special issue, *Los Alamos Science* (1987): 125–130. http://jackman.stanford.edu/mcmc/metropolis1.pdf.

181 **The database came from Shawn Bayern:** "Rock-Paper-Scissors: Humans Versus AI." http://www.essentially.net/rsp.

182 **"A coalition absorbs at least two players":** Von Neumann, J., and Oskar Morgenstern. *Theory of Games and Economic Behavior* (Princeton, NJ: Princeton University Press, 1944).

182 **Parisa Mazrooei and colleagues at the University of Alberta:** Mazrooei, Parisa, Christopher Archibald, and Michael Bowling. "Automating Collusion Detection in Sequential Games." *Association for the Advancement of Artificial Intelligence* (2013). https://webdocs.cs.ualberta.ca/~bowling/papers/13aaai-collusion.pdf.

182 **There are reports of unscrupulous players:** Goldberg, Adrian. "Can the World of Online Poker Chase Out the Cheats?" BBC News, September 12, 2010. http://www.bbc.com/news/uk-11250835.

182 **"In any form of poker":** Dahl, F. "A Reinforcement Learning Algorithm Applied to Simplified Two-Player Texas Hold'em Poker." In *European Conference on Machine Learning 2001, Lecture Notes in Artificial Intelligence 2167*, ed. L. De Raedt and P. Flach (Berlin: Springer-Verlag, 2001).

184 **tweak your tactics as you learn:** Author interview with Tuomas Sandholm, December 2013. Additional details in: Sandholm, T. "Perspectives on Multiagent Learning." *Artificial Intelligence* 171 (2007): 382–391.

184 **Sandholm has been developing "hybrid" bots:** Ganzfried, Sam, and Tuomas Sandholm. "Game Theory-Based Opponent Modeling in Large Imperfect-Information Games." *Proceedings of the 10th International Conference on Autonomous Agents and Multiagent Systems* 2 (2011): 533–540.

185 **professional players Phil Laak and Ali Eslami:** Details of event in: Wise, "Representing Mankind"; and Harris, Martin. "Laak-Eslami

Team Defeats Polaris in Man-Machine Poker Championship." PokerNews, July 25, 2007. http://www.pokernews.com/news/2007/07/laak-eslami-team-defeats-polaris-man-machine-poker-champions.htm.

186 **there was a second man-machine competition:** Details of event in: Harris, Martin. "Polaris 2.0 Defeats Stoxpoker Team in Man-Machine Poker Championship." PokerNews, July 10, 2008. http://www.pokernews.com/news/2008/07/man-machine-II-poker-championship-polaris-defeats-stoxpoker-.htm; and Johnson, R. Colin. "AI Beats Human Poker Champions." EETimes, July 7, 2008. http://www.eetimes.com/document.asp?doc_id=1168863.

187 **Using the regret minimization approach:** Author interview with Michael Johanson, April 2014.

188 **With a nod to the group's checkers research:** Bowling, Michael, Neil Burch, Michael Johanson, and Oskari Tammelin. "Heads-Up Limit Hold'em Poker Is Solved." *Science* 347, no. 6218 (2015): 145–149. doi:10.1126/science.1259433.

189 **"It would attack the mystique":** Author interview with Michael Johanson, April 2014.

190 **Watson found the short clues the most difficult:** Sutton, John D. "Behind-the-Scenes with IBM's 'Jeopardy!' Computer, Watson." CNN, February 7, 2011. http://www.cnn.com/2011/TECH/innovation/02/07/watson.ibm.jeopardy/.

190 **people might be especially good at sizing up others:** Wright, G. R., C. J. Berry, and G. Bird. "'You Can't Kid a Kidder': Association Between Production and Detection of Deception in an Interactive Deception Task." *Frontiers in Human Neuroscience* 6 (2012): 87. doi:10.3389/fnhum.2012.00087.

191 **In a 2006 survey spanning fifty-eight countries:** Global Deception Research Team. "A World of Lies." *Journal of Cross-Cultural Psychology* 37, no. 1 (2006): 60–74. doi:10.1177/0022022105282295.

191 **There's no evidence that liars avert their gaze:** DePaulo, B. M., J. J. Lindsay, B. E. Malone, L. Muhlenbruck, K. Charlton, and H. Cooper. "Cues to Deception." *Psychological Bulletin* 129, no. 1 (2003): 74–118.

191 **In a 2010 study:** Schlicht, E. J., S. Shimojo, C. F. Camerer, P. Battaglia, and K. Nakayama. "Human Wagering Behavior Depends on Opponents' Faces." *PLoS ONE* 5, no. 7 (2010): e11663.

192 **When Matt Mazur decided to build a poker bot:** Author interview with Matt Mazur, August 2014. Additional details from his blog posts (http://www.mattmazur.com).

Chapter 8

197 **Hundreds of cameras cling:** Author experience.

197 **casinos' definition of such cheating:** History of surveillance in: Hicks, Jesse. "Not in My House: How Vegas Casinos Wage a War on Cheating." *The Verge*, January 14, 2014. http://www.theverge.com/2013/1/14/3857842/las-vegas-casino-security-versus-cheating-technology.

198 **Unlawful Internet Gambling Enforcement Act:** Unlawful Internet Gambling Enforcement Act of 2006, 31 U.S.C. 5361–5366, §5362.

198 **That included Lawrence DiCristina:** Details of DiCristina case from: Weinstein, Jack. *Memorandum, Order & Judgment, United States of America against Lawrence DiCristina*. 11-CR-414. August 2012. http://jurist.org/paperchase/103482098-U-S-vs-DiCristina-Opinion-08–21–2012.pdf.

200 **airport operator William McBoyle helped arrange:** *McBoyle v. U.S.* 1930 10CIR 118, 43 F.2d 273.

201 **The conviction was reversed:** Paraphrased from original comment in *McBoyle v. U.S.* 1930: "When a rule of conduct is laid down in words that evoke in the common mind only the picture of vehicles moving on land, the statute should not be extended to aircraft simply because it may seem to us that a similar policy applies, or upon the speculation that, if the legislature had thought of it, very likely broader words would have been used."

201 **Rather, the state law meant:** Brennan, John. "U.S. Supreme Court Declines to Take DiCristina Poker Case; Reminder of Challenge Faced by NJ Sports Betting Advocates." NorthJersey.com, February 24, 2014. http://blog.northjersey.com/meadowlandsmatters/7891/u-s-supreme-court-declines-to-take-dicristina-poker-case-reminder-of-challenge-faced-by-nj-sports-betting-advocates/.

202 **"There may be such a thing as habitual luck":** Ulam, S. M. *Adventures of a Mathematician* (Oakland: University of California Press, 1991).

202 **"Chance favours the prepared mind":** Quoted in: Weiss, R. A. "HIV and the Naked Ape." In *Serendipity: Fortune and the Prepared Mind*,

ed. M. De Rond and I. Morley (Cambridge: Cambridge University Press, 2010). Originally said during a lecture at University of Lille, 1854.

203 **Matthew Salganik and colleagues at Columbia University:** Salganik, M. J., P. S. Dodds, and D. J. Watts. "Experimental Study of Inequality and Unpredictability in an Artificial Cultural Market." *Science* 311 (2006): 854–856.

203 **"Fame has much less to do . . .":** Dodds, Peter Sheridan. "Homo Narrativus and the Trouble with Fame." *Nautilus*, September 5, 2013. http://nautil.us/issue/5/fame/homo-narrativus-and-the-trouble-with-fame.

204 **"Consider a set of funds with no skill":** Roulston, Mark, and David Hand. "Blinded by Optimism" (working paper, Winton Capital Management, December 2013). https://www.wintoncapital.com/assets/Documents/BlindedbyOptimism.pdf?1398870164.

205 **hockey analyst Brian King suggested a way:** Charron, Cam. "Analytics Mailbag: Save Percentages, PDO, and Repeatability." TheLeafsNation.com. May 27, 2014. http://theleafsnation.com/2014/5/27/analytics-mailbag-save-percentages-pdo-and-repeatability.

205 **The statistic, later dubbed PDO:** Details on PDO and NHL statistics given in: Weissbock, Joshua, Herna Viktor, and Diana Inkpen. "Use of Performance Metrics to Forecast Success in the National Hockey League" (paper presented at the European Conference on Machine Learning and Principles and Practice of Knowledge Discovery in Databases, Prague, September 23–27, 2013).

205 **England had the lowest PDO:** Burn-Murdoch, John. "Were England the Uunluckiest Team in the World Cup Group Stages?" FT Data Blog. 29 June 2014. http://blogs.ft.com/ftdata/2014/06/29/were-england-the-unluckiest-team-in-the-world-cup-group-stages/.

206 **Cambridge college spent on wine:** "In Vino Veritas, Redux." *The Economist*, February 5, 2014. http://www.economist.com/blogs/freeexchange/2014/02/correlation-and-causation-0.

207 **topped the wine list with a spend of £338,559:** Simons, John. "Wages Not Wine: Booze Hound Colleges Spend £3 million on Wine." *Tab* (Cambridge, England), January 22, 2014. http://thetab.com/uk/cambridge/2014/01/22/booze-hound-colleges-spend-3-million-on-wine-32441.

207 **Countries that consume lots of chocolate:** Messerli, F. H. "Chocolate Consumption, Cognitive Function, and Nobel Laureates." *New England Journal of Medicine* 367 (2012): 1562–1564. doi:10.1056/NEJMon1211064.

207 **When ice cream sales rise in New York City:** Peters, Justin. "When Ice Cream Sales Rise, So Do Homicides. Coincidence, or Will Your Next Cone Murder You?" Crime (blog), *Slate*, July 9, 2013. http://www.slate.com/blogs/crime/2013/07/09/warm_weather_homicide_rates_when_ice_cream_sales_rise_homicides_rise_coincidence.html.

209 **When Manchester City won the league:** Lewis, Tim. "How Computer Analysts Took Over at Britain's Top Football Clubs." *The Observer*, March 9, 2014. http://www.theguardian.com/football/2014/mar/09/premier-league-football-clubs-computer-analysts-managers-data-winning.

209 **Roberto Martinez, manager of Everton soccer club:** Ibid.

209 **When Picasso worked his "Bull" lithographs:** Details of bull given in: Lavin, Irving. "Picasso's Lithograph(s) 'The Bull(s)' and the History of Art in Reverse" *Art Without History*, 75th Annual Meeting, College Art Association of America, February 12–14, 1987.

210 **Einstein once said of scientific models:** Quoted by Sugihara, George. "On Early Warning Signs." *Seed Magazine*, May 2013. http://seedmagazine.com/content/article/on_early_warning_signs/.

211 **"The best material model of a cat":** Widely attributed to Wiener. Quote appears in: Rosenblueth, Arturo, and Norbert Wiener. "The Role of Models in Science." *Philosophy of Science* 12, no. 4 (1945): 316–321.

211 **In 1947, *Time* magazine published:** Chapin, R. M. "Communist Contagion." *Time*, April 1946. http://claver.gprep.org/fac/sjochs/communist-contagion-map.htm.

211 **a piece called "Europe from Moscow":** Chapin, R. M. "Europe from Moscow." *Time*, March 1952.

212 **When the pair bet together:** Borel, Émile. "A Propos d'Un Traite de Probabilities. Revue Philosophique." 1924. Quoted in Ellsberg, Daniel. *Risk, Ambiguity, and Decision* (New York: Garland Publishing, 2001).

213 **"How to Gamble If You Must":** Details of the course in: Bernstein, J. *Physicists on Wall Street and Other Essays on Science and Society* (New York: Springer, 2008).

213 **The MIT students therefore worked:** Details of the strategy given in: Mezrich, Ben. *Bringing Down the House: The Inside Story of Six MIT Students Who Took Vegas for Millions* (New York: Simon and Schuster, 2003).

214 **They later discovered the janitor:** Locker story given in: Ball, Janet. "How a Team of Students Beat the Casinos." BBC News Magazine, May 26, 2014. http://www.bbc.com/news/magazine-27519748.

214 **"I know very few people":** Author interview with Richard Munchkin, August 2013.

214 **In 2012, PhD student Will Ma:** Details and quotes from author interview with Will Ma, September 2014.

215 **Courses teaching the science of gambling:** The York University course was Bethune 1800: Mathematics of Gambling, taught in 2009□10, and the Emory course was MATH 190–000: Freshman Seminar: Math: Sports, Games & Gambling, taught in Fall 2012. Further details: http://garsia.math.yorku.ca/~zabrocki/bethune1800fw0910/ and http://college.emory.

216 **"That way of thinking about the world":** Author interview with Ruth Bolton, February 2013.

217 **"It is remarkable that a science":** Quoted widely, but originally given in: Laplace, P. S. *Théorie Analytique des Probabilitiés* (Paris: Courcier, 1812).

218 **"It wasn't as though streetwise Las Vegas gamblers":** Author interview with Bill Benter, July 2013.

INDEX

ADAM KUCHARSKI is an assistant professor in mathematical modeling at the London School of Hygiene and Tropical Medicine and an award-winning science writer. He studied at the University of Warwick before completing a PhD in mathematics at the University of Cambridge. The winner of the 2012 Wellcome Trust Science Writing Prize, Kucharski lives in London.

THE SONG OF THE LARK

WILLA CATHER

EDITED WITH
AN INTRODUCTION AND NOTES
BY SHERRILL HARBISON

PENGUIN BOOKS

PENGUIN BOOKS
Published by the Penguin Group
Penguin Group (USA) Inc., 375 Hudson Street, New York, New York 10014, U.S.A.
Penguin Group (Canada), 90 Eglinton Avenue East, Suite 700, Toronto,
Ontario, Canada M4P 2Y3 (a division of Pearson Penguin Canada Inc.)
Penguin Books Ltd, 80 Strand, London WC2R 0RL, England
Penguin Ireland, 25 St Stephen's Green, Dublin 2, Ireland (a division of Penguin Books Ltd)
Penguin Group (Australia), 250 Camberwell Road, Camberwell,
Victoria 3124, Australia (a division of Pearson Australia Group Pty Ltd)
Penguin Books India Pvt Ltd, 11 Community Centre, Panchsheel Park, New Delhi – 110 017, India
Penguin Group (NZ), cnr Airborne and Rosedale Roads,
Albany, Auckland 1310, New Zealand (a division of Pearson New Zealand Ltd)
Penguin Books (South Africa) (Pty) Ltd, 24 Sturdee Avenue,
Rosebank, Johannesburg 2196, South Africa

Penguin Books Ltd, Registered Offices: 80 Strand, London WC2R 0RL, England

First published in the United States of America by
Houghton Mifflin Company 1915
This edition with an introduction and notes by Sherrill Harbison
published in Penguin Books 1999

7 9 10 8

LIBRARY OF CONGRESS CATALOGING-IN-PUBLICATION DATA
Cather, Willa, 1873–1947.
The song of the lark / Willa Cather ; edited with an introduction
and notes by Sherrill Harbison.
p. cm.—(Penguin twentieth-century classics)
ISBN 0 14 11.8104 4
1. Women singers—United States—Fiction. 2. Young women—
Colorado—Fiction. 3. Opera—United States—Fiction.
I. Harbison, Sherrill. II. Title. III. Series.
PS3505.A87S6 1999
813'.52—dc21 98-41698

Printed in the United States of America
Set in Stempel Garamond

INTRODUCTION

The Song of the Lark is a *Künstlerroman*, the story of an artist's growth and development from childhood to maturity. More particularly—and decidedly more rarely—it is a *female Künstlerroman*, one in which the male characters are satellites and willing servants to a woman's career.

Thea Kronborg is a Scandinavian-American singer who works her way up from the dusty desert town of Moonstone, Colorado, to the boards of the Metropolitan Opera House. Although Willa Cather herself was not a musician—she neither sang nor played a musical instrument—the portions of the novel covering childhood, apprenticeship, and artistic awakening in the western landscape are frankly autobiographical. Its final section, dealing with Thea's professional life, is drawn largely from the career of the Wagnerian soprano Olive Fremstad, who was the kind of artist Willa Cather still aspired to be. For although Cather was forty-two when *The Song of the Lark* was published in 1915, it was only the third of her twelve novels, and thus belongs to the early stage of her distinguished literary career.

Willa Cather (1873–1947) was born in the Blue Ridge Mountains of northwestern Virginia. At the age of nine she moved with her family to Nebraska, joining three generations of Cathers who had relocated to the frontier in hopes that the arid western climate might thwart a rising toll of tuberculosis in the family.

The child was deeply affected by being plucked from the lush Appalachian hills and "hurled out" into a country as flat and "bare as a piece of sheet iron." The sameness and instability of the prairie landscape—an endless sea of churning grasses without firm contours or familiar landmarks—threatened her with "a kind of erasure of personality," a sense of drowning or "going under."[1] She conquered her fear by learning to read botanical and other fine distinctions in the terrain, and by listening to the stories of old-time pioneers, who had trekked for months across the treacherous grasses. They became her heroes, and although she gradually learned to love the land they had tamed, for the rest of her life the prairie's overwhelming scale afflicted her with an unsettling combination of homesickness and agoraphobia.

Willa was the oldest of seven children to grow up in a tiny frame house in the booming new railroad town of Red Cloud. She was happiest in the company of her brothers, with whom she rode the range, explored the muddy Republican River banks, and imagined faraway places. She was insatiably curious, theatrical, and a voracious reader, and her parents, recognizing her unusual intensity, gave her her own small room in the unheated attic, to which she could retreat from the boisterous family.

Even in adolescence Cather was unintimidated by convention, and counted among her friends adults considered unusual or eccentric in the bustling frontier town—educated men and impractical dreamers who admired her qualities and encouraged her interests. With an elderly Englishman named William Ducker she studied Latin, discussed religion and ethics, dissected animals, and conducted scientific experiments. From a Herr Schindelmeisser, an itinerant, alcoholic musician, she learned stories about music and musical life, but never—despite his best efforts—to play the piano.

She was also passionately interested in the neighboring Scandinavian and Bohemian farm women, homesick immigrants like herself, whose stories in broken English first awakened her imagination to an older world beyond the sea. "I have never found any intellectual excitement any more intense," she recalled to an interviewer in 1913, "than I used to feel when I spent a morning with one of those old farm women. I used to ride home in the most unreasonable state of excitement; I always felt as if they had told me so much more than they said—as if I had actually got inside another person's skin." The excitement of merging with another personality left a deep impression, one she later recognized as *artistic,* and "if one begins that early," she explained, "it is the story of the maneating tiger all over again—no other adventure ever carries one quite so far."[2]

At first, however, Cather planned to be a doctor, not an artist, and in her early teens apprenticed herself to the town's physician, Dr. G. E. McKeeby, with whom she rode about the countryside assisting with house calls on prairie dugouts. In these years she signed her name as "William Cather, M.D.," cropped her hair, and wore men's clothing, a practice she continued through her first year at the University of Nebraska, which she entered at age sixteen.

At the end of that first year, 1891, Cather was surprised by the publication of her freshman theme on Thomas Carlyle, submitted by her professor without her knowledge to the *Nebraska State Journal*. It was a passionate, romantic defense of artistic genius, arguing that because great artists are divinely inspired, Carlyle's irregular personal behavior deserved special consideration. "The wife of an artist," Cather maintained, "must always be a secondary consideration with him; she should realize that from the outset," because "art of every kind is an exacting master, more so even than Jehovah."[3] She later admitted that the exhilaration of seeing her name in print for the first time had made her change her professional plans.

As a fledgling writer Cather joined the professional leagues immediately, becoming not only a founding and contributing editor of the campus literary magazine, but also a book and drama critic for the *Nebraska State Journal* by her junior year. Her enthusiasm for drama and music only grew with exposure, and as a graduation present to herself she traveled in 1895 to Chicago, where the Metropolitan Opera was on tour. During her one-week stay there, she saw all five productions—one Gounod, one Meyerbeer, and three Verdi operas, with casts including Nellie Melba, Jean de Reszke, and Lillian Nordica. It was her first experience of professional opera and her first trip out of Nebraska, and she left the city dazzled, exhausted, and addicted.

In 1896, at the age of twenty-two, Cather accepted an offer in Pittsburgh to edit a new magazine, *The Home Monthly*, whose pages she helped to fill with her own pseudonymous stories. Later she broke the gender barrier by becoming a news reporter for the Pittsburgh *Daily Leader*. At no time did she abandon arts reviews, which she wrote both for the Pittsburgh audience and as a field correspondent for the Lincoln, Nebraska, *Courier*. Arts reviewing gave her free access to theaters and concert halls, as well as opportunities to interview such great performers as Lillian Nordica and Helena Modjeska—heady fodder for a young woman stagestruck since childhood.

It was in Pittsburgh, where the Metropolitan Opera company appeared in 1897 and 1900, that Cather first saw the Wagner operas that would affect her artistic taste so strongly. "When Wagner called his goddess women down out of Walhalla," she wrote in her 1900 review, "they relegated the fragile heroines of the old Italian operas to oblivion." If audiences had lost patience even with Mozart, it was "all because that malicious man Wagner has so stung the palate that all other styles seem insipid."[4]

Keeping to the grueling schedule of daily journalism was not conducive to Cather's creative work, however, and between 1899 and 1905 she tried high-school teaching so her summers could be free. In 1906, after publishing several stories in *McClure's*, she was recruited by the publisher himself to come to New York and work for his magazine. His bait: a promise to publish her first collection of stories (issued as *The Troll Garden* in 1905).

McClure's was then the nation's most prominent muckraking magazine, achieving fame through exposés of graft and corruption by star reporters Lincoln Steffens and Ida Tarbell. Its publisher, Samuel S. McClure, was a brilliant, erratic man for whom Cather would come to feel profound affection and loyalty, but leaving Pittsburgh to work for him must have been personally painful, as well as an obvious professional decision.

Cather was by then deeply committed to Isabelle McClung, the handsome daughter of a prominent Pittsburgh judge, who enjoyed challenging the city's Presbyterian propriety by mixing with artists. The two women's friendship had deepened to love, and in 1901 McClung rescued Cather from her noisy boardinghouse and brought her home to live. In the following six years Cather enjoyed for the first time the comforts and social connections that come with wealth. She was happy living in a family setting again, and was able to re-create her childhood nest in a private attic study. In 1902 she and Isabelle traveled to Europe. But Cather was ambitious, and had long contended that the demands of intimate relationships were a hindrance to the serious artist. Although Isabelle remained her lifelong romance, after Cather left her for *McClure's*, they never lived together again.

Working for Sam McClure, who made her the magazine's managing editor in 1908, was more demanding than anything Cather had done before, and she wrote very little in the next three years. It was in her editorial capacity, however, that she met and befriended Elizabeth Shepley Sergeant, a gifted young journalist whose letters and valuable memoir of Cather supply important information about the making of *The Song of the Lark*.

It was not until 1911 that Cather had saved enough money to resign her managing editor's post and devote herself to writing, and by that time she felt intellectually dulled and dispirited from sifting through mountains of mediocre prose, fatigued and irritable from running damage control for her energetic and irascible boss. To recuperate, in the spring of 1912 she made a holiday visit to her brother Douglass, a railroad worker based in Winslow, Arizona. It was a trip that would have a profound impact on her life, and on this novel.

In the desert country of New Mexico and Arizona, Cather discovered a new world of bracing air and dazzling color, of jagged canyons, painted hills, and desert flowers. Here too was history—a record of geological and human time even more exciting, she wrote to Elizabeth Sergeant, than she had encountered in Europe. She and Douglass hiked and scaled rock cliffs to explore the ancient cliff dwellers' ruins in the sequestered Walnut Canyon. When her brother was on duty she soaked up trail lore from his railroad buddies, or collected Spanish and Indian legends from another of his friends, a Catholic priest. She also consorted and danced with the local Mexicans, whose beauty and musicality delighted her—she was particularly infatuated with a graceful "young Antinous of a singer" called Julio, who serenaded her nightly and took her to see the Painted Desert.[5] The combination of physical challenge, scenery, history, and romance was exhilarating, and after two months in the Southwest she returned to New York as fresh and ebullient as a child.

Later that same year Cather published her first novel, *Alexander's Bridge*—a somewhat contrived Jamesian story set in urban Boston and London. The following year she "hit home pasture" with *O Pioneers!*, a story of Scandinavian and Bohemian immigrant families in Nebraska.[6] She had seen the world; she could now come home to the parish.[7]

In June 1913, with *O Pioneers!* still in press, Cather met a steamer bringing Elizabeth Sergeant home from France. Before the two women's greetings were over, Sergeant realized her friend was in the grip of a new enthusiasm. Her memoir reports:

> The word "Fremstad" had already appeared in her letters. Fremstad, Fremstad, wonderful Fremstad. . . . Nothing, nothing, Willa

> murmured in the cab—still quite unaware of my Provençal daze . . . could equal the bliss of entering into the very skin of another human being. Did I not agree? And if this skin were Scandinavian—then what?[8]

Olive Fremstad had electrified New York audiences in the role of Richard Wagner's most extreme and exotic female character, the schizoid temptress Kundry of *Parsifal*, and Willa Cather was one of those most smitten. In the decade (1903–14) of the singer's meteoric career at the Metropolitan Opera,[9] Cather had closely watched what she called "the rapid crystallization of ideas as definite, as significant, as profound as Wagner's own."[10] For over a year she had been entertaining an idea for a novel about an opera singer, and now, in preparation, was writing an article for *McClure's*—a triple profile of three reigning divas, Louise Homer, Geraldine Farrar, and Olive Fremstad.

She already knew that Fremstad interested her most, but after meeting and interviewing her, Cather found that her feelings about the singer grew deeper and more complex. Only Fremstad, Cather concluded, was equipped with the talent, intelligence, and dedication to reach what Geraldine Farrar had termed "the frozen heights" of perfection.

Cather wrote Sergeant that Fremstad's presence left her profoundly moved, "choked up by things unutterable."[11] From watching Fremstad both on- and offstage, Cather recognized how completely the singer lived through her roles, how little of the person was left for anything else. As an artist both equipped and determined to reach the "frozen heights," Fremstad had only one great passion: her art. "Work is the only thing that interests her," Cather reported; the singer "says she has tried this and that thing . . . which, from a distance, seemed beautiful; but that art is the only thing that *remains* beautiful."[12]

It was also clear that the single-minded pursuit of beauty destined this kind of artist to spend most of her time in personal and intellectual solitude. "We are born alone, we make our way alone, we die alone," Fremstad had confided, and the greatest rewards of an artist's quest are also enjoyed alone. "My work is only for serious people. If you ever really find anything in art, it is so subtle and beautiful that—well, you need never be afraid anyone will take it away from you, for the chances are that nobody will ever know you've got it."[13]

These convictions, issuing from the mouth of a stunning Scandinavian singer, set off a creative inner explosion in Cather. Here was a meeting of minds: well before encountering Fremstad, Cather herself had protested (to the puzzled Elizabeth Sergeant) that while most artists "almost desperately" linked themselves to lovers, spouses, children, "as if to avoid being devoured by art," Cather herself "*wanted* to be so devoured."[14] Cather understood that lovers or children could never fully share such an impersonal mission, nor would they be satisfied with the husk of humanity left after the artist's vitality was drained away. Although personal attachments would always be alluring in times of loneliness, they could never satisfy this artist's ambition or fulfill her dreams.

Cather watched Fremstad perform every role in her repertoire. The two women became friends—Fremstad cheering and tending Cather when she was hospitalized with a nasty infection; Cather spending a week visiting Fremstad's primitive cabin retreat in Maine. The singer's companion, Mary Watkins, also warmed to Cather, finding her a welcome ally during bouts of Madame's mercurial temper.[15]

The shared belief that art was an exacting master was not all the two women had in common. Like Cather, Fremstad had been an immigrant to the prairie in her adolescence. Born in Stockholm in 1871 and raised in Oslo, she moved to a frontier settlement in Minnesota at about the same time the Cather family arrived in Nebraska. Young Olava (changed to Olive by her American neighbors) had already won singing competitions in Norway; on the prairie her musical talents were pressed into serving at church and revival meetings conducted by her father, a Methodist minister, and into providing piano lessons to the local population.[16]

Like Cather growing up in Red Cloud, Fremstad had artistic gifts that were out of scale with her surroundings; yet those surroundings still left their mark on her, making her seem strangely familiar to Cather. She wrote Sergeant that Fremstad had the same kind of suspicious, defiant, far-seeing pioneer eyes Cather remembered from the Scandinavian farm women she had known as a child.[17] In fact, Fremstad's habits and manner reminded Cather of her own fictional Swedish heroine, Alexandra Bergson of *O Pioneers!*—Alexandra with a voice.

To Willa Cather, the combination of artist and Scandinavian immigrant suggested adventure of the highest order. Toward Olive Fremstad, Cather's feelings were clearly both identification and infatuation, a mature combination of emotions she had first experienced for the rugged farm women on the prairie. The great "intellectual excitement" now before her would be to consummate this attraction, to merge Fremstad and herself in the character of Thea Kronborg, the opera singer–heroine of *The Song of the Lark*.

Thea, like Olive Fremstad, comes of Swedish-Norwegian stock and is the musical daughter of a Methodist minister, obliged to give lessons and play and sing at prayer meetings and revivals. Most of the other details of her youth, however, are drawn directly from Willa Cather's own experience. Like Cather, Thea is one of seven children growing up in an overcrowded little house in Moonstone, a small western town that closely resembles Red Cloud. Her parents recognize and respect their daughter's unusual gifts, but her more conventional siblings and neighbors think of Thea as spoiled, rebellious, and stuck-up. Her refuge is a tiny room in the high-windowed gable end of the attic, a rose-papered bower where she can read, write, and dream in peace.

Thea's closest friends are a handful of adult men who appreciate her qualities and are themselves restless or unhappy in Moonstone. All have counterparts in Cather's life: the chivalrous, self-educated railroad brakeman Ray Kennedy, who loves to explore cliff ruins, combines features of her brother Douglass and his railroading friends. Professor Wunsch, the romantic, alcoholic piano teacher, is a sympathetic portrait of Herr Schindelmeisser, while the wild mandolin player Spanish Johnny was inspired by the musical Mexicans Cather had met in Arizona. Thea's most important childhood friend, the town physician, Dr. Howard Archie, was modeled on Dr. McKeeby.

The scenes conveying the sensual and imaginative richness of Thea's childhood and youth are among the most memorable ones in the novel. Physical memories—hauling her baby brother in his wagon on hot, dusty afternoons, stretching out her growing body in her moonlit room, running about in the dazzling sand hills, feeling a hot brick seep warmth between her blankets on icy winter nights, and the noise and danger of the railroad—merge with intellectual discoveries about music, human aspirations, moral riddles, and small-town disappointments and jealousies. The plan of the novel reinforces the reader's experience of it: these same vivid after-impressions, realized viscerally by the child, are the sensual memories that sustain the mature artist in her professional solitude. As she later tries to explain to Dr. Archie, "A child's attitude toward everything is an artist's attitude. I am more or less of an artist now, but then I was nothing else" (381).

When Thea leaves home to study music in Chicago, she manages to remain oblivious to the city itself, with its bustling crowds, brilliant shops, and obnoxious loitering men. What grips her imagination is a Jules Breton painting in the Art Institute called *The Song of the Lark*, depicting a peasant girl standing in a field, arrested by the song of a meadowlark. The image reinforces an even more important revelation in the concert hall, when Dvořák's *New World* Symphony reveals to Thea a link between the landscape in her memory and the musician she wants to become. From that moment she understands what she wants, and she leaves determined that "as long as she lived that ecstasy was going to be hers. She would live for it, work for it, die for it; but she was going to have it, time after time, height after height" (171–72).

Thea's first teacher in Chicago, a sensitive, one-eyed Hungarian violinist named Harsanyi, discovers her voice, and steers her away from piano to the voice teacher Madison Bowers. But Thea's demands and ambition are beyond Bowers's reach or interest, and his cynicism and slovenly standards make her depressed and surly. She finds both a champion and a romantic interest in the musical dilettante Fred Ottenburg, whose wealth, social connections, and enthusiastic promotion of her talent recall Isabelle McClung's role in Cather's own life.

Thea's full artistic awakening does not take place in the cold gray canyons of Chicago, where she labors at her music lessons, but in a brilliant desert canyon where Fred sends her to rest and recuperate. There she comes upon an isolated gorge sheltering silent prehistoric ruins and spends weeks lying alone there on the sunbaked rock ledges and in the shade of ancient pueblo rooms. Enfolded in the shelter of the canyon she sheds restrictive clothing and mental debris, bathes naked in the stream at its base, naps under an Indian blanket, and opens every pore until her body becomes completely receptive, a vehicle of sensation. Thus poised, she suddenly recognizes the spiritual connection between the shards of ancient Indian pottery she finds by the stream—vessels designed to bear life-giving water—and her own throat, a vessel which carries song: "what was any art but an effort to make a sheath, a mould in which to imprison for a moment the shining, elusive element which is life itself? . . . In singing, one made a vessel of one's throat and nostrils and held it on one's breath, caught the stream in a scale of natural intervals" (255).

This epiphany takes place in a setting once described by the critic Ellen Moers as "the most thoroughly elaborated female landscape in literature"[18]—a canyon of shelved pink cliffs, "lightly fringed with *piñons* and dwarf cedars" (249). The swift trajectory of an eagle swooping in and out of the gorge, coinciding with Fred's visit to her refuge, reinforces its erotic suggestiveness. But Thea's is not an ordinary desire, nor an ordinary fecundity. She sees her body as a vessel for aesthetic, spiritual aspiration and achievement, not for human reproduction. As she watches the eagle mount and soar above the canyon rim, she springs to her feet and hails the golden bird: "Endeavor, achievement, desire, glorious striving of human art! From a cleft in the heart of the world she saluted it" (269).

From here on, Thea views her vessel—her body—as part of the sacred order of things. Her daily bath in the stream in Panther Canyon "came to have a ceremonial gravity. The atmosphere of the canyon was ritualistic" (254). To Thea, the vessel bearing spiritual gifts deserves to be treated *sacramentally*. Becoming an artist means being able simultaneously to abandon her body to sensuous experience and to control that experience, keeping it from contamination. "The condition every art requires," Cather would later explain, is "freedom from adulteration and from the intrusion of foreign matter,"[19] and at this point, Thea severs those human ties that threaten to compromise her.

The mature Thea Kronborg we meet in the novel's last section is ten years older than the young woman who came of age in Panther Canyon. She has returned from study and successful performances in Germany, and is now a reigning soprano at the Metropolitan Opera. This is a changed Thea, one who speaks noticeably in the experienced voice of Olive Fremstad; Cather's own authorial voice now blends with those of Thea's many admirers.

The diva Thea Kronborg—whose first name means "gift of God" and surname means "crown fortress"—is presented as a woman both blessed and isolated by her divine gift. Her professional crown is won by the resolute defense of her person as the vehicle through which that gift can be perfected and returned. Her family, mentors, and suitors serve Thea the woman only as they serve Thea the artist. She is completely obsessed with the intellectual and physical rewards of her craft. Her regimen is grueling, and her exacting standards make her arrogant and lonely. She is sometimes frightened, and more than once the idea of marrying and being taken care of tempts her. She grieves at the conflict between personal and professional needs, particularly when choosing an important European debut over a journey home to see her dying mother.

But art always comes first. It takes every ounce of her strength, leaving her drained, aged, and often unfit for company. When urged to take more time for her "personal life," she replies, "Your work becomes your personal life. You're not much good until it does (378)."[20] Her work requires the kind of perfect dedication that Nietzsche called *chastity*,[21] and its goal is a paradox, the kind of "sensuous spirituality"[22] which is also the goal of the mystic.

The association of art with religious or spiritual discipline was hardly new to Cather; as we have seen, she had adopted the Romantic view of the artist as mediator between human and divine realms as early as her teens. In her youthful arts reviews she insisted repeatedly that the artist had a sober responsibility to a higher calling, proclaiming in 1895 that "only the veil and the cloister" could keep "the priesthood of art untainted from the world."[23] Given these convictions, it is not surprising that she was attracted to the music of the most influential Romantic artist of his century, Richard Wagner.

Many readers have been puzzled by Cather's taste for Wagner, whom Henry James once called a "ridiculous mixture of Nihilism and bric-a-brac."[24] The composer's crude anti-Semitism and intellectually and morally muddled exhibitionism had compromised his reputation well before he was championed by the Third Reich. Yet Wagner understood better than any composer before or after him the emotional and intellectual potential of opera, and his impact on Western intellectual life—filtered partly through the interdisciplinary Symbolist movement, which adopted him as its patron saint—was pervasive and profound.

Wagner's aesthetic ideas were as complicated and self-contradictory as the man himself, but three interrelated concepts are unmistakably present in Cather's work. One is the influence of Schopenhauer's philosophy, which proposed that since no sensate experience could finally satisfy human cravings, peace could be found only in aesthetic contemplation and/or the renunciation of desire. A second is Wagner's concept of the *Gesamtkunstwerk* (variously translated as "total artwork" or "synthesis of all the arts") and the ideal interrelations among its different formal components. A third is what Carl Dahlhaus calls Wagner's "rigorous artistic morality"—his intolerance of philistinism, and his uncompromising belief in the value of art for its own sake. In an industrial age which had consigned art to a peripheral existence, Wagner promoted the "simultaneously despairing and ecstatic conviction that art is the sole justification for life," elevating it to a religion.[25]

These Wagnerian themes appear throughout Cather's writing, both explicitly (as in her early stories from *The Troll Garden*, where Wagner's music is repeatedly invoked to establish mood and theme) and implicitly (as in her 1922 novel *One of Ours*, in whose structure she deeply embedded the *Parsifal* legend, without ever mentioning either the composer or the opera).[26] Wagner's music is of course much discussed in the final section of *The Song of the Lark,* when Thea performs several Wagnerian roles. But Wagnerian references appear much earlier—starting with the dedication poem to Isabelle McClung—an echo of the passage in "The Prize Song" in *Die Meistersinger von Nürnberg*, in which the singer chooses the woman he wants to marry.[27]

Many other Wagnerian ideas float through the early part of the book—unnamed, like symbolic leitmotifs. As a child, Thea begins secretly to dream of studying music in Germany ("the only place you can really learn") after reading "the strange 'Musical Memories' of the Reverend H. R. Haweis" (58, 75)—a reference surely lost on most of today's readers. Haweis, a British Anglican clergyman and amateur musician, was an ardent Wagnerite who frequently used the operas as subjects of his sermons; over half of *My Musical Memories* (1884) is devoted to his personal reminiscences of the composer and of Bayreuth, and to analyses of the *Ring* and *Parsifal.* Other Wagnerian themes surface later when Thea studies in

Chicago. Her first teacher, the one-eyed, fatherly Harsanyi, plays Wotan to the fearless, rock-climbing valkyrie in Panther Canyon—who will one day sing the role of Wotan's favored daughter Brünnhilde. Thea's lifelong battle against musical mediocrity can be read as a commentary on the song contest in *Die Meistersinger*. Even Thea's first experience of Wagner's music—the Walhalla theme and the famous rainbow bridge that "throb[s] out into the air" in *Götterdammerung*—enters Thea's mind as a leitmotif, fragmentary and tantalizing, "as people hear things in their sleep" (170).

The most important mark of Wagner's influence on this novel is not in its structure or style, but in its philosophy—Cather's idea of the artist, and her conception of the links among art, eroticism, and religion. "Music first came to us," she had proclaimed in 1899, "as a religious chant or a love song," and "through all its evolutions it should always express those two cardinal needs of humanity."[28] Like so many artists of her generation, Cather was bewitched by Wagner's compelling *fusion* of these "two cardinal needs of humanity" in art, and by his efforts to control the risks of ecstasy by subjecting it to artistic form.

Many critics have commented on Cather's tendency to circumvent love stories or to punish erotic relationships between her fictional characters—in most of her novels someone is either humiliated or destroyed by love gone wrong. While this is usually discussed autobiographically, as a sign of Cather's own sexual repression or neurosis, it is Wagner—not Freud—who provides the most useful model for her *artistic* approach to eroticism.[29] A telling example is her early *Troll Garden* story "The Garden Lodge," the tale of a genteel, reserved matron struggling to control her fantasies about a visiting singer with whom she rehearses the Siegmund/Sieglinde love duet in her garden bower. Here the leitmotif is Klingsor's garden from *Parsifal*—a land of sensual delights where diabolical flowers lured chaste knights from their rigorous spiritual path. To Cather, dedicated like Wagner to "the priesthood of art," these dangers were potent and real, and they were epitomized in Klingsor's slave Kundry, the role that made Olive Fremstad famous.

Wagner, however, had found an intellectual way (if not a physical one) around this conflict. In two influential essays, *The Art Work of the Future* (1849) and *Opera and Drama* (1851), he expounded his views on the relationship between art and nature in specifically sexual terms, using procreation as a metaphor for the creative process.

Conventional opera had been exhausted, he argued, because its "component arts" had devolved to empty formulas, had become competitive rather than cooperative. He himself would give birth to a new music drama, born from the "marriage" of two ancient arts—music (the emotional element, the "life-bearing, female organism") joined in "the transports of love" to poetry (the intellectual element, "the fertilizing seed").[30] Neither was adequate by itself. To be complete, "the poetic intent" must be nurtured through the artist's "own necessary essence" and brought forth as "the realizational, redemptive expression of feeling."[31] The "word-poet"'s achievement, in other words, would be measured by what he left *unsaid;* the "tone-poet," or musician, would echo the verbal silences to speak "the unutterable."[32]

Applying the gestation metaphor to artistic creativity had a curious consequence, however: it meant that in the artist, the erotic impulse must be *internalized*—masculine and feminine principles must *merge* in the single person of the creator. The artist thus became *androgynous*—as Wagner put it, "a social being subject to the sexual conditions of both *male* and *female.*"[33] This was an idea that would have important consequences in Cather's conception of Thea Kronborg—and perhaps also for her understanding of herself.

Wagner's use of the sexual metaphor stemmed both from his own erotic disposition and from his view of art as religion. He had recognized early that the most serious craving of his generation was for a new channel for undirected religious emotion, cast adrift when Darwin undermined Biblical teaching about the privileged relation between God and man, and he set about providing such a channel with the *Gesamtkünstwerk*, a wedding of philosophical treatises with mythic structures set to music. It was Wagner who introduced the darkened theater, the hidden orchestra, and who first commanded silence during performances—all strategies designed to give audiences unprecedented intimacy and privacy in which to relate to the sacred mysteries onstage. This was not to be "religious music" but music *as religion*, a transformative experience that would ravish the soul.[34]

But with Wagner nothing was ever simple. He himself was no ascetic, and in *Tristan und Isolde*—his astonishing "revision" of Schopenhauer—he proposed obsessive sexual love as a kind of "redemption" from desire, embracing and flouting the idea of renunciation in the same breath. *Parsifal*, his most ritualistic drama about the battle between sensuality and spirituality, was an arcane mix of Schopenhauerian and Christian ideas which Wagner termed a *Bühnenweihfestspiel*, or "festival of consecration in a theater."[35]

Indeed, it was precisely his slippery mix of sensuality and spirituality that most aroused Wagner's audiences: the libretto spoke of salvation, but the music spoke quite a different language. Strait-laced Victorians listening to the thrusting Venusberg theme, to the long, unclimaxed raptures of *Tristan*'s *Liebestod*, or to Siegmund and Sieglinde's incestuous melting into the spring night found themselves both aroused and frustrated—involved, his biographer Robert Gutman observes, in "a dream of purity and renunciation as they embraced the flesh." The confusion was quite deliberate, produced by the composer's own ambivalence, and it was the secret of Wagner's power.[36]

Wagner had an enormous cult following during the years Cather was coming of age, and most of it was female. As Joseph Horowitz has documented, Wagner spoke particularly to Gilded Age women of passionate sensibility—young girls never instructed about sexuality, neurasthenic housewives shackled by genteel breeding and decorum, and independent, sometimes emotionally isolated or closeted New Women. For them Wagner's erotically charged scores were a source of violent excitation, provoking passionate feelings that felt dangerous but came without the risks of personal relationships.[37]

Willa Cather was emphatically one of their number. A *Troll Garden* story, "The Wagner Matinee," shows her familiarity with the urban Wagner societies specializing in matinee concerts, where women could indulge in their romantic fantasies unescorted if they chose.[38] She was also well acquainted with the Bayreuth Wagner groupies, and in *The Song of the Lark*, casts Fred Ottenburg's mother as one—a member of the flock of "young women who followed Wagner about in his old age. . . . When the composer died [she] took to her bed and saw no one for a week" (237).[39]

Cather was not only a Wagner fan, however; she was also a student, who continued to hone and develop her own Wagnerian aesthetic into the 1920s. Her Fremstad profile makes clear that she understood and emulated the ideal relationship between description and suggestion that Wagner believed essential for drama, which he defined as "an act of music made visible."[40] Wagner understood,

Cather wrote, that the libretto "must retain the simplicity of a legend, that the characters must be indicated rather than actualized." It was the music that made the poem "flower," that drew the legend out "from the low relief of archaic simplicity" to make it "present and passionate and personal."[41] In Olive Fremstad's histrionic minimalism—her ability to convey emotion through music alone, ignoring the melodramatic conventions of her time—Cather felt she had witnessed the perfect realization of the composer's aims.

By 1922—the same year she published *One of Ours*, with its unnamed *Parsifal* motif—Cather had developed a minimalist aesthetic of her own, called "the unfurnished novel" ("novel démeublé"), and she used musical language to describe it:

> Whatever is felt on the page without being specifically named there—that, one might say, is created. It is the inexplicable presence of the thing not named, of the overtone divined by the ear but not heard by it, the verbal mood, the emotional aura of the fact or the thing or the deed, that gives high quality to the novel or the drama, as well as to poetry itself."[42]

Cather's work on *The Song of the Lark* came at the midpoint of her long intellectual relationship with Wagner. This novel is not minimalist and spare, as Fremstad's acting was and as Cather's later books would be; its Wagnerism is expansive and lush, full of gorgeous sensuality, like the operas themselves. And like the operas, too, Cather's text tells about chaste dedication to a spiritual goal, while its "emotional aura"—the "overtone divined by the ear but not heard by it"—speaks a very different language.

There are, for example, no explicit sexual encounters in *The Song of the Lark*: although Thea is always surrounded by admiring men, their lack of prurient interest in her is one of the most remarkable things about them. There is, however, a powerful erotic undercurrent in the book, beginning with Thea's intense physical response in childhood to the natural world. The sensuality is diffused—like music—enveloping all her physical experience, rather than being directed at a particular human object.

The Panther Canyon episode only culminates a lifelong pattern of such responses, all experienced in solitude, and all translated into artistic emotion. On her thirteenth birthday, when she is "shaken with passionate excitement" after an intense music lesson with Wunsch, Thea wanders about, "looking into the yellow prickly-pear blossoms with their thousand stamens," looking at the sand hills "until she wished she *were* a sand hill" (71). At fifteen, after a moving discussion with Dr. Archie on a languid summer night, she lies on the floor by her bedroom window "vibrating with excitement" while "her chest ached and it seemed as if her heart were spreading all over the desert." Thea is not "sentimental" about her mentors, as some young girls might be—she is a Wagnerian androgyne, containing all generativity in herself. To make sure we understand this, Cather's authorial voice intervenes to explain: "There is no work of art so big or so beautiful that it was not once all contained in some youthful body, like this one which lay on the floor in the moonlight, pulsing with ardor and anticipation" (122).

Wunsch, who notices Thea's androgynous quality early, compares her to the prickly-pear blossoms of the desert—"thornier and sturdier than the maiden flowers" he remembered from his youth (87). Years later Fred Ottenburg observes it as a quality of her voice " 'the high voice we dream of; so pure and yet so virile and human' " (349).[43] Indeed, the fact that descriptions of the act of singing—of translating music through the physical body—are among this novel's most erotic passages is highest Wagnerism on Cather's part. Harsanyi, laying his hand on Thea's throat to evaluate the newly discovered voice, responds to the experience like a lover with a virgin: "He loved to hear a big voice throb in a relaxed, natural throat, and he was thinking that no one had ever felt this voice vibrate before" (159–60).[44] When Thea spends an evening singing with her Mexican friends in Moonstone, Cather transposes male and female imagery to describe how, "at the appointed, at the acute, moment, the soprano voice, like a fountain jet, shot up into the light. . . . How it leaped from among those dusky male voices!" (199). In the "female landscape" of Panther Canyon, Thea's dream-rapture is described orgasmically but without climax, recalling the music of *Tristan*: "A song would go through her head all morning, as a spring keeps welling up, and it was like a pleasant sensation indefinitely prolonged" (251).

In *The Art Work of the Future*, Wagner defined true dramatic action as "*a bough from the Tree of Life*" whose fruit had been "*planted in the soil of Art.*"[45] Such artistic procreation is also the theme of Walther's "Prize Song" in *Die Meistersinger*, and Thea Kronborg absorbs this Wagnerian lesson deeply. In perfect accommodation, she reaches the pinnacle of her powers as an artist who communicates the spiritual message of art through her very sensual human body. In her final performance in the novel, a transcendent portrayal of Sieglinde in *Die Walküre*, that body becomes "absolutely the instrument of her idea. Not for nothing had she kept it so severely, kept it filled with such energy and fire. All that deep-rooted vitality flowered in her voice, her face, in her very fingertips. She felt like a tree bursting into bloom" (395).

In an equally Wagnerian gesture, Cather uses theological language to explain how such a thing could happen: Thea had merely "entered into the inheritance that she herself had laid up, into the fullness of the faith she had kept before she knew its name or its meaning" (395).

Cather was pleased with the manuscript she sent off in March 1915 to Ferris Greenslet, her editor at Houghton Mifflin, but, to her consternation, Greenslet wanted changes. The early parts, he felt, were too long; there was too much predictable detail about Thea's student years in Germany and about Dr. Archie's politics. And the book seemed to have a split personality—while the first parts were scrupulously realistic, the last sections were highly romanticized.

Cather's feathers were ruffled. She had been a rigorous student of a singer's career, after all—attending lessons, rehearsals, and performances, consulting with voice teachers, critics, and performers about technical details; she knew what she was doing. She agreed to the cuts about Germany and Dr. Archie, but put her foot down about the rest. Greenslet didn't press further, and the book was issued in October of that year.[46]

Critical response was mostly enthusiastic. Especially important to Cather, Fremstad praised it—particularly the early chapters about childhood and Panther Canyon; the last sections, too, were treated seriously, making it the only novel Fremstad had read about an artist in which there was "something doing" in the character.[47] The influential Germanophile H. L. Mencken declared that *The Song of the Lark* placed Cather in "the small class of American novelists who are seriously to be reckoned with." The public clearly admired the novel's strengths, as it went through nineteen reprintings—more than one per year—between 1915 and 1936.

But there were some demurring voices. Though the quality of characterization was widely praised, most reviewers found the book overwritten. Randolph Bourne, whose influence was nearly as important as Mencken's, ventured that the author would "probably be shocked" if she realized "how sharp were the contrasts between those parts of her book which are built out of her own experience and those which are imagined." The opera singer who so fascinated Cather had been "admired," not "assimilated"; Cather had not made the character fully her own.[48] More seriously still, Cather's British publisher, William Heinemann, turned the book down. While he admired the economy of *O Pioneers!*, Heinemann explained, this was the "distressingly familiar" kind of book that "told everything about everybody," and he felt she had taken a wrong turn.[49]

For a long time Cather felt so personally entangled with her double-portrait-heroine that she could not accept these criticisms, and when she eventually did understand them she seems to have been embarrassed. In 1932 she made two adjustments. She deleted the personal dedicatory poem (but not the dedication itself) to Isabelle McClung.[50] She also provided the book with a new preface (see Appendix), in which she claimed that the book's "chief fault" was in describing "a descending curve; the life of a successful artist in the full tide of achievement is not so interesting as the life of a talented young girl 'fighting her way.' " But to Greenslet she wrote something else: she wanted him to quash rumors that her character had any direct relation to Olive Fremstad. Knowing Fremstad had only helped make her aware of the routine of a singer's life, she protested; no one incident in the book had any basis in Fremstad's life.[51]

This was dissimulation, of course: one of Cather's many attempts, as her biographer James Woodress notes, to revise her own biography for the sake of vanity or privacy.[52] Her major books of the 1920s—*A Lost Lady*, *The Professor's House*, *My Mortal Enemy*, and *Death Comes for the Archbishop*—had all described the "descending curves" of mature protagonists, and brilliantly. The problem was something else: Cather herself had fallen short of the ideal that Thea had reached. Watching Thea perform for the first time, Dr. Archie had been "chagrined" that he could not recognize the person he thought he knew so well—the role "had somehow devoured" his friend, leaving "his personal, proprietary pride in her . . . frozen out" (344). Cather was chagrined by the opposite: her own self-projection via Fremstad into Thea had proved a less successful hybrid, a mixture, as Hermione Lee observes, of wish-fulfillment, hero-worship, and (perhaps worst of all) self-exposure.[53]

In 1937, at the age of sixty-four, Cather revised the novel once more for a new Autograph Edition of her work, and this edition—shorter by 6,900 words—has been the one reissued by Houghton Mifflin ever since. The bulk of the cuts were made in the last three sections, "Dr. Archie's Venture," "Kronborg," and the Epilogue. She deleted some dated references and opinions, and some dialogue that now sounded immature. She cut back further on Dr. Archie's business affairs in Colorado and the discussions of Fred and Aunt Tillie in the Epilogue.[54]

She also made some adjustments in favor of discretion. The description of Thea's accompanist's fussy Greenwich Village apartment is heavily pruned; gone too is a suggestive allusion to his homosexuality. Several of Fred's more sexually suggestive remarks disappear, together with a significant number of rapturous passages about Thea's physical beauty.

The revisions streamlined the final chapters somewhat, tilting the book's emotional weight even more toward Thea's early life. They also represent a mature author's censorship of the buoyant, indiscriminate enthusiasm of her own earlier infatuation with a singer.[55] Fortunately, perhaps, Cather realized she could not fully repair the novel's disjunction without a complete rewriting, so its split personality remains in both versions.

Thea Kronborg, of course, struggles with an artistic problem very similar to Willa Cather's. She, too, must learn to traverse the "break" in the voice, the shift between registers which is every singer's greatest vulnerability. Vulnerability, however, can be an enormously appealing quality in a singer, and some fans value the "break" in the voice as an exciting listening experience—a dangerous, even erotic moment.[56] Fred, who marvels at Thea's handling of Elsa's balcony scene because it lies " 'just on the "break" in the voice' " (348), seems to be aware of this. The "break" in this novel—the place where Cather's and Fremstad's stories imperfectly merge—can be seen as a defect, like the tricky break in the singer's voice. But it is also a sign of a very gifted writer's daring, her willingness to take risks—and that, of course, only adds to the novel's seductive power.

SHERRILL HARBISON
Amherst, Massachusetts
September 1998

NOTES

1. From an interview published in the *Philadelphia Record*, August 10, 1913, included in *Willa Cather in Person: Interviews, Speeches, Letters*, ed. L. Brent Bohlke (Lincoln: University of Nebraska Press, 1986), p. 10.

2. *Willa Cather in Person*, pp. 10–11.

3. "Concerning Thomas Carlyle" was published simultaneously in the campus literary magazine, the *Hesperion*, and in the *Nebraska State Journal* on March 1, 1891. It is reprinted in *The Kingdom of Art: Willa Cather's First Principles and Critical Statements, 1893–1896*, ed. Bernice Slote (Lincoln: University of Nebraska Press, 1966), pp. 421–25.

4. Cather reviewed the Metropolitan tour for the Lincoln *Courier* on May 12, 1900 (p. 11), reprinted in volume 2 of *The World and the Parish: Willa Cather's Articles and Reviews, 1893–1902*, ed. William M. Curtin (Lincoln: University of Nebraska Press, 1970), pp. 655–58.

5. She rhapsodized about Julio in letters to Sergeant. See Elizabeth Shepley Sergeant's *Willa Cather: A Memoir* (1953) (Lincoln: University of Nebraska Press/Bison, 1963), pp. 80–81.

6. In the copy of *O Pioneers!* she presented to her Red Cloud friend Carrie Miner Sherwood, Cather wrote: "This was the first time I walked off on my own two feet—everything before was half real and half imitation of writers whom I admired. In this one I hit home pasture and found that I was Yance Sorgeson [a local Norwegian farmer] and not Henry James." See James Woodress, *Willa Cather: A Literary Life* (Lincoln: University of Nebraska Press, 1987), pp. 239–40.

7. In a 1921 interview in the *Lincoln State Journal* Cather remarked, "I often recall what Sarah Orne Jewett said to me many years ago . . . that a knowledge of the world was needed in order to understand the parish." *Willa Cather in Person*, p. 40.

8. Sergeant, *Willa Cather: A Memoir*, p. 111.

9. From 1903 until her retirement in 1914, Fremstad appeared at the Met as Sieglinde, Brünnhilde, Fricka, Isolde, Brangaene, Venus, Elisabeth, Elsa, and Kundry, as well as in half a dozen non-Wagnerian roles. Her merciless interpretation of Strauss's *Salome*—a character even more depraved than Kundry, and still worse, unredeemed—caused the opera to be shut down after its first performance. For a discussion of the vehement aesthetic and moral debate that took place around this aborted production, see Joseph Horowitz, *Wagner Nights*, pp. 283–85, and John Dizikes, *Opera in America*, pp. 315–16. Interestingly, Cather—who disliked Strauss, comparing him unfavorably to Wagner in "The Case of Richard Strauss" (1904)—never mentioned the furor in her profile.

10. Willa Cather, "Three American Singers," *McClure's Magazine*, December 13, 1913, p. 42.

11. Woodress, *Willa Cather: A Literary Life*, p. 253.

12. Cather, "Three American Singers," p. 42.

13. Ibid.

14. Sergeant, *Willa Cather: A Memoir*, p. 3.

15. Mary Watkins Cushing, *The Rainbow Bridge* (New York: G. P. Putnam's Sons, 1954), p. 244.

16. Ibid., pp. 71–75.

17. Sergeant, *Willa Cather: A Memoir*, pp. 98–99.

18. Ellen Moers, *Literary Women* (Garden City, N.Y.: Doubleday, 1976), p. 252.

19. From Cather's essay "Escapism" (1936), collected in *Willa Cather on Writing*, foreword by Stephen Tennant (Lincoln and London, University of Nebraska Press/Bison, 1988), p. 26.

20. Compare Cather on Fremstad: "with Mme. Fremstad one feels that the idea is always more living than the emotion; perhaps it would be nearer the truth to say that the idea is so intensely experienced that it becomes emotion" ("Three American Singers," p. 46), and Richard Wagner's statement about himself: "I know nothing about the real enjoyment of life: for me the 'enjoyment of life and love' is something I have only imagined, not experienced. So my heart has had to move into my head and my living become artificial; it is only as 'the artist' that I can live, as a 'man' I have been completely absorbed into him." John Deathridge and Carl Dahlhaus, *The New Grove Wagner* (New York and London: W. W. Norton, 1984), p. 94.

21. In the *Genealogy of Morals* (1887), Nietzsche, reflecting on Wagner's interest in asceticism, explained that "there is no necessary antithesis between chastity and sensuality; . . . every authentic heart-felt love transcends this antithesis." *The Philosophy of Nietzsche*, trans. Horace B. Samuel (New York: Modern Library, 1954), pp. 718–19.

22. Richard Giannone describes Cather's artistic sensibility as a "sensuous spirituality," a phrase Cather herself had used in 1897 to describe the music of Massenet. "Willa Cather and the Human Voice," *Five Essays on Willa Cather: The Merrimack Symposium*, ed. John J. Murphy (North Andover, Mass.: Merrimack College, 1974), p. 24.

23. From Cather's discussion of the actress Eleonora Duse, published in the *Nebraska State Journal*, June 16, 1895, p. 12; reprinted in *The World and the Parish*, vol. 1, p. 208.

24. Robert W. Gutman, *Richard Wagner: The Man, His Mind, and His Music* (New York: Harcourt Brace Jovanovich, 1990), p. 403.

25. Deathridge and Dahlhaus, *The New Grove Wagner*, pp. 94–95.

26. In correspondence Cather admitted that she had originally planned to call the final section of *One of Ours* "The Blameless Fool By Pity Enlightened," after *Parsifal*. Woodress, *Willa Cather: A Literary Life*, p. 328.

27. In act 3, scene 2 of *Die Meistersinger,* Walther composes "The Prize Song," with which he hopes to win his beloved Eva:

> *Morgendlich leuchtend in rosigem Schein,*
> *von Blüt' und Duft*
> *geschwellt die Luft,*
> *voll aller Wonnen*
> *nie ersonnen,*
> *ein Garten lud mich ein,*
> *Gast ihm zu zein.*
>
> (Shining in the rosy light of morning,
> the air heavy
> with blossom and scent,
> ful of every
> unthought-of joy,
> a garden invited me
> to be its guest.)

28. From an open letter to the Wagnerian soprano Lillian Nordica, published in the Lincoln *Courier*, December 16, 1899, p. 3. *The World and the Parish*, vol. 2, p. 645.

29. For examples of the autobiographical argument, see Blanche H. Gelfant's "The Forgotten Reaping Hook: Sex in *My Ántonia*" in *Critical Essays on Willa Cather*, ed. John J. Murphy (Boston: G. K. Hall, 1984), pp. 147–64; and Sharon O'Brien, *Willa Cather: The Emerging Voice* (New York and Oxford: Oxford University Press, 1987), chap. 9. Cather was explicit and vehement in her dislike of Freud, however, and would surely have found these interpretations offensive. See Sergeant's *Willa Cather: A Memoir,* pp. 163–64 and 239, and Cather's own 1922 essay "Miss Jewett" in *Not Under Forty* (Lincoln: University of Nebraska Press/Bison, 1988), p. 93.

Surprisingly, the link with Wagner has not been much examined in Cather criticism. The best discussions are in William Blissett's "Wagnerian Fiction in English" (*Criticism* 5 [summer 1963], pp. 239–60); Richard Giannone's *Music in Willa Cather's Fiction* (Lincoln: University of Nebraska Press, 1968); Hermione Lee's *Willa Cather: Double Lives* (New York: Vintage, 1989); and Mary Jane Humphrey's "The White Mulberry Tree as Opera," in *Cather Studies*, vol. 3, ed. Susan J. Rosowski (Lincoln: University of Nebraska Press, 1996), pp. 51–66.

30. Wagner's *Opera and Drama* constitutes vol. 2 of *Richard Wagner's Prose Works*, trans. W. Ashton Ellis (1893; reprint, New York: Broude Brothers, 1966). This concept appears on p. 111. I have here cited Stewart Spencer's translation of a Wagner letter describing his essay, cited in Jean-Jacques Nattiez, *Wagner Androgyne* (Princeton: Princeton University Press, 1993), p. 38.

31. Nattiez, *Wagner Androgyne*, p. 36.

32. Wagner, *Opera and Drama*, pp. 276, 317.

33. Ibid., p. 107; see also Nattiez, *Wagner Androgyne*, p. 41.

34. *The Art Work of the Future*, vol. 1 of *Richard Wagner's Prose Works*, trans. W. Ashton Ellis (1895; reprint, Lincoln: University of Nebraska Press, 1993), pp. 155 and *passim.*

35. Deathridge and Dahlhaus, *The New Grove Wagner*, p. 161.

36. Gutman, *Richard Wagner*, p. 253.

37. Joseph Horowitz, *Wagner Nights: An American History* (Berkeley, Los Angeles, and London: University of California Press, 1994), pp. 215, 226.

38. For a discussion of the popular Wagner matinees, see ibid., pp. 191–92.

39. One such Wagner acolyte was Mabel Dodge Luhan, the *bohémienne* socialite who later would become one of Cather's friends. As a young woman attending the Bayreuth Festival, Luhan sat holding her girlfriends' burning hands in the darkened hall, her heart beating wildly. When people listened to Wagner, she intuited, "they were only listening to their own impatient souls, weary at last of the restraint that had held them." Luhan admitted she left Bayreuth in "a colorless depression"; after such transports, "everything in life seemed flat and hopeless" (Joseph Horowitz, *Wagner Nights*, p. 218). Another visitor, Mark Twain, marveled in 1891 over the Bayreuth "pilgrims" who attended daily "services," returning to their hotels with "heart and soul and . . . body exhausted by long hours of tremendous emotion, . . . in no fit condition to do anything but lie torpid and slowly gather back life and strength for the next service. . . . [A]lways during service, I feel like a heretic in heaven. But by no means do I ever overlook or minify the fact that this is one of the most extraordinary experiences of my life. . . . I have never seen anything so great and fine and real as this devotion" ("At the Shrine of St. Wagner," in *What Is Man? and Other Essays* [New York: Harper, 1917], pp. 226–27.)

40. Deathridge and Dahlhaus, *The New Grove Wagner*, p. 77.

41 Cather, "Three American Singers," p. 46.

42. "The Novel *Démeublé*," in *Willa Cather on Writing*, pp. 41–42.

43. For a discussion of Thea's androgynous voice, see John H. Flanni-

gan, "Thea Kronborg's Vocal Transvestitism: Willa Cather and the 'Voz Contralto,' " *Modern Fiction Studies* 40:4 (Winter 1994): 737–63.

44. For discussions of the physiological similarities between singing and orgasm, see Wayne Koestenbaum, *The Queen's Throat*, chapter 5, and Elizabeth Wood, "Sapphonics," in *Queering the Pitch*, pp. 27–66.

45. Wagner, *The Art Work of the Future*, p. 197.

46. Woodress, *Willa Cather: A Literary Life*, pp. 260–61.

47. Ibid., p. 271.

48. Ibid., p. 274.

49. Willa Cather, "My First Novels (There Were Two)," in *Willa Cather on Writing*, p. 96.

50. By 1932, Cather had adjusted to Isabelle's unexpected and unwelcome marriage to the violinist Jan Hambourg in 1916, the year after *The Song of the Lark* was published; but a poem that was also a proposal was no longer appropriate in the dedication.

51. Woodress, *Willa Cather: A Literary Life*, p. 273.

52. Ibid., p. xiv.

53. Hermione Lee, *Willa Cather: Double Lives* (New York: Vintage Books, 1991), p. 124.

54. The first version of the novel comprised 200,000 words; Houghton Mifflin published it at 146,000 after Cather made the first round of cuts. All but 153 of the 6,900 new cuts and changes were made in the novel's last sections. For more detailed discussion see Robin Heyeck and James Woodress, "Willa Cather's Cuts and Revisions in *The Song of the Lark*," *Modern Fiction Studies* 25 (Winter 1979–80): 651–58.

55. Cather was hardly alone in her fascination with Fremstad. James Gibbons Huneker (1860–1921), distinguished essayist and critic for the New York *Sun* during most of Fremstad's American career, also used her as inspiration for some undistinguished fiction. His 1896 story "The Last of the Valkyries" was collected in *Bedouins* as "Venus or Valkyr" (1926). In his 1920 novel, *Painted Veils*, the soprano Esther Brandes is a composite of Fremstad and soprano Mary Garden (1874–1967). Novelist Gertrude Atherton (1857–1948) based her singer, Marguerite Styr, on Fremstad in *Tower of Ivory* (1910). Marcia Davenport (1903–94) drew a combined portrait of Fremstad and Davenport's mother, the soprano Alma Gluck (1884–1938), in her 1936 novel, *Of Lena Geyer*.

56. For a discussion of the erotic potential of the voice-break and its implications for *The Song of the Lark* see Elizabeth Wood, "Sapphonics," in *Queering the Pitch*, pp. 27–66.

SUGGESTIONS FOR FURTHER READING

WILLA CATHER ON MUSIC AND MUSICIANS

(*Listed in Order of First Uncollected Publication*)

Willa Cather's Collected Short Fiction, 1892–1912. Edited by Virginia Faulkner. Rev. ed. Lincoln: University of Nebraska Press, 1970.

The Kingdom of Art: Willa Cather's First Principles and Critical Statements, 1893–1896. Edited by Bernice Slote. Lincoln: University of Nebraska Press, 1966.

The World and the Parish: Willa Cather's Articles and Reviews, 1893–1902. 2 vols. Edited by William M. Curtin. Lincoln: University of Nebraska Press, 1970.

"The Case of Richard Strauss" (1904). *Prairie Schooner* 55:1–2 (Spring–Summer 1981): 24–33.

The Troll Garden (1905). Afterword by Katherine Anne Porter. New York: Meridian, 1984.

"Three American Singers." *McClure's* (December 1913): 33–48.

Youth and the Bright Medusa. New York: Knopf, 1920.

One of Ours. New York: Knopf, 1922.

"Gertrude Hall's *The Wagnerian Romances*" (1925). In *On Writing: Critical Studies on Writing as an Art*. Foreword by Stephen A. Tennant. Lincoln: University of Nebraska Press, 1988.

My Mortal Enemy. New York: Knopf, 1926.

Uncle Valentine and Other Stories: Willa Cather's Uncollected Short Fiction, 1915–1929. Edited by Bernice Slote. Lincoln: University of Nebraska Press, 1973.

Lucy Gayheart. New York: Knopf, 1935.

Willa Cather in Person: Interviews, Speeches, and Letters. Edited by L. Brent Bohlke. Lincoln: University of Nebraska Press, 1986.

ON WILLA CATHER AND THE SONG OF THE LARK

Bennett, Mildred R. "The Childhood Worlds of Willa Cather." *Great Plains Quarterly* 2:4 (Fall 1982): 204–9.

———. "Willa Cather in Pittsburgh." *Prairie Schooner* 33 (Spring 1959): 64–76.

———. *The World of Willa Cather* (1951). New edition with notes and index. Lincoln, Neb.: Bison, 1961.

Blissett, William. "Wagnerian Fiction in English." *Criticism* 5 (Summer 1963): 239–60.

Brown, E. K., and Leon Edel. *Willa Cather: A Critical Biography.* New York: Knopf, 1953.

Byrne, Kathleen D., and Richard C. Snyder. *Willa Cather in Pittsburgh, 1896–1906.* Pittsburgh: Historical Society of Western Pennsylvania, 1980.

Crane, Joan. *Willa Cather: A Bibliography.* Lincoln: University of Nebraska Press, 1982.

Dubek, Laura. "Rewriting Male Scripts: Willa Cather and *The Song of the Lark.*" *Women's Studies* 23:4 (1994): 293–306.

Duryea, Polly Patricia. "Paintings and Drawings in Willa Cather's Prose: A Catalogue Raisonné." Diss. University of Nebraska, Lincoln, 1993.

Fetterley, Judith. "Willa Cather and the Fiction of Female Development." In *Anxious Power: Reading, Writing, and Ambivalence in Narrative by Women.* Edited by Carol J. Singley and Susan Elizabeth Sweeney. Albany, N.Y.: SUNY Press, 1993.

Flannigan, John H. "Thea Kronborg's Vocal Transvestitism: Willa Cather and the 'Voz Contralto.' " *Modern Fiction Studies* 40:4 (Winter 1994): 737–63.

Foster, Shirley. "The Open Cage: Freedom, Marriage, and the Heroine in Early Twentieth Century American Women's Novels." In *Women's Writing: A Challenge to Theory*, edited by Moira Monteith. (Brighton, England: Harvester Press, 1986).

Giannone, Richard. *Music in Willa Cather's Fiction.* Lincoln: University of Nebraska Press, 1968.

———. "Willa Cather and the Human Voice." In *Five Essays on Willa Cather: The Merrimack Symposium*, edited by John J. Murphy. North Andover, Mass.: Merrimack College, 1974.

Hallgarth, Susan A. "The Woman Who Would Be Artist in *The Song of the Lark* and *Lucy Gayheart.*" In *Willa Cather: Family, Community, and History: The BYU Symposium*, edited by John J. Murphy. Provo, Utah: Brigham Young University Humanities Publication Center, 1990.

Hoover, Sharon. "The Wonderfulness of Thea Kronborg's Voice." *Western American Literature* 30:3 (Fall 1995): 257–74.

Huf, Linda. *A Portrait of the Artist as a Young Woman.* New York: Ungar, 1983.

Humphrey, Mary Jane. "The White Mulberry Tree as Opera." In *Cather Studies*, vol. 3, edited by Susan J. Rosowski. Lincoln: University of Nebraska Press, 1996.

Lee, Hermione. *Willa Cather: Double Lives.* New York: Vintage, 1989.

Leonardi, Susan J. "To Have a Voice: The Politics of the Diva." *Perspectives on Contemporary Literature* 13 (1987): 65–72.

Lewis, Edith. *Willa Cather Living* (1953). Foreword by Marilyn Arnold. Athens: Ohio University Press, 1989.

March, John. *A Reader's Companion to the Fiction of Willa Cather*, edited by Marilyn Arnold. Westport, Conn.: Greenwood Press, 1993.

Martin, Mary Anne. "Music and Willa Cather." In *Willa Cather Yearbook I*, edited by Debbie A. Hansen. Lewiston, Me: Edwin Mellen Press, 1993.

Mencken, H. L. *"Song of the Lark."* In *Willa Cather and Her Critics*, edited by James S. Schroeter. Ithaca: Cornell University Press, 1967.

Moers, Ellen. *Literary Women.* Oxford and New York: Oxford University Press, 1976.

Moseley, Ann. "A New World Symphony: Cultural Pluralism in *The Song of the Lark* and *My Ántonia.*" *Willa Cather Pioneer Memorial Newsletter* 39:1 (Spring 1995): 1 ff.

O'Brien, Sharon. *Willa Cather: The Emerging Voice.* Oxford and New York: Oxford University Press, 1987.

Pannill, Linda. "Willa Cather's Artist Heroines." *Women's Studies* 11 (December 1984): 223–32.

Peck, Demaree. "Thea Kronborg's 'Song of Myself': The Artist's Imaginative Inheritance in *The Song of the Lark.*" *Western American Literature* 26 (1991): 21–38.

Pers, Mona. *Willa Cather's Swedes.* Västerås: Målardalen University Press, 1995.

Randall, John H., III. *The Landscape and the Looking Glass: Willa Cather's Search for Value.* Boston: Houghton Mifflin, 1960.

Rosowski, Susan J. *The Voyage Perilous: Willa Cather's Romanticism.* Lincoln: University of Nebraska Press, 1986.

———. "Willa Cather and the Intimacy of Art, or: In Defense of Privacy." *Willa Cather Pioneer Memorial Newsletter* 36:4 (Winter 1992–93): 47–53.

———. "Willa Cather's Female Landscapes: *The Song of the Lark* and *Lucy Gayheart.*" *Women's Studies* 11:3 (1984): 233–46.

———. "Willa Cather's Women." *Studies in American Fiction* 9:2 (Autumn 1981): 261–75.

———. "Writing Against Silences: Female Adolescent Development in Novels of Willa Cather." *Studies in the Novel* 21:1 (Spring 1989): 60–77.

Roulston, Robert. "The Contrapuntal Complexity of Willa Cather's *The Song of the Lark.*" *Midwest Quarterly* 17:4 (Summer 1976): 350–68.

Schwind, Jean. "Fine and Folk Art in *The Song of the Lark*: Cather's Pictorial Sources." In *Cather Studies*, vol. 1, edited by Susan J. Rosowski. Lincoln: University of Nebraska Press, 1990.

Sergeant, Elizabeth Shepley. *Willa Cather: A Memoir* (1953). Lincoln: University of Nebraska Press/Bison, 1963.

Skaggs, Merrill Maguire. "Key Modulations in Cather's Novels About Music." *Willa Cather Pioneer Memorial Newsletter* 39:2–3 (Summer–Fall 1995): 25–30.

Stouck, David. *Willa Cather's Imagination.* Lincoln: University of Nebraska Press, 1975.

Wasserman, Loretta. "The Lovely Storm: Sexual Initiation in Two Early Willa Cather Novels." *Studies in the Novel* 14:4 (Winter 1982): 348–58.

Wood, Elizabeth. "Sapphonics." In *Queering the Pitch: The New Gay and Lesbian Musicology*, edited by Philip Brett, Elizabeth Wood, and Gary C. Thomas. New York and London: Routledge, 1994.

Woodress, James. *Willa Cather: A Literary Life.* Lincoln: University of Nebraska Press, 1987.

Woodress, James, and Robin Heyeck. "Willa Cather's Cuts and Revisions in *The Song of the Lark.*" *Modern Fiction Studies* 25 (Winter 1979–80): 651–58.

ON PLACE AND CULTURAL CONTEXT

Brownell, Gertrude Hall. *The Wagnerian Romances.* Introduction by Willa Cather. New York: Knopf, 1925.

Colton, Harold Sellers. *Black Sand: Prehistory of Northern Arizona.* Albuquerque: University of New Mexico Press, 1960.

Cushing, Mary Watkins. *The Rainbow Bridge.* New York: Putnam, 1954.

Deathridge, John, and Carl Dahlhaus. *The New Grove Wagner.* New York and London: W. W. Norton, 1984.

Dizikes, John. *Opera in America: A Cultural History.* New Haven: Yale University Press, 1993.

Dorset, Phyllis Flanders. *The New Eldorado: The Story of Colorado's Gold and Silver Rushes.* New York: Macmillan, 1970.

Ellis, Richard N., and Duane A. Smith. *Colorado: A History in Photographs.* Niwot, Colo.: University Press of Colorado, 1991.

Furness, Raymond. *Wagner and Literature.* Manchester, England: Manchester University Press; New York: St. Martin's Press, 1982.

Gutman, Robert W. *Richard Wagner: The Man, His Mind, and His Music.* New York: Harcourt Brace Jovanovich, 1990.

Haweis, Hugh Reginald. *My Musical Memories* (1884). London and New York: Funk & Wagnalls, 1892.

Horowitz, Joseph. *Wagner Nights: An American History.* Berkeley: University of California Press, 1994.

Johnsgard, Paul A. *This Fragile Land: A Natural History of the Nebraska Sandhills.* Lincoln: University of Nebraska Press, 1995.

Koestenbaum, Wayne. *The Queen's Throat: Opera, Homosexuality, and the Mystery of Desire.* New York: Vintage, 1994.

Kramer, Lawrence. *Music and Cultural Practice, 1800–1900.* Los Angeles and Berkeley: University of California Press, 1990.

Krehbiel, Henry Edward. *A Book of Operas.* Garden City, N.Y.: Garden City Publishing Co., 1909.

———. *Studies in Wagnerian Drama.* New York and London: Harper, 1891.

Lister, Robert H., and Florence C. Lister. *Those Who Came Before: Southwestern Archeology in the National Park System.* Tucson: University of Arizona Press, 1983.

May, Henry F. *The End of American Innocence: A Study of the First Years of Our Own Time, 1912–1917.* New York: Knopf, 1959.

Metropolitan Opera Annals. Compiled by William H. Seltsam. New York: H. W. Wilson Co., in association with the Metropolitan Opera Guild, 1947.

Nattiez, Jean-Jacques. *Wagner Androgyne.* Translated by Stewart Spencer. Princeton: Princeton University Press, 1993.

Newman, Ernest. *Wagner Nights.* London: Putnam, 1949.

Noble, David Grant. *Wupatki and Walnut Canyon: New Perspectives on History, Prehistory, and Rock Art.* Santa Fe: School of American Research, 1987.

Shaw, George Bernard. *The Perfect Wagnerite: A Commentary on the Nibelungs' Ring* (1898). New York: Time-Life, 1972.

Showalter, Elaine. *Sexual Anarchy: Gender and Culture at the Fin de Siècle.* New York: Viking, 1990.

Thybony, Scott. *Walnut Canyon National Monument.* 2nd ed. Tucson: Southwest Parks and Monuments Association, 1996.

Twain, Mark. "At the Shrine of St. Wagner." In *What Is Man? and Other Essays.* New York: Harper, 1917.

Viskochill, Larry A., with the assistance of Grant Talbot Dean. *Chicago at the Turn of the Century in Photographs.* New York: Dover, 1984.

Wagner, Richard. *The Art Work of the Future and Other Works.* Translated by W. Ashton Ellis (1895). Vol. 1 of *Richard Wagner's Prose Works.* Lincoln: University of Nebraska Press, 1993.

———. *Opera and Drama.* Translated by W. Ashton Ellis (1893). Vol. 2 of *Richard Wagner's Prose Works.* New York: Broude Brothers, 1966.

Wright, James Edward. *The Politics of Populism: Dissent in Colorado.* New Haven and London: Yale University Press, 1974.

A NOTE ON THE TEXT

This text of *The Song of the Lark* is the original version published by Houghton Mifflin in 1915, reprinted by the University of Nebraska Press in 1978. Cather added a preface in 1932 and revised the book in 1937; the revised edition is the one Houghton Mifflin keeps in print. This edition incorporates the 1932 Preface (Appendix), as well as corrections of typographical errors and irregularities in non-English words and phrases.

THE SONG OF THE LARK

"It was a wond'rous lovely storm that drove me!"
LENAU'S *"Don Juan."*

TO
ISABELLE McCLUNG

On uplands,
At morning,
The world was young, the winds were free;
A garden fair,
In that blue desert air,
Its guest invited me to be.

PART I

FRIENDS OF CHILDHOOD

I

Dr. Howard Archie had just come up from a game of pool with the Jewish clothier and two traveling men who happened to be staying overnight in Moonstone. His offices were in the Duke Block, over the drug store.[1] Larry, the doctor's man, had lit the overhead light in the waiting-room and the double student's lamp on the desk in the study. The isinglass sides of the hard-coal burner were aglow, and the air in the study was so hot that as he came in the doctor opened the door into his little operating-room, where there was no stove. The waiting-room was carpeted and stiffly furnished, something like a country parlor. The study had worn, unpainted floors, but there was a look of winter comfort about it. The doctor's flat-top desk was large and well made; the papers were in orderly piles, under glass weights. Behind the stove a wide bookcase, with double glass doors, reached from the floor to the ceiling. It was filled with medical books of every thickness and color. On the top shelf stood a long row of thirty or forty volumes, bound all alike in dark mottled board covers, with imitation leather backs.

As the doctor in New England villages is proverbially old, so the doctor in small Colorado towns twenty-five years ago[2] was generally young. Dr. Archie was barely thirty. He was tall, with massive shoulders which he held stiffly, and a large, well-shaped head. He was a distinguished-looking man, for that part of the world, at least. There was something individual in the way in which his reddish-brown hair, parted cleanly at the side, bushed over his high forehead. His nose was straight and thick, and his eyes were intelligent. He wore a curly, reddish mustache and an imperial, cut trimly, which made him look a little like the pictures of Napoleon III. His hands were large and well kept, but ruggedly formed, and the backs were shaded with crinkly reddish hair. He wore a blue suit of woolly, wide-waled serge; the traveling men had known at a glance that it was made by a Denver tailor. The doctor was always well dressed.

Dr. Archie turned up the student's lamp and sat down in the swivel chair before his desk. He sat uneasily, beating a tattoo on his knees with his fingers, and looked about him as if he were bored. He glanced at his watch, then absently took from his pocket a bunch of small keys, selected one and looked at it. A contemptuous smile, barely perceptible, played on his lips, but his eyes remained meditative. Behind the door that led into the hall, under his buffalo-skin driving-coat, was a locked cupboard. This the doctor opened mechanically, kicking aside a pile of muddy overshoes. Inside, on the shelves, were whiskey glasses and decanters, lemons, sugar, and bitters. Hearing a step in the empty, echoing hall without, the doctor closed the cupboard again, snapping the Yale lock. The door of the waiting-room opened, a man entered and came on into the consulting-room.

"Good-evening, Mr. Kronborg," said the doctor carelessly. "Sit down."

His visitor was a tall, loosely built man, with a thin brown beard, streaked with gray. He wore a frock coat, a broad-brimmed black hat, a white lawn necktie, and steel-rimmed spectacles. Altogether there was a pretentious and important air about him, as he lifted the skirts of his coat and sat down.

"Good-evening, doctor. Can you step around to the house with me? I think Mrs. Kronborg will need you this evening." This was said with profound gravity and, curiously enough, with a slight embarrassment.

"Any hurry?" the doctor asked over his shoulder as he went into his operating-room.

Mr. Kronborg coughed behind his hand, and contracted his brows. His face threatened at every moment to break into a smile of foolish excitement. He controlled it only by calling upon his habitual pulpit manner. "Well, I think it would be as well to go immediately. Mrs. Kronborg will be more comfortable if you are there. She has been suffering for some time."

The doctor came back and threw a black bag upon his desk. He wrote some instructions for his man on a prescription pad and then drew on his overcoat. "All ready," he announced, putting out his lamp. Mr. Kronborg rose and they tramped through the empty hall and down the stairway to the street. The drug store below was dark, and the saloon next door was just closing. Every other light on Main Street was out.

On either side of the road and at the outer edge of the board sidewalk, the snow had been shoveled into breastworks. The town looked small and black, flattened down in the snow, muffled and all but extinguished. Overhead the stars shone gloriously. It was impossible not to notice them. The air was so clear that the white sand hills to the east of Moonstone gleamed softly. Following the Reverend Mr. Kronborg along the narrow walk, past the little dark, sleeping houses, the doctor looked up at the flashing night and whistled softly. It did seem that people were stupider than they need be; as if on a night like this there ought to be something better to do than to sleep nine hours, or to assist Mrs. Kronborg in functions which she could have performed so admirably unaided. He wished he had gone down to Denver to hear Fay Templeton sing "See-Saw."[3] Then he remembered that he had a personal interest in this family, after all. They turned into another street and saw before them lighted windows; a low story-and-a-half house, with a wing built on at the right and a kitchen addition at the back, everything a little on the slant—roofs, windows, and doors. As they approached the gate, Peter Kronborg's pace grew brisker. His nervous, ministerial cough annoyed the doctor. "Exactly as if he were going to give out a text," he thought. He drew off his glove and felt in his vest pocket. "Have a troche, Kronborg," he said, producing some. "Sent me for samples. Very good for a rough throat."

"Ah, thank you, thank you. I was in something of a hurry. I neglected to put on my overshoes. Here we are, doctor." Kronborg opened his front door—seemed delighted to be at home again.

The front hall was dark and cold; the hatrack was hung with an astonishing number of children's hats and caps and cloaks. They were even piled on the table beneath the hatrack. Under the table was a heap of rubbers and overshoes. While the doctor hung up his coat and hat, Peter Kronborg opened the door into the living-room. A glare of light greeted them, and a rush of hot, stale air, smelling of warming flannels.

At three o'clock in the morning Dr. Archie was in the parlor putting on his cuffs and coat—there was no spare bedroom in that house. Peter Kronborg's seventh child, a boy, was being soothed and cosseted by his aunt, Mrs. Kronborg was asleep, and the doctor was going home. But he wanted first to speak to Kronborg, who, coatless and fluttery, was pouring coal into the kitchen stove. As the doctor crossed the dining-room he paused and listened. From one of the wing rooms, off to the left, he heard rapid, distressed breathing. He went to the kitchen door.

"One of the children sick in there?" he asked, nodding toward the partition.

Kronborg hung up the stove-lifter and dusted his fingers. "It must be Thea. I meant to ask you to look at her. She has a croupy cold. But in my excitement—Mrs. Kronborg is doing finely, eh, doctor? Not many of your patients with such a constitution, I expect."

"Oh, yes. She's a fine mother." The doctor took up the lamp from the kitchen table and unceremoniously went into the wing room. Two chubby little boys were asleep in a double bed, with the coverlids over their noses and their feet drawn up. In a single bed, next to theirs, lay a little girl of eleven, wide awake, two yellow braids sticking up on the pillow behind her. Her face was scarlet and her eyes were blazing.

The doctor shut the door behind him. "Feel pretty sick, Thea?" he asked as he took out his thermometer. "Why didn't you call somebody?"

She looked at him with greedy affection. "I thought you were here," she spoke between quick breaths. "There is a new baby, isn't there? Which?"

"Which?" repeated the doctor.

"Brother or sister?"

He smiled and sat down on the edge of the bed. "Brother," he said, taking her hand. "Open."

"Good. Brothers are better," she murmured as he put the glass tube under her tongue.

"Now, be still, I want to count." Dr. Archie reached for her hand and took out his watch. When he put her hand back under the quilt he went over to one of the windows—they were both tight shut—and lifted it a little way. He reached up and ran his hand along the cold, unpapered wall. "Keep under the covers; I'll come back to you in a moment," he said, bending over the glass lamp with his thermometer. He winked at her from the door before he shut it.

Peter Kronborg was sitting in his wife's room, holding the bundle which contained his son. His air of cheerful importance, his beard and glasses, even his shirt-sleeves, annoyed the doctor. He beckoned Kronborg into the living-room and said sternly:—

"You've got a very sick child in there. Why didn't you call me before? It's pneumonia, and she must have been sick for several days. Put the baby down somewhere, please, and help me make up the bed-lounge here in the parlor. She's got to be in a warm room, and she's got to be quiet. You must keep the other children out. Here, this thing opens up, I see," swinging back the top of the carpet lounge. "We can lift her mattress and carry her in just as she is. I don't want to disturb her more than is necessary."

Kronborg was all concern immediately. The two men took up the mattress and carried the sick child into the parlor. "I'll have to go down to my office to get some medicine, Kronborg. The drug store won't be open. Keep the covers on her. I won't be gone long. Shake down the stove and put on a little coal, but not too much; so it'll catch quickly, I mean. Find an old sheet for me, and put it there to warm."

The doctor caught his coat and hurried out into the dark street. Nobody was stirring yet, and the cold was bitter. He was tired and hungry and in no mild humor. "The idea!" he muttered; "to be such an ass at his age, about the seventh! And to feel no responsibility about the little girl. Silly old goat! The baby would have got into the world somehow; they always do. But a nice little girl like that—she's worth the whole litter. Where she ever got it from—" He turned into the Duke Block and ran up the stairs to his office.

Thea Kronborg, meanwhile, was wondering why she happened to be in the parlor, where nobody but company—usually visiting preachers—ever slept. She had moments of stupor when she did not see anything, and moments of excitement when she felt that something unusual and pleasant was about to happen, when she saw everything clearly in the red light from the isinglass sides of the hard-coal burner—the nickel trimmings on the stove itself, the pictures on the wall, which she thought very beautiful, the flowers on the Brussels carpet, Czerny's "Daily Studies"[4] which stood open on the upright piano. She forgot, for the time being, all about the new baby.

When she heard the front door open, it occurred to her that the pleasant thing which was going to happen was Dr. Archie himself. He came in and warmed his hands at the stove. As he turned to her, she threw herself wearily toward him, half out of her bed. She would have tumbled to the floor had he not caught her. He gave her some medicine and went to the kitchen for something he needed. She drowsed and lost the sense of his being there. When she opened her eyes again, he was kneeling before the stove, spreading something dark and sticky on a white cloth, with a big spoon; batter, perhaps. Presently she felt him taking off her nightgown. He wrapped the hot plaster about her chest. There seemed to be straps which he pinned over her shoulders. Then he took out a thread and needle and began to sew her up in it. That, she felt, was too strange; she must be dreaming anyhow, so she succumbed to her drowsiness.

Thea had been moaning with every breath since the doctor came back, but she did not know it. She did not realize that she was suffering pain. When she was conscious at all, she seemed to be separated from her body; to be perched on top of the piano, or on the hanging lamp, watching the doctor sew her up. It was perplexing and unsatisfactory, like dreaming. She wished she could waken up and see what was going on.

The doctor thanked God that he had persuaded Peter Kronborg to keep out of the way. He could do better by the child if he had her to himself. He had no children of his own. His marriage was a very unhappy one. As he lifted and undressed Thea, he thought to himself what a beautiful thing a little girl's body was,—like a flower. It was so neatly and delicately fashioned, so soft, and so milky white. Thea must have got her hair and her silky skin from her mother. She was a little Swede, through and through. Dr. Archie could not help thinking how he would cherish a little creature like this if she were his. Her hands, so little and hot, so clever, too,—he glanced at the open exercise book on the piano. When he had stitched up the flaxseed jacket,[5] he wiped it neatly about the edges, where the paste had worked out on the skin. He put on her the clean nightgown he had warmed before the fire, and tucked the blankets about her. As he pushed back the hair that had fuzzed down over her eyebrows, he felt her head thoughtfully with the tips of his fingers. No, he couldn't say that it was different from any other child's head, though he believed that there was something very different about her. He looked intently at her wide, flushed face, freckled nose, fierce little mouth, and her delicate, tender chin—the one soft touch in her hard little Scandinavian face, as if some fairy godmother had caressed her there and left a cryptic promise. Her brows were usually drawn together defiantly, but never when she was with Dr. Archie. Her affection for him was prettier than most of the things that went to make up the doctor's life in Moonstone.

The windows grew gray. He heard a tramping on the attic floor, on the back stairs, then cries: "Give me my shirt!" "Where's my other stocking?"

"I'll have to stay till they get off to school," he reflected, "or they'll be in here tormenting her, the whole lot of them."

II

FOR THE NEXT FOUR DAYS it seemed to Dr. Archie that his patient might slip through his hands, do what he might. But she did not. On the contrary, after that she recovered very rapidly. As her father remarked, she must have inherited the "constitution" which he was never tired of admiring in her mother.

One afternoon, when her new brother was a week old, the doctor found Thea very comfortable and happy in her bed in the parlor. The sunlight was pouring in over her shoulders, the baby was asleep on a pillow in a big rocking-chair beside her. Whenever he stirred, she put out her hand and rocked him. Nothing of him was visible but a flushed, puffy forehead and an uncompromisingly big, bald cranium. The door into her mother's room stood open, and Mrs. Kronborg was sitting up in bed darning stockings. She was a short, stalwart woman, with a short neck and a determined-looking head. Her skin was very fair, her face calm and unwrinkled, and her yellow hair, braided down her back as she lay in bed, still looked like a girl's. She was a woman whom Dr. Archie respected; active, practical, unruffled; good-humored, but determined. Exactly the sort of woman to take care of a flighty preacher. She had brought her husband some property, too,—one fourth of her father's broad acres in Nebraska,—but this she kept in her own name. She had profound respect for her husband's erudition and eloquence. She sat under his preaching with deep humility, and was as much taken in by his stiff shirt and white neckties as if she had not ironed them herself by lamplight the night before they appeared correct and spotless in the pulpit. But for all this, she had no confidence in his administration of worldly affairs. She looked to him for morning prayers and grace at table; she expected him to name the babies and to supply whatever parental sentiment there was in the house, to remember birthdays and anniversaries, to point the children to moral and patriotic ideals. It was her work to keep their bodies, their clothes, and their conduct in some sort of order, and this she accomplished with a success that was a source of wonder to her neighbors. As she used to remark, and her husband admiringly to echo, she "had never lost one." With all his flightiness, Peter Kronborg appreciated the matter-of-fact, punctual way in which his wife got her children into the world and along in it. He believed, and he was right in believing, that the sovereign State of Colorado was much indebted to Mrs. Kronborg and women like her.

Mrs. Kronborg believed that the size of every family was decided in heaven. More modern views would not have startled her; they would simply have seemed foolish—thin chatter, like the boasts of the men who built the tower of Babel,[1] or like Axel's plan to breed ostriches in the chicken yard. From what evidence Mrs. Kronborg formed her opinions on this and other matters, it would have been difficult to say, but once formed, they were unchangeable. She would no more have questioned her convictions than she would have questioned revelation. Calm and even-tempered, naturally kind, she was capable of strong prejudices, and she never forgave.

When the doctor came in to see Thea, Mrs. Kronborg was reflecting that the washing was a week behind, and deciding what she had better do about it. The arrival of a new baby meant a revision of her entire domestic schedule, and as she drove her needle along she had been working out new sleeping arrangements and cleaning days. The doctor had entered the house without knocking, after making noise enough in the hall to prepare his patients. Thea was reading, her book propped up before her in the sunlight.

"Mustn't do that; bad for your eyes," he said, as Thea shut the book quickly and slipped it under the covers.

Mrs. Kronborg called from her bed: "Bring the baby here, doctor, and have that chair. She wanted him in there for company."

Before the doctor picked up the baby, he put a yellow paper bag down on Thea's coverlid and winked at her. They had a code of winks and grimaces. When he went in to chat with her mother, Thea opened the bag cautiously, trying to keep it from crackling. She drew out a long bunch of white grapes, with a little of the sawdust in which they had been packed still clinging to them. They were called Malaga grapes in Moonstone, and once or twice during the winter the leading grocer got a keg of them. They were used mainly for table decoration, about Christmas-time. Thea had never had more than one grape at a time before. When the doctor came back she was holding the almost transparent fruit up in the sunlight, feeling the pale-green skins softly with the tips of her fingers. She did not thank him; she only snapped her eyes at him in a special way which he understood, and, when he gave her his hand, put it quickly and shyly under her cheek, as if she were trying to do so without knowing it—and without his knowing it.

Dr. Archie sat down in the rocking-chair. "And how's Thea feeling to-day?"

He was quite as shy as his patient, especially when a third person overheard his conversation. Big and handsome and superior to his fellow townsmen as Dr. Archie was, he was seldom at his ease, and like Peter Kronborg he often dodged behind a professional manner. There was sometimes a contraction of embarrassment and self-consciousness all over his big body, which made him awkward—likely to stumble, to kick up rugs, or to knock over chairs. If any one was very sick, he forgot himself, but he had a clumsy touch in convalescent gossip.

Thea curled up on her side and looked at him with pleasure. "All right. I like to be sick. I have more fun then than other times."

"How's that?"

"I don't have to go to school, and I don't have to practice. I can read all I want to, and have good things,"—she patted the grapes. "I had lots of fun that time I mashed my finger and you wouldn't let Professor Wunsch make me practice. Only I had to do left hand, even then. I think that was mean."

The doctor took her hand and examined the forefinger, where the nail had grown back a little crooked. "You mustn't trim it down close at the corner there, and then it will grow straight. You won't want it crooked when you're a big girl and wear rings and have sweethearts."

She made a mocking little face at him and looked at his new scarf-pin. "That's the prettiest one you ev-*er* had. I wish you'd stay a long while and let me look at it. What is it?"

Dr. Archie laughed. "It's an opal.[2] Spanish Johnny brought it up for me from Chihuahua in his shoe. I had it set in Denver, and I wore it to-day for your benefit."

Thea had a curious passion for jewelry. She wanted every shining stone she saw, and in summer she was always going off into the sand hills to hunt for crystals and agates and bits of pink chalcedony. She had two cigar boxes full of stones that she had found or traded for, and she imagined that they were of enormous value. She was always planning how she would have them set.

"What are you reading?" The doctor reached under the covers and pulled out a book of Byron's poems. "Do you like this?"

She looked confused, turned over a few pages rapidly, and pointed to "My native land, good-night."[3] "That," she said sheepishly.

"How about 'Maid of Athens'?"[4]

She blushed and looked at him suspiciously. "I like 'There was a sound of revelry,' "[5] she muttered.

The doctor laughed and closed the book. It was clumsily bound in padded leather and had been presented to the Reverend Peter Kronborg by his Sunday-School class as an ornament for his parlor table.

"Come into the office some day, and I'll lend you a nice book. You can skip the parts you don't understand. You can read it in vacation. Perhaps you'll be able to understand all of it by then."

Thea frowned and looked fretfully toward the piano. "In vacation I have to practice four hours every day, and then there'll be Thor to take care of." She pronounced it "Tor."

"Thor? Oh, you've named the baby Thor?" exclaimed the doctor.

Thea frowned again, still more fiercely, and said quickly, "That's a nice name, only maybe it's a little—old-fashioned." She was very sensitive about being thought a foreigner, and was proud of the fact that, in town, her father always preached in English; very bookish English, at that, one might add.

Born in an old Scandinavian colony in Minnesota, Peter Kronborg had been sent to a small divinity school in Indiana by the women of a Swedish evangelical mission, who were convinced of his gifts and who skimped and begged and gave church suppers to get the long, lazy youth through the seminary. He could still speak enough Swedish to exhort and to bury the members of his country church out at Copper Hole, and he wielded in his Moonstone pulpit a somewhat pompous English vocabulary he had learned out of books at college. He always spoke of "the infant Saviour," "our Heavenly Father," etc. The poor man had no natural, spontaneous human speech. If he had his sincere moments, they were perforce inarticulate. Probably a good deal of his pretentiousness was due to the fact that he habitually expressed himself in a book-learned language, wholly remote from anything personal, native, or homely. Mrs. Kronborg spoke Swedish to her own sisters and to her sister-in-law Tillie, and colloquial English to her neighbors. Thea, who had a rather sensitive ear, until she went to school never spoke at all, except in monosyllables, and her mother was convinced that she was tongue-tied. She was still inept in speech for a child so intelligent. Her ideas were usually clear, but she seldom attempted to explain them, even at school, where she excelled in "written work" and never did more than mutter a reply.

"Your music professor stopped me on the street to-day and asked me how you were," said the doctor, rising. "He'll be sick himself, trotting around in this slush with no overcoat or overshoes."

"He's poor," said Thea simply.

The doctor sighed. "I'm afraid he's worse than that. Is he always all right when you take your lessons? Never acts as if he'd been drinking?"

Thea looked angry and spoke excitedly. "He knows a lot. More than anybody. I don't care if he does drink; he's old and poor." Her voice shook a little.

Mrs. Kronborg spoke up from the next room. "He's a good teacher, doctor. It's good for us he does drink. He'd never be in a little place like this if he didn't have some weakness. These women that teach music around here don't know nothing. I wouldn't have my child wasting time with them. If Professor Wunsch goes away, Thea'll have nobody to take from. He's careful with his scholars; he don't use bad language. Mrs. Kohler is always present when Thea takes her lesson. It's all right." Mrs. Kronborg spoke calmly and judicially. One could see that she had thought the matter out before.

"I'm glad to hear that, Mrs. Kronborg. I wish we could get the old man off his bottle and keep him tidy. Do you suppose if I gave you an old overcoat you could get him to wear it?" The doctor went to the bedroom door and Mrs. Kronborg looked up from her darning.

"Why, yes, I guess he'd be glad of it. He'll take most anything from me. He won't buy clothes, but I guess he'd wear 'em if he had 'em. I've never had any clothes to give him, having so many to make over for."

"I'll have Larry bring the coat around to-night. You aren't cross with me, Thea?" taking her hand.

Thea grinned warmly. "Not if you give Professor Wunsch a coat—and things," she tapped the grapes significantly. The doctor bent over and kissed her.

III

BEING SICK was all very well, but Thea knew from experience that starting back to school again was attended by depressing difficulties. One Monday morning she got up early with Axel and Gunner, who shared her wing room, and hurried into the back living-room, between the dining-room and the kitchen. There, beside a soft-coal stove, the younger children of the family undressed at night and dressed in the morning. The older daughter, Anna, and the two big boys slept upstairs, where the rooms were theoretically warmed by stovepipes from below. The first (and the worst!) thing that confronted Thea was a suit of clean, prickly red flannel, fresh from the wash. Usually the torment of breaking in a clean suit of flannel came on Sunday, but yesterday, as she was staying in the house, she had begged off. Their winter underwear was a trial to all the children, but it was bitterest to Thea because she happened to have the most sensitive skin. While she was tugging it on, her Aunt Tillie brought in warm water from the boiler and filled the tin pitcher. Thea washed her face, brushed and braided her hair, and got into her blue cashmere dress. Over this she buttoned a long apron, with sleeves, which would not be removed until she put on her cloak to go to school. Gunner and Axel, on the soap box behind the stove, had their usual quarrel about which should wear the tightest stockings, but they exchanged reproaches in low tones, for they were wholesomely afraid of Mrs. Kronborg's rawhide whip. She did not chastise her children often, but she did it thoroughly. Only a somewhat stern system of discipline could have kept any degree of order and quiet in that overcrowded house.

Mrs. Kronborg's children were all trained to dress themselves at the earliest possible age, to make their own beds,—the boys as well as the girls,—to take care of their clothes, to eat what was given them, and to keep out of the way. Mrs. Kronborg would have made a good chess-player; she had a head for moves and positions.

Anna, the elder daughter, was her mother's lieutenant. All the children knew that they must obey Anna, who was an obstinate contender for proprieties and not always fair-minded. To see the young Kronborgs headed for Sunday-School was like watching a military drill. Mrs. Kronborg let her children's minds alone. She did not pry into their thoughts or nag them. She respected them as individuals, and outside of the house they had a great deal of liberty. But their communal life was definitely ordered.

In the winter the children breakfasted in the kitchen; Gus and Charley and Anna first, while the younger children were dressing. Gus was nineteen and was a clerk in a dry-goods store. Charley, eighteen months younger, worked in a feed store. They left the house by the kitchen door at seven o'clock, and then Anna helped her Aunt Tillie get the breakfast for the younger ones. Without the help of this sister-in-law, Tillie Kronborg, Mrs. Kronborg's life would have been a hard one. Mrs. Kronborg often reminded Anna that "no hired help would ever have taken the same interest."

Mr. Kronborg came of a poorer stock than his wife; from a lowly, ignorant family that had lived in a poor part of Sweden. His great-grandfather had gone to Norway to work as a farm laborer and had married a Norwegian girl. This strain of Norwegian blood came out somewhere in each generation of the Kronborgs. The intemperance of one of Peter Kronborg's uncles, and the religious mania of another, had been alike charged to the Norwegian grandmother. Both Peter Kronborg and his sister Tillie were more like the Norwegian root of the family than like the Swedish, and this same Norwegian strain was strong in Thea, though in her it took a very different character.

Tillie was a queer, addle-pated thing, as flighty as a girl at thirty-five, and overweeningly fond of gay clothes—which taste, as Mrs. Kronborg philosophically said, did nobody any harm. Tillie was always cheerful, and her tongue was still for scarcely a minute during the day. She had been cruelly overworked on her father's Minnesota farm when she was a young girl, and she had never been so happy as she was now; had never before, as she said, had such social advantages. She thought her brother the most important man in Moonstone. She never missed a church service, and, much to the embarrassment of the children, she always "spoke a piece" at the Sunday-School concerts. She had a complete set of "Standard Recitations,"[1] which she conned on Sundays. This morning, when

Thea and her two younger brothers sat down to breakfast, Tillie was remonstrating with Gunner because he had not learned a recitation assigned to him for George Washington Day at school. The unmemorized text lay heavily on Gunner's conscience as he attacked his buckwheat cakes and sausage. He knew that Tillie was in the right, and that "when the day came he would be ashamed of himself."

"I don't care," he muttered, stirring his coffee; "they oughtn't to make boys speak. It's all right for girls. They like to show off."

"No showing off about it. Boys ought to like to speak up for their country. And what was the use of your father buying you a new suit, if you're not going to take part in anything?"

"That was for Sunday-School. I'd rather wear my old one, anyhow. Why didn't they give the piece to Thea?" Gunner grumbled.

Tillie was turning buckwheat cakes at the griddle. "Thea can play and sing, she don't need to speak. But you've got to know how to do something, Gunner, that you have. What are you going to do when you git big and want to git into society, if you can't do nothing? Everybody'll say, 'Can you sing? Can you play? Can you speak? Then git right out of society.' An' that's what they'll say to you, Mr. Gunner."

Gunner and Axel grinned at Anna, who was preparing her mother's breakfast. They never made fun of Tillie, but they understood well enough that there were subjects upon which her ideas were rather foolish. When Tillie struck the shallows, Thea was usually prompt in turning the conversation.

"Will you and Axel let me have your sled at recess?" she asked.

"All the time?" asked Gunner dubiously.

"I'll work your examples for you to-night, if you do."

"Oh, all right. There'll be a lot of 'em."

"I don't mind, I can work 'em fast. How about yours, Axel?"

Axel was a fat little boy of seven, with pretty, lazy blue eyes. "I don't care," he murmured, buttering his last buckwheat cake without ambition; "too much trouble to copy 'em down. Jenny Smiley'll let me have hers."

The boys were to pull Thea to school on their sled, as the snow was deep. The three set off together. Anna was now in the high school, and she no longer went with the family party, but walked to school with some of the older girls who were her friends, and wore a hat, not a hood like Thea.

IV

AND IT WAS SUMMER, beautiful Summer!" Those were the closing words of Thea's favorite fairy tale,[1] and she thought of them as she ran out into the world one Saturday morning in May, her music book under her arm. She was going to the Kohlers' to take her lesson, but she was in no hurry.

It was in the summer that one really lived. Then all the little overcrowded houses were opened wide, and the wind blew through them with sweet, earthy smells of garden-planting. The town looked as if it had just been washed. People were out painting their fences. The cottonwood trees were a-flicker with sticky, yellow little leaves, and the feathery tamarisks were in pink bud. With the warm weather came freedom for everybody. People were dug up, as it were. The very old people, whom one had not seen all winter, came out and sunned themselves in the yard. The double windows were taken off the houses, the tormenting flannels in which children had been encased all winter were put away in boxes, and the youngsters felt a pleasure in the cool cotton things next their skin.

Thea had to walk more than a mile to reach the Kohlers' house, a very pleasant mile out of town toward the glittering sand hills,—yellow this morning, with lines of deep violet where the clefts and valleys were. She followed the sidewalk to the depot at the south end of the town; then took the road east to the little group of adobe houses where the Mexicans lived, then dropped into a deep ravine; a dry sand creek, across which the railroad track ran on a trestle. Beyond that gulch, on a little rise of ground that faced the open sandy plain, was the Kohlers' house, where Professor Wunsch lived. Fritz Kohler was the town tailor, one of the first settlers. He had moved there, built a little house and made a garden, when Moonstone was first marked down on the map. He had three sons, but they now worked on the railroad and were stationed in distant cities. One of them had gone to work for the Santa Fé, and lived in New Mexico.

Mrs. Kohler seldom crossed the ravine and went into the town except at Christmas-time, when she had to buy presents and Christmas cards to send to her old friends in Freeport, Illinois. As she did not go to church, she did not possess such a thing as a hat. Year after year she wore the same red hood in winter and a black sunbonnet in summer. She made her own dresses; the skirts came barely to her shoe-tops, and were gathered as full as they could possibly be to the waistband. She preferred men's shoes, and usually wore the cast-offs of one of her sons. She had never learned much English, and her plants and shrubs were her companions. She lived for her men and her garden. Beside that sand gulch, she had tried to reproduce a bit of her own village in the Rhine Valley. She hid herself behind the growth she had fostered, lived under the shade of what she had planted and watered and pruned. In the blaze of the open plain she was stupid and blind like an owl. Shade, shade; that was what she was always planning and making. Behind the high tamarisk hedge, her garden was a jungle of verdure in summer. Above the cherry trees and peach trees and golden plums stood the windmill, with its tank on stilts, which kept all this verdure alive. Outside, the sage-brush grew up to the very edge of the garden, and the sand was always drifting up to the tamarisks.

Every one in Moonstone was astonished when the Kohlers took the wandering music-teacher to live with them. In seventeen years old Fritz had never had a crony, except the harness-maker and Spanish Johnny. This Wunsch came from God knew where,—followed Spanish Johnny into town when that wanderer came back from one of his tramps. Wunsch played in the dance orchestra, tuned pianos, and gave lessons. When Mrs. Kohler rescued him, he was sleeping in a dirty, unfurnished room over one of the saloons, and he had only two shirts in the world. Once he was under her roof, the old woman went at him as she did at her garden. She sewed and washed and mended for him, and made him so clean and respectable that he was able to get a large class of pupils and to rent a piano. As soon as he had money ahead, he sent to the Narrow Gauge lodging-house, in Denver, for a trunkful of music which had been held there for unpaid board. With tears in his eyes the old man—he was not over fifty, but sadly battered—told Mrs. Kohler that he asked nothing better of God than to end his days with her, and to be buried in the garden, under her linden trees. They were not American basswood, but the European linden, which has honey-colored blooms in summer, with a fragrance that surpasses all trees and flowers and drives young people wild with joy.

Thea was reflecting as she walked along that had it not been for Professor Wunsch she might have lived on for years in Moonstone without ever knowing the Kohlers, without ever seeing their garden or the inside of their house. Besides the cuckoo clock,—which was wonderful enough, and which Mrs. Kohler said she kept for "company when she was lonesome,"—the Kohlers had in their house the most wonderful thing Thea had ever seen—but of that later.

Professor Wunsch went to the houses of his other pupils to give them their lessons, but one morning he told Mrs. Kronborg that Thea had talent, and that if she came to him he could teach her in his slippers, and that would be better. Mrs. Kronborg was a strange woman. That word "talent," which no one else in Moonstone, not even Dr. Archie, would have understood, she comprehended perfectly. To any other woman there, it would have meant that a child must have her hair curled every day and must play in public. Mrs. Kronborg knew it meant that Thea must practice four hours a day. A child with talent must be kept at the piano, just as a child with measles must be kept under the blankets. Mrs. Kronborg and her three sisters had all studied piano, and all sang well, but none of them had talent. Their father had played the oboe in an orchestra in Sweden, before he came to America to better his fortunes. He had even known Jenny Lind.[2] A child with talent had to be kept at the piano; so twice a week in summer and once a week in winter Thea went over the gulch to the Kohlers', though the Ladies' Aid Society thought it was not proper for their preacher's daughter to go "where there was so much drinking." Not that the Kohler sons ever so much as looked at a glass of beer. They were ashamed of their old folks and got out into the world as fast as possible; had their clothes made by a Denver tailor and their necks shaved up under their hair and forgot the past. Old Fritz and Wunsch, however, indulged in a friendly bottle pretty often. The two men were like comrades; perhaps the bond between them was the glass wherein lost hopes are found; perhaps it was common memories of another country; perhaps it was the grapevine in the garden—knotty, fibrous shrub, full of homesickness and sentiment, which the Germans have carried around the world with them.

As Thea approached the house she peeped between the pink

sprays of the tamarisk hedge and saw the Professor and Mrs. Kohler in the garden, spading and raking. The garden looked like a relief-map now, and gave no indication of what it would be in August; such a jungle! Pole beans and potatoes and corn and leeks and kale and red cabbage—there would even be vegetables for which there is no American name. Mrs. Kohler was always getting by mail packages of seeds from Freeport and from the old country. Then the flowers! There were big sunflowers for the canary bird, tiger lilies and phlox and zinnias and lady's-slippers and portulaca and hollyhocks,—giant hollyhocks. Besides the fruit trees there was a great umbrella-shaped catalpa, and a balm-of-Gilead, two lindens, and even a ginkgo,—a rigid, pointed tree with leaves shaped like butterflies, which shivered, but never bent to the wind.

This morning Thea saw to her delight that the two oleander trees, one white and one red, had been brought up from their winter quarters in the cellar. There is hardly a German family in the most arid parts of Utah, New Mexico, Arizona, but has its oleander trees. However loutish the American-born sons of the family may be, there was never one who refused to give his muscle to the back-breaking task of getting those tubbed trees down into the cellar in the fall and up into the sunlight in the spring. They may strive to avert the day, but they grapple with the tub at last.

When Thea entered the gate, her professor leaned his spade against the white post that supported the turreted dove-house, and wiped his face with his shirt-sleeve; someway he never managed to have a handkerchief about him. Wunsch was short and stocky, with something rough and bear-like about his shoulders. His face was a dark, bricky red, deeply creased rather than wrinkled, and the skin was like loose leather over his neck band—he wore a brass collar button but no collar. His hair was cropped close; iron-gray bristles on a bullet-like head. His eyes were always suffused and bloodshot. He had a coarse, scornful mouth, and irregular, yellow teeth, much worn at the edges. His hands were square and red, seldom clean, but always alive, impatient, even sympathetic.

"*Morgen,*" he greeted his pupil in a businesslike way, put on a black alpaca coat, and conducted her at once to the piano in Mrs. Kohler's sitting-room. He twirled the stool to the proper height, pointed to it, and sat down in a wooden chair beside Thea.

"The scale of B flat major," he directed, and then fell into an attitude of deep attention. Without a word his pupil set to work.

To Mrs. Kohler, in the garden, came the cheerful sound of effort, of vigorous striving. Unconsciously she wielded her rake more lightly. Occasionally she heard the teacher's voice. "Scale of E minor.... *Weiter, weiter!... Immer* I hear the thumb, like a lame foot. *Weiter... weiter,* once;... *Schön!* The chords, quick!"

The pupil did not open her mouth until they began the second movement of the Clementi sonata,[3] when she remonstrated in low tones about the way he had marked the fingering of a passage.

"It makes no matter what you think," replied her teacher coldly. "There is only one right way. The thumb there. *Ein, zwei, drei, vier,*" etc. Then for an hour there was no further interruption.

At the end of the lesson Thea turned on her stool and leaned her arm on the keyboard. They usually had a little talk after the lesson.

Herr Wunsch grinned. "How soon is it you are free from school? Then we make ahead faster, eh?"

"First week in June. Then will you give me the 'Invitation to the Dance'?"[4]

He shrugged his shoulders. "It makes no matter. If you want him, you play him out of lesson hours."

"All right." Thea fumbled in her pocket and brought out a crumpled slip of paper. "What does this mean, please? I guess it's Latin."

Wunsch blinked at the line penciled on the paper. "Wherefrom you get this?" he asked gruffly.

"Out of a book Dr. Archie gave me to read. It's all English but that. Did you ever see it before?" she asked, watching his face.

"Yes. A long time ago," he muttered, scowling. "Ovidius!" He took a stub of lead pencil from his vest pocket, steadied his hand by a visible effort, and under the words

"Lente currite, lente currite, noctis equi,"[5]

he wrote in a clear, elegant Gothic hand,—

"Go slowly, go slowly, ye steeds of the night."

He put the pencil back in his pocket and continued to stare at the Latin. It recalled the poem, which he had read as a student, and thought very fine. There were treasures of memory which no lodging-house keeper could attach. One carried things about in one's head, long after one's linen could be smuggled out in a tuning-bag. He handed the paper back to Thea. "There is the English, quite elegant," he said, rising.

Mrs. Kohler stuck her head in at the door, and Thea slid off the stool. "Come in, Mrs. Kohler," she called, "and show me the piece-picture."

The old woman laughed, pulled off her big gardening-gloves, and pushed Thea to the lounge before the object of her delight. The "piece-picture," which hung on the wall and nearly covered one whole end of the room, was the handiwork of Fritz Kohler. He had learned his trade under an old-fashioned tailor in Magdeburg who required from each of his apprentices a thesis: that is, before they left his shop, each apprentice had to copy in cloth some well-known German painting, stitching bits of colored stuff together on a linen background; a kind of mosaic. The pupil was allowed to select his subject, and Fritz Kohler had chosen a popular painting of Napoleon's retreat from Moscow.[6] The gloomy Emperor and his staff were represented as crossing a stone bridge, and behind them was the blazing city, the walls and fortresses done in gray cloth with orange tongues of flame darting about the domes and minarets. Napoleon rode his white horse; Murat, in Oriental dress,[7] a bay charger. Thea was never tired of examining this work, of hearing how long it had taken Fritz to make it, how much it had been admired, and what narrow escapes it had had from moths and fire. Silk, Mrs. Kohler explained, would have been much easier to manage than woolen cloth, in which it was often hard to get the right shades. The reins of the horses, the wheels of the spurs, the brooding eyebrows of the Emperor, Murat's fierce mustaches, the great shakos of the Guard, were all worked out with the minutest fidelity. Thea's admiration for this picture had endeared her to Mrs. Kohler. It was now many years since she used to point out its wonders to her own little boys. As Mrs. Kohler did not go to church, she never heard any singing, except the songs that floated over from Mexican Town, and Thea often sang for her after the lesson was over. This morning Wunsch pointed to the piano.

"On Sunday, when I go by the church, I hear you sing something."

Thea obediently sat down on the stool again and began, "*Come, ye Disconsolate.*"[8] Wunsch listened thoughtfully, his hands on his knees. Such a beautiful child's voice! Old Mrs. Kohler's face relaxed in a smile of happiness; she half closed her eyes. A big fly was darting in and out of the window; the sunlight made a golden pool on the rag carpet and bathed the faded cretonne pillows on the lounge, under the piece-picture. *"Earth has no sorrow that Heaven cannot heal,"* the song died away.

"That is a good thing to remember," Wunsch shook himself. "You believe that?" looking quizzically at Thea.

She became confused and pecked nervously at a black key with her middle finger. "I don't know. I guess so," she murmured.

Her teacher rose abruptly. "Remember, for next time, thirds. You ought to get up earlier."

That night the air was so warm that Fritz and Herr Wunsch had their after-supper pipe in the grape arbor, smoking in silence while the sound of fiddles and guitars came across the ravine from Mexican Town. Long after Fritz and his old Paulina had gone to bed, Wunsch sat motionless in the arbor, looking up through the woolly vine leaves at the glittering machinery of heaven.

"Lente currite, noctis equi."

That line awoke many memories. He was thinking of youth; of his own, so long gone by, and of his pupil's, just beginning. He would even have cherished hopes for her, except that he had become superstitious. He believed that whatever he hoped for was destined not to be; that his affection brought ill-fortune, especially to the young; that if he held anything in his thoughts, he harmed it. He had taught in music schools in St. Louis and Kansas City, where the shallowness and complacency of the young misses had maddened him. He had encountered bad manners and bad faith, had been the victim of sharpers of all kinds, was dogged by bad luck. He had played in orchestras that were never paid and wandering opera troupes which disbanded penniless. And there was always the old enemy, more relentless than the others. It was long since he had wished anything or desired anything beyond the necessities of the body. Now that he was tempted to hope for another, he felt alarmed and shook his head.

It was his pupil's power of application, her rugged will, that interested him. He had lived for so long among people whose sole ambition was to get something for nothing that he had learned not to look for seriousness in anything. Now that he by chance encountered it, it recalled standards, ambitions, a society long forgot. What was it she reminded him of? A yellow flower, full of sunlight, perhaps. No; a thin glass full of sweet-smelling, sparkling Moselle wine. He seemed to see such a glass before him in the arbor, to watch the bubbles rising and breaking, like the silent discharge of energy in the nerves and brain, the rapid florescence in young blood—Wunsch felt ashamed and dragged his slippers along the path to the kitchen, his eyes on the ground.

V

THE CHILDREN in the primary grades were sometimes required to make relief maps of Moonstone in sand. Had they used colored sands, as the Navajo medicine men do in their sand mosaics, they could easily have indicated the social classifications of Moonstone, since these conformed to certain topographical boundaries, and every child understood them perfectly.

The main business street ran, of course, through the center of the town. To the west of this street lived all the people who were, as Tillie Kronborg said, "in society." Sylvester Street, the third parallel with Main Street on the west, was the longest in town, and the best dwellings were built along it. Far out at the north end, nearly a mile from the court-house and its cottonwood grove, was Dr. Archie's house, its big yard and garden surrounded by a white paling fence. The Methodist Church was in the center of the town, facing the court-house square. The Kronborgs lived half a mile south of the church, on the long street that stretched out like an arm to the depot settlement. This was the first street west of Main, and was built up only on one side. The preacher's house faced the backs of the brick and frame store buildings and a draw full of sunflowers and scraps of old iron. The sidewalk which ran in front of the Kronborgs' house was the one continuous sidewalk to the depot, and all the train men and roundhouse employees passed the front gate every time they came uptown. Thea and Mrs. Kronborg had many friends among the railroad men, who often paused to chat across the fence, and of one of these we shall have more to say.

In the part of Moonstone that lay east of Main Street, toward the deep ravine which, farther south, wound by Mexican Town, lived all the humbler citizens, the people who voted but did not run for office. The houses were little story-and-a-half cottages, with none of the fussy architectural efforts that marked those on Sylvester

Street. They nestled modestly behind their cottonwoods and Virginia creeper; their occupants had no social pretensions to keep up. There were no half-glass front doors with doorbells, or formidable parlors behind closed shutters. Here the old women washed in the back yard, and the men sat in the front doorway and smoked their pipes. The people on Sylvester Street scarcely knew that this part of the town existed. Thea liked to take Thor and her express wagon and explore these quiet, shady streets, where the people never tried to have lawns or to grow elms and pine trees, but let the native timber have its way and spread in luxuriance. She had many friends there, old women who gave her a yellow rose or a spray of trumpet vine and appeased Thor with a cooky or a doughnut. They called Thea "that preacher's girl," but the demonstrative was misplaced, for when they spoke of Mr. Kronborg they called him "the Methodist preacher."

Dr. Archie was very proud of his yard and garden, which he worked himself. He was the only man in Moonstone who was successful at growing rambler roses, and his strawberries were famous. One morning when Thea was downtown on an errand, the doctor stopped her, took her hand and went over her with a quizzical eye, as he nearly always did when they met.

"You haven't been up to my place to get any strawberries yet, Thea. They're at their best just now. Mrs. Archie doesn't know what to do with them all. Come up this afternoon. Just tell Mrs. Archie I sent you. Bring a big basket and pick till you are tired."

When she got home Thea told her mother that she didn't want to go, because she didn't like Mrs. Archie.

"She is certainly one queer woman," Mrs. Kronborg assented, "but he's asked you so often, I guess you'll have to go this time. She won't bite you."

After dinner Thea took a basket, put Thor in his babybuggy, and set out for Dr. Archie's house at the other end of town. As soon as she came within sight of the house, she slackened her pace. She approached it very slowly, stopping often to pick dandelions and sand-peas for Thor to crush up in his fist.

It was his wife's custom, as soon as Dr. Archie left the house in the morning, to shut all the doors and windows to keep the dust out, and to pull down the shades to keep the sun from fading the carpets. She thought, too, that neighbors were less likely to drop in if the house was closed up. She was one of those people who are stingy without motive or reason, even when they can gain nothing by it. She must have known that skimping the doctor in heat and food made him more extravagant than he would have been had she made him comfortable. He never came home for lunch, because she gave him such miserable scraps and shreds of food. No matter how much milk he bought, he could never get thick cream for his strawberries. Even when he watched his wife lift it from the milk in smooth, ivory-colored blankets, she managed, by some sleight-of-hand, to dilute it before it got to the breakfast table. The butcher's favorite joke was about the kind of meat he sold Mrs. Archie. She felt no interest in food herself, and she hated to prepare it. She liked nothing better than to have Dr. Archie go to Denver for a few days—he often went chiefly because he was hungry—and to be left alone to eat canned salmon and to keep the house shut up from morning until night.

Mrs. Archie would not have a servant because, she said, "they ate too much and broke too much"; she even said they knew too much. She used what mind she had in devising shifts to minimize her housework. She used to tell her neighbors that if there were no men, there would be no housework. When Mrs. Archie was first married, she had been always in a panic for fear she would have children. Now that her apprehensions on that score had grown paler, she was almost as much afraid of having dust in the house as she had once been of having children in it. If dust did not get in, it did not have to be got out, she said. She would take any amount of trouble to avoid trouble. Why, nobody knew. Certainly her husband had never been able to make her out. Such little, mean natures are among the darkest and most baffling of created things. There is no law by which they can be explained. The ordinary incentives of pain and pleasure do not account for their behavior. They live like insects, absorbed in petty activities that seem to have nothing to do with any genial aspect of human life.

Mrs. Archie, as Mrs. Kronborg said, "liked to gad." She liked to have her house clean, empty, dark, locked, and to be out of it—anywhere. A church social, a prayer meeting, a ten-cent show; she seemed to have no preference. When there was nowhere else to go, she used to sit for hours in Mrs. Smiley's millinery and notion store, listening to the talk of the women who came in, watching them while they tried on hats, blinking at them from her corner with her sharp, restless little eyes. She never talked much herself, but she knew all the gossip of the town and she had a sharp ear for racy anecdotes—"traveling men's stories," they used to be called in Moonstone. Her clicking laugh sounded like a typewriting machine in action, and, for very pointed stories, she had a little screech.

Mrs. Archie had been Mrs. Archie for only six years, and when she was Belle White she was one of the "pretty" girls in Lansing, Michigan. She had then a train of suitors. She could truly remind Archie that "the boys hung around her." They did. They thought her very spirited and were always saying, "Oh, that Belle White, she's a case!" She used to play heavy practical jokes which the young men thought very clever. Archie was considered the most promising young man in "the young crowd," so Belle selected him. She let him see, made him fully aware, that she had selected him, and Archie was the sort of boy who could not withstand such enlightenment. Belle's family were sorry for him. On his wedding day her sisters looked at the big, handsome boy—he was twenty-four—as he walked down the aisle with his bride, and then they looked at each other. His besotted confidence, his sober, radiant face, his gentle, protecting arm, made them uncomfortable. Well, they were glad that he was going West at once, to fulfill his doom where they would not be onlookers. Anyhow, they consoled themselves, they had got Belle off their hands.

More than that, Belle seemed to have got herself off her hands. Her reputed prettiness must have been entirely the result of determination, of a fierce little ambition. Once she had married, fastened herself on some one, come to port,—it vanished like the ornamental plumage which drops away from some birds after the mating season. The one aggressive action of her life was over. She began to shrink in face and stature. Of her harum-scarum spirit there was nothing left but the little screech. Within a few years she looked as small and mean as she was.

Thor's chariot crept along. Thea approached the house unwillingly. She didn't care about the strawberries, anyhow. She had come only because she did not want to hurt Dr. Archie's feelings. She not only disliked Mrs. Archie, she was a little afraid of her. While Thea was getting the heavy baby-buggy through the iron gate she heard some one call, "Wait a minute!" and Mrs. Archie came running around the house from the back door, her apron over her head. She came to help with the buggy, because she was afraid the wheels might scratch the paint off the gateposts. She was a skinny little woman with a great pile of frizzy light hair on a small head.

"Dr. Archie told me to come up and pick some strawberries," Thea muttered, wishing she had stayed at home.

Mrs. Archie led the way to the back door, squinting and shading her eyes with her hand. "Wait a minute," she said again, when Thea explained why she had come.

She went into her kitchen and Thea sat down on the porch step. When Mrs. Archie reappeared she carried in her hand a little wooden butter-basket trimmed with fringed tissue paper, which she must have brought home from some church supper. "You'll have to have something to put them in," she said, ignoring the yawning willow basket which stood empty on Thor's feet. "You can have this, and you needn't mind about returning it. You know about not trampling the vines, don't you?"

Mrs. Archie went back into the house and Thea leaned over in the sand and picked a few strawberries. As soon as she was sure that she was not going to cry, she tossed the little basket into the big one and ran Thor's buggy along the gravel walk and out of the gate as fast as she could push it. She was angry, and she was ashamed for Dr. Archie. She could not help thinking how uncomfortable he would be if he ever found out about it. Little things like that were the ones that cut him most. She slunk home by the back way, and again almost cried when she told her mother about it.

Mrs. Kronborg was frying doughnuts for her husband's supper. She laughed as she dropped a new lot into the hot grease. "It's wonderful, the way some people are made," she declared. "But I wouldn't let that upset me if I was you. Think what it would be to live with it all the time. You look in the black pocketbook inside my handbag and take a dime and go downtown and get an ice-cream soda. That'll make you feel better. Thor can have a little of the ice-cream if you feed it to him with a spoon. He likes it, don't you, son?" She stooped to wipe his chin. Thor was only six months old and inarticulate, but it was quite true that he liked ice-cream.

VI

SEEN FROM A BALLOON, Moonstone would have looked like a Noah's ark town set out in the sand and lightly shaded by gray-green tamarisks and cottonwoods. A few people were trying to make soft maples grow in their turfed lawns, but the fashion of planting incongruous trees from the North Atlantic States had not become general then, and the frail, brightly painted desert town was shaded by the light-reflecting, wind-loving trees of the desert, whose roots are always seeking water and whose leaves are always talking about it, making the sound of rain. The long, porous roots of the cottonwood are irrepressible. They break into the wells as rats do into granaries, and thieve the water.

The long street which connected Moonstone with the depot settlement traversed in its course a considerable stretch of rough open country, staked out in lots but not built up at all, a weedy hiatus between the town and the railroad. When you set out along this street to go to the station, you noticed that the houses became smaller and farther apart, until they ceased altogether, and the board sidewalk continued its uneven course through sunflower patches, until you reached the solitary, new brick Catholic Church. The church stood there because the land was given to the parish by the man who owned the adjoining waste lots, in the hope of making them more salable—"Farrier's Addition," this patch of prairie was called in the clerk's office. An eighth of a mile beyond the church was a washout, a deep sand-gully, where the board sidewalk became a bridge for perhaps fifty feet. Just beyond the gully was old Uncle Billy Beemer's grove,—twelve town lots set out in fine, well-grown cottonwood trees, delightful to look upon, or to listen to, as they swayed and rippled in the wind. Uncle Billy had been one of the most worthless old drunkards who ever sat on a store box and told filthy stories. One night he played hide-and-seek with a switch engine and got his sodden brains knocked out. But his grove, the one creditable thing he had ever done in his life, rustled on. Beyond this grove the houses of the depot settlement began, and the naked board walk, that had run in out of the sunflowers, again became a link between human dwellings.

One afternoon, late in the summer, Dr. Howard Archie was fighting his way back to town along this walk through a blinding sandstorm, a silk handkerchief tied over his mouth. He had been to see a sick woman down in the depot settlement, and he was walking because his ponies had been out for a hard drive that morning.

As he passed the Catholic Church he came upon Thea and Thor. Thea was sitting in a child's express wagon, her feet out behind, kicking the wagon along and steering by the tongue. Thor was on her lap and she held him with one arm. He had grown to be a big cub of a baby, with a constitutional grievance, and he had to be continually amused. Thea took him philosophically, and tugged and pulled him about, getting as much fun as she could under her encumbrance. Her hair was blowing about her face, and her eyes were squinting so intently at the uneven board sidewalk in front of her that she did not see the doctor until he spoke to her.

"Look out, Thea. You'll steer that youngster into the ditch."

The wagon stopped. Thea released the tongue, wiped her hot, sandy face, and pushed back her hair. "Oh, no, I won't! I never ran off but once, and then he didn't get anything but a bump. He likes this better than a babybuggy, and so do I."

"Are you going to kick that cart all the way home?"

"Of course. We take long trips; wherever there is a sidewalk. It's no good on the road."

"Looks to me like working pretty hard for your fun. Are you going to be busy to-night? Want to make a call with me? Spanish Johnny's come home again, all used up. His wife sent me word this morning, and I said I'd go over to see him to-night. He's an old chum of yours, isn't he?"

"Oh, I'm glad. She's been crying her eyes out. When did he come?"

"Last night, on Number Six. Paid his fare, they tell me. Too sick to beat it. There'll come a time when that boy won't get back, I'm afraid. Come around to my office about eight o'clock,—and you needn't bring that!"

Thor seemed to understand that he had been insulted, for he

scowled and began to kick the side of the wagon, shouting, "Go-go, go-go!" Thea leaned forward and grabbed the wagon tongue. Dr. Archie stepped in front of her and blocked the way. "Why don't you make him wait? What do you let him boss you like that for?"

"If he gets mad he throws himself, and then I can't do anything with him. When he's mad he's lots stronger than me, aren't you, Thor?" Thea spoke with pride, and the idol was appeased. He grunted approvingly as his sister began to kick rapidly behind her, and the wagon rattled off and soon disappeared in the flying currents of sand.

That evening Dr. Archie was seated in his office, his desk chair tilted back, reading by the light of a hot coal-oil lamp. All the windows were open, but the night was breathless after the sandstorm, and his hair was moist where it hung over his forehead. He was deeply engrossed in his book and sometimes smiled thoughtfully as he read. When Thea Kronborg entered quietly and slipped into a seat, he nodded, finished his paragraph, inserted a bookmark, and rose to put the book back into the case. It was one out of the long row of uniform volumes on the top shelf.

"Nearly every time I come in, when you're alone, you're reading one of those books," Thea remarked thoughtfully. "They must be very nice."

The doctor dropped back into his swivel chair, the mottled volume still in his hand. "They aren't exactly books, Thea," he said seriously. "They're a city."

"A history, you mean?"

"Yes, and no. They're a history of a live city, not a dead one. A Frenchman undertook to write about a whole cityful of people, all the kinds he knew. And he got them nearly all in, I guess. Yes, it's very interesting. You'll like to read it some day, when you're grown up."

Thea leaned forward and made out the title on the back, "A Distinguished Provincial in Paris."[1]

"It doesn't sound very interesting."

"Perhaps not, but it is." The doctor scrutinized her broad face, low enough to be in the direct light from under the green lamp shade. "Yes," he went on with some satisfaction, "I think you'll like them some day. You're always curious about people, and I expect this man knew more about people than anybody that ever lived."

"City people or country people?"

"Both. People are pretty much the same everywhere."

"Oh, no, they're not. The people who go through in the dining-car aren't like us."

"What makes you think they aren't, my girl? Their clothes?"

Thea shook her head. "No, it's something else. I don't know." Her eyes shifted under the doctor's searching gaze and she glanced up at the row of books. "How soon will I be old enough to read them?"

"Soon enough, soon enough, little girl." The doctor patted her hand and looked at her index finger. "The nail's coming all right, isn't it? But I think that man makes you practice too much. You have it on your mind all the time." He had noticed that when she talked to him she was always opening and shutting her hands. "It makes you nervous."

"No, he don't," Thea replied stubbornly, watching Dr. Archie return the book to its niche.

He took up a black leather case, put on his hat, and they went down the dark stairs into the street. The summer moon hung full in the sky. For the time being, it was the great fact in the world. Beyond the edge of the town the plain was so white that every clump of sage stood out distinct from the sand, and the dunes looked like a shining lake. The doctor took off his straw hat and carried it in his hand as they walked toward Mexican Town, across the sand.

North of Pueblo, Mexican settlements were rare in Colorado then. This one had come about accidentally. Spanish Johnny was the first Mexican who came to Moonstone. He was a painter and decorator, and had been working in Trinidad, when Ray Kennedy told him there was a "boom" on in Moonstone, and a good many new buildings were going up. A year after Johnny settled in Moonstone, his cousin, Famos Serreños, came to work in the brickyard; then Serreños' cousins came to help him. During the strike, the master mechanic put a gang of Mexicans to work in the roundhouse. The Mexicans had arrived so quietly, with their blankets and musical instruments, that before Moonstone was awake to the fact, there was a Mexican quarter; a dozen families or more.

As Thea and the doctor approached the 'dobe houses, they heard a guitar, and a rich barytone voice—that of Famos Serreños—singing "La Golandrina."[2] All the Mexican houses had neat little yards, with tamarisk hedges and flowers, and walks bordered with shells or white-washed stones. Johnny's house was dark. His wife,

Mrs. Tellamantez, was sitting on the doorstep, combing her long, blue-black hair. (Mexican women are like the Spartans; when they are in trouble, in love, under stress of any kind, they comb and comb their hair.) She rose without embarrassment or apology, comb in hand, and greeted the doctor.

"Good-evening; will you go in?" she asked in a low, musical voice. "He is in the back room. I will make a light." She followed them indoors, lit a candle and handed it to the doctor, pointing toward the bedroom. Then she went back and sat down on her doorstep.

Dr. Archie and Thea went into the bedroom, which was dark and quiet. There was a bed in the corner, and a man was lying on the clean sheets. On the table beside him was a glass pitcher, half-full of water. Spanish Johnny looked younger than his wife, and when he was in health he was very handsome: slender, gold-colored, with wavy black hair, a round, smooth throat, white teeth, and burning black eyes. His profile was strong and severe, like an Indian's. What was termed his "wildness" showed itself only in his feverish eyes and in the color that burned on his tawny cheeks. That night he was a coppery green, and his eyes were like black holes. He opened them when the doctor held the candle before his face.

"*Mi testa!*" he muttered, "*mi testa,* doctor. *La fiebre!*"[3] Seeing the doctor's companion at the foot of the bed, he attempted a smile. "*Muchacha!*"[4] he exclaimed deprecatingly.

Dr. Archie stuck a thermometer into his mouth. "Now, Thea, you can run outside and wait for me."

Thea slipped noiselessly through the dark house and joined Mrs. Tellamantez. The somber Mexican woman did not seem inclined to talk, but her nod was friendly. Thea sat down on the warm sand, her back to the moon, facing Mrs. Tellamantez on her doorstep, and began to count the moonflowers on the vine that ran over the house. Mrs. Tellamantez was always considered a very homely woman. Her face was of a strongly marked type not sympathetic to Americans. Such long, oval faces, with a full chin, a large, mobile mouth, a high nose, are not uncommon in Spain. Mrs. Tellamantez could not write her name, and could read but little. Her strong nature lived upon itself. She was chiefly known in Moonstone for her forbearance with her incorrigible husband.

Nobody knew exactly what was the matter with Johnny, and everybody liked him. His popularity would have been unusual for a white man, for a Mexican it was unprecedented. His talents were his undoing. He had a high, uncertain tenor voice, and he played the mandolin with exceptional skill. Periodically he went crazy. There was no other way to explain his behavior. He was a clever workman, and, when he worked, as regular and faithful as a burro. Then some night he would fall in with a crowd at the saloon and begin to sing. He would go on until he had no voice left, until he wheezed and rasped. Then he would play his mandolin furiously, and drink until his eyes sank back into his head. At last, when he was put out of the saloon at closing time, and could get nobody to listen to him, he would run away—along the railroad track, straight across the desert. He always managed to get aboard a freight somewhere. Once beyond Denver, he played his way southward from saloon to saloon until he got across the border. He never wrote to his wife; but she would soon begin to get newspapers from La Junta, Albuquerque, Chihuahua, with marked paragraphs announcing that Juan Tellamantez and his wonderful mandolin could be heard at the Jack Rabbit Grill, or the Pearl of Cadiz Saloon. Mrs. Tellamantez waited and wept and combed her hair. When he was completely wrung out and burned up,—all but destroyed,—her Juan always came back to her to be taken care of,—once with an ugly knife wound in the neck, once with a finger missing from his right hand,—but he played just as well with three fingers as he had with four.

Public sentiment was lenient toward Johnny, but everybody was disgusted with Mrs. Tellamantez for putting up with him. She ought to discipline him, people said; she ought to leave him; she had no self-respect. In short, Mrs. Tellamantez got all the blame. Even Thea thought she was much too humble. To-night, as she sat with her back to the moon, looking at the moonflowers and Mrs. Tellamantez's somber face, she was thinking that there is nothing so sad in the world as that kind of patience and resignation. It was much worse than Johnny's craziness. She even wondered whether it did not help to make Johnny crazy. People had no right to be so passive and resigned. She would like to roll over and over in the sand and screech at Mrs. Tellamantez. She was glad when the doctor came out.

The Mexican woman rose and stood respectful and expectant. The doctor held his hat in his hand and looked kindly at her.

"Same old thing, Mrs. Tellamantez. He's no worse than he's been before. I've left some medicine. Don't give him anything but toast water until I see him again. You're a good nurse; you'll get him out." Dr. Archie smiled encouragingly. He glanced about the little garden and wrinkled his brows. "I can't see what makes him behave so. He's killing himself, and he's not a rowdy sort of fellow. Can't you tie him up someway? Can't you tell when these fits are coming on?"

Mrs. Tellamantez put her hand to her forehead. "The saloon, doctor, the excitement; that is what makes him. People listen to him, and it excites him."

The doctor shook his head. "Maybe. He's too much for my calculations. I don't see what he gets out of it."

"He is always fooled,"—the Mexican woman spoke rapidly and tremulously, her long under lip quivering. "He is good at heart, but he has no head. He fools himself. You do not understand in this country, you are progressive. But he has no judgment, and he is fooled." She stooped quickly, took up one of the white conch-shells that bordered the walk, and, with an apologetic inclination of her head, held it to Dr. Archie's ear. "Listen, doctor. You hear something in there? You hear the sea; and yet the sea is very far from here. You have judgment, and you know that. But he is fooled. To him, it is the sea itself. A little thing is big to him." She bent and placed the shell in the white row, with its fellows. Thea took it up softly and pressed it to her own ear. The sound in it startled her; it was like something calling one. So that was why Johnny ran away. There was something awe-inspiring about Mrs. Tellamantez and her shell.

Thea caught Dr. Archie's hand and squeezed it hard as she skipped along beside him back toward Moonstone. She went home, and the doctor went back to his lamp and his book. He never left his office until after midnight. If he did not play whist or pool in the evening, he read. It had become a habit with him to lose himself.

VII

THEA'S TWELFTH BIRTHDAY had passed a few weeks before her memorable call upon Mrs. Tellamantez. There was a worthy man in Moonstone who was already planning to marry Thea as soon as she should be old enough. His name was Ray Kennedy, his age was thirty, and he was conductor on a freight train, his run being from Moonstone to Denver. Ray was a big fellow, with a square, open American face, a rock chin, and features that one would never happen to remember. He was an aggressive idealist, a freethinker, and, like most railroad men, deeply sentimental. Thea liked him for reasons that had to do with the adventurous life he had led in Mexico and the Southwest, rather than for anything very personal. She liked him, too, because he was the only one of her friends who ever took her to the sand hills.[1] The sand hills were a constant tantalization; she loved them better than anything near Moonstone, and yet she could so seldom get to them. The first dunes were accessible enough; they were only a few miles beyond the Kohlers', and she could run out there any day when she could do her practicing in the morning and get Thor off her hands for an afternoon. But the real hills—the Turquoise Hills, the Mexicans called them—were ten good miles away, and one reached them by a heavy, sandy road. Dr. Archie sometimes took Thea on his long drives, but as nobody lived in the sand hills, he never had calls to make in that direction. Ray Kennedy was her only hope of getting there.

This summer Thea had not been to the hills once, though Ray had planned several Sunday expeditions. Once Thor was sick, and once the organist in her father's church was away and Thea had to play the organ for the three Sunday services. But on the first Sunday in September, Ray drove up to the Kronborgs' front gate at nine o'clock in the morning and the party actually set off. Gunner and Axel went with Thea, and Ray had asked Spanish Johnny to come and to bring Mrs. Tellamantez and his mandolin. Ray was artlessly fond of music, especially of Mexican music. He and Mrs. Tellamantez had got up the lunch between them, and they were to make coffee in the desert.

When they left Mexican Town, Thea was on the front seat with Ray and Johnny, and Gunner and Axel sat behind with Mrs. Tellamantez. They objected to this, of course, but there were some things about which Thea would have her own way. "As stubborn as a Finn,"[2] Mrs. Kronborg sometimes said of her, quoting an old Swedish saying. When they passed the Kohlers', old Fritz and Wunsch were cutting grapes at the arbor. Thea gave them a businesslike nod. Wunsch came to the gate and looked after them. He divined Ray Kennedy's hopes, and he distrusted every expedition that led away from the piano. Unconsciously he made Thea pay for frivolousness of this sort.

As Ray Kennedy's party followed the faint road across the sagebrush, they heard behind them the sound of church bells, which gave them a sense of escape and boundless freedom. Every rabbit that shot across the path, every sage hen that flew up by the trail, was like a runaway thought, a message that one sent into the desert. As they went farther, the illusion of the mirage became more instead of less convincing; a shallow silver lake that spread for many miles, a little misty in the sunlight. Here and there one saw reflected the image of a heifer, turned loose to live upon the sparse sandgrass. They were magnified to a preposterous height and looked like mammoths, prehistoric beasts standing solitary in the waters that for many thousands of years actually washed over that desert;—the mirage itself may be the ghost of that long-vanished sea. Beyond the phantom lake lay the line of many-colored hills; rich, sun-baked yellow, glowing turquoise, lavender, purple; all the open, pastel colors of the desert.

After the first five miles the road grew heavier. The horses had to slow down to a walk and the wheels sank deep into the sand, which now lay in long ridges, like waves, where the last high wind had drifted it. Two hours brought the party to Pedro's Cup, named for a Mexican desperado who had once held the sheriff at bay there. The Cup was a great amphitheater, cut out in the hills, its floor smooth and packed hard, dotted with sagebrush and greasewood.

On either side of the Cup the yellow hills ran north and south, with winding ravines between them, full of soft sand which drained down from the crumbling banks. On the surface of this fluid sand, one could find bits of brilliant stone, crystals and agates and onyx, and petrified wood as red as blood. Dried toads and lizards were to be found there, too. Birds, decomposing more rapidly, left only feathered skeletons.

After a little reconnoitering, Mrs. Tellamantez declared that it was time for lunch, and Ray took his hatchet and began to cut greasewood, which burns fiercely in its green state. The little boys dragged the bushes to the spot that Mrs. Tellamantez had chosen for her fire. Mexican women like to cook out of doors.

After lunch Thea sent Gunner and Axel to hunt for agates. "If you see a rattlesnake, run. Don't try to kill it," she enjoined.

Gunner hesitated. "If Ray would let me take the hatchet, I could kill one all right."

Mrs. Tellamantez smiled and said something to Johnny in Spanish.

"Yes," her husband replied, translating, "they say in Mexico, kill a snake but never hurt his feelings. Down in the hot country, *muchacha,*" turning to Thea, "people keep a pet snake in the house to kill rats and mice. They call him the house snake. They keep a little mat for him by the fire, and at night he curl up there and sit with the family, just as friendly!"

Gunner sniffed with disgust. "Well, I think that's a dirty Mexican way to keep house; so there!"

Johnny shrugged his shoulders. "Perhaps," he muttered. A Mexican learns to dive below insults or soar above them, after he crosses the border.

By this time the south wall of the amphitheater cast a narrow shelf of shadow, and the party withdrew to this refuge. Ray and Johnny began to talk about the Grand Canyon and Death Valley, two places much shrouded in mystery in those days, and Thea listened intently. Mrs. Tellamantez took out her drawn-work[3] and pinned it to her knee. Ray could talk well about the large part of the continent over which he had been knocked about, and Johnny was appreciative.

"You been all over, pretty near. Like a Spanish boy," he commented respectfully.

Ray, who had taken off his coat, whetted his pocket-knife

thoughtfully on the sole of his shoe. "I began to browse around early. I had a mind to see something of this world, and I ran away from home before I was twelve. Rustled for myself ever since."

"Ran away?" Johnny looked hopeful. "What for?"

"Couldn't make it go with my old man, and didn't take to farming. There were plenty of boys at home. I wasn't missed."

Thea wriggled down in the hot sand and rested her chin on her arm. "Tell Johnny about the melons, Ray, please do!"

Ray's solid, sunburned cheeks grew a shade redder, and he looked reproachfully at Thea. "You're stuck on that story, kid. You like to get the laugh on me, don't you? That was the finishing split I had with my old man, John. He had a claim along the creek, not far from Denver, and raised a little garden stuff for market. One day he had a load of melons and he decided to take 'em to town and sell 'em along the street, and he made me go along and drive for him. Denver wasn't the queen city it is now, by any means, but it seemed a terrible big place to me; and when we got there, if he didn't make me drive right up Capitol Hill! Pap got out and stopped at folkses houses to ask if they didn't want to buy any melons, and I was to drive along slow. The farther I went the madder I got, but I was trying to look unconscious, when the end-gate came loose and one of the melons fell out and squashed. Just then a swell girl, all dressed up, comes out of one of the big houses and calls out, 'Hello, boy, you're losing your melons!' Some dudes on the other side of the street took their hats off to her and began to laugh. I couldn't stand it any longer. I grabbed the whip and lit into that team, and they tore up the hill like jack-rabbits, them damned melons bouncing out the back every jump, the old man cussin' an' yellin' behind and everybody laughin'. I never looked behind, but the whole of Capitol Hill must have been a mess with them squashed melons. I didn't stop the team till I got out of sight of town. Then I pulled up an' left 'em with a rancher I was acquainted with, and I never went home to get the lickin' that was waitin' for me. I expect it's waitin' for me yet."

Thea rolled over in the sand. "Oh, I wish I could have seen those melons fly, Ray! I'll never see anything as funny as that. Now, tell Johnny about your first job."

Ray had a collection of good stories. He was observant, truthful, and kindly—perhaps the chief requisites in a good story-teller. Occasionally he used newspaper phrases, conscientiously learned in his efforts at self-instruction, but when he talked naturally he was always worth listening to. Never having had any schooling to speak of, he had, almost from the time he first ran away, tried to make good his loss. As a sheep-herder he had worried an old grammar to tatters, and read instructive books with the help of a pocket dictionary. By the light of many camp-fires he had pondered upon Prescott's histories,[4] and the works of Washington Irving,[5] which he bought at a high price from a book-agent. Mathematics and physics were easy for him, but general culture came hard, and he was determined to get it. Ray was a freethinker, and inconsistently believed himself damned for being one. When he was braking, down on the Santa Fé, at the end of his run he used to climb into the upper bunk of the caboose, while a noisy gang played poker about the stove below him, and by the roof-lamp read Robert Ingersoll's speeches[6] and "The Age of Reason."[7]

Ray was a loyal-hearted fellow, and it had cost him a great deal to give up his God. He was one of the step-children of Fortune, and he had very little to show for all his hard work; the other fellow always got the best of it. He had come in too late, or too early, on several schemes that had made money. He brought with him from all his wanderings a good deal of information (more or less correct in itself, but unrelated, and therefore misleading), a high standard of personal honor, a sentimental veneration for all women, bad as well as good, and a bitter hatred of Englishmen. Thea often thought that the nicest thing about Ray was his love for Mexico and the Mexicans, who had been kind to him when he drifted, a homeless boy, over the border. In Mexico, Ray was Señor Ken-áy-dy, and when he answered to that name he was somehow a different fellow. He spoke Spanish fluently, and the sunny warmth of that tongue kept him from being quite as hard as his chin, or as narrow as his popular science.

While Ray was smoking his cigar, he and Johnny fell to talking about the great fortunes that had been made in the Southwest, and about fellows they knew who had "struck it rich."

"I guess you been in on some big deals down there?" Johnny asked trustfully.

Ray smiled and shook his head. "I've been out on some, John. I've never been exactly in on any. So far, I've either held on too long or let go too soon. But mine's coming to me, all right." Ray looked reflective. He leaned back in the shadow and dug out a rest for his elbow in the sand. "The narrowest escape I ever had, was in the Bridal Chamber. If I hadn't let go there, it would have made me rich. That was a close call."

Johnny looked delighted. "You don' say! She was silver mine, I guess?"

"I guess she was! Down at Lake Valley.[8] I put up a few hundred for the prospector, and he gave me a bunch of stock. Before we'd got anything out of it, my brother-in-law died of the fever in Cuba. My sister was beside herself to get his body back to Colorado to bury him. Seemed foolish to me, but she's the only sister I got. It's expensive for dead folks to travel, and I had to sell my stock in the mine to raise the money to get Elmer on the move. Two months afterward, the boys struck that big pocket in the rock, full of virgin silver. They named her the Bridal Chamber. It wasn't ore, you remember. It was pure, soft metal you could have melted right down into dollars. The boys cut it out with chisels. If old Elmer hadn't played that trick on me, I'd have been in for about fifty thousand. That was a close call, Spanish."

"I recollec'. When the pocket gone, the town go bust."

"You bet. Higher'n a kite. There was no vein, just a pocket in the rock that had sometime or another got filled up with molten silver. You'd think there would be more somewhere about, but *nada.* There's fools digging holes in that mountain yet."

When Ray had finished his cigar, Johnny took his mandolin and began Kennedy's favorite, "Ultimo Amor." It was now three o'clock in the afternoon, the hottest hour in the day. The narrow shelf of shadow had widened until the floor of the amphitheater was marked off in two halves, one glittering yellow, and one purple. The little boys had come back and were making a robbers' cave to enact the bold deeds of Pedro the bandit. Johnny, stretched gracefully on the sand, passed from "Ultimo Amor" to "Fluvia de Oro," and then to "Noches de Algeria," playing languidly.

Every one was busy with his own thoughts. Mrs. Tellamantez was thinking of the square in the little town in which she was born; of the white churchsteps, with people genuflecting as they passed, and the round-topped acacia trees, and the band playing in the plaza. Ray Kennedy was thinking of the future, dreaming the large Western dream of easy money, of a fortune kicked up somewhere in the hills,—an oil well, a gold mine, a ledge of copper. He always told himself, when he accepted a cigar from a newly married railroad man, that he knew enough not to marry until he had found his ideal, and could keep her like a queen. He believed that in the yellow head over there in the sand he had found his ideal, and that by the time she was old enough to marry, he would be able to keep her like a queen. He would kick it up from somewhere, when he got loose from the railroad.

Thea, stirred by tales of adventure, of the Grand Canyon and Death Valley, was recalling a great adventure of her own. Early in the summer her father had been invited to conduct a reunion of old frontiersmen, up in Wyoming, near Laramie, and he took Thea along with him to play the organ and sing patriotic songs. There they stayed at the house of an old ranchman who told them about a ridge up in the hills called Laramie Plain, where the wagon-trails of the Forty-niners and the Mormons[9] were still visible. The old man even volunteered to take Mr. Kronborg up into the hills to see this place, though it was a very long drive to make in one day. Thea had begged frantically to go along, and the old rancher, flattered by her rapt attention to his stories, had interceded for her. They set out from Laramie before daylight, behind a strong team of mules. All the way there was much talk of the Forty-niners. The old rancher had been a teamster in a freight train that used to crawl back and forth across the plains between Omaha and Cherry Creek, as Denver was then called, and he had met many a wagon train bound for California. He told of Indians and buffalo, thirst and slaughter, wanderings in snowstorms, and lonely graves in the desert.

The road they followed was a wild and beautiful one. It led up and up, by granite rocks and stunted pines, around deep ravines and echoing gorges. The top of the ridge, when they reached it, was a great flat plain, strewn with white boulders, with the wind howling over it. There was not one trail, as Thea had expected; there were a score; deep furrows, cut in the earth by heavy wagon wheels, and now grown over with dry, whitish grass. The furrows ran side by side; when one trail had been worn too deep, the next party had abandoned it and made a new trail to the right or left. They were, indeed, only old wagon ruts, running east and west, and grown over with grass. But as Thea ran about among the white stones, her skirts blowing this way and that, the wind brought to her eyes tears that might have come anyway. The old rancher picked up an iron oxshoe from one of the furrows and gave it to her for a keepsake. To the west one could see range after range of blue mountains, and at last the snowy range, with its white, windy peaks, the clouds caught here and there on their spurs. Again and again Thea had to hide her face from the cold for a moment. The wind never slept on this plain, the old man said. Every little while eagles flew over.

Coming up from Laramie, the old man had told them that he was in Brownsville, Nebraska, when the first telegraph wires were put across the Missouri River, and that the first message that ever crossed the river was "Westward the course of Empire takes its way."[10] He had been in the room when the instrument began to click, and all the men there had, without thinking what they were doing, taken off their hats, waiting bareheaded to hear the message translated. Thea remembered that message when she sighted down the wagon tracks toward the blue mountains. She told herself she would never, never forget it. The spirit of human courage seemed to live up there with the eagles. For long after, when she was moved by a Fourth-of-July oration, or a band, or a circus parade, she was apt to remember that windy ridge.

To-day she went to sleep while she was thinking about it. When Ray wakened her, the horses were hitched to the wagon and Gunner and Axel were begging for a place on the front seat. The air had cooled, the sun was setting, and the desert was on fire. Thea contentedly took the back seat with Mrs. Tellamantez. As they drove homeward the stars began to come out, pale yellow in a yellow sky, and Ray and Johnny began to sing one of those railroad ditties that are usually born on the Southern Pacific and run the length of the Santa Fé and the "Q" system before they die to give place to a new one. This was a song about a Greaser[11] dance, the refrain being something like this:—

> "Pedró, Pedró, swing high, swing low,
> And it's allamand left again;
> For there's boys that's bold and there's some that's cold,
> But the góld boys come from Spain,
> Oh, the góld boys come from Spain!"

VIII

WINTER WAS LONG in coming that year. Throughout October the days were bathed in sunlight and the air was clear as crystal. The town kept its cheerful summer aspect, the desert glistened with light, the sand hills every day went through magical changes of color. The scarlet sage bloomed late in the front yards, the cottonwood leaves were bright gold long before they fell, and it was not until November that the green on the tamarisks began to cloud and fade. There was a flurry of snow about Thanksgiving, and then December came on warm and clear.

Thea had three music pupils now, little girls whose mothers declared that Professor Wunsch was "much too severe." They took their lessons on Saturday, and this, of course, cut down her time for play. She did not really mind this because she was allowed to use the money—her pupils paid her twenty-five cents a lesson—to fit up a little room for herself upstairs in the half-story. It was the end room of the wing, and was not plastered, but was snugly lined with soft pine. The ceiling was so low that a grown person could reach it with the palm of the hand, and it sloped down on either side. There was only one window, but it was a double one and went to the floor. In October, while the days were still warm, Thea and Tillie papered the room, walls and ceiling in the same paper, small red and brown roses on a yellowish ground. Thea bought a brown cotton carpet, and her big brother, Gus, put it down for her one Sunday. She made white cheesecloth curtains and hung them on a tape. Her mother gave her an old walnut dresser with a broken mirror, and she had her own dumpy walnut single bed, and a blue washbowl and pitcher which she had drawn at a church fair lottery. At the head of her bed she had a tall round wooden hat-crate, from the clothing store. This, standing on end and draped with cretonne, made a fairly steady table for her lantern. She was not allowed to take a lamp upstairs, so Ray Kennedy gave her a railroad lantern by which she could read at night.

In winter this loft room of Thea's was bitterly cold, but against her mother's advice—and Tillie's—she always left her window open a little way. Mrs. Kronborg declared that she "had no patience with American physiology," though the lessons about the injurious effects of alcohol and tobacco were well enough for the boys. Thea asked Dr. Archie about the window, and he told her that a girl who sang must always have plenty of fresh air, or her voice would get husky, and that the cold would harden her throat. The important thing, he said, was to keep your feet warm. On very cold nights Thea always put a brick in the oven after supper, and when she went upstairs she wrapped it in an old flannel petticoat and put it in her bed. The boys, who would never heat bricks for themselves, sometimes carried off Thea's, and thought it a good joke to get ahead of her.

When Thea first plunged in between her red blankets, the cold sometimes kept her awake for a good while, and she comforted herself by remembering all she could of "Polar Explorations," a fat, calf-bound volume her father had bought from a book-agent, and by thinking about the members of Greely's party:[1] how they lay in their frozen sleeping-bags, each man hoarding the warmth of his own body and trying to make it last as long as possible against the on-coming cold that would be everlasting. After half an hour or so, a warm wave crept over her body and round, sturdy legs; she glowed like a little stove with the warmth of her own blood, and the heavy quilts and red blankets grew warm wherever they touched her, though her breath sometimes froze on the coverlid. Before daylight, her internal fires went down a little, and she often wakened to find herself drawn up into a tight ball, somewhat stiff in the legs. But that made it all the easier to get up.

The acquisition of this room was the beginning of a new era in Thea's life. It was one of the most important things that ever happened to her. Hitherto, except in summer, when she could be out of doors, she had lived in constant turmoil; the family, the day school, the Sunday-School. The clamor about her drowned the voice within herself. In the end of the wing, separated from the other upstairs sleeping-rooms by a long, cold, unfinished lumber room, her mind worked better. She thought things out more clearly. Pleasant plans and ideas occurred to her which had never come before. She had certain thoughts which were like companions, ideas which were like older and wiser friends. She left them there in the morning, when she finished dressing in the cold, and at night, when she came up with her lantern and shut the door after a busy day, she found them awaiting her. There was no possible way of heating the room, but that was fortunate, for otherwise it would have been occupied by one of her older brothers.

From the time when she moved up into the wing, Thea began to live a double life. During the day, when the hours were full of tasks, she was one of the Kronborg children, but at night she was a different person. On Friday and Saturday nights she always read for a long while after she was in bed. She had no clock, and there was no one to nag her.

Ray Kennedy, on his way from the depot to his boardinghouse, often looked up and saw Thea's light burning when the rest of the house was dark, and felt cheered as by a friendly greeting. He was a faithful soul, and many disappointments had not changed his nature. He was still, at heart, the same boy who, when he was sixteen, had settled down to freeze with his sheep in a Wyoming blizzard, and had been rescued only to play the losing game of fidelity to other charges.

Ray had no very clear idea of what might be going on in Thea's head, but he knew that something was. He used to remark to Spanish Johnny, "That girl is developing something fine." Thea was patient with Ray, even in regard to the liberties he took with her name. Outside the family, every one in Moonstone, except Wunsch and Dr. Archie, called her "Thee-a,"[2] but this seemed cold and distant to Ray, so he called her "Thee." Once, in a moment of exasperation, Thea asked him why he did this, and he explained that he once had a chum, Theodore, whose name was always abbreviated thus, and that since he was killed down on the Santa Fé, it seemed natural to call somebody "Thee." Thea sighed and submitted. She was always helpless before homely sentiment and usually changed the subject.

It was the custom for each of the different Sunday-Schools in Moonstone to give a concert on Christmas Eve. But this year all the churches were to unite and give, as was announced from the pulpits, "a semi-sacred concert of picked talent" at the opera house. The Moonstone Orchestra, under the direction of Professor Wunsch, was to play, and the most talented members of each Sunday-School were to take part in the programme. Thea was put down by the committee "for instrumental." This made her indignant, for the vocal numbers were always more popular. Thea went to the president of the committee and demanded hotly if her rival, Lily Fisher, were going to sing. The president was a big, florid, powdered woman, a fierce W.C.T.U.[3] worker, one of Thea's natural enemies. Her name was Johnson; her husband kept the livery stable, and she was called Mrs. Livery Johnson, to distinguish her from other families of the same surname. Mrs. Johnson was a prominent Baptist, and Lily Fisher was the Baptist prodigy. There was a not very Christian rivalry between the Baptist Church and Mr. Kronborg's church.

When Thea asked Mrs. Johnson whether her rival was to be allowed to sing, Mrs. Johnson, with an eagerness which told how she had waited for this moment, replied that "Lily was going to recite to be obliging, and to give other children a chance to sing." As she delivered this thrust, her eyes glittered more than the Ancient Mariner's,[4] Thea thought. Mrs. Johnson disapproved of the way in which Thea was being brought up, of a child whose chosen associates were Mexicans and sinners, and who was, as she pointedly put it, "bold with men." She so enjoyed an opportunity to rebuke Thea, that, tightly corseted as she was, she could scarcely control her breathing, and her lace and her gold watch chain rose and fell "with short, uneasy motion." Frowning, Thea turned away and walked slowly homeward. She suspected guile. Lily Fisher was the most stuck-up doll in the world, and it was certainly not like her to recite to be obliging. Nobody who could sing ever recited, because the warmest applause always went to the singers.

However, when the programme was printed in the Moonstone *Gleam*, there it was: "Instrumental solo, Thea Kronborg. Recitation, Lily Fisher."

Because his orchestra was to play for the concert, Mr. Wunsch imagined that he had been put in charge of the music, and he became arrogant. He insisted that Thea should play a "Ballade" by Reinecke.[5] When Thea consulted her mother, Mrs. Kronborg agreed with her that the "Ballade" would "never take" with a Moonstone audience. She advised Thea to play "something with variations," or, at least, "The Invitation to the Dance."

"It makes no matter what they like," Wunsch replied to Thea's entreaties. "It is time already that they learn something."

Thea's fighting powers had been impaired by an ulcerated tooth and consequent loss of sleep, so she gave in. She finally had the molar pulled, though it was a second tooth and should have been saved. The dentist was a clumsy, ignorant country boy, and Mr. Kronborg would not hear of Dr. Archie's taking Thea to a dentist in Denver, though Ray Kennedy said he could get a pass for her. What with the pain of the tooth, and family discussions about it, with trying to make Christmas presents and to keep up her school work and practicing, and giving lessons on Saturdays, Thea was fairly worn out.

On Christmas Eve she was nervous and excited. It was the first time she had ever played in the opera house, and she had never before had to face so many people. Wunsch would not let her play with her notes, and she was afraid of forgetting. Before the concert began, all the participants had to assemble on the stage and sit there to be looked at. Thea wore her white summer dress and a blue sash, but Lily Fisher had a new pink silk, trimmed with white swansdown.

The hall was packed. It seemed as if every one in Moonstone was there, even Mrs. Kohler, in her hood, and old Fritz. The seats were wooden kitchen chairs, numbered, and nailed to long planks which held them together in rows. As the floor was not raised, the chairs were all on the same level. The more interested persons in the audience peered over the heads of the people in front of them to get a good view of the stage. From the platform Thea picked out many friendly faces. There was Dr. Archie, who never went to church entertainments; there was the friendly jeweler who ordered her music for her,—he sold accordions and guitars as well as watches,—and the druggist who often lent her books, and her favorite teacher from the school. There was Ray Kennedy, with a party of freshly barbered railroad men he had brought along with him. There was Mrs. Kronborg with all the children, even Thor, who had been brought out in a new white plush coat. At the back of the hall sat a little group of Mexicans, and among them Thea caught the gleam of Spanish Johnny's white teeth, and of Mrs. Tellamantez's lustrous, smoothly coiled black hair.

After the orchestra played "Selections from Erminie,"[6] and the Baptist preacher made a long prayer, Tillie Kronborg came on with a highly colored recitation, "The Polish Boy."[7] When it was over every one breathed more freely. No committee had the courage to leave Tillie off a programme. She was accepted as a trying feature of every entertainment. The Progressive Euchre Club[8] was the only social organization in the town that entirely escaped Tillie. After Tillie sat down, the Ladies' Quartette sang, "Beloved, it is Night,"[9] and then it was Thea's turn.

The "Ballade" took ten minutes, which was five minutes too long. The audience grew restive and fell to whispering. Thea could hear Mrs. Livery Johnson's bracelets jangling as she fanned herself, and she could hear her father's nervous, ministerial cough. Thor behaved better than any one else. When Thea bowed and returned to her seat at the back of the stage there was the usual applause, but it was vigorous only from the back of the house where the Mexicans sat, and from Ray Kennedy's *claqueurs.* Any one could see that a good-natured audience had been bored.

Because Mr. Kronborg's sister was on the programme, it had also been necessary to ask the Baptist preacher's wife's cousin to sing. She was a "deep alto" from McCook, and she sang, "Thy Sentinel Am I."[10] After her came Lily Fisher. Thea's rival was also a blonde, but her hair was much heavier than Thea's, and fell in long round curls over her shoulders. She was the angel-child of the Baptists, and looked exactly like the beautiful children on soap calendars. Her pink-and-white face, her set smile of innocence, were surely born of a color-press. She had long, drooping eyelashes, a little pursed-up mouth, and narrow, pointed teeth, like a squirrel's.

Lily began:—

> "*Rock of Ages, cleft for me,* carelessly the maiden sang."[11]

Thea drew a long breath. That was the game; it was a recitation and a song in one. Lily trailed the hymn through half a dozen verses with great effect. The Baptist preacher had announced at the beginning of the concert that "owing to the length of the programme, there would be no encores." But the applause which followed Lily to her seat was such an unmistakable expression of enthusiasm that Thea had to admit Lily was justified in going back. She was attended this time by Mrs. Livery Johnson herself, crimson with triumph and gleaming-eyed, nervously rolling and unrolling a sheet of music. She took off her bracelets and played Lily's accompaniment. Lily had the effrontery to come out with, "She sang the song of Home, Sweet Home, the song that touched my heart."[12] But this did not surprise Thea; as Ray said later in the evening, "the cards had been stacked against her from the beginning." The next issue of the *Gleam* correctly stated that "unquestionably the honors of the evening must be accorded to Miss Lily Fisher." The Baptists had everything their own way.

After the concert Ray Kennedy joined the Kronborgs' party and walked home with them. Thea was grateful for his silent sympathy, even while it irritated her. She inwardly vowed that she would never take another lesson from old Wunsch. She wished that her father would not keep cheerfully singing, "When Shepherds Watched,"[13] as he marched ahead, carrying Thor. She felt that silence would become the Kronborgs for a while. As a family, they somehow seemed a little ridiculous, trooping along in the starlight. There were so many of them, for one thing. Then Tillie was so absurd. She was giggling and talking to Anna just as if she had not made, as even Mrs. Kronborg admitted, an exhibition of herself.

When they got home, Ray took a box from his overcoat pocket and slipped it into Thea's hand as he said good-night. They all hurried in to the glowing stove in the parlor. The sleepy children were sent to bed. Mrs. Kronborg and Anna stayed up to fill the stockings.

"I guess you're tired, Thea. You needn't stay up." Mrs. Kronborg's clear and seemingly indifferent eye usually measured Thea pretty accurately.

Thea hesitated. She glanced at the presents laid out on the dining-room table, but they looked unattractive. Even the brown plush monkey she had bought for Thor with such enthusiasm seemed to have lost his wise and humorous expression. She murmured, "All right," to her mother, lit her lantern, and went upstairs.

Ray's box contained a hand-painted white satin fan, with pond lilies—an unfortunate reminder. Thea smiled grimly and tossed it into her upper drawer. She was not to be consoled by toys. She undressed quickly and stood for some time in the cold, frowning in the broken looking-glass at her flaxen pig-tails, at her white neck and arms. Her own broad, resolute face set its chin at her, her eyes flashed into her own defiantly. Lily Fisher was pretty, and she was willing to be just as big a fool as people wanted her to be. Very well; Thea Kronborg wasn't. She would rather be hated than be stupid, any day. She popped into bed and read stubbornly at a queer paper book the drug-store man had given her because he couldn't sell it. She had trained herself to put her mind on what she was doing, otherwise she would have come to grief with her complicated daily schedule. She read, as intently as if she had not been flushed with anger, the strange "Musical Memories" of the Reverend H. R. Haweis.[14] At last she blew out the lantern and went to sleep. She had many curious dreams that night. In one of them Mrs. Tellamantez held her shell to Thea's ear, and she heard the roaring, as before, and distant voices calling, "Lily Fisher! Lily Fisher!"

IX

Mr. Kronborg considered Thea a remarkable child; but so were all his children remarkable. If one of the business men downtown remarked to him that he "had a mighty bright little girl, there," he admitted it, and at once began to explain what a "long head for business" his son Gus had, or that Charley was "a natural electrician," and had put in a telephone from the house to the preacher's study behind the church.

Mrs. Kronborg watched her daughter thoughtfully. She found her more interesting than her other children, and she took her more seriously, without thinking much about why she did so. The other children had to be guided, directed, kept from conflicting with one another. Charley and Gus were likely to want the same thing, and to quarrel about it. Anna often demanded unreasonable service from her older brothers; that they should sit up until after midnight to bring her home from parties when she did not like the youth who had offered himself as her escort; or that they should drive twelve miles into the country, on a winter night, to take her to a ranch dance, after they had been working hard all day. Gunner often got bored with his own clothes or stilts or sled, and wanted Axel's. But Thea, from the time she was a little thing, had her own routine. She kept out of every one's way, and was hard to manage only when the other children interfered with her. Then there was trouble indeed: bursts of temper which used to alarm Mrs. Kronborg. "You ought to know enough to let Thea alone. She lets you alone," she often said to the other children.

One may have staunch friends in one's own family, but one seldom has admirers. Thea, however, had one in the person of her addle-pated aunt, Tillie Kronborg. In older countries, where dress and opinions and manners are not so thoroughly standardized as in our own West, there is a belief that people who are foolish about the more obvious things of life are apt to have peculiar insight into what lies beyond the obvious. The old woman who can never learn not to put the kerosene can on the stove, may yet be able to tell fortunes, to persuade a backward child to grow, to cure warts, or to tell people what to do with a young girl who has gone melancholy. Tillie's mind was a curious machine; when she was awake it went round like a wheel when the belt has slipped off, and when she was asleep she dreamed follies. But she had intuitions. She knew, for instance, that Thea was different from the other Kronborgs, worthy though they all were. Her romantic imagination found possibilities in her niece. When she was sweeping or ironing, or turning the ice-cream freezer at a furious rate, she often built up brilliant futures for Thea, adapting freely the latest novel she had read.

Tillie made enemies for her niece among the church people because, at sewing societies and church suppers, she sometimes spoke vauntingly, with a toss of her head, just as if Thea's "wonderfulness" were an accepted fact in Moonstone, like Mrs. Archie's stinginess, or Mrs. Livery Johnson's duplicity. People declared that, on this subject, Tillie made them tired.

Tillie belonged to a dramatic club that once a year performed in the Moonstone Opera House such plays as "Among the Breakers," and "The Veteran of 1812."[1] Tillie played character parts, the flirtatious old maid or the spiteful *intrigante.* She used to study her parts up in the attic at home. While she was committing the lines, she got Gunner or Anna to hold the book for her, but when she began "to bring out the expression," as she said, she used, very timorously, to ask Thea to hold the book. Thea was usually—not always—agreeable about it. Her mother had told her that, since she had some influence with Tillie, it would be a good thing for them all if she could tone her down a shade and "keep her from taking on any worse than need be." Thea would sit on the foot of Tillie's bed, her feet tucked under her, and stare at the silly text. "I wouldn't make so much fuss, there, Tillie," she would remark occasionally; "I don't see the point in it"; or, "What do you pitch your voice so high for? It don't carry half as well."

"I don't see how it comes Thea is so patient with Tillie," Mrs. Kronborg more than once remarked to her husband. "She ain't patient with most people, but it seems like she's got a peculiar patience for Tillie."

Tillie always coaxed Thea to go "behind the scenes" with her when the club presented a play, and help her with her make-up. Thea hated it, but she always went. She felt as if she had to do it. There was something in Tillie's adoration of her that compelled her. There was no family impropriety that Thea was so much ashamed of as Tillie's "acting" and yet she was always being dragged in to assist her. Tillie simply had her, there. She didn't know why, but it was so. There was a string in her somewhere that Tillie could pull; a sense of obligation to Tillie's misguided aspirations. The saloon-keepers had some such feeling of responsibility toward Spanish Johnny.

The dramatic club was the pride of Tillie's heart, and her enthusiasm was the principal factor in keeping it together. Sick or well, Tillie always attended rehearsals, and was always urging the young people, who took rehearsals lightly, to "stop fooling and begin now." The young men—bank clerks, grocery clerks, insurance agents—played tricks, laughed at Tillie, and "put it up on each other" about seeing her home; but they often went to tiresome rehearsals just to oblige her. They were good-natured young fellows. Their trainer and stage-manager was young Upping, the jeweler who ordered Thea's music for her. Though barely thirty, he had followed half a dozen professions, and had once been a violinist in the orchestra of the Andrews Opera Company, then well known in little towns throughout Colorado and Nebraska.

By one amazing indiscretion Tillie very nearly lost her hold upon the Moonstone Drama Club. The club had decided to put on "The Drummer Boy of Shiloh,"[2] a very ambitious undertaking because of the many supers needed and the scenic difficulties of the act which took place in Andersonville Prison. The members of the club consulted together in Tillie's absence as to who should play the part of the drummer boy. It must be taken by a very young person, and village boys of that age are self-conscious and are not apt at memorizing. The part was a long one, and clearly it must be given to a girl. Some members of the club suggested Thea Kronborg, others advocated Lily Fisher. Lily's partisans urged that she was much prettier than Thea, and had a much "sweeter disposition." Nobody denied these facts. But there was nothing in the least boyish about Lily, and she sang all songs and played all parts alike. Lily's simper was popular, but it seemed not quite the right thing for the heroic drummer boy.

Upping, the trainer, talked to one and another: "Lily's all right for girl parts," he insisted, "but you've got to get a girl with some ginger in her for this. Thea's got the voice, too. When she sings, 'Just Before the Battle, Mother,'[3] she'll bring down the house."

When all the members of the club had been privately consulted, they announced their decision to Tillie at the first regular meeting that was called to cast the parts. They expected Tillie to be overcome with joy, but, on the contrary, she seemed embarrassed. "I'm afraid Thea hasn't got time for that," she said jerkily. "She is always so busy with her music. Guess you'll have to get somebody else."

The club lifted its eyebrows. Several of Lily Fisher's friends coughed. Mr. Upping flushed. The stout woman who always played the injured wife called Tillie's attention to the fact that this would be a fine opportunity for her niece to show what she could do. Her tone was condescending.

Tillie threw up her head and laughed; there was something sharp and wild about Tillie's laugh—when it was not a giggle. "Oh, I guess Thea hasn't got time to do any showing off. Her time to show off ain't come yet. I expect she'll make us all sit up when it does. No use asking her to take the part. She'd turn her nose up at it. I guess they'd be glad to get her in the Denver Dramatics, if they could."

The company broke up into groups and expressed their amazement. Of course all Swedes were conceited, but they would never have believed that all the conceit of all the Swedes put together would reach such a pitch as this. They confided to each other that Tillie was "just a little off, on the subject of her niece," and agreed that it would be as well not to excite her further. Tillie got a cold reception at rehearsals for a long while afterward, and Thea had a crop of new enemies without even knowing it.

X

WUNSCH AND OLD FRITZ and Spanish Johnny celebrated Christmas together, so riotously that Wunsch was unable to give Thea her lesson the next day. In the middle of the vacation week Thea went to the Kohlers' through a soft, beautiful snowstorm. The air was a tender blue-gray, like the color on the doves that flew in and out of the white dove-house on the post in the Kohlers' garden. The sand hills looked dim and sleepy. The tamarisk hedge was full of snow, like a foam of blossoms drifted over it. When Thea opened the gate, old Mrs. Kohler was just coming in from the chicken yard, with five fresh eggs in her apron and a pair of old top-boots on her feet. She called Thea to come and look at a bantam egg, which she held up proudly. Her bantam hens were remiss in zeal, and she was always delighted when they accomplished anything. She took Thea into the sitting-room, very warm and smelling of food, and brought her a plateful of little Christmas cakes, made according to old and hallowed formulæ, and put them before her while she warmed her feet. Then she went to the door of the kitchen stairs and called: "Herr Wunsch, Herr Wunsch!"

Wunsch came down wearing an old wadded jacket, with a velvet collar. The brown silk was so worn that the wadding stuck out almost everywhere. He avoided Thea's eyes when he came in, nodded without speaking, and pointed directly to the piano stool. He was not so insistent upon the scales as usual, and throughout the little sonata of Mozart's she was studying, he remained languid and absent-minded. His eyes looked very heavy, and he kept wiping them with one of the new silk handkerchiefs Mrs. Kohler had given him for Christmas. When the lesson was over he did not seem inclined to talk. Thea, loitering on the stool, reached for a tattered book she had taken off the music-rest when she sat down. It was a very old

Leipsic[1] edition of the piano score of Gluck's "Orpheus."[2] She turned over the pages curiously.

"Is it nice?" she asked.

"It is the most beautiful opera ever made," Wunsch declared solemnly. "You know the story, eh? How, when she die, Orpheus went down below for his wife?"

"Oh, yes, I know. I didn't know there was an opera about it, though. Do people sing this now?"

"*Aber ja!* What else? You like to try? See." He drew her from the stool and sat down at the piano. Turning over the leaves to the third act, he handed the score to Thea. "Listen, I play it through and you get the *rhythmus. Eins, zwei, drei, vier.*" He played through Orpheus' lament, then pushed back his cuffs with awakening interest and nodded at Thea. "Now, *vom blatt, mit mir.*"[3]

"Ach, ich habe sie verloren,
All' mein Glück ist nun dahin."[4]

Wunsch sang the aria with much feeling. It was evidently one that was very dear to him.

"*Noch einmal,* alone, yourself." He played the introductory measures, then nodded at her vehemently, and she began:—

"Ach, ich habe sie verloren."

When she finished, Wunsch nodded again. "*Schön,*" he muttered as he finished the accompaniment softly. He dropped his hands on his knees and looked up at Thea. "That is very fine, eh? There is no such beautiful melody in the world. You can take the book for one week and learn something, to pass the time. It is good to know—always. *Euridice, Eu—ri—di—ce, weh dass ich auf Erden bin!*"[5] he sang softly, playing the melody with his right hand.

Thea, who was turning over the pages of the third act, stopped and scowled at a passage. The old German's blurred eyes watched her curiously.

"For what do you look so, *immer?*"[6] puckering up his own face. "You see something a little difficult, may-be, and you make such a face like it was an enemy."

Thea laughed, disconcerted. "Well, difficult things are enemies, aren't they? When you have to get them?"

Wunsch lowered his head and threw it up as if he were butting something. "Not at all! By no means." He took the book from her and looked at it. "Yes, that is not so easy, there. This is an old book. They do not print it so now any more, I think. They leave it out, may-be. Only one woman could sing that good."

Thea looked at him in perplexity.

Wunsch went on. "It is written for alto, you see. A woman sings the part, and there was only one to sing that good in there. You understand? Only one!"[7] He glanced at her quickly and lifted his red forefinger upright before her eyes.

Thea looked at the finger as if she were hypnotized. "Only one?" she asked breathlessly; her hands, hanging at her sides, were opening and shutting rapidly.

Wunsch nodded and still held up that compelling finger. When he dropped his hands, there was a look of satisfaction in his face.

"Was she very great?"

Wunsch nodded.

"Was she beautiful?"

"*Aber gar nicht!* Not at all. She was ugly; big mouth, big teeth, no figure, nothing at all," indicating a luxuriant bosom by sweeping his hands over his chest. "A pole, a post! But for the voice—*ach!* She have something in there, behind the eyes," tapping his temples.

Thea followed all his gesticulations intently. "Was she German?"

"No, *Spanisch.*" He looked down and frowned for a moment. "*Ach,* I tell you, she look like the Frau Tellamantez, some-thing. Long face, long chin, and ugly al-so."

"Did she die a long while ago?"

"Die? I think not. I never hear, anyhow. I guess she is alive somewhere in the world; Paris, may-be. But old, of course. I hear her when I was a youth. She is too old to sing now any more."

"Was she the greatest singer you ever heard?"

Wunsch nodded gravely. "Quite so. She was the most—" he hunted for an English word, lifted his hand over his head and snapped his fingers noiselessly in the air, enunciating fiercely, "*künst-ler-isch!*"[8] The word seemed to glitter in his uplifted hand, his voice was so full of emotion.

Wunsch rose from the stool and began to button his wadded jacket, preparing to return to his half-heated room in the loft. Thea regretfully put on her cloak and hood and set out for home.

When Wunsch looked for his score late that afternoon, he found that Thea had not forgotten to take it with her. He smiled his loose, sarcastic smile, and thoughtfully rubbed his stubbly chin with his red fingers. When Fritz came home in the early blue twilight the snow was flying faster, Mrs. Kohler was cooking *Hasenpfeffer*[9] in the kitchen, and the professor was seated at the piano, playing the Gluck, which he knew by heart. Old Fritz took off his shoes quietly behind the stove and lay down on the lounge before his masterpiece, where the firelight was playing over the walls of Moscow. He listened, while the room grew darker and the windows duller. Wunsch always came back to the same thing:—

"*Ach, ich habe sie verloren,*

. . . . .

Euridice, Euridice!"

From time to time Fritz sighed softly. He, too, had lost a Euridice.

XI

ONE SATURDAY, late in June, Thea arrived early for her lesson. As she perched herself upon the piano stool,—a wobbly, old-fashioned thing that worked on a creaky screw,—she gave Wunsch a side glance, smiling. "You must not be cross to me to-day. This is my birthday."

"So?" he pointed to the keyboard.

After the lesson they went out to join Mrs. Kohler, who had asked Thea to come early, so that she could stay and smell the linden bloom. It was one of those still days of intense light, when every particle of mica in the soil flashed like a little mirror, and the glare from the plain below seemed more intense than the rays from above. The sand ridges ran glittering gold out to where the mirage licked them up, shining and steaming like a lake in the tropics. The sky looked like blue lava, forever incapable of clouds,—a turquoise bowl that was the lid of the desert. And yet within Mrs. Kohler's green patch the water dripped, the beds had all been hosed, and the air was fresh with rapidly evaporating moisture.

The two symmetrical linden trees were the proudest things in the garden. Their sweetness embalmed all the air. At every turn of the paths,—whether one went to see the hollyhocks or the bleeding heart, or to look at the purple morning-glories that ran over the bean-poles,—wherever one went, the sweetness of the lindens struck one afresh and one always came back to them. Under the round leaves, where the waxen yellow blossoms hung, bevies of wild bees were buzzing. The tamarisks were still pink, and the flower-beds were doing their best in honor of the linden festival. The white dove-house was shining with a fresh coat of paint, and the pigeons were crooning contentedly, flying down often to drink at the drip from the water tank. Mrs. Kohler, who was transplanting pansies, came up with her trowel and told Thea it was lucky to have your birthday when the lindens were in bloom, and that she must go and look at the sweet peas. Wunsch accompanied her, and as they walked between the flower-beds he took Thea's hand.

"Es flüstern und sprechen die Blumen,"—

he muttered. "You know that von Heine? *Im leuchtenden Sommermorgen?*"[1] He looked down at Thea and softly pressed her hand.

"No, I don't know it. What does *flüstern* mean?"

"*Flüstern?*—to whisper. You must begin now to know such things. That is necessary. How many birthdays?"

"Thirteen. I'm in my 'teens now. But how can I know words like that? I only know what you say at my lessons. They don't teach German at school. How can I learn?"

"It is always possible to learn when one likes," said Wunsch. His words were peremptory, as usual, but his tone was mild, even confidential. "There is always a way. And if some day you are going to sing, it is necessary to know well the German language."

Thea stooped over to pick a leaf of rosemary. How did Wunsch know that, when the very roses on her wall-paper had never heard it? "But am I going to?" she asked, still stooping.

"That is for you to say," returned Wunsch coldly. "You would better marry some *Jacob* here and keep the house for him, may-be? That is as one desires."

Thea flashed up at him a clear, laughing look. "No, I don't want to do that. You know," she brushed his coat-sleeve quickly with her yellow head. "Only how can I learn anything here? It's so far from Denver."

Wunsch's loose lower lip curled in amusement. Then, as if he suddenly remembered something, he spoke seriously. "Nothing is far and nothing is near, if one desires. The world is little, people are little, human life is little. There is only one big thing—desire. And before it, when it is big, all is little. It brought Columbus across the sea in a little boat, *und so weiter.*"[2] Wunsch made a grimace, took his pupil's hand and drew her toward the grape arbor. "Hereafter I will more speak to you in German. Now, sit down and I will teach you for your birthday that little song. Ask me the words you do not know already. Now: *Im leuchtenden Sommermorgen.*"

Thea memorized quickly because she had the power of listening intently. In a few moments she could repeat the eight lines for him. Wunsch nodded encouragingly and they went out of the arbor into the sunlight again. As they went up and down the gravel paths between the flowerbeds, the white and yellow butterflies kept darting before them, and the pigeons were washing their pink feet at the drip and crooning in their husky bass. Over and over again Wunsch made her say the lines to him. "You see it is nothing. If you learn a great many of the *Lieder*, you will know the German language already. *Weiter, nun.*"[3] He would incline his head gravely and listen.

"*Im leuchtenden Sommermorgen*
Geh' ich im Garten herum;
Es flüstern und sprechen die Blumen,
Ich aber, ich wandte stumm.

"*Es flüstern und sprechen die Blumen*
Und schau'n mitleidig mich an:
'*Sei unserer Schwester nicht böse,*
Du trauriger, blasser Mann!' "

(In the soft-shining summer morning
I wandered the garden within.
The flowers they whispered and murmured,
But I, I wandered dumb.

The flowers they whisper and murmur,
And me with compassion they scan:
"Oh, be not harsh to our sister,
Thou sorrowful, death-pale man!")

Wunsch had noticed before that when his pupil read anything in verse the character of her voice changed altogether; it was no longer the voice which spoke the speech of Moonstone. It was a soft, rich contralto, and she read quietly; the feeling was in the voice itself, not indicated by emphasis or change of pitch. She repeated the little verses musically, like a song, and the entreaty of the flowers was even softer than the rest, as the shy speech of flowers might be, and she ended with the voice suspended, almost with a rising inflection. It was a nature-voice, Wunsch told himself, breathed from the creature and apart from language, like the sound of the wind in the trees, or the murmur of water.

"What is it the flowers mean when they ask him not to be harsh to their sister, eh?" he asked, looking down at her curiously and wrinkling his dull red forehead.

Thea glanced at him in surprise. "I suppose he thinks they are asking him not to be harsh to his sweetheart—or some girl they remind him of."

"And why *trauriger, blasser Mann?*"

They had come back to the grape arbor, and Thea picked out a sunny place on the bench, where a tortoise-shell cat was stretched at full length. She sat down, bending over the cat and teasing his whiskers. "Because he had been awake all night, thinking about her, wasn't it? Maybe that was why he was up so early."

Wunsch shrugged his shoulders. "If he think about her all night already, why do you say the flowers remind him?"

Thea looked up at him in perplexity. A flash of comprehension lit her face and she smiled eagerly. "Oh, I didn't mean 'remind' in that way! I didn't mean they brought her to his mind! I meant it was only when he came out in the morning, that she seemed to him like that,—like one of the flowers."

"And before he came out, how did she seem?"

This time it was Thea who shrugged her shoulders. The warm smile left her face. She lifted her eyebrows in annoyance and looked off at the sand hills.

Wunsch persisted. "Why you not answer me?"

"Because it would be silly. You are just trying to make me say things. It spoils things to ask questions."

Wunsch bowed mockingly; his smile was disagreeable. Suddenly his face grew grave, grew fierce, indeed. He pulled himself up from his clumsy stoop and folded his arms. "But it is necessary to know if you know somethings. Somethings cannot be taught. If you not know in the beginning, you not know in the end. For a singer there must be something in the inside from the beginning. I shall not be long in this place, may-be, and I like to know. Yes,"—he ground his heel in the gravel,—"yes, when you are barely six, you must know that already. That is the beginning of all things; *der Geist, die Phantasie.*[4] It must be in the baby, when it makes its first cry, like *der Rhythmus*, or it is not to be. You have some voice already, and if in the beginning, when you are with things-to-play, you know that what you will not tell me, then you can learn to sing, may-be."

Wunsch began to pace the arbor, rubbing his hands together. The dark flush of his face had spread up under the iron-gray bristles on his head. He was talking to himself, not to Thea. Insidious power of the linden bloom! "Oh, much you can learn! *Aber nicht die Americanischen Fräulein.*[5] They have nothing inside them," striking his chest with both fists. "They are like the ones in the *Märchen*,[6] a grinning face and hollow in the insides. Something they can learn, oh, yes, may-be! But the secret—what make the rose to red, the sky to blue, the man to love—*in der Brust, in der Brust* it is, *und ohne dieses giebt es keine Kunst, giebt es keine Kunst!*"[7] He threw up his square hand and shook it, all the fingers apart and wagging. Purple and breathless he went out of the arbor and into the house, without saying good-bye. These outbursts frightened Wunsch. They were always harbingers of ill.

Thea got her music-book and stole quietly out of the garden. She did not go home, but wandered off into the sand dunes, where the prickly pear was in blossom and the green lizards were racing each other in the glittering light. She was shaken by a passionate excitement. She did not altogether understand what Wunsch was talking about; and yet, in a way she knew. She knew, of course, that there was something about her that was different. But it was more like a friendly spirit than like anything that was a part of herself. She brought everything to it, and it answered her; happiness consisted of that backward and forward movement of herself. The something came and went, she never knew how. Sometimes she hunted for it and could not find it; again, she lifted her eyes from a book, or stepped out of doors, or wakened in the morning, and it was there,—under her cheek, it usually seemed to be, or over her breast,—a kind of warm sureness. And when it was there, everything was more interesting and beautiful, even people. When this companion was with her, she could get the most wonderful things out of Spanish Johnny, or Wunsch, or Dr. Archie.

On her thirteenth birthday she wandered for a long while about the sand ridges, picking up crystals and looking into the yellow prickly-pear blossoms with their thousand stamens. She looked at the sand hills until she wished she *were* a sand hill. And yet she knew that she was going to leave them all behind some day. They would be changing all day long, yellow and purple and lavender, and she would not be there. From that day on, she felt there was a secret between her and Wunsch. Together they had lifted a lid, pulled out a drawer, and looked at something. They hid it away and never spoke of what they had seen; but neither of them forgot it.

XII

ONE JULY NIGHT, when the moon was full, Dr. Archie was coming up from the depot, restless and discontented, wishing there were something to do. He carried his straw hat in his hand, and kept brushing his hair back from his forehead with a purposeless, unsatisfied gesture. After he passed Uncle Billy Beemer's cottonwood grove, the sidewalk ran out of the shadow into the white moonlight and crossed the sand gully on high posts, like a bridge. As the doctor approached this trestle, he saw a white figure, and recognized Thea Kronborg. He quickened his pace and she came to meet him.

"What are you doing out so late, my girl?" he asked as he took her hand.

"Oh, I don't know. What do people go to bed so early for? I'd like to run along before the houses and screech at them. Isn't it glorious out here?"

The young doctor gave a melancholy laugh and pressed her hand.

"Think of it," Thea snorted impatiently. "Nobody up but us and the rabbits! I've started up half a dozen of 'em. Look at that little one down there now,"—she stooped and pointed. In the gully below them there was, indeed, a little rabbit with a white spot of a tail, crouching down on the sand, quite motionless. It seemed to be lapping up the moonlight like cream. On the other side of the walk, down in the ditch, there was a patch of tall, rank sunflowers, their shaggy leaves white with dust. The moon stood over the cottonwood grove. There was no wind, and no sound but the wheezing of an engine down on the tracks.

"Well, we may as well watch the rabbits." Dr. Archie sat down on the sidewalk and let his feet hang over the edge. He pulled out a smooth linen handkerchief that smelled of German cologne water.

"Well, how goes it? Working hard? You must know about all Wunsch can teach you by this time."

Thea shook her head. "Oh, no, I don't, Dr. Archie. He's hard to get at, but he's been a real musician in his time. Mother says she believes he's forgotten more than the music-teachers down in Denver ever knew."

"I'm afraid he won't be around here much longer," said Dr. Archie. "He's been making a tank of himself lately. He'll be pulling his freight one of these days. That's the way they do, you know. I'll be sorry on your account." He paused and ran his fresh handkerchief over his face. "What the deuce are we all here for anyway, Thea?" he said abruptly.

"On earth, you mean?" Thea asked in a low voice.

"Well, primarily, yes. But secondarily, why are we in Moonstone? It isn't as if we'd been born here. You were, but Wunsch wasn't, and I wasn't. I suppose I'm here because I married as soon as I got out of medical school and had to get a practice quick. If you hurry things, you always get left in the end. I don't learn anything here, and as for the people— In my own town in Michigan, now, there were people who liked me on my father's account, who had even known my grandfather. That meant something. But here it's all like the sand: blows north one day and south the next. We're all a lot of gamblers without much nerve, playing for small stakes. The railroad is the one real fact in this country. That has to be; the world has to be got back and forth. But the rest of us are here just because it's the end of a run and the engine has to have a drink. Some day I'll get up and find my hair turning gray, and I'll have nothing to show for it."

Thea slid closer to him and caught his arm. "No, no. I won't let you get gray. You've got to stay young for me. I'm getting young now, too."

Archie laughed. "Getting?"

"Yes. People aren't young when they're children. Look at Thor, now; he's just a little old man. But Gus has a sweetheart, and he's young!"

"Something in that!" Dr. Archie patted her head, and then felt the shape of her skull gently, with the tips of his fingers. "When you were little, Thea, I used always to be curious about the shape of your head. You seemed to have more inside it than most youngsters. I haven't examined it for a long time. Seems to be the usual shape, but uncommonly hard, some how. What are you going to do with yourself, anyway?"

"I don't know."

"Honest, now?" He lifted her chin and looked into her eyes.

Thea laughed and edged away from him.

"You've got something up your sleeve, haven't you? Anything you like; only don't marry and settle down here without giving yourself a chance, will you?"

"Not much. See, there's another rabbit!"

"That's all right about the rabbits, but I don't want you to get tied up. Remember that."

Thea nodded. "Be nice to Wunsch, then. I don't know what I'd do if he went away."

"You've got older friends than Wunsch here, Thea."

"I know." Thea spoke seriously and looked up at the moon, propping her chin on her hand. "But Wunsch is the only one that can teach me what I want to know. I've got to learn to do something well, and that's the thing I can do best."

"Do you want to be a music-teacher?"

"Maybe, but I want to be a good one. I'd like to go to Germany to study, some day. Wunsch says that's the best place,—the only place you can really learn." Thea hesitated and then went on nervously, "I've got a book that says so, too. It's called 'My Musical Memories.' It made me want to go to Germany even before Wunsch said anything. Of course it's a secret. You're the first one I've told."

Dr. Archie smiled indulgently. "That's a long way off. Is that what you've got in your hard noddle?" He put his hand on her hair, but this time she shook him off.

"No, I don't think much about it. But you talk about going, and a body has to have something to go *to!*"

"That's so." Dr. Archie sighed. "You're lucky if you have. Poor Wunsch, now, he hasn't. What do such fellows come out here for? He's been asking me about my mining stock, and about mining towns. What would he do in a mining town? He wouldn't know a piece of ore if he saw one. He's got nothing to sell that a mining town wants to buy. Why don't those old fellows stay at home? We won't need them for another hundred years. An engine wiper can get a job, but a piano player! Such people can't make good."

"My grandfather Alstrom was a musician, and he made good."

Dr. Archie chuckled. "Oh, a Swede can make good anywhere, at anything! You've got that in your favor, miss. Come, you must be getting home."

Thea rose. "Yes, I used to be ashamed of being a Swede, but I'm not any more. Swedes are kind of common, but I think it's better to be *something.*"

"It surely is! How tall you are getting. You come above my shoulder now."

"I'll keep on growing, don't you think? I particularly want to be tall. Yes, I guess I must go home. I wish there'd be a fire."

"A fire?"

"Yes, so the fire-bell would ring and the roundhouse whistle would blow, and everybody would come running out. Sometime I'm going to ring the fire-bell myself and stir them all up."

"You'd be arrested."

"Well, that would be better than going to bed."

"I'll have to lend you some more books."

Thea shook herself impatiently. "I can't read every night."

Dr. Archie gave one of his low, sympathetic chuckles as he opened the gate for her. "You're beginning to grow up, that's what's the matter with you. I'll have to keep an eye on you. Now you'll have to say good-night to the moon."

"No, I won't. I sleep on the floor now, right in the moonlight. My window comes down to the floor, and I can look at the sky all night."

She shot round the house to the kitchen door, and Dr. Archie watched her disappear with a sigh. He thought of the hard, mean, frizzy little woman who kept his house for him; once the belle of a Michigan town, now dry and withered up at thirty. "If I had a daughter like Thea to watch," he reflected, "I wouldn't mind anything. I wonder if all of my life's going to be a mistake just because I made a big one then? Hardly seems fair."

Howard Archie was "respected" rather than popular in Moonstone. Everyone recognized that he was a good physician, and a progressive Western town likes to be able to point to a handsome, well-set-up, well-dressed man among its citizens. But a great many people thought Archie "distant," and they were right. He had the uneasy manner of a man who is not among his own kind, and who has not seen enough of the world to feel that all people are in some sense his own kind. He knew that every one was curious about his wife, that she played a sort of character part in Moonstone, and that people made fun of her, not very delicately. Her own friends—most of them women who were distasteful to Archie—liked to ask her to contribute to church charities, just to see how mean she could be. The little, lop-sided cake at the church supper, the cheapest pin-cushion, the skimpiest apron at the bazaar, were always Mrs. Archie's contribution.

All this hurt the doctor's pride. But if there was one thing he had learned, it was that there was no changing Belle's nature. He had married a mean woman; and he must accept the consequences. Even in Colorado he would have had no pretext for divorce, and, to do him justice, he had never thought of such a thing. The tenets of the Presbyterian Church in which he had grown up, though he had long ceased to believe in them, still influenced his conduct and his conception of propriety. To him there was something vulgar about divorce. A divorced man was a disgraced man; at least, he had exhibited his hurt, and made it a matter for common gossip. Respectability was so necessary to Archie that he was willing to pay a high price for it. As long as he could keep up a decent exterior, he could manage to get on; and if he could have concealed his wife's littleness from all his friends, he would scarcely have complained. He was more afraid of pity than he was of any unhappiness. Had there been another woman for whom he cared greatly, he might have had plenty of courage; but he was not likely to meet such a woman in Moonstone.

There was a puzzling timidity in Archie's make-up. The thing that held his shoulders stiff, that made him resort to a mirthless little laugh when he was talking to dull people, that made him sometimes stumble over rugs and carpets, had its counterpart in his mind. He had not the courage to be an honest thinker. He could comfort himself by evasions and compromises. He consoled himself for his own marriage by telling himself that other people's were not much better. In his work he saw pretty deeply into marital relations in Moonstone, and he could honestly say that there were not many of his friends whom he envied. Their wives seemed to suit them well enough, but they would never have suited him.

Although Dr. Archie could not bring himself to regard marriage merely as a social contract, but looked upon it as somehow made sacred by a church in which he did not believe,—as a physician he knew that a young man whose marriage is merely nominal must yet go on living his life. When he went to Denver or to Chicago, he drifted about in careless company where gayety and good-humor can be bought, not because he had any taste for such society, but because he honestly believed that anything was better than divorce. He often told himself that "hanging and wiving go by destiny." If wiving went badly with a man,—and it did oftener than not,—then he must do the best he could to keep up appearances and help the tradition of domestic happiness along. The Moonstone gossips, assembled in Mrs. Smiley's millinery and notion store, often discussed Dr. Archie's politeness to his wife, and his pleasant manner of speaking about her. "Nobody has ever got a thing out of him yet," they agreed. And it was certainly not because no one had ever tried.

When he was down in Denver, feeling a little jolly, Archie could forget how unhappy he was at home, and could even make himself believe that he missed his wife. He always bought her presents, and would have liked to send her flowers if she had not repeatedly told him never to send her anything but bulbs,—which did not appeal to him in his expansive moments. At the Denver Athletic Club banquets, or at dinner with his colleagues at the Brown Palace Hotel, he sometimes spoke sentimentally about "little Mrs. Archie," and he always drank the toast "to our wives, God bless them!" with gusto.

The determining factor about Dr. Archie was that he was romantic. He had married Belle White because he was romantic—too romantic to know anything about women, except what he wished them to be, or to repulse a pretty girl who had set her cap for him. At medical school, though he was a rather wild boy in behavior, he had always disliked coarse jokes and vulgar stories. In his old Flint's Physiology there was still a poem he had pasted there when he was a student; some verses by Dr. Oliver Wendell Holmes[1] about the ideals of the medical profession. After so much and such disillusioning experience with it, he still had a romantic feeling about the human body; a sense that finer things dwelt in it than could be explained by anatomy. He never jested about birth or death or marriage, and did not like to hear other doctors do it. He was a good nurse, and had a reverence for the bodies of women and children. When he was tending them, one saw him at his best. Then his constraint and self-consciousness fell away from him. He was easy, gentle, competent, master of himself and of other people. Then the idealist in him was not afraid of being discovered and ridiculed.

In his tastes, too, the doctor was romantic. Though he read

Balzac all the year through, he still enjoyed the Waverley Novels[2] as much as when he had first come upon them, in thick leather-bound volumes, in his grandfather's library. He nearly always read Scott on Christmas and holidays, because it brought back the pleasures of his boyhood so vividly. He liked Scott's women. Constance de Beverley and the minstrel girl in "The Fair Maid of Perth,"[3] not the Duchesse de Langeais,[4] were his heroines. But better than anything that ever got from the heart of a man into printer's ink, he loved the poetry of Robert Burns.[5] "Death and Dr. Hornbook" and "The Jolly Beggars," Burns's "Reply to his Tailor," he often read aloud to himself in his office, late at night, after a glass of hot toddy. He used to read "Tam o'Shanter" to Thea Kronborg, and he got her some of the songs, set to the old airs for which they were written. He loved to hear her sing them. Sometimes when she sang, "Oh, wert thou in the cauld blast," the doctor and even Mr. Kronborg joined in. Thea never minded if people could not sing; she directed them with her head and somehow carried them along. When her father got off the pitch she let her own voice out and covered him.

XIII

AT THE BEGINNING OF JUNE, when school closed, Thea had told Wunsch that she didn't know how much practicing she could get in this summer because Thor had his worst teeth still to cut.

"My God! all last summer he was doing that!" Wunsch exclaimed furiously.

"I know, but it takes them two years, and Thor is slow," Thea answered reprovingly.

The summer went well beyond her hopes, however. She told herself that it was the best summer of her life, so far. Nobody was sick at home, and her lessons were uninterrupted. Now that she had four pupils of her own and made a dollar a week, her practicing was regarded more seriously by the household. Her mother had always arranged things so that she could have the parlor four hours a day in summer. Thor proved a friendly ally. He behaved handsomely about his molars, and never objected to being pulled off into remote places in his cart. When Thea dragged him over the hill and made a camp under the shade of a bush or a bank, he would waddle about and play with his blocks, or bury his monkey in the sand and dig him up again. Sometimes he got into the cactus and set up a howl, but usually he let his sister read peacefully, while he coated his hands and face, first with an all-day sucker and then with gravel.

Life was pleasant and uneventful until the first of September, when Wunsch began to drink so hard that he was unable to appear when Thea went to take her mid-week lesson, and Mrs. Kohler had to send her home after a tearful apology. On Saturday morning she set out for the Kohlers' again, but on her way, when she was crossing the ravine, she noticed a woman sitting at the bottom of the gulch, under the railroad trestle. She turned from her path and saw that it was Mrs. Tellamantez, and she seemed to be doing drawn-work. Then Thea noticed that there was something beside her, covered up with a purple and yellow Mexican blanket. She ran up the gulch and called to Mrs. Tellamantez. The Mexican woman held up a warning finger. Thea glanced at the blanket and recognized a square red hand which protruded. The middle finger twitched slightly.

"Is he hurt?" she gasped.

Mrs. Tellamantez shook her head. "No; very sick. He knows nothing," she said quietly, folding her hands over her drawn-work.

Thea learned that Wunsch had been out all night, that this morning Mrs. Kohler had gone to look for him and found him under the trestle covered with dirt and cinders. Probably he had been trying to get home and had lost his way. Mrs. Tellamantez was watching beside the unconscious man while Mrs. Kohler and Johnny went to get help.

"You better go home now, I think," said Mrs. Tellamantez, in closing her narration.

Thea hung her head and looked wistfully toward the blanket.

"Couldn't I just stay till they come?" she asked. "I'd like to know if he's very bad."

"Bad enough," sighed Mrs. Tellamantez, taking up her work again.

Thea sat down under the narrow shade of one of the trestle posts and listened to the locusts rasping in the hot sand while she watched Mrs. Tellamantez evenly draw her threads. The blanket looked as if it were over a heap of bricks.

"I don't see him breathing any," she said anxiously.

"Yes, he breathes," said Mrs. Tellamantez, not lifting her eyes.

It seemed to Thea that they waited for hours. At last they heard voices, and a party of men came down the hill and up the gulch. Dr. Archie and Fritz Kohler came first; behind were Johnny and Ray, and several men from the roundhouse. Ray had the canvas litter that was kept at the depot for accidents on the road. Behind them trailed half a dozen boys who had been hanging round the depot.

When Ray saw Thea, he dropped his canvas roll and hurried forward. "Better run along home, Thee. This is ugly business." Ray was indignant that anybody who gave Thea music lessons should behave in such a manner.

Thea resented both his proprietary tone and his superior virtue.

"I won't. I want to know how bad he is. I'm not a baby!" she exclaimed indignantly, stamping her foot into the sand.

Dr. Archie, who had been kneeling by the blanket, got up and came toward Thea, dusting his knees. He smiled and nodded confidentially. "He'll be all right when we get him home. But he wouldn't want you to see him like this, poor old chap! Understand? Now, skip!"

Thea ran down the gulch and looked back only once, to see them lifting the canvas litter with Wunsch upon it, still covered with the blanket.

The men carried Wunsch up the hill and down the road to the Kohlers'. Mrs. Kohler had gone home and made up a bed in the sitting-room, as she knew the litter could not be got round the turn in the narrow stairway. Wunsch was like a dead man. He lay unconscious all day. Ray Kennedy stayed with him till two o'clock in the afternoon, when he had to go out on his run. It was the first time he had ever been inside the Kohlers' house, and he was so much impressed by Napoleon that the piece-picture formed a new bond between him and Thea.

Dr. Archie went back at six o'clock, and found Mrs. Kohler and Spanish Johnny with Wunsch, who was in a high fever, muttering and groaning.

"There ought to be some one here to look after him to-night, Mrs. Kohler," he said. "I'm on a confinement case, and I can't be here, but there ought to be somebody. He may get violent."

Mrs. Kohler insisted that she could always do anything with Wunsch, but the doctor shook his head and Spanish Johnny grinned. He said he would stay. The doctor laughed at him. "Ten fellows like you couldn't hold him, Spanish, if he got obstreperous; an Irishman would have his hands full. Guess I'd better put the soft pedal on him." He pulled out his hypodermic.

Spanish Johnny stayed, however, and the Kohlers went to bed. At about two o'clock in the morning Wunsch rose from his ignominious cot. Johnny, who was dozing on the lounge, awoke to find the German standing in the middle of the room in his undershirt and drawers, his arms bare, his heavy body seeming twice its natural girth. His face was snarling and savage, and his eyes were crazy. He had risen to avenge himself, to wipe out his shame, to destroy his enemy. One look was enough for Johnny. Wunsch raised a chair threateningly, and Johnny, with the lightness of a *picador,* darted under the missile and out of the open window. He shot across the gully to get help, meanwhile leaving the Kohlers to their fate.

Fritz, upstairs, heard the chair crash upon the stove. Then he heard doors opening and shutting, and some one stumbling about in the shrubbery of the garden. He and Paulina sat up in bed and held a consultation. Fritz slipped from under the covers, and going cautiously over to the window, poked out his head. Then he rushed to the door and bolted it.

"*Mein Gott,* Paulina," he gasped, "he has the axe, he will kill us!"

"The dresser," cried Mrs. Kohler; "push the dresser before the door. *Ach,* if you had your rabbit gun, now!"

"It is in the barn," said Fritz sadly. "It would do no good; he would not be afraid of anything now. Stay you in the bed, Paulina." The dresser had lost its casters years ago, but he managed to drag it in front of the door. "He is in the garden. He makes nothing. He will get sick again, may-be."

Fritz went back to bed and his wife pulled the quilt over him and made him lie down. They heard stumbling in the garden again, then a smash of glass.

"*Ach, das Mistbeet!*"[1] gasped Paulina, hearing her hot-bed shivered. "The poor soul, Fritz, he will cut himself. *Ach!* what is that?" They both sat up in bed. "*Wieder! Ach!* What is he doing?"

The noise came steadily, a sound of chopping. Paulina tore off her night-cap. "*Die Bäume, die Bäume!* He is cutting our trees, Fritz!" Before her husband could prevent her, she had sprung from the bed and rushed to the window. "*Der Taubenschlag! Gerechter Himmel,*[2] he is chopping the dove-house down!"

Fritz reached her side before she had got her breath again, and poked his head out beside hers. There, in the faint starlight, they saw a bulky man, barefoot, half dressed, chopping away at the white post that formed the pedestal of the dove-house. The startled pigeons were croaking and flying about his head, even beating their wings in his face, so that he struck at them furiously with the axe. In a few seconds there was a crash, and Wunsch had actually felled the dove-house.

"Oh, if only it is not the trees next!" prayed Paulina. "The dove-house you can make new again, but not *die Bäume.*"

They watched breathlessly. In the garden below Wunsch stood in the attitude of a woodman, contemplating the fallen cote. Suddenly he threw the axe over his shoulder and went out of the front gate toward the town.

"The poor soul, he will meet his death!" Mrs. Kohler wailed. She ran back to her feather bed and hid her face in the pillow.

Fritz kept watch at the window. "No, no, Paulina," he called presently; "I see lanterns coming. Johnny must have gone for somebody. Yes, four lanterns, coming along the gulch. They stop; they must have seen him already. Now they are under the hill and I cannot see them, but I think they have him. They will bring him back. I must dress and go down." He caught his trousers and began pulling them on by the window. "Yes, here they come, half a dozen men. And they have tied him with a rope, Paulina!"

"*Ach,* the poor man! To be led like a cow," groaned Mrs. Kohler. "Oh, it is good that he has no wife!" She was reproaching herself for nagging Fritz when he drank himself into foolish pleasantry or mild sulks, and felt that she had never before appreciated her blessings.

Wunsch was in bed for ten days, during which time he was gossiped about and even preached about in Moonstone. The Baptist preacher took a shot at the fallen man from his pulpit, Mrs. Livery Johnson nodding approvingly from her pew. The mothers of Wunsch's pupils sent him notes informing him that their daughters would discontinue their music-lessons. The old maid who had rented him her piano sent the town dray for her contaminated instrument, and ever afterward declared that Wunsch had ruined its tone and scarred its glossy finish. The Kohlers were unremitting in their kindness to their friend. Mrs. Kohler made him soups and broths without stint, and Fritz repaired the dove-house and mounted it on a new post, lest it might be a sad reminder.

As soon as Wunsch was strong enough to sit about in his slippers and wadded jacket, he told Fritz to bring him some stout thread from the shop. When Fritz asked what he was going to sew, he produced the tattered score of "Orpheus" and said he would like to fix it up for a little present. Fritz carried it over to the shop and stitched it into pasteboards, covered with dark suiting-cloth. Over the stitches he glued a strip of thin red leather which he got from his friend, the harness-maker. After Paulina had cleaned the pages with fresh bread, Wunsch was amazed to see what a fine book he had. It opened stiffly, but that was no matter.

Sitting in the arbor one morning, under the ripe grapes and the brown, curling leaves, with a pen and ink on the bench beside him and the Gluck score on his knee, Wunsch pondered for a long while. Several times he dipped the pen in the ink, and then put it back again in the cigar box in which Mrs. Kohler kept her writing utensils. His thoughts wandered over a wide territory; over many countries and many years. There was no order or logical sequence in his ideas. Pictures came and went without reason. Faces, mountains, rivers, autumn days in other vineyards far away. He thought of a *Fuszreise*[3] he had made through the Hartz Mountains in his student days; of the innkeeper's pretty daughter who had lighted his pipe for him in the garden one summer evening, of the woods above Wiesbaden, haymakers on an island in the river. The roundhouse whistle woke him from his reveries. Ah, yes, he was in Moonstone, Colorado. He frowned for a moment and looked at the book on his knee. He had thought of a great many appropriate things to write in it, but suddenly he rejected all of them, opened the book, and at the top of the much-engraved title-page he wrote rapidly in purple ink:—

Einst, O Wunder!—
A. Wunsch.

Moonstone, Colo.
September 30, 18—

Nobody in Moonstone ever found what Wunsch's first name was. That "A" may have stood for Adam, or August, or even Amadeus; he got very angry if any one asked him. He remained A. Wunsch to the end of his chapter there. When he presented this score to Thea, he told her that in ten years she would either know what the inscription meant, or she would not have the least idea, in which case it would not matter.

When Wunsch began to pack his trunk, both the Kohlers were very unhappy. He said he was coming back some day, but that for the present, since he had lost all his pupils, it would be better for him to try some "new town." Mrs. Kohler darned and mended all his clothes, and gave him two new shirts she had made for Fritz.

Fritz made him a new pair of trousers and would have made him an overcoat but for the fact that overcoats were so easy to pawn.

Wunsch would not go across the ravine to the town until he went to take the morning train for Denver. He said that after he got to Denver he would "look around." He left Moonstone one bright October morning, without telling any one good-bye. He bought his ticket and went directly into the smoking-car. When the train was beginning to pull out, he heard his name called frantically, and looking out of the window he saw Thea Kronborg standing on the siding, bareheaded and panting. Some boys had brought word to school that they saw Wunsch's trunk going over to the station, and Thea had run away from school. She was at the end of the station platform, her hair in two braids, her blue gingham dress wet to the knees because she had run across lots through the weeds. It had rained during the night, and the tall sunflowers behind her were fresh and shining.

"Good-bye, Herr Wunsch, good-bye!" she called, waving to him.

He thrust his head out at the car window and called back, "*Leben sie wohl, leben sie wohl, mein Kind!*"[4] He watched her until the train swept around the curve beyond the roundhouse, and then sank back into his seat, muttering, "She had been running. Ah, she will run a long way; they cannot stop her!"

What was it about the child that one believed in? Was it her dogged industry, so unusual in this free-and-easy country? Was it her imagination? More likely it was because she had both imagination and a stubborn will, curiously balancing and interpenetrating each other. There was something unconscious and unawakened about her, that tempted curiosity. She had a kind of seriousness that he had not met with in a pupil before. She hated difficult things, and yet she could never pass one by. They seemed to challenge her; she had no peace until she mastered them. She had the power to make a great effort, to lift a weight heavier than herself. Wunsch hoped he would always remember her as she stood by the track, looking up at him; her broad eager face, so fair in color, with its high cheekbones, yellow eyebrows and greenish-hazel eyes. It was a face full of light and energy, of the unquestioning hopefulness of first youth. Yes, she was like a flower full of sun, but not the soft German flowers of his childhood. He had it now, the comparison he had absently reached for before: she was like the yellow prickly-pear blossoms that open there in the desert; thornier and sturdier than the maiden flowers he remembered; not so sweet, but wonderful.

That night Mrs. Kohler brushed away many a tear as she got supper and set the table for two. When they sat down, Fritz was more silent than usual. People who have lived long together need a third at table: they know each other's thoughts so well that they have nothing left to say. Mrs. Kohler stirred and stirred her coffee and clattered the spoon, but she had no heart for her supper. She felt, for the first time in years, that she was tired of her own cooking. She looked across the glass lamp at her husband and asked him if the butcher liked his new overcoat, and whether he had got the shoulders right in a ready-made suit he was patching over for Ray Kennedy. After supper Fritz offered to wipe the dishes for her, but she told him to go about his business, and not to act as if she were sick or getting helpless.

When her work in the kitchen was all done, she went out to cover the oleanders against frost, and to take a last look at her chickens. As she came back from the hen-house she stopped by one of the linden trees and stood resting her hand on the trunk. He would never come back, the poor man; she knew that. He would drift on from new town to new town, from catastrophe to catastrophe. He would hardly find a good home for himself again. He would die at last in some rough place, and be buried in the desert or on the wild prairie, far enough from any linden tree!

Fritz, smoking his pipe on the kitchen doorstep, watched his Paulina and guessed her thoughts. He, too, was sorry to lose his friend. But Fritz was getting old; he had lived a long while and had learned to lose without struggle.

XIV

"MOTHER," said Peter Kronborg to his wife one morning about two weeks after Wunsch's departure, "how would you like to drive out to Copper Hole with me to-day?"

Mrs. Kronborg said she thought she would enjoy the drive. She put on her gray cashmere dress and gold watch and chain, as befitted a minister's wife, and while her husband was dressing she packed a black oilcloth satchel with such clothing as she and Thor would need overnight.

Copper Hole was a settlement fifteen miles northwest of Moonstone where Mr. Kronborg preached every Friday evening. There was a big spring there and a creek and a few irrigating ditches. It was a community of discouraged agriculturists who had disastrously experimented with dry farming. Mr. Kronborg always drove out one day and back the next, spending the night with one of his parishioners. Often, when the weather was fine, his wife accompanied him. To-day they set out from home after the midday meal, leaving Tillie in charge of the house. Mrs. Kronborg's maternal feeling was always garnered up in the baby, whoever the baby happened to be. If she had the baby with her, the others could look out for themselves. Thor, of course, was not, accurately speaking, a baby any longer. In the matter of nourishment he was quite independent of his mother, though this independence had not been won without a struggle. Thor was conservative in all things, and the whole family had anguished with him when he was being weaned. Being the youngest, he was still the baby for Mrs. Kronborg, though he was nearly four years old and sat up boldly on her lap this afternoon, holding on to the ends of the lines and shouting "'mup, 'mup, horsey." His father watched him affectionately and hummed hymn tunes in the jovial way that was sometimes such a trial to Thea.

Mrs. Kronborg was enjoying the sunshine and the brilliant sky and all the faintly marked features of the dazzling, monotonous landscape. She had a rather unusual capacity for getting the flavor of places and of people. Although she was so enmeshed in family cares most of the time, she could emerge serene when she was away from them. For a mother of seven, she had a singularly unprejudiced point of view. She was, moreover, a fatalist, and as she did not attempt to direct things beyond her control, she found a good deal of time to enjoy the ways of man and nature.

When they were well upon their road, out where the first lean pasture lands began and the sand grass made a faint showing between the sagebushes, Mr. Kronborg dropped his tune and turned to his wife. "Mother, I've been thinking about something."

"I guessed you had. What is it?" She shifted Thor to her left knee, where he would be more out of the way.

"Well, it's about Thea. Mr. Follansbee came to my study at the church the other day and said they would like to have their two girls take lessons of Thea. Then I sounded Miss Meyers" (Miss Meyers was the organist in Mr. Kronborg's church) "and she said there was a good deal of talk about whether Thea wouldn't take over Wunsch's pupils. She said if Thea stopped school she wouldn't wonder if she could get pretty much all Wunsch's class. People think Thea knows about all Wunsch could teach."

Mrs. Kronborg looked thoughtful. "Do you think we ought to take her out of school so young?"

"She is young, but next year would be her last year anyway. She's far along for her age. And she can't learn much under the principal we've got now, can she?"

"No, I'm afraid she can't," his wife admitted. "She frets a good deal and says that man always has to look in the back of the book for the answers. She hates all that diagramming they have to do, and I think myself it's a waste of time."

Mr. Kronborg settled himself back into the seat and slowed the mare to a walk. "You see, it occurs to me that we might raise Thea's prices, so it would be worth her while. Seventy-five cents for hour lessons, fifty cents for half-hour lessons. If she got, say two thirds of Wunsch's class, that would bring her in upwards of ten dollars a week. Better pay than teaching a country school, and there would be more work in vacation than in winter. Steady work twelve months in the year; that's an advantage. And she'd be living at home, with no expenses."

"There'd be talk if you raised her prices," said Mrs. Kronborg dubiously.

"At first there would. But Thea is so much the best musician in town that they'd all come into line after a while. A good many people in Moonstone have been making money lately, and have bought new pianos. There were ten new pianos shipped in here from Denver in the last year. People ain't going to let them stand idle; too much money invested. I believe Thea can have as many scholars as she can handle, if we set her up a little."

"How set her up, do you mean?" Mrs. Kronborg felt a certain reluctance about accepting this plan, though she had not yet had time to think out her reasons.

"Well, I've been thinking for some time we could make good use of another room. We couldn't give up the parlor to her all the time. If we built another room on the ell and put the piano in there, she could give lessons all day long and it wouldn't bother us. We could build a clothes-press in it, and put in a bed-lounge and a dresser and let Anna have it for her sleeping-room. She needs a place of her own, now that she's beginning to be dressy."

"Seems like Thea ought to have the choice of the room, herself," said Mrs. Kronborg.

"But, my dear, she don't want it. Won't have it. I sounded her coming home from church on Sunday; asked her if she would like to sleep in a new room, if we built on. She fired up like a little wildcat and said she'd made her own room all herself, and she didn't think anybody ought to take it away from her."

"She don't mean to be impertinent, father. She's made decided that way, like my father." Mrs. Kronborg spoke warmly. "I never have any trouble with the child. I remember my father's ways and go at her carefully. Thea's all right."

Mr. Kronborg laughed indulgently and pinched Thor's full cheek. "Oh, I didn't mean anything against your girl, mother! She's all right, but she's a little wild-cat, just the same. I think Ray Kennedy's planning to spoil a born old maid."

"Huh! She'll get something a good sight better than Ray Kennedy, you see! Thea's an awful smart girl. I've seen a good many girls take music lessons in my time, but I ain't seen one that took to it so. Wunsch said so, too. She's got the making of something in her."

"I don't deny that, and the sooner she gets at it in a businesslike way, the better. She's the kind that takes responsibility, and it'll be good for her."

Mrs. Kronborg was thoughtful. "In some ways it will, maybe. But there's a good deal of strain about teaching youngsters, and she's always worked so hard with the scholars she has. I've often listened to her pounding it into 'em. I don't want to work her too hard. She's so serious that she's never had what you might call any real childhood. Seems like she ought to have the next few years sort of free and easy. She'll be tied down with responsibilities soon enough."

Mr. Kronborg patted his wife's arm. "Don't you believe it, mother. Thea is not the marrying kind. I've watched 'em. Anna will marry before long and make a good wife, but I don't see Thea bringing up a family. She's got a good deal of her mother in her, but she hasn't got all. She's too peppery and too fond of having her own way. Then she's always got to be ahead in everything. That kind make good church-workers and missionaries and school teachers, but they don't make good wives. They fret all their energy away, like colts, and get cut on the wire."

Mrs. Kronborg laughed. "Give me the graham crackers I put in your pocket for Thor. He's hungry. You're a funny man, Peter. A body wouldn't think, to hear you, you was talking about your own daughters. I guess you see through 'em. Still, even if Thea ain't apt to have children of her own, I don't know as that's a good reason why she should wear herself out on other people's."

"That's just the point, mother. A girl with all that energy has got to do something, same as a boy, to keep her out of mischief. If you don't want her to marry Ray, let her do something to make herself independent."

"Well, I'm not against it. It might be the best thing for her. I wish I felt sure she wouldn't worry. She takes things hard. She nearly cried herself sick about Wunsch's going away. She's the smartest child of 'em all, Peter, by a long ways."

Peter Kronborg smiled. "There you go, Anna. That's you all over again. Now, I have no favorites; they all have their good points. But you," with a twinkle, "always did go in for brains."

Mrs. Kronborg chuckled as she wiped the cracker crumbs from Thor's chin and fists. "Well, you're mighty conceited, Peter! But I don't know as I ever regretted it. I prefer having a family of my own to fussing with other folks' children, that's the truth."

Before the Kronborgs reached Copper Hole, Thea's destiny was pretty well mapped out for her. Mr. Kronborg was always delighted to have an excuse for enlarging the house.

Mrs. Kronborg was quite right in her conjecture that there would be unfriendly comment in Moonstone when Thea raised her prices for music-lessons. People said she was getting too conceited for anything. Mrs. Livery Johnson put on a new bonnet and paid up all her back calls to have the pleasure of announcing in each parlor she entered that her daughters, at least, would "never pay professional prices to Thea Kronborg."

Thea raised no objection to quitting school. She was now in the "high room," as it was called, in next to the highest class, and was studying geometry and beginning Cæsar. She no longer recited her lessons to the teacher she liked, but to the Principal, a man who belonged, like Mrs. Livery Johnson, to the camp of Thea's natural enemies. He taught school because he was too lazy to work among grown-up people, and he made an easy job of it. He got out of real work by inventing useless activities for his pupils, such as the "tree-diagramming system." Thea had spent hours making trees out of "Thanatopsis," Hamlet's soliloquy, Cato on "Immortality."[1] She agonized under this waste of time, and was only too glad to accept her father's offer of liberty.

So Thea left school the first of November. By the first of January she had eight one-hour pupils and ten half-hour pupils, and there would be more in the summer. She spent her earnings generously. She bought a new Brussels carpet for the parlor, and a rifle for Gunner and Axel, and an imitation tiger-skin coat and cap for Thor. She enjoyed being able to add to the family possessions, and thought Thor looked quite as handsome in his spots as the rich children she had seen in Denver. Thor was most complacent in his conspicuous apparel. He could walk anywhere by this time—though he always preferred to sit, or to be pulled in his cart. He was a blissfully lazy child, and had a number of long, dull plays, such as making nests for his china duck and waiting for her to lay him an egg. Thea thought him very intelligent, and she was proud that he was so big and burly. She found him restful, loved to hear him call her "sitter," and really liked his companionship, especially when she was tired.

On Saturday, for instance, when she taught from nine in the morning until five in the afternoon, she liked to get off in a corner with Thor after supper, away from all the bathing and dressing and joking and talking that went on in the house, and ask him about his duck, or hear him tell one of his rambling stories.

XV

BY THE TIME Thea's fifteenth birthday came round, she was established as a music teacher in Moonstone. The new room had been added to the house early in the spring, and Thea had been giving her lessons there since the middle of May. She liked the personal independence which was accorded her as a wage-earner. The family questioned her comings and goings very little. She could go buggy-riding with Ray Kennedy, for instance, without taking Gunner or Axel. She could go to Spanish Johnny's and sing part songs with the Mexicans, and nobody objected.

Thea was still under the first excitement of teaching, and was terribly in earnest about it. If a pupil did not get on well, she fumed and fretted. She counted until she was hoarse. She listened to scales in her sleep. Wunsch had taught only one pupil seriously, but Thea taught twenty. The duller they were, the more furiously she poked and prodded them. With the little girls she was nearly always patient, but with pupils older than herself, she sometimes lost her temper. One of her mistakes was to let herself in for a calling-down from Mrs. Livery Johnson. That lady appeared at the Kronborgs' one morning and announced that she would allow no girl to stamp her foot at her daughter Grace. She added that Thea's bad manners with the older girls were being talked about all over town, and that if her temper did not speedily improve she would lose all her advanced pupils. Thea was frightened. She felt she could never bear the disgrace, if such a thing happened. Besides, what would her father say, after he had gone to the expense of building an addition to the house? Mrs. Johnson demanded an apology to Grace. Thea said she was willing to make it. Mrs. Johnson said that hereafter, since she had taken lessons of the best piano teacher in Grinnell, Iowa, she herself would decide what pieces Grace should study. Thea readily consented to that, and Mrs. Johnson rustled away to tell a neighbor woman that Thea Kronborg could be meek enough when you went at her right.

Thea was telling Ray about this unpleasant encounter as they were driving out to the sand hills the next Sunday.

"She was stuffing you, all right, Thee," Ray reassured her. "There's no general dissatisfaction among your scholars. She just wanted to get in a knock. I talked to the piano tuner the last time he was here, and he said all the people he tuned for expressed themselves very favorably about your teaching. I wish you didn't take so much pains with them, myself."

"But I have to, Ray. They're all so dumb. They've got no ambition," Thea exclaimed irritably. "Jenny Smiley is the only one who isn't stupid. She can read pretty well, and she has such good hands. But she don't care a rap about it. She has no pride."

Ray's face was full of complacent satisfaction as he glanced sidewise at Thea, but she was looking off intently into the mirage, at one of those mammoth cattle that are nearly always reflected there. "Do you find it easier to teach in your new room?" he asked.

"Yes; I'm not interrupted so much. Of course, if I ever happen to want to practice at night, that's always the night Anna chooses to go to bed early."

"It's a darned shame, Thee, you didn't cop that room for yourself. I'm sore at the *padre* about that. He ought to give you that room. You could fix it up so pretty."

"I didn't want it, honest I didn't. Father would have let me have it. I like my own room better. Somehow I can think better in a little room. Besides, up there I am away from everybody, and I can read as late as I please and nobody nags me."

"A growing girl needs lots of sleep," Ray providently remarked.

Thea moved restlessly on the buggy cushions. "They need other things more," she muttered. "Oh, I forgot. I brought something to show you. Look here, it came on my birthday. Wasn't it nice of him to remember?" She took from her pocket a postcard, bent in the middle and folded, and handed it to Ray. On it was a white dove, perched on a wreath of very blue forget-me-nots, and "Birthday Greetings" in gold letters. Under this was written, "From A. Wunsch."

Ray turned the card over, examined the postmark, and then began to laugh.

"Concord, Kansas. He has my sympathy!"

"Why, is that a poor town?"

"It's the jumping-off place, no town at all. Some houses dumped down in the middle of a cornfield. You get lost in the corn. Not even a saloon to keep things going; sell whiskey without a license at the butcher shop, beer on ice with the liver and beefsteak. I wouldn't stay there over Sunday for a ten-dollar bill."

"Oh, dear! What do you suppose he's doing there? Maybe he just stopped off there a few days to tune pianos," Thea suggested hopefully.

Ray gave her back the card. "He's headed in the wrong direction. What does he want to get back into a grass country for? Now, there are lots of good live towns down on the Santa Fé, and everybody down there is musical. He could always get a job playing in saloons if he was dead-broke. I've figured out that I've got no years of my life to waste in a Methodist country where they raise pork."

"We must stop on our way back and show this card to Mrs. Kohler. She misses him so."

"By the way, Thee, I hear the old woman goes to church every Sunday to hear you sing. Fritz tells me he has to wait till two o'clock for his Sunday dinner these days. The church people ought to give you credit for that, when they go for you."

Thea shook her head and spoke in a tone of resignation. "They'll always go for me, just as they did for Wunsch. It wasn't because he drank they went for him; not really. It was something else."

"You want to salt your money down, Thee, and go to Chicago and take some lessons. Then you come back, and wear a long feather and high heels and put on a few airs, and that'll fix 'em. That's what they like."

"I'll never have money enough to go to Chicago. Mother meant to lend me some, I think, but now they've got hard times back in Nebraska, and her farm don't bring her in anything. Takes all the tenant can raise to pay the taxes. Don't let's talk about that. You promised to tell me about the play you went to see in Denver."

Any one would have liked to hear Ray's simple and clear account of the performance he had seen at the Tabor Grand Opera House[1]—Maggie Mitchell in *Little Barefoot*[2]—and any one would have liked to watch his kind face. Ray looked his best out of doors, when his thick red hands were covered by gloves, and the dull red of his sunburned face somehow seemed right in the light and wind. He looked better, too, with his hat on; his hair was thin and dry, with no particular color or character, "regular Willy-boy hair," as he himself described it. His eyes were pale beside the reddish bronze of his skin. They had the faded look often seen in the eyes of men who have lived much in the sun and wind and who have been accustomed to train their vision upon distant objects.

Ray realized that Thea's life was dull and exacting, and that she missed Wunsch. He knew she worked hard, that she put up with a great many little annoyances, and that her duties as a teacher separated her more than ever from the boys and girls of her own age. He did everything he could to provide recreation for her. He brought her candy and magazines and pineapples—of which she was very fond—from Denver, and kept his eyes and ears open for anything that might interest her. He was, of course, living for Thea. He had thought it all out carefully and had made up his mind just when he would speak to her. When she was seventeen, then he would tell her his plan and ask her to marry him. He would be willing to wait two, or even three years, until she was twenty, if she thought best. By that time he would surely have got in on something: copper, oil, gold, silver, sheep,—something.

Meanwhile, it was pleasure enough to feel that she depended on him more and more, that she leaned upon his steady kindness. He never broke faith with himself about her; he never hinted to her of his hopes for the future, never suggested that she might be more intimately confidential with him, or talked to her of the thing he thought about so constantly. He had the chivalry which is perhaps the proudest possession of his race. He had never embarrassed her by so much as a glance. Sometimes, when they drove out to the sand hills, he let his left arm lie along the back of the buggy seat, but it never came any nearer to Thea than that, never touched her. He often turned to her a face full of pride, and frank admiration, but his glance was never so intimate or so penetrating as Dr. Archie's. His blue eyes were clear and shallow, friendly, uninquiring. He rested Thea because he was so different; because, though he often told her interesting things, he never set lively fancies going in her head; because he never misunderstood her, and because he never, by any chance, for a single instant, understood her! Yes, with Ray she was safe; by him she would never be discovered!

XVI

THE PLEASANTEST EXPERIENCE Thea had that summer was a trip that she and her mother made to Denver in Ray Kennedy's caboose. Mrs. Kronborg had been looking forward to this excursion for a long while, but as Ray never knew at what hour his freight would leave Moonstone, it was difficult to arrange. The call-boy was as likely to summon him to start on his run at twelve o'clock midnight as at twelve o'clock noon. The first week in June started out with all the scheduled trains running on time, and a light freight business. Tuesday evening Ray, after consulting with the dispatcher, stopped at the Kronborgs' front gate to tell Mrs. Kronborg—who was helping Tillie water the flowers—that if she and Thea could be at the depot at eight o'clock the next morning, he thought he could promise them a pleasant ride and get them into Denver before nine o'clock in the evening. Mrs. Kronborg told him cheerfully, across the fence, that she would "take him up on it," and Ray hurried back to the yards to scrub out his car.

The one complaint Ray's brakemen had to make of him was that he was too fussy about his caboose. His former brakeman had asked to be transferred because, he said, "Kennedy was as fussy about his car as an old maid about her bird-cage." Joe Giddy, who was braking with Ray now, called him "the bride," because he kept the caboose and bunks so clean.

It was properly the brakeman's business to keep the car clean, but when Ray got back to the depot, Giddy was nowhere to be found. Muttering that all his brakemen seemed to consider him "easy," Ray went down to his car alone. He built a fire in the stove and put water on to heat while he got into his overalls and jumper. Then he set to work with a scrubbing-brush and plenty of soap and "cleaner." He scrubbed the floor and seats, blacked the stove, put clean sheets on the bunks, and then began to demolish Giddy's picture gallery. Ray found that his brakemen were likely to have what he termed "a taste for the nude in art," and Giddy was no exception. Ray took down half a dozen girls in tights and ballet skirts,—premiums for cigarette coupons,—and some racy calendars advertising saloons and sporting clubs, which had cost Giddy both time and trouble; he even removed Giddy's particular pet, a naked girl lying on a couch with her knee carelessly poised in the air. Underneath the picture was printed the title, "The Odalisque."[1] Giddy was under the happy delusion that this title meant something wicked,—there was a wicked look about the consonants,—but Ray, of course, had looked it up, and Giddy was indebted to the dictionary for the privilege of keeping his lady. If "odalisque" had been what Ray called an objectionable word, he would have thrown the picture out in the first place. Ray even took down a picture of Mrs. Langtry in evening dress, because it was entitled the "Jersey Lily,"[2] and because there was a small head of Edward VII, then Prince of Wales, in one corner. Albert Edward's conduct was a popular subject of discussion among railroad men in those days, and as Ray pulled the tacks out of this lithograph he felt more indignant with the English than ever. He deposited all these pictures under the mattress of Giddy's bunk, and stood admiring his clean car in the lamplight; the walls now exhibited only a wheatfield, advertising agricultural implements, a map of Colorado, and some pictures of race-horses and hunting-dogs. At this moment Giddy, freshly shaved and shampooed, his shirt shining with the highest polish known to Chinese laundrymen, his straw hat tipped over his right eye, thrust his head in at the door.

"What in hell—" he brought out furiously. His good-humored, sunburned face seemed fairly to swell with amazement and anger.

"That's all right, Giddy," Ray called in a conciliatory tone. "Nothing injured. I'll put 'em all up again as I found 'em. Going to take some ladies down in the car to-morrow."

Giddy scowled. He did not dispute the propriety of Ray's measures, if there were to be ladies on board, but he felt injured. "I suppose you'll expect me to behave like a Y.M.C.A. secretary," he growled. "I can't do my work and serve tea at the same time."

"No need to have a tea-party," said Ray with determined cheerfulness. "Mrs. Kronborg will bring the lunch, and it will be a darned good one."

Giddy lounged against the car, holding his cigar between two thick fingers. "Then I guess she'll get it," he observed knowingly. "I don't think your musical friend is much on the grub-box. Has to keep her hands white to tickle the ivories." Giddy had nothing against Thea, but he felt cantankerous and wanted to get a rise out of Kennedy.

"Every man to his own job," Ray replied agreeably, pulling his white shirt on over his head.

Giddy emitted smoke disdainfully. "I suppose so. The man that gets her will have to wear an apron and bake the pancakes. Well, some men like to mess about the kitchen." He paused, but Ray was intent on getting into his clothes as quickly as possible. Giddy thought he could go a little further. "Of course, I don't dispute your right to haul women in this car if you want to; but personally, so far as I'm concerned, I'd a good deal rather drink a can of tomatoes and do without the women *and* their lunch. I was never much enslaved to hard-boiled eggs, anyhow."

"You'll eat 'em to-morrow, all the same." Ray's tone had a steely glitter as he jumped out of the car, and Giddy stood aside to let him pass. He knew that Kennedy's next reply would be delivered by hand. He had once seen Ray beat up a nasty fellow for insulting a Mexican woman who helped about the grub-car in the work train, and his fists had worked like two steel hammers. Giddy wasn't looking for trouble.

At eight o'clock the next morning Ray greeted his ladies and helped them into the car. Giddy had put on a clean shirt and yellow pigskin gloves and was whistling his best. He considered Kennedy a fluke as a ladies' man, and if there was to be a party, the honors had to be done by some one who wasn't a blacksmith at small-talk. Giddy had, as Ray sarcastically admitted, "a local reputation as a jollier," and he was fluent in gallant speeches of a not too-veiled nature. He insisted that Thea should take his seat in the cupola, opposite Ray's, where she could look out over the country. Thea told him, as she clambered up, that she cared a good deal more about riding in that seat than about going to Denver. Ray was never so companionable and easy as when he sat chatting in the lookout of his little house on wheels. Good stories came to him, and interesting recollections. Thea had a great respect for the reports he had to write out, and for the telegrams that were handed to him at stations;

for all the knowledge and experience it must take to run a freight train.

Giddy, down in the car, in the pauses of his work, made himself agreeable to Mrs. Kronborg.

"It's a great rest to be where my family can't get at me, Mr. Giddy," she told him. "I thought you and Ray might have some housework here for me to look after, but I couldn't improve any on this car."

"Oh, we like to keep her neat," returned Giddy glibly, winking up at Ray's expressive back. "If you want to see a clean ice-box, look at this one. Yes, Kennedy always carries fresh cream to eat on his oatmeal. I'm not particular. The tin cow's good enough for me."

"Most of you boys smoke so much that all victuals taste alike to you," said Mrs. Kronborg. "I've got no religious scruples against smoking, but I couldn't take as much interest cooking for a man that used tobacco. I guess it's all right for bachelors who have to eat round."

Mrs. Kronborg took off her hat and veil and made herself comfortable. She seldom had an opportunity to be idle, and she enjoyed it. She could sit for hours and watch the sage-hens fly up and the jack-rabbits dart away from the track, without being bored. She wore a tan bombazine dress, made very plainly, and carried a roomy, worn, mother-of-the-family handbag.

Ray Kennedy always insisted that Mrs. Kronborg was "a fine-looking lady," but this was not the common opinion in Moonstone. Ray had lived long enough among the Mexicans to dislike fussiness, to feel that there was something more attractive in ease of manner than in absent-minded concern about hairpins and dabs of lace. He had learned to think that the way a woman stood, moved, sat in her chair, looked at you, was more important than the absence of wrinkles from her skirt. Ray had, indeed, such unusual perceptions in some directions, that one could not help wondering what he would have been if he had ever, as he said, had "half a chance."

He was right; Mrs. Kronborg was a fine-looking woman. She was short and square, but her head was a real head, not a mere jerky termination of the body. It had some individuality apart from hats and hairpins. Her hair, Moonstone women admitted, would have been very pretty "on anybody else." Frizzy bangs were worn then, but Mrs. Kronborg always dressed her hair in the same way, parted in the middle, brushed smoothly back from her low, white forehead, pinned loosely on the back of her head in two thick braids. It was growing gray about the temples, but after the manner of yellow hair it seemed only to have grown paler there, and had taken on a color like that of English primroses. Her eyes were clear and untroubled; her face smooth and calm, and, as Ray said, "strong."

Thea and Ray, up in the sunny cupola, were laughing and talking. Ray got great pleasure out of seeing her face there in the little box where he so often imagined it. They were crossing a plateau where great red sandstone boulders lay about, most of them much wider at the top than at the base, so that they looked like great toadstools.

"The sand has been blowing against them for a good many hundred years," Ray explained, directing Thea's eyes with his gloved hand. "You see the sand blows low, being so heavy, and cuts them out underneath. Wind and sand are pretty high-class architects. That's the principle of most of the Cliff-Dweller remains down at Canyon de Chelly.[3] The sandstorms had dug out big depressions in the face of a cliff, and the Indians built their houses back in that depression."

"You told me that before, Ray, and of course you know. But the geography says their houses were cut out of the face of the living rock, and I like that better."

Ray sniffed. "What nonsense does get printed! It's enough to give a man disrespect for learning. How could them Indians cut houses out of the living rock, when they knew nothing about the art of forging metals?" Ray leaned back in his chair, swung his foot, and looked thoughtful and happy. He was in one of his favorite fields of speculation, and nothing gave him more pleasure than talking these things over with Thea Kronborg. "I'll tell you, Thee, if those old fellows had learned to work metals once, your ancient Egyptians and Assyrians wouldn't have beat them very much. Whatever they did do, they did well. Their masonry's standing there to-day, the corners as true as the Denver Capitol. They were clever at most everything but metals; and that one failure kept them from getting across. It was the quicksand that swallowed 'em up, as a race. I guess civilization proper began when men mastered metals."

Ray was not vain about his bookish phrases. He did not use them to show off, but because they seemed to him more adequate than colloquial speech. He felt strongly about these things, and groped for words, as he said, "to express himself." He had the lamentable American belief that "expression" is obligatory. He still carried in his trunk, among the unrelated possessions of a railroad man, a notebook on the title-page of which was written "Impressions on First Viewing the Grand Canyon, Ray H. Kennedy." The pages of that book were like a battlefield; the laboring author had fallen back from metaphor after metaphor, abandoned position after position. He would have admitted that the art of forging metals was nothing to this treacherous business of recording impressions, in which the material you were so full of vanished mysteriously under your striving hand. "Escaping steam!" he had said to himself, the last time he tried to read that notebook.

Thea didn't mind Ray's travel-lecture expressions. She dodged them, unconsciously, as she did her father's professional palaver. The light in Ray's pale-blue eyes and the feeling in his voice more than made up for the stiffness of his language.

"Were the Cliff-Dwellers really clever with their hands, Ray, or do you always have to make allowance and say, 'That was pretty good for an Indian'?" she asked.

Ray went down into the car to give some instructions to Giddy. "Well," he said when he returned, "about the aborigines: once or twice I've been with some fellows who were cracking burial mounds. Always felt a little ashamed of it, but we did pull out some remarkable things. We got some pottery out whole; seemed pretty fine to me. I guess their women were their artists. We found lots of old shoes and sandals made out of yucca fiber, neat and strong; and feather blankets, too."

"Feather blankets? You never told me about them."

"Didn't I? The old fellows—or the squaws—wove a close netting of yucca fiber, and then tied on little bunches of down feathers, overlapping, just the way feathers grow on a bird. Some of them were feathered on both sides. You can't get anything warmer than that, now, can you?—or prettier. What I like about those old aborigines is, that they got all their ideas from nature."

Thea laughed. "That means you're going to say something about girls' wearing corsets. But some of your Indians flattened their babies' heads, and that's worse than wearing corsets."

"Give me an Indian girl's figure for beauty," Ray insisted. "And a girl with a voice like yours ought to have plenty of lung-action. But you know my sentiments on that subject. I was going to tell you about the handsomest thing we ever looted out of those burial mounds. It was on a woman, too, I regret to say. She was preserved as perfect as any mummy that ever came out of the pyramids. She had a big string of turquoises around her neck, and she was wrapped in a fox-fur cloak, lined with little yellow feathers that must have come off wild canaries. Can you beat that, now? The fellow that claimed it sold it to a Boston man for a hundred and fifty dollars."

Thea looked at him admiringly. "Oh, Ray, and didn't you get anything off her, to remember her by, even? She must have been a princess."

Ray took a wallet from the pocket of the coat that was hanging beside him, and drew from it a little lump wrapped in worn tissue paper. In a moment a stone, soft and blue as a robin's egg, lay in the hard palm of his hand. It was a turquoise, rubbed smooth in the Indian finish, which is so much more beautiful than the incongruous high polish the white man gives that tender stone. "I got this from her necklace. See the hole where the string went through? You know how the Indians drill them? Work the drill with their teeth. You like it, don't you? They're just right for you. Blue and yellow are the Swedish colors." Ray looked intently at her head, bent over his hand, and then gave his whole attention to the track.

"I'll tell you, Thee," he began after a pause, "I'm going to form a camping party one of these days and persuade your *padre* to take you and your mother down to that country, and we'll live in the rock houses—they're as comfortable as can be—and start the cook fires up in 'em once again. I'll go into the burial mounds and get you more keepsakes than any girl ever had before." Ray had planned such an expedition for his wedding journey, and it made his heart thump to see how Thea's eyes kindled when he talked about it. "I've learned more down there about what makes history," he went on, "than in all the books I've ever read. When you sit in the sun and let your heels hang out of a doorway that drops a thousand feet, ideas come to you. You begin to feel what the human race has been up against from the beginning. There's something mighty elevating about those old habitations. You feel like it's up to you to do your best, on account of those fellows having it so hard. You feel like you owed them something."

At Wassiwappa, Ray got instructions to sidetrack until Thirty-six went by. After reading the message, he turned to his guests. "I'm afraid this will hold us up about two hours, Mrs. Kronborg, and we won't get into Denver till near midnight."

"That won't trouble me," said Mrs. Kronborg contentedly. "They know me at the Y.W.C.A., and they'll let me in any time of night. I came to see the country, not to make time. I've always wanted to get out at this white place and look around, and now I'll have a chance. What makes it so white?"

"Some kind of chalky rock." Ray sprang to the ground and gave Mrs. Kronborg his hand. "You can get soil of any color in Colorado; match most any ribbon."

While Ray was getting his train on to a side track, Mrs. Kronborg strolled off to examine the post-office and station house; these, with the water tank, made up the town. The station agent "batched" and raised chickens. He ran out to meet Mrs. Kronborg, clutched at her feverishly, and began telling her at once how lonely he was and what bad luck he was having with his poultry. She went to his chicken yard with him, and prescribed for gapes.[4]

Wassiwappa seemed a dreary place enough to people who looked for verdure, a brilliant place to people who liked color. Beside the station house there was a blue-grass plot, protected by a red plank fence, and six fly-bitten box-elder trees, not much larger than bushes, were kept alive by frequent hosings from the water plug. Over the windows some dusty morning-glory vines were trained on strings. All the country about was broken up into low chalky hills, which were so intensely white, and spotted so evenly with sage, that they looked like white leopards crouching. White dust powdered everything, and the light was so intense that the station agent usually wore blue glasses. Behind the station there was a water course, which roared in flood time, and a basin in the soft white rock where a pool of alkali water flashed in the sun like a mirror. The agent looked almost as sick as his chickens, and Mrs. Kronborg at once invited him to lunch with her party. He had, he confessed, a distaste for his own cooking, and lived mainly on soda crackers and canned beef. He laughed apologetically when Mrs. Kronborg said she guessed she'd look about for a shady place to eat lunch.

She walked up the track to the water tank, and there, in the narrow shadows cast by the uprights on which the tank stood, she found two tramps. They sat up and stared at her, heavy with sleep. When she asked them where they were going, they told her "to the coast." They rested by day and traveled by night; walked the ties unless they could steal a ride, they said; adding that "these Western roads were getting strict." Their faces were blistered, their eyes blood-shot, and their shoes looked fit only for the trash pile.

"I suppose you're hungry?" Mrs. Kronborg asked. "I suppose you both drink?" she went on thoughtfully, not censoriously.

The huskier of the two hoboes, a bushy, bearded fellow, rolled his eyes and said, "I wonder?" But the other, who was old and spare, with a sharp nose and watery eyes, sighed. "Some has one affliction, some another," he said.

Mrs. Kronborg reflected. "Well," she said at last, "you can't get liquor here, anyway. I am going to ask you to vacate, because I want to have a little picnic under this tank for the freight crew that brought me along. I wish I had lunch enough to provide you, but I ain't. The station agent says he gets his provisions over there at the post-office store, and if you are hungry you can get some canned stuff there." She opened her handbag and gave each of the tramps a half-dollar.

The old man wiped his eyes with his forefinger. "Thank 'ee, ma'am. A can of tomatters will taste pretty good to me. I wasn't always walkin' ties; I had a good job in Cleveland before—"

The hairy tramp turned on him fiercely. "Aw, shut up on that, grandpaw! Ain't you got no gratitude? What do you want to hand the lady that fur?"

The old man hung his head and turned away. As he went off, his comrade looked after him and said to Mrs. Kronborg: "It's true, what he says. He had a job in the car shops; but he had bad luck." They both limped away toward the store, and Mrs. Kronborg sighed. She was not afraid of tramps. She always talked to them, and never turned one away. She hated to think how many of them there were, crawling along the tracks over that vast country.

Her reflections were cut short by Ray and Giddy and Thea, who came bringing the lunch box and water bottles. Although there was not shadow enough to accommodate all the party at once, the air under the tank was distinctly cooler than the surrounding air, and the drip made a pleasant sound in that breathless noon. The station agent ate as if he had never been fed before, apologizing every time he took another piece of fried chicken. Giddy was unabashed before the devilled eggs of which he had spoken so scornfully last night. After lunch the men lit their pipes and lay back against the uprights that supported the tank.

"This is the sunny side of railroading, all right," Giddy drawled luxuriously.

"You fellows grumble too much," said Mrs. Kronborg as she corked the pickle jar. "Your job has its drawbacks, but it don't tie you down. Of course there's the risk; but I believe a man's watched over, and he can't be hurt on the railroad or anywhere else if it's intended he shouldn't be."

Giddy laughed. "Then the trains must be operated by fellows the Lord has it in for, Mrs. Kronborg. They figure it out that a railroad man's only due to last eleven years; then it's his turn to be smashed."

"That's a dark Providence, I don't deny," Mrs. Kronborg admitted. "But there's lots of things in life that's hard to understand."

"I guess!" murmured Giddy, looking off at the spotted white hills.

Ray smoked in silence, watching Thea and her mother clear away the lunch. He was thinking that Mrs. Kronborg had in her face the same serious look that Thea had; only hers was calm and satisfied, and Thea's was intense and questioning. But in both it was a large kind of look, that was not all the time being broken up and convulsed by trivial things. They both carried their heads like Indian women, with a kind of noble unconsciousness. He got so tired of women who were always nodding and jerking; apologizing, deprecating, coaxing, insinuating with their heads.

When Ray's party set off again that afternoon the sun beat fiercely into the cupola, and Thea curled up in one of the seats at the back of the car and had a nap.

As the short twilight came on, Giddy took a turn in the cupola, and Ray came down and sat with Thea on the rear platform of the caboose and watched the darkness come in soft waves over the plain. They were now about thirty miles from Denver, and the mountains looked very near. The great toothed wall behind which the sun had gone down now separated into four distinct ranges, one behind the other. They were a very pale blue, a color scarcely stronger than wood smoke, and the sunset had left bright streaks in the snow-filled gorges. In the clear, yellow-streaked sky the stars were coming out, flickering like newly lighted lamps, growing steadier and more golden as the sky darkened and the land beneath them fell into complete shadow. It was a cool, restful darkness that was not black or forbidding, but somehow open and free;

the night of high plains where there is no moistness or mistiness in the atmosphere.

Ray lit his pipe. "I never get tired of them old stars, Thee. I miss 'em up in Washington and Oregon where it's misty. Like 'em best down in Mother Mexico, where they have everything their own way. I'm not for any country where the stars are dim." Ray paused and drew on his pipe. "I don't know as I ever really noticed 'em much till that first year I herded sheep up in Wyoming. That was the year the blizzard caught me."

"And you lost all your sheep, didn't you, Ray?" Thea spoke sympathetically. "Was the man who owned them nice about it?"

"Yes, he was a good loser. But I didn't get over it for a long while. Sheep are so damned resigned. Sometimes, to this day, when I'm dog-tired, I try to save them sheep all night long. It comes kind of hard on a boy when he first finds out how little he is, and how big everything else is."

Thea moved restlessly toward him and dropped her chin on her hand, looking at a low star that seemed to rest just on the rim of the earth. "I don't see how you stood it. I don't believe I could. I don't see how people can stand it to get knocked out, anyhow!" She spoke with such fierceness that Ray glanced at her in surprise. She was sitting on the floor of the car, crouching like a little animal about to spring.

"No occasion for you to see," he said warmly. "There'll always be plenty of other people to take the knocks for you."

"That's nonsense, Ray." Thea spoke impatiently and leaned lower still, frowning at the red star. "Everybody's up against it for himself, succeeds or fails—himself."

"In one way, yes," Ray admitted, knocking the sparks from his pipe out into the soft darkness that seemed to flow like a river beside the car. "But when you look at it another way, there are a lot of halfway people in this world who help the winners win, and the failers fail. If a man stumbles, there's plenty of people to push him down. But if he's like 'the youth who bore,'[5] those same people are foreordained to help him along. They may hate to, worse than blazes, and they may do a lot of cussin' about it, but they have to help the winners and they can't dodge it. It's a natural law, like what keeps the big clock up there going, little wheels and big, and no mix-up." Ray's hand and his pipe were suddenly outlined against the sky. "Ever occur to you, Thee, that they have to be on time close enough to *make time?* The Dispatcher up there must have a long head." Pleased with his similitude, Ray went back to the lookout. Going into Denver, he had to keep a sharp watch.

Giddy came down, cheerful at the prospect of getting into port, and singing a new topical ditty that had come up from the Santa Fé by way of La Junta. Nobody knows who makes these songs; they seem to follow events automatically. Mrs. Kronborg made Giddy sing the whole twelve verses of this one, and laughed until she wiped her eyes. The story was that of Katie Casey, head dining-room girl at Winslow, Arizona, who was unjustly discharged by the Harvey House manager.[6] Her suitor, the yardmaster, took the switchmen out on a strike until she was reinstated. Freight trains from the east and the west piled up at Winslow until the yards looked like a log-jam. The division superintendent, who was in California, had to wire instructions for Katie Casey's restoration before he could get his trains running. Giddy's song told all this with much detail, both tender and technical, and after each of the dozen verses came the refrain:—

"Oh, who would think that Katie Casey owned the Santa Fé?
But it really looks that way,
The dispatcher's turnin' gray,
All the crews is off their pay;
She can hold the freight from Albuquerq' to Needles any day;
The division superintendent, he come home from Monterey,
Just to see if things was pleasin' Katie Ca—a—a—sey."

Thea laughed with her mother and applauded Giddy. Everything was so kindly and comfortable; Giddy and Ray, and their hospitable little house, and the easy-going country, and the stars. She curled up on the seat again with that warm, sleepy feeling of the friendliness of the world—which nobody keeps very long, and which she was to lose early and irrevocably.

XVII

THE SUMMER FLEW BY. Thea was glad when Ray Kennedy had a Sunday in town and could take her driving. Out among the sand hills she could forget the "new room" which was the scene of wearing and fruitless labor. Dr. Archie was away from home a good deal that year. He had put all his money into mines above Colorado Springs, and he hoped for great returns from them.

In the fall of that year, Mr. Kronborg decided that Thea ought to show more interest in church work. He put it to her frankly, one night at supper, before the whole family. "How can I insist on the other girls in the congregation being active in the work, when one of my own daughters manifests so little interest?"

"But I sing every Sunday morning, and I have to give up one night a week to choir practice," Thea declared rebelliously, pushing back her plate with an angry determination to eat nothing more.

"One night a week is not enough for the pastor's daughter," her father replied. "You won't do anything in the sewing society, and you won't take part in the Christian Endeavor or the Band of Hope.[1] Very well, you must make it up in other ways. I want some one to play the organ and lead the singing at prayer-meeting this winter. Deacon Potter told me some time ago that he thought there would be more interest in our prayer-meetings if we had the organ. Miss Meyers don't feel that she can play on Wednesday nights. And there ought to be somebody to start the hymns. Mrs. Potter is getting old, and she always starts them too high. It won't take much of your time, and it will keep people from talking."

This argument conquered Thea, though she left the table sullenly. The fear of the tongue, that terror of little towns, is usually felt more keenly by the minister's family than by other households. Whenever the Kronborgs wanted to do anything, even to buy a new carpet, they had to take counsel together as to whether people would talk. Mrs. Kronborg had her own conviction that people talked when they felt like it, and said what they chose, no matter how the minister's family conducted themselves. But she did not impart these dangerous ideas to her children. Thea was still under the belief that public opinion could be placated; that if you clucked often enough, the hens would mistake you for one of themselves.

Mrs. Kronborg did not have any particular zest for prayer-meetings, and she stayed at home whenever she had a valid excuse. Thor was too old to furnish such an excuse now, so every Wednesday night, unless one of the children was sick, she trudged off with Thea, behind Mr. Kronborg. At first Thea was terribly bored. But she got used to prayer-meeting, got even to feel a mournful interest in it.

The exercises were always pretty much the same. After the first hymn her father read a passage from the Bible, usually a Psalm. Then there was another hymn, and then her father commented upon the passage he had read and, as he said, "applied the Word to our necessities." After a third hymn, the meeting was declared open, and the old men and women took turns at praying and talking. Mrs. Kronborg never spoke in meeting. She told people firmly that she had been brought up to keep silent and let the men talk, but she gave respectful attention to the others, sitting with her hands folded in her lap.

The prayer-meeting audience was always small. The young and energetic members of the congregation came only once or twice a year, "to keep people from talking." The usual Wednesday night gathering was made up of old women, with perhaps six or eight old men, and a few sickly girls who had not much interest in life; two of them, indeed, were already preparing to die. Thea accepted the mournfulness of the prayer-meetings as a kind of spiritual discipline, like funerals. She always read late after she went home and felt a stronger wish than usual to live and to be happy.

The meetings were conducted in the Sunday-School room, where there were wooden chairs instead of pews; an old map of Palestine hung on the wall, and the bracket lamps gave out only a dim light. The old women sat motionless as Indians in their shawls and bonnets; some of them wore long black mourning veils. The old men drooped in their chairs. Every back, every face, every head said "resignation." Often there were long silences, when you could hear nothing but the crackling of the soft coal in the stove and the muffled cough of one of the sick girls.

There was one nice old lady,—tall, erect, self-respecting, with a delicate white face and a soft voice. She never whined, and what she said was always cheerful, though she spoke so nervously that Thea knew she dreaded getting up, and that she made a real sacrifice to, as she said, "testify to the goodness of her Saviour." She was the mother of the girl who coughed, and Thea used to wonder how she explained things to herself. There was, indeed, only one woman who talked because she was, as Mr. Kronborg said, "tonguey." The others were somehow impressive. They told about the sweet thoughts that came to them while they were at their work; how, amid their household tasks, they were suddenly lifted by the sense of a divine Presence. Sometimes they told of their first conversion, of how in their youth that higher Power had made itself known to them. Old Mr. Carsen, the carpenter, who gave his services as janitor to the church, used often to tell how, when he was a young man and a scoffer, bent on the destruction of both body and soul, his Saviour had come to him in the Michigan woods and had stood, it seemed to him, beside the tree he was felling; and how he dropped his axe and knelt in prayer "to Him who died for us upon the tree." Thea always wanted to ask him more about it; about his mysterious wickedness, and about the vision.

Sometimes the old people would ask for prayers for their absent children. Sometimes they asked their brothers and sisters in Christ to pray that they might be stronger against temptations. One of the sick girls used to ask them to pray that she might have more faith in the times of depression that came to her, "when all the way before seemed dark." She repeated that husky phrase so often, that Thea always remembered it.

One old woman, who never missed a Wednesday night, and who nearly always took part in the meeting, came all the way up from the depot settlement. She always wore a black crocheted "fascinator" over her thin white hair, and she made long, tremulous prayers, full of railroad terminology. She had six sons in the service of different railroads, and she always prayed "for the boys on the road, who know not at what moment they may be cut off. When, in Thy divine wisdom, their hour is upon them, may they, O our Heavenly Father, see only white lights along the road to Eternity." She used to speak, too, of "the engines that race with death"; and though she looked so old and little when she was on her knees, and her voice was so shaky, her prayers had a thrill of speed and danger in them; they made one think of the deep black canyons, the slender trestles, the pounding trains. Thea liked to look at her sunken eyes that seemed full of wisdom, at her black thread gloves, much too long in the fingers and so meekly folded one over the other. Her face was brown, and worn away as rocks are worn by water. There are many ways of describing that color of age, but in reality it is not like parchment, or like any of the things it is said to be like. That brownness and that texture of skin are found only in the faces of old human creatures, who have worked hard and who have always been poor.

One bitterly cold night in December the prayer-meeting seemed to Thea longer than usual. The prayers and the talks went on and on. It was as if the old people were afraid to go out into the cold, or were stupefied by the hot air of the room. She had left a book at home that she was impatient to get back to. At last the Doxology[2] was sung, but the old people lingered about the stove to greet each other, and Thea took her mother's arm and hurried out to the frozen sidewalk, before her father could get away. The wind was whistling up the street and whipping the naked cottonwood trees against the telegraph poles and the sides of the houses. Thin snow clouds were flying overhead, so that the sky looked gray, with a dull phosphorescence. The icy streets and the shingle roofs of the houses were gray, too. All along the street, shutters banged or windows rattled, or gates wobbled, held by their latch but shaking on loose hinges. There was not a cat or a dog in Moonstone that night that was not given a warm shelter; the cats under the kitchen stove, the dogs in barns or coal-sheds. When Thea and her mother reached home, their mufflers were covered with ice, where their breath had frozen. They hurried into the house and made a dash for the parlor and the hard-coal burner, behind which Gunner was sitting on a stool, reading his Jules Verne[3] book. The door stood open into the dining-room, which was heated from the parlor. Mr. Kronborg always had a lunch when he came home from prayer-meeting, and his pumpkin pie and milk were set out on the dining-table. Mrs. Kronborg said she thought she felt hungry, too, and asked Thea if she didn't want something to eat.

"No, I'm not hungry, mother. I guess I'll go upstairs."

"I expect you've got some book up there," said Mrs. Kronborg, bringing out another pie. "You'd better bring it down here and read. Nobody'll disturb you, and it's terrible cold up in that loft."

Thea was always assured that no one would disturb her if she read downstairs, but the boys talked when they came in, and her father fairly delivered discourses after he had been renewed by half a pie and a pitcher of milk.

"I don't mind the cold. I'll take a hot brick up for my feet. I put one in the stove before I left, if one of the boys hasn't stolen it. Good-night, mother." Thea got her brick and lantern, and dashed upstairs through the windy loft. She undressed at top speed and got into bed with her brick. She put a pair of white knitted gloves on her hands, and pinned over her head a piece of soft flannel that had been one of Thor's long petticoats when he was a baby. Thus equipped, she was ready for business. She took from her table a thick paper-backed volume, one of the "line" of paper novels the druggist kept to sell to traveling men. She had bought it, only yesterday, because the first sentence interested her very much, and because she saw, as she glanced over the pages, the magical names of two Russian cities. The book was a poor translation of "Anna Karenina."[4] Thea opened it at a mark, and fixed her eyes intently upon the small print. The hymns, the sick girl, the resigned black figures were forgotten. It was the night of the ball in Moscow.

Thea would have been astonished if she could have known how, years afterward, when she had need of them, those old faces were to come back to her, long after they were hidden away under the earth; that they would seem to her then as full of meaning, as mysteriously marked by Destiny, as the people who danced the mazurka under the elegant Korsunsky.

XVIII

MR. KRONBORG WAS TOO FOND of his ease and too sensible to worry his children much about religion. He was more sincere than many preachers, but when he spoke to his family about matters of conduct it was usually with a regard for keeping up appearances. The church and church work were discussed in the family like the routine of any other business. Sunday was the hard day of the week with them, just as Saturday was the busy day with the merchants on Main Street. Revivals were seasons of extra work and pressure, just as threshing-time was on the farms. Visiting elders had to be lodged and cooked for, the folding-bed in the parlor was let down, and Mrs. Kronborg had to work in the kitchen all day long and attend the night meetings.

During one of these revivals Thea's sister Anna professed religion with, as Mrs. Kronborg said, "a good deal of fluster." While Anna was going up to the mourners' bench nightly and asking for the prayers of the congregation, she disseminated general gloom throughout the household, and after she joined the church she took on an air of "set-apartness" that was extremely trying to her brothers and her sister, though they realized that Anna's sanctimoniousness was perhaps a good thing for their father. A preacher ought to have one child who did more than merely acquiesce in religious observances, and Thea and the boys were glad enough that it was Anna and not one of themselves who assumed this obligation.

"Anna, she's American," Mrs. Kronborg used to say. The Scandinavian mould of countenance, more or less marked in each of the other children, was scarcely discernible in her, and she looked enough like other Moonstone girls to be thought pretty. Anna's nature was conventional, like her face. Her position as the minister's eldest daughter was important to her, and she tried to live up to it. She read sentimental religious story-books and emulated the spiritual struggles and magnanimous behavior of their persecuted heroines. Everything had to be interpreted for Anna. Her opinions about the smallest and most commonplace things were gleaned from the Denver papers, the church weeklies, from sermons and Sunday-School addresses. Scarcely anything was attractive to her in its natural state—indeed, scarcely anything was decent until it was clothed by the opinion of some authority. Her ideas about habit, character, duty, love, marriage, were grouped under heads, like a book of popular quotations, and were totally unrelated to the emergencies of human living. She discussed all these subjects with other Methodist girls of her age. They would spend hours, for instance, in deciding what they would or would not tolerate in a suitor or a husband, and the frailties of masculine nature were too often a subject of discussion among them. In her behavior Anna was a harmless girl, mild except where her prejudices were concerned, neat and industrious, with no graver fault than priggishness; but her mind had really shocking habits of classification. The wickedness of Denver and of Chicago, and even of Moonstone, occupied her thoughts too much. She had none of the delicacy that goes with a nature of warm impulses, but the kind of fishy curiosity which justifies itself by an expression of horror.

Thea, and all Thea's ways and friends, seemed indecorous to Anna. She not only felt a grave social discrimination against the Mexicans; she could not forget that Spanish Johnny was a drunkard and that "nobody knew what he did when he ran away from home." Thea pretended, of course, that she liked the Mexicans because they were fond of music; but every one knew that music was nothing very real, and that it did not matter in a girl's relations with people. What was real, then, and what did matter? Poor Anna!

Anna approved of Ray Kennedy as a young man of steady habits and blameless life, but she regretted that he was an atheist, and that he was not a passenger conductor with brass buttons on his coat. On the whole, she wondered what such an exemplary young man found to like in Thea. Dr. Archie she treated respectfully because of his position in Moonstone, but she *knew* he had kissed the Mexican barytone's pretty daughter, and she had a whole *dossier* of evidence about his behavior in his hours of relaxation in Denver. He was "fast," and it was because he was "fast" that Thea liked him. Thea always liked that kind of people. Dr. Archie's whole manner with

Thea, Anna often told her mother, was too free. He was always putting his hand on Thea's head, or holding her hand while he laughed and looked down at her. The kindlier manifestations of human nature (about which Anna sang and talked, in the interests of which she went to conventions and wore white ribbons) were never realities to her after all. She did not believe in them. It was only in attitudes of protest or reproof, clinging to the cross, that human beings could be even temporarily decent.

Preacher Kronborg's secret convictions were very much like Anna's. He believed that his wife was absolutely good, but there was not a man or woman in his congregation whom he trusted all the way.

Mrs. Kronborg, on the other hand, was likely to find something to admire in almost any human conduct that was positive and energetic. She could always be taken in by the stories of tramps and runaway boys. She went to the circus and admired the bareback riders, who were "likely good enough women in their way." She admired Dr. Archie's fine physique and well-cut clothes as much as Thea did, and said she "felt it was a privilege to be handled by such a gentleman when she was sick."

Soon after Anna became a church member she began to remonstrate with Thea about practicing—playing "secular music"—on Sunday. One Sunday the dispute in the parlor grew warm and was carried to Mrs. Kronborg in the kitchen. She listened judicially and told Anna to read the chapter about how Naaman the leper was permitted to bow down in the house of Rimmon.[1] Thea went back to the piano, and Anna lingered to say that, since she was in the right, her mother should have supported her.

"No," said Mrs. Kronborg, rather indifferently, "I can't see it that way, Anna. I never forced you to practice, and I don't see as I should keep Thea from it. I like to hear her, and I guess your father does. You and Thea will likely follow different lines, and I don't see as I'm called upon to bring you up alike."

Anna looked meek and abused. "Of course all the church people must hear her. Ours is the only noisy house on this street. You hear what she's playing now, don't you?"

Mrs. Kronborg rose from browning her coffee. "Yes; it's the Blue Danube waltzes.[2] I'm familiar with 'em. If any of the church people come at you, you just send 'em to me. I ain't afraid to speak out on occasion, and I wouldn't mind one bit telling the Ladies' Aid a few things about standard composers." Mrs. Kronborg smiled, and added thoughtfully, "No, I wouldn't mind that one bit."

Anna went about with a reserved and distant air for a week, and Mrs. Kronborg suspected that she held a larger place than usual in her daughter's prayers; but that was another thing she didn't mind.

Although revivals were merely a part of the year's work, like examination week at school, and although Anna's piety impressed her very little, a time came when Thea was perplexed about religion. A scourge of typhoid broke out in Moonstone and several of Thea's schoolmates died of it. She went to their funerals, saw them put into the ground, and wondered a good deal about them. But a certain grim incident, which caused the epidemic, troubled her even more than the death of her friends.

Early in July, soon after Thea's fifteenth birthday, a particularly disgusting sort of tramp came into Moonstone in an empty box car. Thea was sitting in the hammock in the front yard when he first crawled up to the town from the depot, carrying a bundle wrapped in dirty ticking under one arm, and under the other a wooden box with rusty screening nailed over one end. He had a thin, hungry face covered with black hair. It was just before suppertime when he came along, and the street smelled of fried potatoes and fried onions and coffee. Thea saw him sniffing the air greedily and walking slower and slower. He looked over the fence. She hoped he would not stop at their gate, for her mother never turned any one away, and this was the dirtiest and most utterly wretched-looking tramp she had ever seen. There was a terrible odor about him, too. She caught it even at that distance, and put her handkerchief to her nose. A moment later she was sorry, for she knew that he had noticed it. He looked away and shuffled a little faster.

A few days later Thea heard that the tramp had camped in an empty shack over on the east edge of town, beside the ravine, and was trying to give a miserable sort of show there. He told the boys who went to see what he was doing, that he had traveled with a circus. His bundle contained a filthy clown's suit, and his box held half a dozen rattlesnakes.

Saturday night, when Thea went to the butcher shop to get the chickens for Sunday, she heard the whine of an accordion and saw a crowd before one of the saloons. There she found the tramp, his bony body grotesquely attired in the clown's suit, his face shaved and painted white,—the sweat trickling through the paint and washing it away,—and his eyes wild and feverish. Pulling the accordion in and out seemed to be almost too great an effort for him, and he panted to the tune of "Marching through Georgia."[3] After a considerable crowd had gathered, the tramp exhibited his box of snakes, announced that he would now pass the hat, and that when the onlookers had contributed the sum of one dollar, he would eat "one of these living reptiles." The crowd began to cough and murmur, and the saloon keeper rushed off for the marshal, who arrested the wretch for giving a show without a license and hurried him away to the calaboose.

The calaboose stood in a sunflower patch,—an old hut with a barred window and a padlock on the door. The tramp was utterly filthy and there was no way to give him a bath. The law made no provision to grub-stake vagrants, so after the constable had detained the tramp for twenty-four hours, he released him and told him to "get out of town, and get quick." The fellow's rattlesnakes had been killed by the saloon keeper. He hid in a box car in the freight yard, probably hoping to get a ride to the next station, but he was found and put out. After that he was seen no more. He had disappeared and left no trace except an ugly, stupid word, chalked on the black paint of the seventy-five-foot standpipe which was the reservoir for the Moonstone water-supply; the same word, in another tongue, that the French soldier shouted at Waterloo to the English officer who bade the Old Guard surrender;[4] a comment on life which the defeated, along the hard roads of the world, sometimes bawl at the victorious.

A week after the tramp excitement had passed over, the city water began to smell and to taste. The Kronborgs had a well in their back yard and did not use city water, but they heard the complaints of their neighbors. At first people said that the town well was full of rotting cottonwood roots, but the engineer at the pumping-station convinced the mayor that the water left the well untainted. Mayors reason slowly, but, the well being eliminated, the official mind had to travel toward the standpipe—there was no other track for it to go in. The standpipe amply rewarded investigation. The tramp had got even with Moonstone. He had climbed the standpipe by the handholds and let himself down into seventy-five feet of cold water, with his shoes and hat and roll of ticking. The city council had a mild panic and passed a new ordinance about tramps. But the fever had already broken out, and several adults and half a dozen children died of it.

Thea had always found everything that happened in Moonstone exciting, disasters particularly so. It was gratifying to read sensational Moonstone items in the Denver paper. But she wished she had not chanced to see the tramp as he came into town that evening, sniffing the supper-laden air. His face remained unpleasantly clear in her memory, and her mind struggled with the problem of his behavior as if it were a hard page in arithmetic. Even when she was practicing, the drama of the tramp kept going on in the back of her head, and she was constantly trying to make herself realize what pitch of hatred or despair could drive a man to do such a hideous thing. She kept seeing him in his bedraggled clown suit, the white paint on his roughly shaven face, playing his accordion before the saloon. She had noticed his lean body, his high, bald forehead that sloped back like a curved metal lid. How could people fall so far out of fortune? She tried to talk to Ray Kennedy about her perplexity, but Ray would not discuss things of that sort with her. It was in his sentimental conception of women that they should be deeply religious, though men were at liberty to doubt and finally to deny. A picture called "The Soul Awakened,"[5] popular in Moonstone parlors, pretty well interpreted Ray's idea of woman's spiritual nature.

One evening when she was haunted by the figure of the tramp, Thea went up to Dr. Archie's office. She found him sewing up two bad gashes in the face of a little boy who had been kicked by a mule. After the boy had been bandaged and sent away with his father, Thea helped the doctor wash and put away the surgical instruments. Then she dropped into her accustomed seat beside his desk and began to talk about the tramp. Her eyes were hard and green with excitement, the doctor noticed.

"It seems to me, Dr. Archie, that the whole town's to blame. I'm to blame, myself. I know he saw me hold my nose when he went by. Father's to blame. If he believes the Bible, he ought to have gone to the calaboose and cleaned that man up and taken care of him. That's what I can't understand; do people believe the Bible, or don't they? If the next life is all that matters, and we're put here to get ready for it, then why do we try to make money, or learn things, or have a good time? There's not one person in Moonstone that really lives the way the New Testament says. Does it matter, or don't it?"

Dr. Archie swung round in his chair and looked at her, honestly and leniently. "Well, Thea, it seems to me like this. Every people has had its religion. All religions are good, and all are pretty much alike. But I don't see how we could live up to them in the sense you mean. I've thought about it a good deal, and I can't help feeling that while we are in this world we have to live for the best things of this world, and those things are material and positive. Now, most religions are passive, and they tell us chiefly what we should not do." The doctor moved restlessly, and his eyes hunted for something along the opposite wall: "See here, my girl, take out the years of early childhood and the time we spend in sleep and dull old age, and we only have about twenty able, waking years. That's not long enough to get acquainted with half the fine things that have been done in the world, much less to do anything ourselves. I think we ought to keep the Commandments and help other people all we can; but the main thing is to live those twenty splendid years; to do all we can and enjoy all we can."

Dr. Archie met his little friend's searching gaze, the look of acute inquiry which always touched him.

"But poor fellows like that tramp—" she hesitated and wrinkled her forehead.

The doctor leaned forward and put his hand protectingly over hers, which lay clenched on the green felt desktop. "Ugly accidents happen, Thea; always have and always will. But the failures are swept back into the pile and forgotten. They don't leave any lasting scar in the world, and they don't affect the future. The things that last are the good things. The people who forge ahead and do something, they really count." He saw tears on her cheeks, and he remembered that he had never seen her cry before, not even when she crushed her finger when she was little. He rose and walked to the window, came back and sat down on the edge of his chair.

"Forget the tramp, Thea. This is a great big world, and I want you to get about and see it all. You're going to Chicago some day, and do something with that fine voice of yours. You're going to be a number one musician and make us proud of you. Take Mary Anderson,[6] now; even the tramps are proud of her. There isn't a tramp along the 'Q' system who hasn't heard of her. We all like people who do things, even if we only see their faces on a cigar-box lid."

They had a long talk. Thea felt that Dr. Archie had never let himself out to her so much before. It was the most grown-up conversation she had ever had with him. She left his office happy, flattered and stimulated. She ran for a long while about the white, moonlit streets, looking up at the stars and the bluish night, at the quiet houses sunk in black shade, the glittering sand hills. She loved the familiar trees, and the people in those little houses, and she loved the unknown world beyond Denver. She felt as if she were being pulled in two, between the desire to go away forever and the desire to stay forever. She had only twenty years—no time to lose.

Many a night that summer she left Dr. Archie's office with a desire to run and run about those quiet streets until she wore out her shoes, or wore out the streets themselves; when her chest ached and it seemed as if her heart were spreading all over the desert. When she went home, it was not to go to sleep. She used to drag her mattress beside her low window and lie awake for a long while, vibrating with excitement, as a machine vibrates from speed. Life rushed in upon her through that window—or so it seemed. In reality, of course, life rushes from within, not from without. There is no work of art so big or so beautiful that it was not once all contained in some youthful body, like this one which lay on the floor in the moonlight, pulsing with ardor and anticipation. It was on such nights that Thea Kronborg learned the thing that old Dumas[7] meant when he told the Romanticists that to make a drama he needed but one passion and four walls.

XIX

IT IS WELL for its peace of mind that the traveling public takes railroads so much for granted. The only men who are incurably nervous about railway travel are the railroad operatives. A railroad man never forgets that the next run may be his turn.

On a single-track road, like that upon which Ray Kennedy worked, the freight trains make their way as best they can between passenger trains. Even when there is such a thing as a freight time-schedule, it is merely a form. Along the one track dozens of fast and slow trains dash in both directions, kept from collision only by the brains in the dispatcher's office. If one passenger train is late, the whole schedule must be revised in an instant; the trains following must be warned, and those moving toward the belated train must be assigned new meeting-places.

Between the shifts and modifications of the passenger schedule, the freight trains play a game of their own. They have no right to the track at any given time, but are supposed to be on it when it is free, and to make the best time they can between passenger trains. A freight train, on a single-track road, gets anywhere at all only by stealing bases.

Ray Kennedy had stuck to the freight service, although he had had opportunities to go into the passenger service at higher pay. He always regarded railroading as a temporary makeshift, until he "got into something," and he disliked the passenger service. No brass buttons for him, he said; too much like a livery. While he was railroading he would wear a jumper, thank you!

The wreck that "caught" Ray was a very commonplace one; nothing thrilling about it, and it got only six lines in the Denver papers. It happened about daybreak one morning, only thirty-two miles from home.

At four o'clock in the morning Ray's train had stopped to take water at Saxony, having just rounded the long curve which lies south of that station. It was Joe Giddy's business to walk back along the curve about three hundred yards and put out torpedoes to warn any train which might be coming up from behind—a freight crew is not notified of trains following, and the brakeman is supposed to protect his train. Ray was so fussy about the punctilious observance of orders that almost any brakeman would take a chance once in a while, from natural perversity.

When the train stopped for water that morning, Ray was at the desk in his caboose, making out his report. Giddy took his torpedoes, swung off the rear platform, and glanced back at the curve. He decided that he would not go back to flag this time. If anything was coming up behind, he could hear it in plenty of time. So he ran forward to look after a hot journal that had been bothering him. In a general way, Giddy's reasoning was sound. If a freight train, or even a passenger train, had been coming up behind them, he could have heard it in time. But as it happened, a light engine, which made no noise at all, was coming,—ordered out to help with the freight that was piling up at the other end of the division. This engine got no warning, came round the curve, struck the caboose, went straight through it, and crashed into the heavy lumber car ahead.

The Kronborgs were just sitting down to breakfast, when the night telegraph operator dashed into the yard at a run and hammered on the front door. Gunner answered the knock, and the telegraph operator told him he wanted to see his father a minute, quick. Mr. Kronborg appeared at the door, napkin in hand. The operator was pale and panting.

"Fourteen was wrecked down at Saxony this morning," he shouted, "and Kennedy's all broke up. We're sending an engine down with the doctor, and the operator at Saxony says Kennedy wants you to come along with us and bring your girl." He stopped for breath.

Mr. Kronborg took off his glasses and began rubbing them with his napkin.

"Bring—I don't understand," he muttered. "How did this happen?"

"No time for that, sir. Getting the engine out now. Your girl, Thea. You'll surely do that for the poor chap. Everybody knows he thinks the world of her." Seeing that Mr. Kronborg showed no indication of having made up his mind, the operator turned to Gunner. "Call your sister, kid. I'm going to ask the girl herself," he blurted out.

"Yes, yes, certainly. Daughter," Mr. Kronborg called. He had somewhat recovered himself and reached to the hall hatrack for his hat.

Just as Thea came out on the front porch, before the operator had had time to explain to her, Dr. Archie's ponies came up to the gate at a brisk trot. Archie jumped out the moment his driver stopped the team and came up to the bewildered girl without so much as saying good-morning to any one. He took her hand with the sympathetic, reassuring graveness which had helped her at more than one hard time in her life. "Get your hat, my girl. Kennedy's hurt down the road, and he wants you to run down with me. They'll have a car for us. Get into my buggy, Mr. Kronborg. I'll drive you down, and Larry can come for the team."

The driver jumped out of the buggy and Mr. Kronborg and the doctor got in. Thea, still bewildered, sat on her father's knee. Dr. Archie gave his ponies a smart cut with the whip.

When they reached the depot, the engine, with one car attached, was standing on the main track. The engineer had got his steam up, and was leaning out of the cab impatiently. In a moment they were off. The run to Saxony took forty minutes. Thea sat still in her seat while Dr. Archie and her father talked about the wreck. She took no part in the conversation and asked no questions, but occasionally she looked at Dr. Archie with a frightened, inquiring glance, which he answered by an encouraging nod. Neither he nor her father said anything about how badly Ray was hurt. When the engine stopped near Saxony, the main track was already cleared. As they got out of the car, Dr. Archie pointed to a pile of ties.

"Thea, you'd better sit down here and watch the wreck crew while your father and I go up and look Kennedy over. I'll come back for you when I get him fixed up."

The two men went off up the sand gulch, and Thea sat down and looked at the pile of splintered wood and twisted iron that had lately been Ray's caboose. She was frightened and absent-minded. She felt that she ought to be thinking about Ray, but her mind kept racing off to all sorts of trivial and irrelevant things. She wondered whether Grace Johnson would be furious when she came to take her music lesson and found nobody there to give it to her; whether she had forgotten to close the piano last night and whether Thor would get into the new room and mess the keys all up with his sticky fingers; whether Tillie would go upstairs and make her bed for her. Her mind worked fast, but she could fix it upon nothing. The grasshoppers, the lizards, distracted her attention and seemed more real to her than poor Ray.

On their way to the sand bank where Ray had been carried, Dr. Archie and Mr. Kronborg met the Saxony doctor. He shook hands with them.

"Nothing you can do, doctor. I couldn't count the fractures. His back's broken, too. He wouldn't be alive now if he weren't so confoundedly strong, poor chap. No use bothering him. I've given him morphia, one and a half, in eighths."

Dr. Archie hurried on. Ray was lying on a flat canvas litter, under the shelter of a shelving bank, lightly shaded by a slender cottonwood tree. When the doctor and the preacher approached, he looked at them intently.

"Didn't—" he closed his eyes to hide his bitter disappointment.

Dr. Archie knew what was the matter. "Thea's back there, Ray. I'll bring her as soon as I've had a look at you."

Ray looked up. "You might clean me up a trifle, doc. Won't need you for anything else, thank you all the same."

However little there was left of him, that little was certainly Ray Kennedy. His personality was as positive as ever, and the blood and dirt on his face seemed merely accidental, to have nothing to do with the man himself. Dr. Archie told Mr. Kronborg to bring a pail of water, and he began to sponge Ray's face and neck. Mr. Kronborg stood by, nervously rubbing his hands together and trying to think of something to say. Serious situations always embarrassed him and made him formal, even when he felt real sympathy.

"In times like this, Ray," he brought out at last, crumpling up his handkerchief in his long fingers,—"in times like this, we don't want to forget the Friend that sticketh closer than a brother."

Ray looked up at him; a lonely, disconsolate smile played over his mouth and his square cheeks. "Never mind about all that, *padre*," he said quietly. "Christ and me fell out long ago."

There was a moment of silence. Then Ray took pity on Mr. Kronborg's embarrassment. "You go back for the little girl, *padre*. I want a word with the doc in private."

Ray talked to Dr. Archie for a few moments, then stopped suddenly, with a broad smile. Over the doctor's shoulder he saw Thea coming up the gulch, in her pink chambray dress, carrying her sunhat by the strings. Such a yellow head! He often told himself that he "was perfectly foolish about her hair." The sight of her, coming, went through him softly, like the morphia. "There she is," he whispered. "Get the old preacher out of the way, doc. I want to have a little talk with her."

Dr. Archie looked up. Thea was hurrying and yet hanging back. She was more frightened than he had thought she would be. She had gone with him to see very sick people and had always been steady and calm. As she came up, she looked at the ground, and he could see that she had been crying.

Ray Kennedy made an unsuccessful effort to put out his hand. "Hello, little kid, nothing to be afraid of. Darned if I don't believe they've gone and scared you! Nothing to cry about. I'm the same old goods, only a little dented. Sit down on my coat there, and keep me company. I've got to lay still a bit."

Dr. Archie and Mr. Kronborg disappeared. Thea cast a timid glance after them, but she sat down resolutely and took Ray's hand.

"You ain't scared now, are you?" he asked affectionately. "You were a regular brick to come, Thee. Did you get any breakfast?"

"No, Ray, I'm not scared. Only I'm dreadful sorry you're hurt, and I can't help crying."

His broad, earnest face, languid from the opium and smiling with such simple happiness, reassured her. She drew nearer to him and lifted his hand to her knee. He looked at her with his clear, shallow blue eyes. How he loved everything about that face and head! How many nights in his cupola, looking up the track, he had seen that face in the darkness; through the sleet and snow, or in the soft blue air when the moonlight slept on the desert.

"You needn't bother to talk, Thee. The doctor's medicine makes me sort of dopey. But it's nice to have company. Kind of cozy, don't you think? Pull my coat under you more. It's a darned shame I can't wait on you."

"No, no, Ray. I'm all right. Yes, I like it here. And I guess you ought not to talk much, ought you? If you can sleep, I'll stay right here, and be awful quiet. I feel just as much at home with you as ever, now."

That simple, humble, faithful something in Ray's eyes went

straight to Thea's heart. She did feel comfortable with him, and happy to give him so much happiness. It was the first time she had ever been conscious of that power to bestow intense happiness by simply being near any one. She always remembered this day as the beginning of that knowledge. She bent over him and put her lips softly to his cheek.

Ray's eyes filled with light. "Oh, do that again, kid!" he said impulsively. Thea kissed him on the forehead, blushing faintly. Ray held her hand fast and closed his eyes with a deep sigh of happiness. The morphia and the sense of her nearness filled him with content. The gold mine, the oil well, the copper ledge—all pipe dreams, he mused, and this was a dream, too. He might have known it before. It had always been like that; the things he admired had always been away out of his reach: a college education, a gentleman's manner, an Englishman's accent—things over his head. And Thea was farther out of his reach than all the rest put together. He had been a fool to imagine it, but he was glad he had been a fool. She had given him one grand dream. Every mile of his run, from Moonstone to Denver, was painted with the colors of that hope. Every cactus knew about it. But now that it was not to be, he knew the truth. Thea was never meant for any rough fellow like him—hadn't he really known that all along, he asked himself? She wasn't meant for common men. She was like wedding cake, a thing to dream on. He raised his eyelids a little. She was stroking his hand and looking off into the distance. He felt in her face that look of unconscious power that Wunsch had seen there. Yes, she was bound for the big terminals of the world; no way stations for her. His lids drooped. In the dark he could see her as she would be after a while; in a box at the Tabor Grand in Denver, with diamonds on her neck and a tiara in her yellow hair, with all the people looking at her through their opera-glasses, and a United States Senator, maybe, talking to her. "Then you'll remember me!" He opened his eyes, and they were full of tears.

Thea leaned closer. "What did you say, Ray? I couldn't hear."

"Then you'll remember me," he whispered.

The spark in his eye, which is one's very self, caught the spark in hers that was herself, and for a moment they looked into each other's natures. Thea realized how good and how great-hearted he was, and he realized about her many things. When that elusive spark of personality retreated in each of them, Thea still saw in his wet eyes her own face, very small, but much prettier than the cracked glass at home had ever shown it. It was the first time she had seen her face in that kindest mirror a woman can ever find.

Ray had felt things in that moment when he seemed to be looking into the very soul of Thea Kronborg. Yes, the gold mine, the oil well, the copper ledge, they'd all got away from him, as things will; but he'd backed a winner once in his life! With all his might he gave his faith to the broad little hand he held. He wished he could leave her the rugged strength of his body to help her through with it all. He would have liked to tell her a little about his old dream,—there seemed long years between him and it already,—but to tell her now would somehow be unfair; wouldn't be quite the straightest thing in the world. Probably she knew, anyway. He looked up quickly. "You know, don't you, Thee, that I think you are just the finest thing I've struck in this world?"

The tears ran down Thea's cheeks. "You're too good to me, Ray. You're a lot too good to me," she faltered.

"Why, kid," he murmured, "everybody in this world's going to be good to you!"

Dr. Archie came to the gulch and stood over his patient. "How's it going?"

"Can't you give me another punch with your pacifier, doc? The little girl had better run along now." Ray released Thea's hand. "See you later, Thee."

She got up and moved away aimlessly, carrying her hat by the strings. Ray looked after her with the exaltation born of bodily pain and said between his teeth, "Always look after that girl, doc. She's a queen!"

Thea and her father went back to Moonstone on the one-o'clock passenger. Dr. Archie stayed with Ray Kennedy until he died, late in the afternoon.

XX

On Monday morning, the day after Ray Kennedy's funeral, Dr. Archie called at Mr. Kronborg's study, a little room behind the church. Mr. Kronborg did not write out his sermons, but spoke from notes jotted upon small pieces of cardboard in a kind of shorthand of his own. As sermons go, they were not worse than most. His conventional rhetoric pleased the majority of his congregation, and Mr. Kronborg was generally regarded as a model preacher. He did not smoke, he never touched spirits. His indulgence in the pleasures of the table was an endearing bond between him and the women of his congregation. He ate enormously, with a zest which seemed incongruous with his spare frame.

This morning the doctor found him opening his mail and reading a pile of advertising circulars with deep attention.

"Good-morning, Mr. Kronborg," said Dr. Archie, sitting down. "I came to see you on business. Poor Kennedy asked me to look after his affairs for him. Like most railroad men he spent his wages, except for a few investments in mines which don't look to me very promising. But his life was insured for six hundred dollars in Thea's favor."

Mr. Kronborg wound his feet about the standard of his desk-chair. "I assure you, doctor, this is a complete surprise to me."

"Well, it's not very surprising to me," Dr. Archie went on. "He talked to me about it the day he was hurt. He said he wanted the money to be used in a particular way, and in no other." Dr. Archie paused meaningly.

Mr. Kronborg fidgeted. "I am sure Thea would observe his wishes in every respect."

"No doubt; but he wanted me to see that you agreed to his plan. It seems that for some time Thea has wanted to go away to study music. It was Kennedy's wish that she should take this money and go to Chicago this winter. He felt that it would be an advantage to her in a business way: that even if she came back here to teach, it would give her more authority and make her position here more comfortable."

Mr. Kronborg looked a little startled. "She is very young," he hesitated; "she is barely seventeen. Chicago is a long way from home. We would have to consider. I think, Dr. Archie, we had better consult Mrs. Kronborg."

"I think I can bring Mrs. Kronborg around, if I have your consent. I've always found her pretty level-headed. I have several old classmates practicing in Chicago. One is a throat specialist. He has a good deal to do with singers. He probably knows the best piano teachers and could recommend a boarding-house where music students stay. I think Thea needs to get among a lot of young people who are clever like herself. Here she has no companions but old fellows like me. It's not a natural life for a young girl. She'll either get warped, or wither up before her time. If it will make you and Mrs. Kronborg feel any easier, I'll be glad to take Thea to Chicago and see that she gets started right. This throat man I speak of is a big fellow in his line, and if I can get him interested, he may be able to put her in the way of a good many things. At any rate, he'll know the right teachers. Of course, six hundred dollars won't take her very far, but even half the winter there would be a great advantage. I think Kennedy sized the situation up exactly."

"Perhaps; I don't doubt it. You are very kind, Dr. Archie." Mr. Kronborg was ornamenting his desk-blotter with hieroglyphics. "I should think Denver might be better. There we could watch over her. She is very young."

Dr. Archie rose. "Kennedy didn't mention Denver. He said Chicago, repeatedly. Under the circumstances, it seems to me we ought to try to carry out his wishes exactly, if Thea is willing."

"Certainly, certainly. Thea is conscientious. She would not waste her opportunities." Mr. Kronborg paused. "If Thea were your own daughter, doctor, would you consent to such a plan, at her present age?"

"I most certainly should. In fact, if she were my daughter, I'd have sent her away before this. She's a most unusual child, and she's only wasting herself here. At her age she ought to be learning, not teaching. She'll never learn so quickly and easily as she will right now."

"Well, doctor, you had better talk it over with Mrs. Kronborg. I make it a point to defer to her wishes in such matters. She understands all her children perfectly. I may say that she has all a mother's insight, and more."

Dr. Archie smiled. "Yes, and then some. I feel quite confident about Mrs. Kronborg. We usually agree. Good-morning."

Dr. Archie stepped out into the hot sunshine and walked rapidly toward his office, with a determined look on his face. He found his waiting-room full of patients, and it was one o'clock before he had dismissed the last one. Then he shut his door and took a drink before going over to the hotel for his lunch. He smiled as he locked his cupboard. "I feel almost as gay as if I were going to get away for a winter myself," he thought.

Afterward Thea could never remember much about that summer, or how she lived through her impatience. She was to set off with Dr. Archie on the fifteenth of October, and she gave lessons until the first of September. Then she began to get her clothes ready, and spent whole afternoons in the village dressmaker's stuffy, littered little sewing-room. Thea and her mother made a trip to Denver to buy the materials for her dresses. Ready-made clothes for girls were not to be had in those days. Miss Spencer, the dressmaker, declared that she could do handsomely by Thea if they would only let her carry out her own ideas. But Mrs. Kronborg and Thea felt that Miss Spencer's most daring productions might seem out of place in Chicago, so they restrained her with a firm hand. Tillie, who always helped Mrs. Kronborg with the family sewing, was for letting Miss Spencer challenge Chicago on Thea's person. Since Ray Kennedy's death, Thea had become more than ever one of Tillie's heroines. Tillie swore each of her friends to secrecy, and, coming home from church or leaning over the fence, told them the most touching stories about Ray's devotion, and how Thea would "never get over it."

Tillie's confidences stimulated the general discussion of Thea's venture. This discussion went on, upon front porches and in back yards, pretty much all summer. Some people approved of Thea's going to Chicago, but most people did not. There were others who changed their minds about it every day.

Tillie said she wanted Thea to have a ball dress "above all things." She bought a fashion book especially devoted to evening clothes and looked hungrily over the colored plates, picking out costumes that would be becoming to "a blonde." She wanted Thea to have all the gay clothes she herself had always longed for; clothes she often told herself she needed "to recite in."

"Tillie," Thea used to cry impatiently, "can't you see that if Miss Spencer tried to make one of those things, she'd make me look like a circus girl? Anyhow, I don't know anybody in Chicago. I won't be going to parties."

Tillie always replied with a knowing toss of her head, "You see! You'll be in society before you know it. There ain't many girls as accomplished as you."

On the morning of the fifteenth of October the Kronborg family, all of them but Gus, who couldn't leave the store, started for the station an hour before train time. Charley had taken Thea's trunk and telescope[1] to the depot in his delivery wagon early that morning. Thea wore her new blue serge traveling-dress, chosen for its serviceable qualities. She had done her hair up carefully, and had put a pale-blue ribbon around her throat, under a little lace collar that Mrs. Kohler had crocheted for her. As they went out of the gate, Mrs. Kronborg looked her over thoughtfully. Yes, that blue ribbon went very well with the dress, and with Thea's eyes. Thea had a rather unusual touch about such things, she reflected comfortably. Tillie always said that Thea was "so indifferent to dress," but her mother noticed that she usually put her clothes on well. She felt the more at ease about letting Thea go away from home, because she had good sense about her clothes and never tried to dress up too much. Her coloring was so individual, she was so unusually fair, that in the wrong clothes she might easily have been "conspicuous."

It was a fine morning, and the family set out from the house in good spirits. Thea was quiet and calm. She had forgotten nothing, and she clung tightly to her handbag, which held her trunk-key and all of her money that was not in an envelope pinned to her chemise. Thea walked behind the others, holding Thor by the hand, and this time she did not feel that the procession was too long. Thor was uncommunicative that morning, and would only talk about how he would rather get a sand bur in his toe every day than wear shoes and stockings. As they passed the cottonwood grove where Thea often used to bring him in his cart, she asked him who would take him for nice long walks after sister went away.

"Oh, I can walk in our yard," he replied unappreciatively. "I guess I can make a pond for my duck."

Thea leaned down and looked into his face. "But you won't forget about sister, will you?" Thor shook his head. "And won't you be glad when sister comes back and can take you over to Mrs. Kohler's to see the pigeons?"

"Yes, I'll be glad. But I'm going to have a pigeon my own self."

"But you haven't got any little house for one. Maybe Axel would make you a little house."

"Oh, her can live in the barn, her can," Thor drawled indifferently.

Thea laughed and squeezed his hand. She always liked his sturdy matter-of-factness. Boys ought to be like that, she thought.

When they reached the depot, Mr. Kronborg paced the platform somewhat ceremoniously with his daughter. Any member of his flock would have gathered that he was giving her good counsel about meeting the temptations of the world. He did, indeed, begin to admonish her not to forget that talents come from our Heavenly Father and are to be used for His glory, but he cut his remarks short and looked at his watch. He believed that Thea was a religious girl, but when she looked at him with that intent, that passionately inquiring gaze which used to move even Wunsch, Mr. Kronborg suddenly felt his eloquence fail. Thea was like her mother, he reflected; you couldn't put much sentiment across with her. As a usual thing, he liked girls to be a little more responsive. He liked them to blush at his compliments; as Mrs. Kronborg candidly said, "Father could be very soft with the girls." But this morning he was thinking that hard-headedness was a reassuring quality in a daughter who was going to Chicago alone.

Mr. Kronborg believed that big cities were places where people went to lose their identity and to be wicked. He himself, when he was a student at the Seminary—he coughed and opened his watch again. He knew, of course, that a great deal of business went on in Chicago, that there was an active Board of Trade, and that hogs and cattle were slaughtered there. But when, as a young man, he had stopped over in Chicago, he had not interested himself in the commercial activities of the city. He remembered it as a place full of cheap shows and dance halls and boys from the country who were behaving disgustingly.

Dr. Archie drove up to the station about ten minutes before the train was due. His man tied the ponies and stood holding the doctor's alligator-skin bag—very elegant, Thea thought it. Mrs. Kronborg did not burden the doctor with warnings and cautions. She said again that she hoped he could get Thea a comfortable place to stay, where they had good beds, and she hoped the landlady would be a woman who'd had children of her own. "I don't go much on old maids looking after girls," she remarked as she took a pin out of her own hat and thrust it into Thea's blue turban. "You'll be sure to lose your hatpins on the train, Thea. It's better to have an extra one in case." She tucked in a little curl that had escaped from Thea's careful twist. "Don't forget to brush your dress often, and pin it up to the curtains of your berth to-night, so it won't wrinkle. If you get it wet, have a tailor press it before it draws."

She turned Thea about by the shoulders and looked her over a last time. Yes, she looked very well. She wasn't pretty, exactly,—her face was too broad and her nose was too big. But she had that lovely skin, and she looked fresh and sweet. She had always been a sweet-smelling child. Her mother had always liked to kiss her, when she happened to think of it.

The train whistled in, and Mr. Kronborg carried the canvas "telescope" into the car. Thea kissed them all good-bye. Tillie cried, but she was the only one who did. They all shouted things up at the closed window of the Pullman car, from which Thea looked down at them as from a frame, her face glowing with excitement, her turban a little tilted in spite of three hatpins. She had already taken off her new gloves to save them. Mrs. Kronborg reflected that she would never see just that same picture again, and as Thea's car slid off along the rails, she wiped a tear from her eye. "She won't come back a little girl," Mrs. Kronborg said to her husband as they turned to go home. "Anyhow, she's been a sweet one."

While the Kronborg family were trooping slowly homeward, Thea was sitting in the Pullman, her telescope in the seat beside her, her handbag tightly gripped in her fingers. Dr. Archie had gone into the smoker. He thought she might be a little tearful, and that it would be kinder to leave her alone for a while. Her eyes did fill once, when she saw the last of the sand hills and realized that she was going to leave them behind for a long while. They always made her think of Ray, too. She had had such good times with him out there.

But, of course, it was herself and her own adventure that mattered to her. If youth did not matter so much to itself, it would never have the heart to go on. Thea was surprised that she did not feel a deeper sense of loss at leaving her old life behind her. It seemed, on the contrary, as she looked out at the yellow desert speeding by, that she had left very little. Everything that was essential seemed to be right there in the car with her. She lacked nothing. She even felt more compact and confident than usual. She was all there, and something else was there, too,—in her heart, was it, or under her cheek? Anyhow, it was about her somewhere, that warm sureness, that sturdy little companion with whom she shared a secret.

When Dr. Archie came in from the smoker, she was sitting still, looking intently out of the window and smiling, her lips a little parted, her hair in a blaze of sunshine. The doctor thought she was the prettiest thing he had ever seen, and very funny, with her telescope and big handbag. She made him feel jolly, and a little mournful, too. He knew that the splendid things of life are few, after all, and so very easy to miss.

PART II

THE SONG OF THE LARK

I

THEA AND DR. ARCHIE had been gone from Moonstone four days. On the afternoon of the nineteenth of October they were in a street-car, riding through the depressing, unkept wastes of North Chicago,[1] on their way to call upon the Reverend Lars Larsen, a friend to whom Mr. Kronborg had written. Thea was still staying at the rooms of the Young Women's Christian Association, and was miserable and homesick there. The housekeeper watched her in a way that made her uncomfortable. Things had not gone very well, so far. The noise and confusion of a big city tired and disheartened her. She had not had her trunk sent to the Christian Association rooms because she did not want to double cartage charges, and now she was running up a bill for storage on it. The contents of her gray telescope were becoming untidy, and it seemed impossible to keep one's face and hands clean in Chicago. She felt as if she were still on the train, traveling without enough clothes to keep clean. She wanted another nightgown, and it did not occur to her that she could buy one. There were other clothes in her trunk that she needed very much, and she seemed no nearer a place to stay than when she arrived in the rain, on that first disillusioning morning.

Dr. Archie had gone at once to his friend Hartley Evans, the throat specialist, and had asked him to tell him of a good piano teacher and direct him to a good boarding-house. Dr. Evans said he could easily tell him who was the best piano teacher in Chicago, but that most students' boarding-houses were "abominable places, where girls got poor food for body and mind." He gave Dr. Archie several addresses, however, and the doctor went to look the places over. He left Thea in her room, for she seemed tired and was not at all like herself. His inspection of boarding-houses was not encouraging. The only place that seemed to him at all desirable was full, and the mistress of the house could not give Thea a room in which she could have a piano. She said Thea might use the piano in her parlor; but when Dr. Archie went to look at the parlor he found a girl talking to a young man on one of the corner sofas. Learning that the boarders received all their callers there, he gave up that house, too, as hopeless.

So when they set out to make the acquaintance of Mr. Larsen on the afternoon he had appointed, the question of a lodging was still undecided. The Swedish Reform Church was in a sloughy, weedy district, near a group of factories. The church itself was a very neat little building. The parsonage, next door, looked clean and comfortable, and there was a well-kept yard about it, with a picket fence. Thea saw several little children playing under a swing, and wondered why ministers always had so many. When they rang at the parsonage door, a capable-looking Swedish servant girl answered the bell and told them that Mr. Larsen's study was in the church, and that he was waiting for them there.

Mr. Larsen received them very cordially. The furniture in his study was so new and the pictures were so heavily framed, that Thea thought it looked more like the waiting-room of the fashionable Denver dentist to whom Dr. Archie had taken her that summer, than like a preacher's study. There were even flowers in a glass vase on the desk. Mr. Larsen was a small, plump man, with a short, yellow beard, very white teeth, and a little turned-up nose on which he wore gold-rimmed eye-glasses. He looked about thirty-five, but he was growing bald, and his thin hair was parted above his left ear and brought up over the bare spot on the top of his head. He looked cheerful and agreeable. He wore a blue coat and no cuffs.

After Dr. Archie and Thea sat down on a slippery leather couch, the minister asked for an outline of Thea's plans. Dr. Archie explained that she meant to study piano with Andor Harsanyi; that they had already seen him, that Thea had played for him and he said he would be glad to teach her.

Mr. Larsen lifted his pale eyebrows and rubbed his plump white hands together. "But he is a concert pianist already. He will be very expensive."

"That's why Miss Kronborg wants to get a church position if possible. She has not money enough to see her through the winter. There's no use her coming all the way from Colorado and studying with a second-rate teacher. My friends here tell me Harsanyi is the best."

"Oh, very likely! I have heard him play with Thomas. You Western people do things on a big scale. There are half a dozen teachers that I should think— However, you know what you want." Mr. Larsen showed his contempt for such extravagant standards by a shrug. He felt that Dr. Archie was trying to impress him. He had succeeded, indeed, in bringing out the doctor's stiffest manner. Mr. Larsen went on to explain that he managed the music in his church himself, and drilled his choir, though the tenor was the official choirmaster. Unfortunately there were no vacancies in his choir just now. He had his four voices, very good ones. He looked away from Dr. Archie and glanced at Thea. She looked troubled, even a little frightened when he said this, and drew in her lower lip. She, certainly, was not pretentious, if her protector was. He continued to study her. She was sitting on the lounge, her knees far apart, her gloved hands lying stiffly in her lap, like a country girl. Her turban, which seemed a little too big for her, had got tilted in the wind,—it was always windy in that part of Chicago,—and she looked tired. She wore no veil, and her hair, too, was the worse for the wind and dust. When he said he had all the voices he required, he noticed that her gloved hands shut tightly. Mr. Larsen reflected that she was not, after all, responsible for the lofty manner of her father's physician; that she was not even responsible for her father, whom he remembered as a tiresome fellow. As he watched her tired, worried face, he felt sorry for her.

"All the same, I would like to try your voice," he said, turning pointedly away from her companion. "I am interested in voices. Can you sing to the violin?"

"I guess so," Thea replied dully. "I don't know. I never tried."

Mr. Larsen took his violin out of the case and began to tighten the keys. "We might go into the lecture-room and see how it goes. I can't tell much about a voice by the organ. The violin is really the proper instrument to try a voice." He opened a door at the back of his study, pushed Thea gently through it, and looking over his shoulder to Dr. Archie said, "Excuse us, sir. We will be back soon."

Dr. Archie chuckled. All preachers were alike, officious and on their dignity; liked to deal with women and girls, but not with men. He took up a thin volume from the minister's desk. To his amusement it proved to be a book of "Devotional and Kindred Poems; by Mrs. Aurelia S. Larsen." He looked them over, thinking that the world changed very little. He could remember when the wife of his father's minister had published a volume of verses, which all the church members had to buy and all the children were encouraged to read. His grandfather had made a face at the book and said, "Puir body!" Both ladies seemed to have chosen the same subjects, too: Jephthah's Daughter, Rizpah, David's Lament for Absalom, etc.[2] The doctor found the book very amusing.

The Reverend Lars Larsen was a reactionary Swede. His father came to Iowa in the sixties, married a Swedish girl who was ambitious, like himself, and they moved to Kansas and took up land under the Homestead Act.[3] After that, they bought land and leased it from the Government, acquired land in every possible way. They worked like horses, both of them; indeed, they would never have used any horse-flesh they owned as they used themselves. They reared a large family and worked their sons and daughters as mercilessly as they worked themselves; all of them but Lars. Lars was the fourth son, and he was born lazy. He seemed to bear the mark of overstrain on the part of his parents. Even in his cradle he was an example of physical inertia; anything to lie still. When he was a growing boy his mother had to drag him out of bed every morning, and he had to be driven to his chores. At school he had a model "attendance record," because he found getting his lessons easier than farm work. He was the only one of the family who went through the high school, and by the time he graduated he had already made up his mind to study for the ministry, because it seemed to him the least laborious of all callings. In so far as he could see, it was the only business in which there was practically no competition, in which a man was not all the time pitted against other men who were willing to work themselves to death. His father stubbornly opposed Lars's plan, but after keeping the boy at home for a year and finding how useless he was on the farm, he sent him to a theological seminary—as much to conceal his laziness from the neighbors as because he did not know what else to do with him.

Larsen, like Peter Kronborg, got on well in the ministry, because he got on well with the women. His English was no worse than that of most young preachers of American parentage, and he made the most of his skill with the violin. He was supposed to exert a very desirable influence over young people and to stimulate their interest in church work. He married an American girl, and when his father died he got his share of the property—which was very considerable. He invested his money carefully and was that rare thing, a preacher of independent means. His white, well-kept hands were his result,—the evidence that he had worked out his life successfully in the way that pleased him. His Kansas brothers hated the sight of his hands.

Larsen liked all the softer things of life,—in so far as he knew about them. He slept late in the morning, was fussy about his food, and read a great many novels, preferring sentimental ones. He did not smoke, but he ate a great deal of candy "for his throat," and always kept a box of chocolate drops in the upper right-hand drawer of his desk. He always bought season tickets for the symphony concerts, and he played his violin for women's culture clubs. He did not wear cuffs, except on Sunday, because he believed that a free wrist facilitated his violin practice. When he drilled his choir he always held his hand with the little and index fingers curved higher than the other two, like a noted German conductor he had seen. On the whole, the Reverend Larsen was not an insincere man; he merely spent his life resting and playing, to make up for the time his forebears had wasted grubbing in the earth. He was simple-hearted and kind; he enjoyed his candy and his children and his sacred cantatas. He could work energetically at almost any form of play.

Dr. Archie was deep in "The Lament of Mary Magdalen," when Mr. Larsen and Thea came back to the study. From the minister's expression he judged that Thea had succeeded in interesting him.

Mr. Larsen seemed to have forgotten his hostility toward him, and addressed him frankly as soon as he entered. He stood holding his violin, and as Thea sat down he pointed to her with his bow:—

"I have just been telling Miss Kronborg that though I cannot promise her anything permanent, I might give her something for the next few months. My soprano is a young married woman and is temporarily indisposed.[4] She would be glad to be excused from her duties for a while. I like Miss Kronborg's singing very much, and I think she would benefit by the instruction in my choir. Singing here might very well lead to something else. We pay our soprano only eight dollars a Sunday, but she always gets ten dollars for singing at funerals. Miss Kronborg has a sympathetic voice, and I think there would be a good deal of demand for her at funerals. Several American churches apply to me for a soloist on such occasions, and I could help her to pick up quite a little money that way."

This sounded lugubrious to Dr. Archie, who had a physician's dislike of funerals, but he tried to accept the suggestion cordially.

"Miss Kronborg tells me she is having some trouble getting located," Mr. Larsen went on with animation, still holding his violin. "I would advise her to keep away from boarding-houses altogether. Among my parishioners there are two German women, a mother and daughter. The daughter is a Swede by marriage, and clings to the Swedish Church. They live near here, and they rent some of their rooms. They have now a large room vacant, and have asked me to recommend some one. They have never taken boarders, but Mrs. Lorch, the mother, is a good cook,—at least, I am always glad to take supper with her,—and I think I could persuade her to let this young woman partake of the family table. The daughter, Mrs. Andersen, is musical, too, and sings in the Mozart Society. I think they might like to have a music student in the house. You speak German, I suppose?" he turned to Thea.

"Oh, no; a few words. I don't know the grammar," she murmured.

Dr. Archie noticed that her eyes looked alive again, not frozen as they had looked all morning. "If this fellow can help her, it's not for me to be stand-offish," he said to himself.

"Do you think you would like to stay in such a quiet place, with old-fashioned people?" Mr. Larsen asked. "I shouldn't think you could find a better place to work, if that's what you want."

"I think mother would like to have me with people like that," Thea replied. "And I'd be glad to settle down most anywhere. I'm losing time."

"Very well, there's no time like the present. Let us go to see Mrs. Lorch and Mrs. Andersen."

The minister put his violin in its case and caught up a black-and-white checked traveling-cap that he wore when he rode his high Columbia wheel.[5] The three left the church together.

II

So Thea did not go to a boarding-house after all. When Dr. Archie left Chicago she was comfortably settled with Mrs. Lorch, and her happy reunion with her trunk somewhat consoled her for his departure.

Mrs. Lorch and her daughter lived half a mile from the Swedish Reform Church, in an old square frame house, with a porch supported by frail pillars, set in a damp yard full of big lilac bushes. The house, which had been left over from country times, needed paint badly, and looked gloomy and despondent among its smart Queen Anne neighbors. There was a big back yard with two rows of apple trees and a grape arbor, and a warped walk, two planks wide, which led to the coal bins at the back of the lot. Thea's room was on the second floor, overlooking this back yard, and she understood that in the winter she must carry up her own coal and kindling from the bin. There was no furnace in the house, no running water except in the kitchen, and that was why the room rent was small. All the rooms were heated by stoves, and the lodgers pumped the water they needed from the cistern under the porch, or from the well at the entrance of the grape arbor. Old Mrs. Lorch could never bring herself to have costly improvements made in her house; indeed she had very little money. She preferred to keep the house just as her husband built it, and she thought her way of living good enough for plain people.

Thea's room was large enough to admit a rented upright piano without crowding. It was, the widowed daughter said, "a double room that had always before been occupied by two gentlemen"; the piano now took the place of a second occupant. There was an ingrain carpet on the floor, green ivy leaves on a red ground, and clumsy, old-fashioned walnut furniture. The bed was very wide, and the mattress thin and hard. Over the fat pillows were "shams" embroidered in Turkey red, each with a flowering scroll—one with "Gute' Nacht," the other with "Guten Morgen." The dresser was so big that Thea wondered how it had ever been got into the house and up the narrow stairs. Besides an old horsehair armchair, there were two low plush "spring-rockers," against the massive pedestals of which one was always stumbling in the dark. Thea sat in the dark a good deal those first weeks, and sometimes a painful bump against one of those brutally immovable pedestals roused her temper and pulled her out of a heavy hour. The wall-paper was brownish yellow, with blue flowers. When it was put on, the carpet, certainly, had not been consulted. There was only one picture on the wall when Thea moved in: a large colored print of a brightly lighted church in a snow-storm, on Christmas Eve, with greens hanging about the stone doorway and arched windows. There was something warm and home-like about this picture, and Thea grew fond of it. One day, on her way into town to take her lesson, she stopped at a bookstore and bought a photograph of the Naples bust of Julius Cæsar. This she had framed, and hung it on the big bare wall behind her stove. It was a curious choice, but she was at the age when people do inexplicable things. She had been interested in Cæsar's "Commentaries" when she left school to begin teaching, and she loved to read about great generals; but these facts would scarcely explain her wanting that grim bald head to share her daily existence. It seemed a strange freak, when she bought so few things, and when she had, as Mrs. Andersen said to Mrs. Lorch, "no pictures of the composers at all."

Both the widows were kind to her, but Thea liked the mother better. Old Mrs. Lorch was fat and jolly, with a red face, always shining as if she had just come from the stove, bright little eyes, and hair of several colors. Her own hair was one cast of iron-gray, her switch another, and her false front still another. Her clothes always smelled of savory cooking, except when she was dressed for church or *Kaffeeklatsch*,[1] and then she smelled of bay rum or of the lemon-verbena sprig which she tucked inside her puffy black kid glove. Her cooking justified all that Mr. Larsen had said of it, and Thea had never been so well nourished before.

The daughter, Mrs. Andersen,—Irene, her mother called her,—was a different sort of woman altogether. She was perhaps forty years old, angular, big-boned, with large, thin features, light-blue eyes, and dry, yellow hair, the bang tightly frizzed. She was pale, anæmic, and sentimental. She had married the youngest son of a rich, arrogant Swedish family who were lumber merchants in St. Paul. There she dwelt during her married life. Oscar Andersen was a strong, full-blooded fellow who had counted on a long life and had been rather careless about his business affairs. He was killed by the explosion of a steam boiler in the mills, and his brothers managed to prove that he had very little stock in the big business. They had strongly disapproved of his marriage and they agreed among themselves that they were entirely justified in defrauding his widow, who, they said, "would only marry again and give some fellow a good thing of it." Mrs. Andersen would not go to law with the family that had always snubbed and wounded her—she felt the humiliation of being thrust out more than she felt her impoverishment; so she went back to Chicago to live with her widowed mother on an income of five hundred a year. This experience had given her sentimental nature an incurable hurt. Something withered away in her. Her head had a downward droop; her step was soft and apologetic, even in her mother's house, and her smile had the sickly, uncertain flicker that so often comes from a secret humiliation. She was affable and yet shrinking, like one who has come down in the world, who has known better clothes, better carpets, better people, brighter hopes. Her husband was buried in the Andersen lot in St. Paul, with a locked iron fence around it. She had to go to his eldest brother for the key when she went to say good-bye to his grave. She clung to the Swedish Church because it had been her husband's church.

As her mother had no room for her household belongings, Mrs. Andersen had brought home with her only her bedroom set, which now furnished her own room at Mrs. Lorch's. There she spent most of her time, doing fancywork or writing letters to sympathizing German friends in St. Paul, surrounded by keepsakes and photographs of the burly Oscar Andersen. Thea, when she was admitted to this room, and shown these photographs, found herself wondering, like the Andersen family, why such a lusty, gay-looking fellow ever thought he wanted this pallid, long-cheeked woman, whose manner was always that of withdrawing, and who must have been rather thin-blooded even as a girl.

Mrs. Andersen was certainly a depressing person. It sometimes annoyed Thea very much to hear her insinuating knock on the door, her flurried explanation of why she had come, as she backed toward the stairs. Mrs. Andersen admired Thea greatly. She thought it a distinction to be even a "temporary soprano"—Thea called herself so quite seriously—in the Swedish Church. She also thought it distinguished to be a pupil of Harsanyi's. She considered Thea very handsome, very Swedish, very talented. She fluttered about the upper floor when Thea was practicing. In short, she tried to make a heroine of her, just as Tillie Kronborg had always done, and Thea was conscious of something of the sort. When she was working and heard Mrs. Andersen tip-toeing past her door, she used to shrug her shoulders and wonder whether she was always to have a Tillie diving furtively about her in some disguise or other.

At the dressmaker's Mrs. Andersen recalled Tillie even more painfully. After her first Sunday in Mr. Larsen's choir, Thea saw that she must have a proper dress for morning service. Her Moonstone party dress might do to wear in the evening, but she must have one frock that could stand the light of day. She, of course, knew nothing about Chicago dressmakers, so she let Mrs. Andersen take her to a German woman whom she recommended warmly. The German dressmaker was excitable and dramatic. Concert dresses, she said, were her specialty. In her fitting-room there were photographs of singers in the dresses she had made them for this or that *Sängerfest.*[2] She and Mrs. Andersen together achieved a costume which would have warmed Tillie Kronborg's heart. It was clearly intended for a woman of forty, with violent tastes. There seemed to be a piece of every known fabric in it somewhere. When it came home, and was spread out on her huge bed, Thea looked it over and told herself candidly that it was "a horror." However, her money was gone, and there was nothing to do but make the best of the dress. She never wore it except, as she said, "to sing in," as if it were an unbecoming uniform. When Mrs. Lorch and Irene told her that she "looked like a little bird-of-Paradise in it," Thea shut her teeth and repeated to herself words she had learned from Joe Giddy and Spanish Johnny.

In these two good women Thea found faithful friends, and in their house she found the quiet and peace which helped her to support the great experiences of that winter.

III

ANDOR HARSANYI HAD NEVER had a pupil in the least like Thea Kronborg. He had never had one more intelligent, and he had never had one so ignorant. When Thea sat down to take her first lesson from him, she had never heard a work by Beethoven or a composition by Chopin. She knew their names vaguely. Wunsch had been a musician once, long before he wandered into Moonstone, but when Thea awoke his interest there was not much left of him. From him Thea had learned something about the works of Gluck and Bach, and he used to play her some of the compositions of Schumann. In his trunk he had a mutilated score of the F sharp minor sonata, which he had heard Clara Schumann play at a festival in Leipsic. Though his powers of execution were at such a low ebb, he used to play at this sonata for his pupil and managed to give her some idea of its beauty. When Wunsch was a young man, it was still daring to like Schumann; enthusiasm for his work was considered an expression of youthful waywardness. Perhaps that was why Wunsch remembered him best. Thea studied some of the *Kinderszenen*[1] with him, as well as some little sonatas by Mozart and Clementi. But for the most part Wunsch stuck to Czerny and Hummel.[2]

Harsanyi found in Thea a pupil with sure, strong hands, one who read rapidly and intelligently, who had, he felt, a richly gifted nature. But she had been given no direction, and her ardor was unawakened. She had never heard a symphony orchestra. The literature of the piano was an undiscovered world to her. He wondered how she had been able to work so hard when she knew so little of what she was working toward. She had been taught according to the old Stuttgart method;[3] stiff back, stiff elbows, a very formal position of the hands. The best thing about her preparation was that she had developed an unusual power of work. He noticed at once her way of charging at difficulties. She ran to meet them as if they were foes she had long been seeking, seized them as if they were destined for her and she for them. Whatever she did well, she took for granted. Her eagerness aroused all the young Hungarian's chivalry. Instinctively one went to the rescue of a creature who had so much to overcome and who struggled so hard. He used to tell his wife that Miss Kronborg's hour took more out of him than half a dozen other lessons. He usually kept her long over time; he changed her lessons about so that he could do so, and often gave her time at the end of the day, when he could talk to her afterward and play for her a little from what he happened to be studying. It was always interesting to play for her. Sometimes she was so silent that he wondered, when she left him, whether she had got anything out of it. But a week later, two weeks later, she would give back his idea again in a way that set him vibrating.

All this was very well for Harsanyi; an interesting variation in the routine of teaching. But for Thea Kronborg, that winter was almost beyond enduring. She always remembered it as the happiest and wildest and saddest of her life. Things came too fast for her; she had not had enough preparation. There were times when she came home from her lesson and lay upon her bed hating Wunsch and her family, hating a world that had let her grow up so ignorant; when she wished that she could die then and there, and be born over again to begin anew. She said something of this kind once to her teacher, in the midst of a bitter struggle. Harsanyi turned the light of his wonderful eye upon her—poor fellow, he had but one, though that was set in such a handsome head—and said slowly: "Every artist makes himself born. It is very much harder than the other time, and longer. Your mother did not bring anything into the world to play piano. That you must bring into the world yourself."

This comforted Thea temporarily, for it seemed to give her a chance. But a great deal of the time she was comfortless. Her letters to Dr. Archie were brief and businesslike. She was not apt to chatter much, even in the stimulating company of people she liked, and to chatter on paper was simply impossible for her. If she tried to write him anything definite about her work, she immediately scratched it out as being only partially true, or not true at all. Nothing that she could say about her studies seemed unqualifiedly true, once she put it down on paper.

Late one afternoon, when she was thoroughly tired and wanted to struggle on into the dusk, Harsanyi, tired too, threw up his hands and laughed at her. “Not to-day, Miss Kronborg. That sonata will keep; it won’t run away. Even if you and I should not waken up to-morrow, it will be there.”

Thea turned to him fiercely. “No, it isn’t here unless I have it—not for me,” she cried passionately. “Only what I hold in my two hands is there for me!”

Harsanyi made no reply. He took a deep breath and sat down again. “The second movement now, quietly, with the shoulders relaxed.”

There were hours, too, of great exaltation; when she was at her best and became a part of what she was doing and ceased to exist in any other sense. There were other times when she was so shattered by ideas that she could do nothing worth while; when they trampled over her like an army and she felt as if she were bleeding to death under them. She sometimes came home from a late lesson so exhausted that she could eat no supper. If she tried to eat, she was ill afterward. She used to throw herself upon the bed and lie there in the dark, not thinking, not feeling, but evaporating. That same night, perhaps, she would waken up rested and calm, and as she went over her work in her mind, the passages seemed to become something of themselves, to take a sort of pattern in the darkness. She had never learned to work away from the piano until she came to Harsanyi, and it helped her more than anything had ever helped her before.

She almost never worked now with the sunny, happy contentment that had filled the hours when she worked with Wunsch—“like a fat horse turning a sorgum mill,” she said bitterly to herself. Then, by sticking to it, she could always do what she set out to do. Now, everything that she really wanted was impossible; a *cantabile* like Harsanyi’s, for instance, instead of her own cloudy tone. No use telling her she might have it in ten years. She wanted it now. She wondered how she had ever found other things interesting: books, “Anna Karenina”—all that seemed so unreal and on the outside of things. She was not born a musician, she decided; there was no other way of explaining it.

Sometimes she got so nervous at the piano that she left it, and snatching up her hat and cape went out and walked, hurrying through the streets like Christian fleeing from the City of Destruction.[4] And while she walked she cried. There was scarcely a street in the neighborhood that she had not cried up and down before that winter was over. The thing that used to lie under her cheek, that sat so warmly over her heart when she glided away from the sand hills that autumn morning, was far from her. She had come to Chicago to be with it, and it had deserted her, leaving in its place a painful longing, an unresigned despair.

Harsanyi knew that his interesting pupil—"the savage blonde," one of his male students called her—was sometimes very unhappy. He saw in her discontent a curious definition of character. He would have said that a girl with so much musical feeling, so intelligent, with good training of eye and hand, would, when thus suddenly introduced to the great literature of the piano, have found boundless happiness. But he soon learned that she was not able to forget her own poverty in the richness of the world he opened to her. Often when he played to her, her face was the picture of restless misery. She would sit crouching forward, her elbows on her knees, her brows drawn together and her gray-green eyes smaller than ever, reduced to mere pin-points of cold, piercing light. Sometimes, while she listened, she would swallow hard, two or three times, and look nervously from left to right, drawing her shoulders together. "Exactly," he thought, "as if she were being watched, or as if she were naked and heard some one coming."

On the other hand, when she came several times to see Mrs. Harsanyi and the two babies, she was like a little girl, jolly and gay and eager to play with the children, who loved her. The little daughter, Tanya, liked to touch Miss Kronborg's yellow hair and pat it, saying, "Dolly, dolly," because it was of a color much oftener seen on dolls than on people. But if Harsanyi opened the piano and sat down to play, Miss Kronborg gradually drew away from the children, retreated to a corner and became sullen or troubled. Mrs. Harsanyi noticed this, also, and thought it very strange behavior.

Another thing that puzzled Harsanyi was Thea's apparent lack of curiosity. Several times he offered to give her tickets to concerts, but she said she was too tired or that it "knocked her out to be up late." Harsanyi did not know that she was singing in a choir, and had often to sing at funerals, neither did he realize how much her work with him stirred her and exhausted her. Once, just as she was leaving his studio, he called her back and told her he could give her some tickets that had been sent him for Emma Juch[5] that evening.

Thea fingered the black wool on the edge of her plush cape and replied, "Oh, thank you, Mr. Harsanyi, but I have to wash my hair to-night."

Mrs. Harsanyi liked Miss Kronborg thoroughly. She saw in her the making of a pupil who would reflect credit upon Harsanyi. She felt that the girl could be made to look strikingly handsome, and that she had the kind of personality which takes hold of audiences. Moreover, Miss Kronborg was not in the least sentimental about her husband. Sometimes from the show pupils one had to endure a good deal. "I like that girl," she used to say, when Harsanyi told her of one of Thea's *gaucheries.* "She doesn't sigh every time the wind blows. With her one swallow doesn't make a summer."

Thea told them very little about herself. She was not naturally communicative, and she found it hard to feel confidence in new people. She did not know why, but she could not talk to Harsanyi as she could to Dr. Archie, or to Johnny and Mrs. Tellamantez. With Mr. Larsen she felt more at home, and when she was walking she sometimes stopped at his study to eat candy with him or to hear the plot of the novel he happened to be reading.

One evening toward the middle of December Thea was to dine with the Harsanyis. She arrived early, to have time to play with the children before they went to bed. Mrs. Harsanyi took her into her own room and helped her take off her country "fascinator" and her clumsy plush cape. Thea had bought this cape at a big department store and had paid $16.50 for it. As she had never paid more than ten dollars for a coat before, that seemed to her a large price. It was very heavy and not very warm, ornamented with a showy pattern in black disks, and trimmed around the collar and the edges with some kind of black wool that "crocked" badly in snow or rain. It was lined with a cotton stuff called "farmer's satin." Mrs. Harsanyi was one woman in a thousand. As she lifted this cape from Thea's shoulders and laid it on her white bed, she wished that her husband did not have to charge pupils like this one for their lessons. Thea wore her Moonstone party dress, white organdie, made with a "V" neck and elbow sleeves, and a blue sash. She looked very pretty in it, and around her throat she had a string of pink coral and tiny white shells that Ray once brought her from Los Angeles. Mrs. Harsanyi noticed that she wore high heavy shoes which needed blacking. The choir in Mr. Larsen's church stood behind a railing, so Thea did not pay much attention to her shoes.

"You have nothing to do to your hair," Mrs. Harsanyi said kindly, as Thea turned to the mirror. "However it happens to lie, it's always pretty. I admire it as much as Tanya does."

Thea glanced awkwardly away from her and looked stern, but Mrs. Harsanyi knew that she was pleased. They went into the living-room, behind the studio, where the two children were playing on the big rug before the coal grate. Andor, the boy, was six, a sturdy, handsome child, and the little girl was four. She came tripping to meet Thea, looking like a little doll in her white net dress—her mother made all her clothes. Thea picked her up and hugged her. Mrs. Harsanyi excused herself and went to the dining-room. She kept only one maid and did a good deal of the housework herself, besides cooking her husband's favorite dishes for him. She was still under thirty, a slender, graceful woman, gracious, intelligent, and capable. She adapted herself to circumstances with a well-bred ease which solved many of her husband's difficulties, and kept him, as he said, from feeling cheap and down at the heel. No musician ever had a better wife. Unfortunately her beauty was of a very frail and impressionable kind, and she was beginning to lose it. Her face was too thin now, and there were often dark circles under her eyes.

Left alone with the children, Thea sat down on Tanya's little chair—she would rather have sat on the floor, but was afraid of rumpling her dress—and helped them play "cars" with Andor's iron railway set. She showed him new ways to lay his tracks and how to make switches, set up his Noah's ark village for stations and packed the animals in the open coal cars to send them to the stockyards. They worked out their shipment so realistically that when Andor put the two little reindeer into the stock car, Tanya snatched them out and began to cry, saying she wasn't going to have all their animals killed.

Harsanyi came in, jaded and tired, and asked Thea to go on with her game, as he was not equal to talking much before dinner. He sat down and made pretense of glancing at the evening paper, but he soon dropped it. After the railroad began to grow tiresome, Thea went with the children to the lounge in the corner, and played for them the game with which she used to amuse Thor for hours together behind the parlor stove at home, making shadow pictures against the wall with her hands. Her fingers were very supple, and she could make a duck and a cow and a sheep and a fox and a rabbit and even an elephant. Harsanyi, from his low chair, watched them, smiling. The boy was on his knees, jumping up and down with the excitement of guessing the beasts, and Tanya sat with her feet tucked under her and clapped her frail little hands. Thea's profile, in the lamplight, teased his fancy. Where had he seen a head like it before?

When dinner was announced, little Andor took Thea's hand and walked to the dining-room with her. The children always had dinner with their parents and behaved very nicely at table. "Mamma," said Andor seriously as he climbed into his chair and tucked his napkin into the collar of his blouse, "Miss Kronborg's hands are every kind of animal there is."

His father laughed. "I wish somebody would say that about my hands, Andor."

When Thea dined at the Harsanyis before, she noticed that there was an intense suspense from the moment they took their places at the table until the master of the house had tasted the soup. He had a theory that if the soup went well, the dinner would go well; but if the soup was poor, all was lost. To-night he tasted his soup and smiled, and Mrs. Harsanyi sat more easily in her chair and turned her attention to Thea. Thea loved their dinner table, because it was lighted by candles in silver candle-sticks, and she had never seen a table so lighted anywhere else. There were always flowers, too. To-night there was a little orange tree, with oranges on it, that one of Harsanyi's pupils had sent him at Thanksgiving time. After Harsanyi had finished his soup and a glass of red Hungarian wine, he lost his fagged look and became cordial and witty. He persuaded Thea to drink a little wine to-night. The first time she dined with them, when he urged her to taste the glass of sherry beside her plate, she astonished them by telling them that she "never drank."

Harsanyi was then a man of thirty-two. He was to have a very brilliant career, but he did not know it then. Theodore Thomas was perhaps the only man in Chicago who felt that Harsanyi might have a great future. Harsanyi belonged to the softer Slavic type, and was more like a Pole than a Hungarian. He was tall, slender, active, with sloping, graceful shoulders and long arms. His head was very fine, strongly and delicately modelled, and, as Thea put it, "so independent." A lock of his thick brown hair usually hung over his forehead. His eye was wonderful; full of light and fire when he was interested, soft and thoughtful when he was tired or melancholy. The meaning and power of two very fine eyes must all have gone into this one—the right one, fortunately, the one next his audience when he played. He believed that the glass eye which gave one side of his face such a dull, blind look, had ruined his career, or rather had made a career impossible for him. Harsanyi lost his eye when he was twelve years old, in a Pennsylvania mining town where explosives happened to be kept too near the frame shanties in which the company packed newly arrived Hungarian families.

His father was a musician and a good one, but he had cruelly over-worked the boy; keeping him at the piano for six hours a day and making him play in cafés and dance halls for half the night. Andor ran away and crossed the ocean with an uncle, who smuggled him through the port as one of his own many children. The explosion in which Andor was hurt killed a score of people, and he was thought lucky to get off with an eye. He still had a clipping from a Pittsburgh paper, giving a list of the dead and injured. He appeared as "Harsanyi, Andor, left eye and slight injuries about the head." That was his first American "notice"; and he kept it. He held no grudge against the coal company; he understood that the accident was merely one of the things that are bound to happen in the general scramble of American life, where every one comes to grab and takes his chance.

While they were eating dessert, Thea asked Harsanyi if she could change her Tuesday lesson from afternoon to morning. "I have to be at a choir rehearsal in the afternoon, to get ready for the Christmas music, and I expect it will last until late."

Harsanyi put down his fork and looked up. "A choir rehearsal? You sing in a church?"

"Yes. A little Swedish church, over on the North side."

"Why did you not tell us?"

"Oh, I'm only a temporary. The regular soprano is not well."

"How long have you been singing there?"

"Ever since I came. I had to get a position of some kind," Thea explained, flushing, "and the preacher took me on. He runs the choir himself. He knew my father, and I guess he took me to oblige."

Harsanyi tapped the tablecloth with the ends of his fingers. "But why did you never tell us? Why are you so reticent with us?"

Thea looked shyly at him from under her brows. "Well, it's certainly not very interesting. It's only a little church. I only do it for business reasons."

"What do you mean? Don't you like to sing? Don't you sing well?"

"I like it well enough, but, of course, I don't know anything about singing. I guess that's why I never said anything about it. Anybody that's got a voice can sing in a little church like that."

Harsanyi laughed softly—a little scornfully, Thea thought. "So you have a voice, have you?"

Thea hesitated, looked intently at the candles and then at Harsanyi. "Yes," she said firmly; "I have got some, anyway."

"Good girl," said Mrs. Harsanyi, nodding and smiling at Thea. "You must let us hear you sing after dinner."

This remark seemingly closed the subject, and when the coffee was brought they began to talk of other things. Harsanyi asked Thea how she happened to know so much about the way in which freight trains are operated, and she tried to give him some idea of how the people in little desert towns live by the railway and order their lives by the coming and going of the trains. When they left the dining-room the children were sent to bed and Mrs. Harsanyi took Thea into the studio. She and her husband usually sat there in the evening.

Although their apartment seemed so elegant to Thea, it was small and cramped. The studio was the only spacious room. The Harsanyis were poor, and it was due to Mrs. Harsanyi's good management that their lives, even in hard times, moved along with dignity and order. She had long ago found out that bills or debts of any kind frightened her husband and crippled his working power. He said they were like bars on the windows, and shut out the future; they meant that just so many hundred dollars' worth of his life was debilitated and exhausted before he got to it. So Mrs. Harsanyi saw to it that they never owed anything. Harsanyi was not extravagant, though he was sometimes careless about money. Quiet and order and his wife's good taste were the things that meant most to him. After these, good food, good cigars, a little good wine. He wore his clothes until they were shabby, until his wife had to ask the tailor to come to the house and measure him for new ones. His neckties she usually made herself, and when she was in shops she always kept her eye open for silks in very dull or pale shades, grays and olives, warm blacks and browns.

When they went into the studio Mrs. Harsanyi took up her embroidery and Thea sat down beside her on a low stool, her hands clasped about her knees. While his wife and his pupil talked, Harsanyi sank into a *chaise longue* in which he sometimes snatched a few moments' rest between his lessons, and smoked. He sat well out of the circle of the lamplight, his feet to the fire. His feet were slender and well shaped, always elegantly shod. Much of the grace of his movements was due to the fact that his feet were almost as sure and flexible as his hands. He listened to the conversation with amusement. He admired his wife's tact and kindness with crude young people; she taught them so much without seeming to be instructing. When the clock struck nine, Thea said she must be going home.

Harsanyi rose and flung away his cigarette. "Not yet. We have just begun the evening. Now you are going to sing for us. I have been waiting for you to recover from dinner. Come, what shall it be?" he crossed to the piano.

Thea laughed and shook her head, locking her elbows still tighter about her knees. "Thank you, Mr. Harsanyi, but if you really make me sing, I'll accompany myself. You couldn't stand it to play the sort of things I have to sing."

As Harsanyi still pointed to the chair at the piano, she left her stool and went to it, while he returned to his *chaise longue*. Thea looked at the keyboard uneasily for a moment, then she began "Come, ye Disconsolate," the hymn Wunsch had always liked to hear her sing. Mrs. Harsanyi glanced questioningly at her husband, but he was looking intently at the toes of his boots, shading his forehead with his long white hand. When Thea finished the hymn she did not turn around, but immediately began "The Ninety and Nine."[6] Mrs. Harsanyi kept trying to catch her husband's eye, but his chin only sank lower on his collar.

> "There were ninety and nine that safely lay
> In the shelter of the fold,
> But one was out on the hills away,
> Far off from the gates of gold."

Harsanyi looked at her, then back at the fire.

> "Rejoice, for the Shepherd has found his sheep."

Thea turned on the chair and grinned. "That's about enough, isn't it? That song got me my job. The preacher said it was sympathetic," she minced the word, remembering Mr. Larsen's manner.

Harsanyi drew himself up in his chair, resting his elbows on the low arms. "Yes? That is better suited to your voice. Your upper tones are good, above G. I must teach you some songs. Don't you know anything—pleasant?"

Thea shook her head ruefully. "I'm afraid I don't. Let me see— Perhaps," she turned to the piano and put her hands on the keys. "I used to sing this for Mr. Wunsch a long while ago. It's for contralto, but I'll try it." She frowned at the keyboard a moment, played the few introductory measures, and began

"Ach, ich habe sie verloren,"

She had not sung it for a long time, and it came back like an old friendship. When she finished, Harsanyi sprang from his chair and dropped lightly upon his toes, a kind of *entre-chat* that he sometimes executed when he formed a sudden resolution, or when he was about to follow a pure intuition, against reason. His wife said that when he gave that spring he was shot from the bow of his ancestors, and now when he left his chair in that manner she knew he was intensely interested. He went quickly to the piano.

"Sing that again. There is nothing the matter with your low voice, my girl. I will play for you. Let your voice out." Without looking at her he began the accompaniment. Thea drew back her shoulders, relaxed them instinctively, and sang.

When she finished the aria, Harsanyi beckoned her nearer. "Sing *ah—ah* for me, as I indicate." He kept his right hand on the keyboard and put his left to her throat, placing the tips of his delicate fingers over her larynx. "Again,—until your breath is gone.—Trill between the two tones, always; good! Again; excellent!—Now up,—stay there. E and F. Not so good, is it? F is always a hard one.—Now, try the half-tone.—That's right, nothing difficult about it.—Now, pianissimo, *ah—ah.* Now, swell it, *ah—ah.*—Again, follow my hand.—Now, carry it down.—Anybody ever tell you anything about your breathing?"

"Mr. Larsen says I have an unusually long breath," Thea replied with spirit.

Harsanyi smiled. "So you have, so you have. That was what I meant. Now, once more; carry it up and then down, *ah—ah.*" He put his hand back to her throat and sat with his head bent, his one eye closed. He loved to hear a big voice throb in a relaxed, natural throat, and he was thinking that no one had ever felt this voice vibrate before. It was like a wild bird that had flown into his studio on Middleton Street from goodness knew how far! No one knew that it had come, or even that it existed; least of all the strange, crude girl in whose throat it beat its passionate wings. What a simple thing it was, he reflected; why had he never guessed it before? Everything about her indicated it,—the big mouth, the wide jaw and chin, the strong white teeth, the deep laugh. The machine was so simple and strong, seemed to be so easily operated. She sang from the bottom of herself. Her breath came from down where her laugh came from, the deep laugh which Mrs. Harsanyi had once called "the laugh of the people." A relaxed throat, a voice that lay on the breath, that had never been forced off the breath; it rose and fell in the air-column like the little balls which are put to shine in the jet of a fountain. The voice did not thin as it went up; the upper tones were as full and rich as the lower, produced in the same way and as unconsciously, only with deeper breath.

At last Harsanyi threw back his head and rose. "You must be tired, Miss Kronborg."

When she replied, she startled him; he had forgotten how hard and full of burs her speaking voice was. "No," she said, "singing never tires me."

Harsanyi pushed back his hair with a nervous hand. "I don't know much about the voice, but I shall take liberties and teach you some good songs. I think you have a very interesting voice."

"I'm glad if you like it. Good-night, Mr. Harsanyi." Thea went with Mrs. Harsanyi to get her wraps.

When Mrs. Harsanyi came back to her husband, she found him walking restlessly up and down the room.

"Don't you think her voice wonderful, dear?" she asked.

"I scarcely know what to think. All I really know about that girl is that she tires me to death. We must not have her often. If I did not have my living to make, then—" he dropped into a chair and closed his eyes. "How tired I am. What a voice!"

IV

After that evening Thea's work with Harsanyi changed somewhat. He insisted that she should study some songs with him, and after almost every lesson he gave up half an hour of his own time to practicing them with her. He did not pretend to know much about voice production, but so far, he thought, she had acquired no really injurious habits. A healthy and powerful organ had found its own method, which was not a bad one. He wished to find out a good deal before he recommended a vocal teacher. He never told Thea what he thought about her voice, and made her general ignorance of anything worth singing his pretext for the trouble he took. That was in the beginning. After the first few lessons his own pleasure and hers were pretext enough. The singing came at the end of the lesson hour, and they both treated it as a form of relaxation.

Harsanyi did not say much even to his wife about his discovery. He brooded upon it in a curious way. He found that these unscientific singing lessons stimulated him in his own study. After Miss Kronborg left him he often lay down in his studio for an hour before dinner, with his head full of musical ideas, with an effervescence in his brain which he had sometimes lost for weeks together under the grind of teaching. He had never got so much back for himself from any pupil as he did from Miss Kronborg. From the first she had stimulated him; something in her personality invariably affected him. Now that he was feeling his way toward her voice, he found her more interesting than ever before. She lifted the tedium of the winter for him, gave him curious fancies and reveries. Musically, she was sympathetic to him. Why all this was true, he never asked himself. He had learned that one must take where and when one can the mysterious mental irritant that rouses one's imagination; that it is not to be had by order. She often wearied him, but she never bored him. Under her crudeness and brusque hardness, he felt there was a nature quite different, of which he never got so much as a hint except when she was at the piano, or when she sang. It was toward this hidden creature that he was trying, for his own pleasure, to find his way. In short, Harsanyi looked forward to his hour with Thea for the same reason that poor Wunsch had sometimes dreaded his; because she stirred him more than anything she did could adequately explain.

One afternoon Harsanyi, after the lesson, was standing by the window putting some collodion on a cracked finger, and Thea was at the piano trying over "Die Lorelei"[1] which he had given her last week to practice. It was scarcely a song which a singing master would have given her, but he had his own reasons. How she sang it mattered only to him and to her. He was playing his own game now, without interference; he suspected that he could not do so always.

When she finished the song, she looked back over her shoulder at him and spoke thoughtfully. "That wasn't right, at the end, was it?"

"No, that should be an open, flowing tone, something like this,"—he waved his fingers rapidly in the air. "You get the idea?"

"No, I don't. Seems a queer ending, after the rest."

Harsanyi corked his little bottle and dropped it into the pocket of his velvet coat. "Why so? Shipwrecks come and go, *Märchen* come and go, but the river keeps right on. There you have your open, flowing tone."

Thea looked intently at the music. "I see," she said dully. "Oh, I see!" she repeated quickly and turned to him a glowing countenance. "It is the river.—Oh, yes, I get it now!" She looked at him but long enough to catch his glance, then turned to the piano again. Harsanyi was never quite sure where the light came from when her face suddenly flashed out at him in that way. Her eyes were too small to account for it, though they glittered like green ice in the sun. At such moments her hair was yellower, her skin whiter, her cheeks pinker, as if a lamp had suddenly been turned up inside of her. She went at the song again:

> *"Ich weiss nicht, was soll es bedeuten,*
> *Das ich so traurig bin."*[2]

A kind of happiness vibrated in her voice. Harsanyi noticed how much and how unhesitatingly she changed her delivery of the whole song, the first part as well as the last. He had often noticed that she could not think a thing out in passages. Until she saw it as a whole, she wandered like a blind man surrounded by torments. After she once had her "revelation," after she got the idea that to her—not always to him—explained everything, then she went forward rapidly. But she was not always easy to help. She was sometimes impervious to suggestion; she would stare at him as if she were deaf and ignore everything he told her to do. Then, all at once, something would happen in her brain and she would begin to do all that he had been for weeks telling her to do, without realizing that he had ever told her.

To-night Thea forgot Harsanyi and his finger. She finished the song only to begin it with fresh enthusiasm.

> *"Und das hat mit ihrem singen*
> *Die Lorelei getlan."*[3]

She sat there singing it until the darkening room was so flooded with it that Harsanyi threw open a window.

"You really must stop it, Miss Kronborg. I shan't be able to get it out of my head to-night."

Thea laughed tolerantly as she began to gather up her music. "Why, I thought you had gone, Mr. Harsanyi. I like that song."

That evening at dinner Harsanyi sat looking intently into a glass of heavy yellow wine; boring into it, indeed, with his one eye, when his face suddenly broke into a smile.

"What is it, Andor?" his wife asked.

He smiled again, this time at her, and took up the nutcrackers and a Brazil nut. "Do you know," he said in a tone so intimate and confidential that he might have been speaking to himself,—"do you know, I like to see Miss Kronborg get hold of an idea. In spite of being so talented, she's not quick. But when she does get an idea, it fills her up to the eyes. She had my room so reeking of a song this afternoon that I couldn't stay there."

Mrs. Harsanyi looked up quickly, " 'Die Lorelei,' you mean? One couldn't think of anything else anywhere in the house. I thought she was possessed. But don't you think her voice is wonderful sometimes?"

Harsanyi tasted his wine slowly. "My dear, I've told you before that I don't know what I think about Miss Kronborg, except that

I'm glad there are not two of her. I sometimes wonder whether she is not glad. Fresh as she is at it all, I've occasionally fancied that, if she knew how, she would like to—diminish." He moved his left hand out into the air as if he were suggesting a *diminuendo* to an orchestra.

V

By the first of February Thea had been in Chicago almost four months, and she did not know much more about the city than if she had never quitted Moonstone. She was, as Harsanyi said, incurious. Her work took most of her time, and she found that she had to sleep a good deal. It had never before been so hard to get up in the morning. She had the bother of caring for her room, and she had to build her fire and bring up her coal. Her routine was frequently interrupted by a message from Mr. Larsen summoning her to sing at a funeral. Every funeral took half a day, and the time had to be made up. When Mrs. Harsanyi asked her if it did not depress her to sing at funerals, she replied that she "had been brought up to go to funerals and didn't mind."

Thea never went into shops unless she had to, and she felt no interest in them. Indeed, she shunned them, as places where one was sure to be parted from one's money in some way. She was nervous about counting her change, and she could not accustom herself to having her purchases sent to her address. She felt much safer with her bundles under her arm.

During this first winter Thea got no city consciousness. Chicago was simply a wilderness through which one had to find one's way. She felt no interest in the general briskness and zest of the crowds. The crash and scramble of that big, rich, appetent Western city she did not take in at all, except to notice that the noise of the drays and street-cars tired her. The brilliant window displays, the splendid furs and stuffs, the gorgeous flower-shops, the gay candy-shops, she scarcely noticed. At Christmas-time she did feel some curiosity about the toy-stores, and she wished she held Thor's little mittened fist in her hand as she stood before the windows. The jewelers' windows, too, had a strong attraction for her—she had always liked bright stones. When she went into the city she used to brave the bit-ing lake winds and stand gazing in at the displays of diamonds and pearls and emeralds; the tiaras and necklaces and earrings, on white velvet. These seemed very well worth while to her, things worth coveting.

Mrs. Lorch and Mrs. Andersen often told each other it was strange that Miss Kronborg had so little initiative about "visiting points of interest." When Thea came to live with them she had expressed a wish to see two places: Montgomery Ward and Company's big mail-order store, and the packing-houses, to which all the hogs and cattle that went through Moonstone were bound. One of Mrs. Lorch's lodgers worked in a packing-house, and Mrs. Andersen brought Thea word that she had spoken to Mr. Eckman and he would gladly take her to Packing-town. Eckman was a toughish young Swede, and he thought it would be something of a lark to take a pretty girl through the slaughter-houses. But he was disappointed. Thea neither grew faint nor clung to the arm he kept offering her. She asked innumerable questions and was impatient because he knew so little of what was going on outside of his own department. When they got off the street-car and walked back to Mrs. Lorch's house in the dusk, Eckman put her hand in his overcoat pocket—she had no muff—and kept squeezing it ardently until she said, "Don't do that; my ring cuts me." That night he told his roommate that he "could have kissed her as easy as rolling off a log, but she wasn't worth the trouble." As for Thea, she had enjoyed the afternoon very much, and wrote her father a brief but clear account of what she had seen.

One night at supper Mrs. Andersen was talking about the exhibit of students' work she had seen at the Art Institute that afternoon. Several of her friends had sketches in the exhibit. Thea, who always felt that she was behindhand in courtesy to Mrs. Andersen, thought that here was an opportunity to show interest without committing herself to anything. "Where is that, the Institute?" she asked absently.

Mrs. Andersen clasped her napkin in both hands. "The Art Institute? Our beautiful Art Institute on Michigan Avenue? Do you mean to say you have never visited it?"

"Oh, is it the place with the big lions out in front? I remember; I saw it when I went to Montgomery Ward's. Yes, I thought the lions were beautiful."

"But the pictures! Didn't you visit the galleries?"

"No. The sign outside said it was a pay-day. I've always meant to go back, but I haven't happened to be down that way since."

Mrs. Lorch and Mrs. Andersen looked at each other. The old mother spoke, fixing her shining little eyes upon Thea across the table. "Ah, but Miss Kronborg, there are old masters! Oh, many of them, such as you could not see anywhere out of Europe."

"And Corots," breathed Mrs. Andersen, tilting her head feelingly. "Such examples of the Barbizon school!"[1] This was meaningless to Thea, who did not read the art columns of the Sunday *Inter-Ocean*[2] as Mrs. Andersen did.

"Oh, I'm going there some day," she reassured them. "I like to look at oil paintings."

One bleak day in February, when the wind was blowing clouds of dirt like a Moonstone sandstorm, dirt that filled your eyes and ears and mouth, Thea fought her way across the unprotected space in front of the Art Institute and into the doors of the building. She did not come out again until the closing hour. In the street-car, on the long cold ride home, while she sat staring at the waistcoat buttons of a fat strap-hanger, she had a serious reckoning with herself. She seldom thought about her way of life, about what she ought or ought not to do; usually there was but one obvious and important thing to be done. But that afternoon she remonstrated with herself severely. She told herself that she was missing a great deal; that she ought to be more willing to take advice and to go to see things. She was sorry that she had let months pass without going to the Art Institute. After this she would go once a week.

The Institute proved, indeed, a place of retreat, as the sand hills or the Kohlers' garden used to be; a place where she could forget Mrs. Andersen's tiresome overtures of friendship, the stout contralto in the choir whom she so unreasonably hated, and even, for a little while, the torment of her work. That building was a place in which she could relax and play, and she could hardly ever play now. On the whole, she spent more time with the casts than with the pictures. They were at once more simple and more perplexing; and some way they seemed more important, harder to overlook. It never occurred to her to buy a catalogue, so she called most of the casts by names she made up for them. Some of them she knew; the Dying Gladiator she had read about in "Childe Harold" almost as long ago as she could remember; he was strongly associated with Dr. Archie and childish illnesses. The Venus di Milo puzzled her;

she could not see why people thought her so beautiful. She told herself over and over that she did not think the Apollo Belvedere "at all handsome." Better than anything else she liked a great equestrian statue[3] of an evil, cruel-looking general with an unpronounceable name. She used to walk round and round this terrible man and his terrible horse, frowning at him, brooding upon him, as if she had to make some momentous decision about him.

The casts, when she lingered long among them, always made her gloomy. It was with a lightening of the heart, a feeling of throwing off the old miseries and old sorrows of the world, that she ran up the wide staircase to the pictures. There she liked best the ones that told stories. There was a painting by Gérôme called "The Pasha's Grief"[4] which always made her wish for Gunner and Axel. The Pasha was seated on a rug, beside a green candle almost as big as a telegraph pole, and before him was stretched his dead tiger, a splendid beast, and there were pink roses scattered about him. She loved, too, a picture of some boys bringing in a newborn calf on a litter,[5] the cow walking beside it and licking it. The Corot which hung next to this painting she did not like or dislike; she never saw it.

But in that same room there was a picture—oh, that was the thing she ran upstairs so fast to see! That was her picture. She imagined that nobody cared for it but herself, and that it waited for her. That was a picture indeed. She liked even the name of it, "The Song of the Lark."[6] The flat country, the early morning light, the wet fields, the look in the girl's heavy face—well, they were all hers, anyhow, whatever was there. She told herself that that picture was "right." Just what she meant by this, it would take a clever person to explain. But to her the word covered the almost boundless satisfaction she felt when she looked at the picture.

Before Thea had any idea how fast the weeks were flying, before Mr. Larsen's "permanent" soprano had returned to her duties, spring came; windy, dusty, strident, shrill; a season almost more violent in Chicago than the winter from which it releases one, or the heat to which it eventually delivers one. One sunny morning the apple trees in Mrs. Lorch's back yard burst into bloom, and for the first time in months Thea dressed without building a fire. The morning shone like a holiday, and for her it was to be a holiday. There was in the air that sudden, treacherous softness which makes the Poles who work in the packing-houses get drunk. At such times beauty is necessary, and in Packingtown there is no place to get it except at the saloons, where one can buy for a few hours the illusion of comfort, hope, love,—whatever one most longs for.

Harsanyi had given Thea a ticket for the symphony concert that afternoon, and when she looked out at the white apple trees her doubts as to whether she ought to go vanished at once. She would make her work light that morning, she told herself. She would go to the concert full of energy. When she set off, after dinner, Mrs. Lorch, who knew Chicago weather, prevailed upon her to take her cape. The old lady said that such sudden mildness, so early in April, presaged a sharp return of winter, and she was anxious about her apple trees.

The concert began at two-thirty, and Thea was in her seat in the Auditorium[7] at ten minutes after two—a fine seat in the first row of the balcony, on the side, where she could see the house as well as the orchestra. She had been to so few concerts that the great house, the crowd of people, and the lights, all had a stimulating effect. She was surprised to see so many men in the audience, and wondered how they could leave their business in the afternoon. During the first number Thea was so much interested in the orchestra itself, in the men, the instruments, the volume of sound, that she paid little attention to what they were playing. Her excitement impaired her power of listening. She kept saying to herself, "Now I must stop this foolishness and listen; I may never hear this again"; but her mind was like a glass that is hard to focus. She was not ready to listen until the second number, Dvořák's Symphony in E minor, called on the programme, "From the New World."[8] The first theme had scarcely been given out when her mind became clear; instant composure fell upon her, and with it came the power of concentration. This was music she could understand, music from the New World indeed! Strange how, as the first movement went on, it brought back to her that high tableland above Laramie; the grass-grown wagon trails, the far-away peaks of the snowy range, the wind and the eagles, that old man and the first telegraph message.

When the first movement ended, Thea's hands and feet were cold as ice. She was too much excited to know anything except that she wanted something desperately, and when the English horns gave out the theme of the Largo, she knew that what she wanted was exactly that. Here were the sand hills, the grasshoppers and locusts, all the things that wakened and chirped in the early morning; the reaching and reaching of high plains, the immeasurable yearning of all flat lands. There was home in it, too; first memories, first mornings long ago; the amazement of a new soul in a new world; a soul new and yet old, that had dreamed something despairing, something glorious, in the dark before it was born; a soul obsessed by what it did not know, under the cloud of a past it could not recall.

If Thea had had much experience in concert-going, and had known her own capacity, she would have left the hall when the symphony was over. But she sat still, scarcely knowing where she was, because her mind had been far away and had not yet come back to her. She was startled when the orchestra began to play again—the entry of the gods into Walhalla. She heard it as people hear things in their sleep. She knew scarcely anything about the Wagner operas. She had a vague idea that "Rhinegold" was about the strife between gods and men;[9] she had read something about it in Mr. Haweis's book long ago. Too tired to follow the orchestra with much understanding, she crouched down in her seat and closed her eyes. The cold, stately measures of the Walhalla music rang out, far away; the rainbow bridge throbbed out into the air, under it the wailing of the Rhine daughters and the singing of the Rhine. But Thea was sunk in twilight; it was all going on in another world. So it happened that with a dull, almost listless ear she heard for the first time that troubled music, ever-darkening, ever-brightening, which was to flow through so many years of her life.

When Thea emerged from the concert hall, Mrs. Lorch's predictions had been fulfilled. A furious gale was beating over the city from Lake Michigan. The streets were full of cold, hurrying, angry people, running for street-cars and barking at each other. The sun was setting in a clear, windy sky, that flamed with red as if there were a great fire somewhere on the edge of the city. For almost the first time Thea was conscious of the city itself, of the congestion of life all about her, of the brutality and power of those streams that flowed in the streets, threatening to drive one under. People jostled her, ran into her, poked her aside with their elbows, uttering angry exclamations. She got on the wrong car and was roughly ejected by the conductor at a windy corner, in front of a saloon. She stood there dazed and shivering. The cars passed, screaming as they rounded curves, but either they were full to the doors, or were bound for places where she did not want to go. Her hands were so cold that she took off her tight kid gloves. The street lights began to gleam in the dusk. A young man came out of the saloon and stood eyeing her questioningly while he lit a cigarette. "Looking for a friend to-night?" he asked. Thea drew up the collar of her cape and walked on a few paces. The young man shrugged his shoulders and drifted away.

Thea came back to the corner and stood there irresolutely. An old man approached her. He, too, seemed to be waiting for a car. He wore an overcoat with a black fur collar, his gray mustache was waxed into little points, and his eyes were watery. He kept thrusting his face up near hers. Her hat blew off and he ran after it—a stiff, pitiful skip he had—and brought it back to her. Then, while she was pinning her hat on, her cape blew up, and he held it down for her, looking at her intently. His face worked as if he were going to cry or were frightened. He leaned over and whispered something to her. It struck her as curious that he was really quite timid, like an old beggar. "Oh, let me *alone!*" she cried miserably between her teeth. He vanished, disappeared like the Devil in a play. But in the mean time something had got away from her; she could not remember how the violins came in after the horns, just there. When her cape blew up, perhaps—Why did these men torment her? A cloud of dust blew in her face and blinded her. There was some power abroad in the world bent upon taking away from her that feeling with which she had come out of the concert hall. Everything seemed to sweep down on her to tear it out from under her cape. If one had that, the world became one's enemy; people, buildings, wagons, cars, rushed at one to crush it under, to make one let go of it. Thea glared round her at the crowds, the ugly, sprawling streets, the long lines of lights, and she was not crying now. Her eyes were brighter than even Harsanyi had ever seen them. All these things and people were no longer remote and negligible; they had to be met, they were lined up against her, they were there to take something from her. Very well; they should never have it. They might trample her to death, but they should never have it. As long as she lived that ecstasy was going to be hers. She would live for it, work for it, die for it; but she was going to have it, time after time, height after height. She could hear the crash of the orchestra again, and she rose on the brasses. She would have it, what the trumpets were singing! She would have it, have it,—it! Under the old cape she pressed her hands upon her heaving bosom, that was a little girl's no longer.

VI

One afternoon in April, Theodore Thomas,[1] the conductor of the Chicago Symphony Orchestra, had turned out his desk light and was about to leave his office in the Auditorium Building, when Harsanyi appeared in the doorway. The conductor welcomed him with a hearty hand-grip and threw off the overcoat he had just put on. He pushed Harsanyi into a chair and sat down at his burdened desk, pointing to the piles of papers and railway folders upon it.

"Another tour, clear to the coast. This traveling is the part of my work that grinds me, Andor. You know what it means: bad food, dirt, noise, exhaustion for the men and for me. I'm not so young as I once was. It's time I quit the highway. This is the last tour, I swear!"

"Then I'm sorry for the 'highway.' I remember when I first heard you in Pittsburgh, long ago. It was a life-line you threw me. It's about one of the people along your highway that I've come to see you. Whom do you consider the best teacher for voice in Chicago?"

Mr. Thomas frowned and pulled his heavy mustache. "Let me see; I suppose on the whole Madison Bowers is the best. He's intelligent, and he had good training. I don't like him."

Harsanyi nodded. "I thought there was no one else. I don't like him, either, so I hesitated. But I suppose he must do, for the present."

"Have you found anything promising? One of your own students?"

"Yes, sir. A young Swedish girl from somewhere in Colorado. She is very talented, and she seems to me to have a remarkable voice."

"High voice?"

"I think it will be; though her low voice has a beautiful quality, very individual. She has had no instruction in voice at all, and I shrink from handing her over to anybody; her own instinct about it has been so good. It is one of those voices that manages itself easily, without thinning as it goes up; good breathing and perfect relaxation. But she must have a teacher, of course. There is a break in the middle voice, so that the voice does not all work together; an unevenness."

Thomas looked up. "So? Curious; that cleft often happens with the Swedes. Some of their best singers have had it. It always reminds me of the space you so often see between their front teeth. Is she strong physically?"

Harsanyi's eye flashed. He lifted his hand before him and clenched it. "Like a horse, like a tree! Every time I give her a lesson, I lose a pound. She goes after what she wants."

"Intelligent, you say? Musically intelligent?"

"Yes; but no cultivation whatever. She came to me like a fine young savage, a book with nothing written in it. That is why I feel the responsibility of directing her." Harsanyi paused and crushed his soft gray hat over his knee. "She would interest you, Mr. Thomas," he added slowly. "She has a quality—very individual."

"Yes; the Scandinavians are apt to have that, too. She can't go to Germany, I suppose?"

"Not now, at any rate. She is poor."

Thomas frowned again. "I don't think Bowers a really first-rate man. He's too petty to be really first-rate; in his nature, I mean. But I dare say he's the best you can do, if you can't give her time enough yourself."

Harsanyi waved his hand. "Oh, the time is nothing—she may have all she wants. But I cannot teach her to sing."

"Might not come amiss if you made a musician of her, however," said Mr. Thomas dryly.

"I have done my best. But I can only play with a voice, and this is not a voice to be played with. I think she will be a musician, whatever happens. She is not quick, but she is solid, real; not like these others. My wife says that with that girl one swallow does not make a summer."

Mr. Thomas laughed. "Tell Mrs. Harsanyi that her remark conveys something to me. Don't let yourself get too much interested. Voices are so often disappointing; especially women's voices. So much chance about it, so many factors."

"Perhaps that is why they interest one. All the intelligence and talent in the world can't make a singer. The voice is a wild thing. It can't be bred in captivity. It is a sport, like the silver fox. It happens."

Mr. Thomas smiled into Harsanyi's gleaming eye. "Why haven't you brought her to sing for me?"

"I've been tempted to, but I knew you were driven to death, with this tour confronting you."

"Oh, I can always find time to listen to a girl who has a voice, if she means business. I'm sorry I'm leaving so soon. I could advise you better if I had heard her. I can sometimes give a singer suggestions. I've worked so much with them."

"You're the only conductor I know who is not snobbish about singers." Harsanyi spoke warmly.

"Dear me, why should I be? They've learned from me, and I've learned from them." As they rose, Thomas took the younger man affectionately by the arm. "Tell me about that wife of yours. Is she well, and as lovely as ever? And such fine children! Come to see me oftener, when I get back. I miss it when you don't."

The two men left the Auditorium Building together. Harsanyi walked home. Even a short talk with Thomas always stimulated him. As he walked he was recalling an evening they once spent together in Cincinnati.

Harsanyi was the soloist at one of Thomas's concerts there, and after the performance the conductor had taken him off to a *Rathskeller* where there was excellent German cooking, and where the proprietor saw to it that Thomas had the best wines procurable. Thomas had been working with the great chorus of the Festival Association[2] and was speaking of it with enthusiasm when Harsanyi asked him how it was that he was able to feel such an interest in choral directing and in voices generally. Thomas seldom spoke of his youth or his early struggles, but that night he turned back the pages and told Harsanyi a long story.

He said he had spent the summer of his fifteenth year wandering about alone in the South, giving violin concerts in little towns. He traveled on horseback. When he came into a town, he went about all day tacking up posters announcing his concert in the evening. Before the concert, he stood at the door taking in the admission money until his audience had arrived, and then he went on the platform and played. It was a lazy, hand-to-mouth existence, and

Thomas said he must have got to like that easy way of living and the relaxing Southern atmosphere. At any rate, when he got back to New York in the fall, he was rather torpid; perhaps he had been growing too fast. From this adolescent drowsiness the lad was awakened by two voices, by two women who sang in New York in 1851,—Jenny Lind and Henrietta Sontag.[3] They were the first great artists he had ever heard, and he never forgot his debt to them.

As he said, "It was not voice and execution alone. There was a greatness about them. They were great women, great artists. They opened a new world to me." Night after night he went to hear them, striving to reproduce the quality of their tone upon his violin. From that time his idea about strings was completely changed, and on his violin he tried always for the singing, vibrating tone, instead of the loud and somewhat harsh tone then prevalent among even the best German violinists. In later years he often advised violinists to study singing, and singers to study violin. He told Harsanyi that he got his first conception of tone quality from Jenny Lind.

"But, of course," he added, "the great thing I got from Lind and Sontag was the indefinite, not the definite, thing. For an impressionable boy, their inspiration was incalculable. They gave me my first feeling for the Italian style—but I could never say how much they gave me. At that age, such influences are actually creative. I always think of my artistic consciousness as beginning then."

All his life Thomas did his best to repay what he felt he owed to the singer's art. No man could get such singing from choruses, and no man worked harder to raise the standard of singing in schools and churches and choral societies.

VII

ALL THROUGH THE LESSON Thea had felt that Harsanyi was restless and abstracted. Before the hour was over, he pushed back his chair and said resolutely, "I am not in the mood, Miss Kronborg. I have something on my mind, and I must talk to you. When do you intend to go home?"

Thea turned to him in surprise. "The first of June, about. Mr. Larsen will not need me after that, and I have not much money ahead. I shall work hard this summer, though."

"And to-day is the first of May; May-day." Harsanyi leaned forward, his elbows on his knees, his hands locked between them. "Yes, I must talk to you about something. I have asked Madison Bowers to let me bring you to him on Thursday, at your usual lesson-time. He is the best vocal teacher in Chicago, and it is time you began to work seriously with your voice."

Thea's brow wrinkled. "You mean take lessons of Bowers?"

Harsanyi nodded, without lifting his head.

"But I can't, Mr. Harsanyi. I haven't got the time, and, besides—" she blushed and drew her shoulders up stiffly—"besides, I can't afford to pay two teachers." Thea felt that she had blurted this out in the worst possible way, and she turned back to the keyboard to hide her chagrin.

"I know that. I don't mean that you shall pay two teachers. After you go to Bowers you will not need me. I need scarcely tell you that I shan't be happy at losing you."

Thea turned to him, hurt and angry. "But I don't want to go to Bowers. I don't want to leave you. What's the matter? Don't I work hard enough? I'm sure you teach people that don't try half as hard."

Harsanyi rose to his feet. "Don't misunderstand me, Miss Kronborg. You interest me more than any pupil I have. I have been thinking for months about what you ought to do, since that night when you first sang for me." He walked over to the window, turned, and came toward her again. "I believe that your voice is worth all that you can put into it. I have not come to this decision rashly. I have studied you, and I have become more and more convinced, against my own desires. I cannot make a singer of you, so it was my business to find a man who could. I have even consulted Theodore Thomas about it."

"But suppose I don't want to be a singer? I want to study with you. What's the matter? Do you really think I've no talent? Can't I be a pianist?"

Harsanyi paced up and down the long rug in front of her. "My girl, you are very talented. You could be a pianist, a good one. But the early training of a pianist, such a pianist as you would want to be, must be something tremendous. He must have had no other life than music. At your age he must be the master of his instrument. Nothing can ever take the place of that first training. You know very well that your technique is good, but it is not remarkable. It will never overtake your intelligence. You have a fine power of work, but you are not by nature a student. You are not by nature, I think, a pianist. You would never find yourself. In the effort to do so, I'm afraid your playing would become warped, eccentric." He threw back his head and looked at his pupil intently with that one eye which sometimes seemed to see deeper than any two eyes, as if its singleness gave it privileges.[1] "Oh, I have watched you very carefully, Miss Kronborg. Because you had had so little and had yet done so much for yourself, I had a great wish to help you. I believe that the strongest need of your nature is to find yourself, to emerge *as* yourself. Until I heard you sing I wondered how you were to do this, but it has grown clearer to me every day."

Thea looked away toward the window with hard, narrow eyes. "You mean I can be a singer because I haven't brains enough to be a pianist."

"You have brains enough and talent enough. But to do what you will want to do, it takes more than these—it takes vocation. Now, I think you have vocation, but for the voice, not for the piano. If you knew,"—he stopped and sighed,—"if you knew how fortunate I sometimes think you. With the voice the way is so much shorter, the rewards are more easily won. In your voice I think Nature herself did for you what it would take you many years to do at the piano. Perhaps you were not born in the wrong place after all. Let us talk frankly now. We have never done so before, and I have respected your reticence. What you want more than anything else in the world is to be an artist; is that true?"

She turned her face away from him and looked down at the keyboard. Her answer came in a thickened voice. "Yes, I suppose so."

"When did you first feel that you wanted to be an artist?"

"I don't know. There was always—something."

"Did you never think that you were going to sing?"

"Yes."

"How long ago was that?"

"Always, until I came to you. It was you who made me want to play piano." Her voice trembled. "Before, I tried to think I did, but I was pretending."

Harsanyi reached out and caught the hand that was hanging at her side. He pressed it as if to give her something. "Can't you see, my dear girl, that was only because I happened to be the first artist you have ever known? If I had been a trombone player, it would have been the same; you would have wanted to play trombone. But all the while you have been working with such good-will, something has been struggling against me. See, here we were, you and I and this instrument,"—he tapped the piano,—"three good friends, working so hard. But all the while there was something fighting us: your gift, and the woman you were meant to be. When you find your way to that gift and to that woman, you will be at peace. In the beginning it was an artist that you wanted to be; well, you may be an artist, always."

Thea drew a long breath. Her hands fell in her lap. "So I'm just where I began. No teacher, nothing done. No money."

Harsanyi turned away. "Feel no apprehension about the money, Miss Kronborg. Come back in the fall and we shall manage that. I shall even go to Mr. Thomas if necessary. This year will not be lost. If you but knew what an advantage this winter's study, all your study of the piano, will give you over most singers. Perhaps things have come out better for you than if we had planned them knowingly."

"You mean they have *if* I can sing."

Thea spoke with a heavy irony, so heavy, indeed, that it was coarse. It grated upon Harsanyi because he felt that it was not sincere, an awkward affectation.

He wheeled toward her. "Miss Kronborg, answer me this. *You know that you can sing*, do you not? You have always known it. While we worked here together you sometimes said to yourself, 'I have something you know nothing about; I could surprise you.' Is that also true?"

Thea nodded and hung her head.

"Why were you not frank with me? Did I not deserve it?"

She shuddered. Her bent shoulders trembled. "I don't know," she muttered. "I didn't mean to be like that. I couldn't. I can't. It's different."

"You mean it is very personal?" he asked kindly.

She nodded. "Not at church or funerals, or with people like Mr. Larsen. But with you it was—personal. I'm not like you and Mrs. Harsanyi. I come of rough people. I'm rough. But I'm independent, too. It was—all I had. There is no use my talking, Mr. Harsanyi. I can't tell you."

"You needn't tell me. I know. Every artist knows." Harsanyi stood looking at his pupil's back, bent as if she were pushing something, at her lowered head. "You can sing for those people because with them you do not commit yourself. But the reality, one cannot uncover *that* until one is sure. One can fail one's self, but one must not live to see that fail; better never reveal it. Let me help you to make yourself sure of it. That I can do better than Bowers."

Thea lifted her face and threw out her hands.

Harsanyi shook his head and smiled. "Oh, promise nothing! You will have much to do. There will not be voice only, but French, German, Italian. You will have work enough. But sometimes you will need to be understood; what you never show to any one will need companionship. And then you must come to me." He peered into her face with that searching, intimate glance. "You know what I mean, the thing in you that has no business with what is little, that will have to do only with beauty and power."

Thea threw out her hands fiercely, as if to push him away. She made a sound in her throat, but it was not articulate. Harsanyi took one of her hands and kissed it lightly upon the back. His salute was one of greeting, not of farewell, and it was for some one he had never seen.

When Mrs. Harsanyi came in at six o'clock, she found her husband sitting listlessly by the window. "Tired?" she asked.

"A little. I've just got through a difficulty. I've sent Miss Kronborg away; turned her over to Bowers, for voice."

"Sent Miss Kronborg away? Andor, what is the matter with you?"

"It's nothing rash. I've known for a long while I ought to do it. She is made for a singer, not a pianist."

Mrs. Harsanyi sat down on the piano chair. She spoke a little bitterly: "How can you be sure of that? She was, at least, the best you had. I thought you meant to have her play at your students' recital next fall. I am sure she would have made an impression. I could have dressed her so that she would have been very striking. She had so much individuality."

Harsanyi bent forward, looking at the floor. "Yes, I know. I shall miss her, of course."

Mrs. Harsanyi looked at her husband's fine head against the gray window. She had never felt deeper tenderness for him than she did at that moment. Her heart ached for him. "You will never get on, Andor," she said mournfully.

Harsanyi sat motionless. "No, I shall never get on," he repeated quietly. Suddenly he sprang up with that light movement she knew so well, and stood in the window, with folded arms. "But some day I shall be able to look her in the face and laugh because I did what I could for her. I believe in her. She will do nothing common. She is uncommon, in a common, common world. That is what I get out of it. It means more to me than if she played at my concert and brought me a dozen pupils. All this drudgery will kill me if once in a while I cannot hope something, for somebody! If I cannot sometimes see a bird fly and wave my hand to it."

His tone was angry and injured. Mrs. Harsanyi understood that this was one of the times when his wife was a part of the drudgery, of the "common, common world." He had let something he cared for go, and he felt bitterly about whatever was left. The mood would pass, and he would be sorry. She knew him. It wounded her, of course, but that hurt was not new. It was as old as her love for him. She went out and left him alone.

VIII

ONE WARM DAMP JUNE NIGHT the Denver Express was speeding westward across the earthy-smelling plains of Iowa. The lights in the day-coach were turned low and the ventilators were open, admitting showers of soot and dust upon the occupants of the narrow green plush chairs which were tilted at various angles of discomfort. In each of these chairs some uncomfortable human being lay drawn up, or stretched out, or writhing from one position to another. There were tired men in rumpled shirts, their necks bare and their suspenders down; old women with their heads tied up in black handkerchiefs; bedraggled young women who went to sleep while they were nursing their babies and forgot to button up their dresses; dirty boys who added to the general discomfort by taking off their boots. The brakeman, when he came through at midnight, sniffed the heavy air disdainfully and looked up at the ventilators. As he glanced down the double rows of contorted figures, he saw one pair of eyes that were wide open and bright, a yellow head that was not overcome by the stupefying heat and smell in the car. "There's a girl for you," he thought as he stopped by Thea's chair.

"Like to have the window up a little?" he asked.

Thea smiled up at him, not misunderstanding his friendliness. "The girl behind me is sick; she can't stand a draft. What time is it, please?"

He took out his open-faced watch and held it before her eyes with a knowing look. "In a hurry?" he asked. "I'll leave the end door open and air you out. Catch a wink; the time'll go faster."

Thea nodded good-night to him and settled her head back on her pillow, looking up at the oil lamps. She was going back to Moonstone for her summer vacation, and she was sitting up all night in a day-coach because that seemed such an easy way to save money. At her age discomfort was a small matter, when one made five dollars a day by it. She had confidently expected to sleep after the car got quiet, but in the two chairs behind her were a sick girl and her mother, and the girl had been coughing steadily since ten o'clock. They had come from somewhere in Pennsylvania, and this was their second night on the road. The mother said they were going to Colorado "for her daughter's lungs." The daughter was a little older than Thea, perhaps nineteen, with patient dark eyes and curly brown hair. She was pretty in spite of being so sooty and travel-stained. She had put on an ugly figured satine kimono over her loosened clothes. Thea, when she boarded the train in Chicago, happened to stop and plant her heavy telescope on this seat. She had not intended to remain there, but the sick girl had looked up at her with an eager smile and said, "Do sit there, miss. I'd so much rather not have a gentleman in front of me."

After the girl began to cough there were no empty seats left, and if there had been Thea could scarcely have changed without hurting her feelings. The mother turned on her side and went to sleep; she was used to the cough. But the girl lay wide awake, her eyes fixed on the roof of the car, as Thea's were. The two girls must have seen very different things there.

Thea fell to going over her winter in Chicago. It was only under unusual or uncomfortable conditions like these that she could keep her mind fixed upon herself or her own affairs for any length of time. The rapid motion and the vibration of the wheels under her seemed to give her thoughts rapidity and clearness. She had taken twenty very expensive lessons from Madison Bowers, but she did not yet know what he thought of her or of her ability. He was different from any man with whom she had ever had to do. With her other teachers she had felt a personal relation; but with him she did not. Bowers was a cold, bitter, avaricious man, but he knew a great deal about voices. He worked with a voice as if he were in a laboratory, conducting a series of experiments. He was conscientious and industrious, even capable of a certain cold fury when he was working with an interesting voice, but Harsanyi declared that he had the soul of a shrimp, and could no more make an artist than a throat specialist could. Thea realized that he had taught her a great deal in twenty lessons.

Although she cared so much less for Bowers than for Harsanyi, Thea was, on the whole, happier since she had been studying with him than she had been before. She had always told herself that she studied piano to fit herself to be a music teacher. But she never asked herself why she was studying voice. Her voice, more than any other part of her, had to do with that confidence, that sense of wholeness and inner well-being that she had felt at moments ever since she could remember.

Of this feeling Thea had never spoken to any human being until that day when she told Harsanyi that "there had always been—something." Hitherto she had felt but one obligation toward it—secrecy; to protect it even from herself. She had always believed that by doing all that was required of her by her family, her teachers, her pupils, she kept that part of herself from being caught up in the meshes of common things. She took it for granted that some day, when she was older, she would know a great deal more about it. It was as if she had an appointment to meet the rest of herself sometime, somewhere. It was moving to meet her and she was moving to meet it. That meeting awaited her, just as surely as, for the poor girl in the seat behind her, there awaited a hole in the earth, already dug.

For Thea, so much had begun with a hole in the earth. Yes, she reflected, this new part of her life had all begun that morning when she sat on the clay bank beside Ray Kennedy, under the flickering shade of the cottonwood tree. She remembered the way Ray had looked at her that morning. Why had he cared so much? And Wunsch, and Dr. Archie, and Spanish Johnny, why had they? It was something that had to do with her that made them care, but it was not she. It was something they believed in, but it was not she. Perhaps each of them concealed another person in himself, just as she did. Why was it that they seemed to feel and to hunt for a second person in her and not in each other? Thea frowned up at the dull lamp in the roof of the car. What if one's second self could somehow speak to all these second selves? What if one could bring them out, as whiskey did Spanish Johnny's? How deep they lay, these second persons, and how little one knew about them, except to guard them fiercely. It was to music, more than to anything else, that these hidden things in people responded. Her mother—even her mother had something of that sort which replied to music.

Thea found herself listening for the coughing behind her and not hearing it. She turned cautiously and looked back over the head-rest of her chair. The poor girl had fallen asleep. Thea looked at her intently. Why was she so afraid of men? Why did she shrink into herself and avert her face whenever a man passed her chair? Thea thought she knew; of course, she knew. How horrible to waste away like that, in the time when one ought to be growing fuller and stronger and rounder every day. Suppose there were such a dark hole open for her, between to-night and that place where she was to meet herself? Her eyes narrowed. She put her hand on her breast and felt how warm it was; and within it there was a full, powerful pulsation. She smiled—though she was ashamed of it—with the natural contempt of strength for weakness, with the sense of physical security which makes the savage merciless. Nobody could die while they felt like that inside. The springs there were wound so tight that it would be a long while before there was any slack in them. The life in there was rooted deep. She was going to have a few things before she died. She realized that there were a great many trains dashing east and west on the face of the continent that night, and that they all carried young people who meant to have things. But the difference was that *she was going to get them!* That was all. Let people try to stop her! She glowered at the rows of feckless bodies that lay sprawled in the chairs. Let them try it once! Along with the yearning that came from some deep part of her, that was selfless and exalted, Thea had a hard kind of cockiness, a determination to get ahead. Well, there are passages in life when that fierce, stubborn self-assertion will stand its ground after the nobler feeling is overwhelmed and beaten under.

Having told herself once more that she meant to grab a few things, Thea went to sleep.

She was wakened in the morning by the sunlight, which beat fiercely through the glass of the car window upon her face. She made herself as clean as she could, and while the people all about her were getting cold food out of their lunch-baskets she escaped into the dining-car. Her thrift did not go to the point of enabling her to carry a lunch-basket. At that early hour there were few people in the dining-car. The linen was white and fresh, the darkies were trim and smiling, and the sunlight gleamed pleasantly upon the silver and the glass water-bottles. On each table there was a slender vase with a single pink rose in it. When Thea sat down she looked into her rose and thought it the most beautiful thing in the world; it was wide open, recklessly offering its yellow heart, and there were drops of water on the petals. All the future was in that rose, all that one would like to be. The flower put her in an absolutely regal mood. She had a whole pot of coffee, and scrambled eggs with chopped ham, utterly disregarding the astonishing price they cost. She had faith enough in what she could do, she told herself, to have eggs if she wanted them. At the table opposite her sat a man and his wife and little boy—Thea classified them as being "from the East." They spoke in that quick, sure staccato, which Thea, like Ray Kennedy, pretended to scorn and secretly admired. People who could use words in that confident way, and who spoke them elegantly, had a great advantage in life, she reflected. There were so many words which she could not pronounce in speech as she had to do in singing. Language was like clothes; it could be a help to one, or it could give one away. But the most important thing was that one should not pretend to be what one was not.

When she paid her check she consulted the waiter. "Waiter, do you suppose I could buy one of those roses? I'm out of the day-coach, and there is a sick girl in there. I'd like to take her a cup of coffee and one of those flowers."

The waiter liked nothing better than advising travelers less sophisticated than himself. He told Thea there were a few roses left in the icebox and he would get one. He took the flower and the coffee into the day-coach. Thea pointed out the girl, but she did not accompany him. She hated thanks and never received them gracefully. She stood outside on the platform to get some fresh air into her lungs. The train was crossing the Platte River now, and the sunlight was so intense that it seemed to quiver in little flames on the glittering sandbars, the scrub willows, and the curling, fretted shallows.

Thea felt that she was coming back to her own land. She had often heard Mrs. Kronborg say that she "believed in immigration," and so did Thea believe in it. This earth seemed to her young and fresh and kindly, a place where refugees from old, sad countries were given another chance. The mere absence of rocks gave the soil a kind of amiability and generosity, and the absence of natural boundaries gave the spirit a wider range. Wire fences might mark the end of a man's pasture, but they could not shut in his thoughts as mountains and forests can. It was over flat lands like this, stretching out to drink the sun, that the larks sang—and one's heart sang there, too. Thea was glad that this was her country, even if one did not learn to speak elegantly there. It was, somehow, an honest country, and there was a new song in that blue air which had never been sung in the world before. It was hard to tell about it, for it had nothing to do with words; it was like the light of the desert at noon, or the smell of the sagebrush after rain; intangible but powerful. She had the sense of going back to a friendly soil, whose friendship was somehow going to strengthen her; a naïve, generous country that gave one its joyous force, its large-hearted, childlike power to love, just as it gave one its coarse, brilliant flowers.

As she drew in that glorious air Thea's mind went back to Ray Kennedy. He, too, had that feeling of empire; as if all the Southwest really belonged to him because he had knocked about over it so much, and knew it, as he said, "like the blisters on his own hands." That feeling, she reflected, was the real element of companionship between her and Ray. Now that she was going back to Colorado, she realized this as she had not done before.

IX

THEA REACHED MOONSTONE in the late afternoon, and all the Kronborgs were there to meet her except her two older brothers. Gus and Charley were young men now, and they had declared at noon that it would "look silly if the whole bunch went down to the train." "There's no use making a fuss over Thea just because she's been to Chicago," Charley warned his mother. "She's inclined to think pretty well of herself, anyhow, and if you go treating her like company, there'll be no living in the house with her." Mrs. Kronborg simply leveled her eyes at Charley, and he faded away, muttering. She had, as Mr. Kronborg always said with an inclination of his head, good control over her children. Anna, too, wished to absent herself from the party, but in the end her curiosity got the better of her. So when Thea stepped down from the porter's stool, a very creditable Kronborg representation was grouped on the platform to greet her. After they had all kissed her (Gunner and Axel shyly), Mr. Kronborg hurried his flock into the hotel omnibus, in which they were to be driven ceremoniously home, with the neighbors looking out of their windows to see them go by.

All the family talked to her at once, except Thor,—impressive in new trousers,—who was gravely silent and who refused to sit on Thea's lap. One of the first things Anna told her was that Maggie Evans, the girl who used to cough in prayer meeting, died yesterday, and had made a request that Thea sing at her funeral.

Thea's smile froze. "I'm not going to sing at all this summer, except my exercises. Bowers says I taxed my voice last winter, singing at funerals so much. If I begin the first day after I get home, there'll be no end to it. You can tell them I caught cold on the train, or something."

Thea saw Anna glance at their mother. Thea remembered having seen that look on Anna's face often before, but she had never thought anything about it because she was used to it. Now she realized that the look was distinctly spiteful, even vindictive. She suddenly realized that Anna had always disliked her.

Mrs. Kronborg seemed to notice nothing, and changed the trend of the conversation, telling Thea that Dr. Archie and Mr. Upping, the jeweler, were both coming in to see her that evening, and that she had asked Spanish Johnny to come, because he had behaved well all winter and ought to be encouraged.

The next morning Thea wakened early in her own room up under the eaves and lay watching the sunlight shine on the roses of her wall-paper. She wondered whether she would ever like a plastered room as well as this one lined with scantlings. It was snug and tight, like the cabin of a little boat. Her bed faced the window and stood against the wall, under the slant of the ceiling. When she went away she could just touch the ceiling with the tips of her fingers; now she could touch it with the palm of her hand. It was so little that it was like a sunny cave, with roses running all over the roof. Through the low window, as she lay there, she could watch people going by on the farther side of the street; men, going downtown to open their stores. Thor was over there, rattling his express wagon along the sidewalk. Tillie had put a bunch of French pinks in a tumbler of water on her dresser, and they gave out a pleasant perfume. The blue jays were fighting and screeching in the cottonwood tree outside her window, as they always did, and she could hear the old Baptist deacon across the street calling his chickens, as she had heard him do every summer morning since she could remember. It was pleasant to waken up in that bed, in that room, and to feel the brightness of the morning, while light quivered about the low, papered ceiling in golden spots, refracted by the broken mirror and the glass of water that held the pinks. *"Im leuchtenden Sommermorgen"*; those lines, and the face of her old teacher, came back to Thea, floated to her out of sleep, perhaps. She had been dreaming something pleasant, but she could not remember what. She would go to call upon Mrs. Kohler to-day, and see the pigeons washing their pink feet in the drip under the water tank, and flying about their house that was sure to have a fresh coat of white paint on it for summer. On the way home she would stop to see Mrs. Tellamantez. On Sunday she would coax Gunner to take her out to the sand hills. She had missed them in Chicago; had been homesick for their brilliant morning gold and for their soft colors at evening. The Lake, somehow, had never taken their place.

While she lay planning, relaxed in warm drowsiness, she heard a knock at her door. She supposed it was Tillie, who sometimes fluttered in on her before she was out of bed to offer some service which the family would have ridiculed. But instead, Mrs. Kronborg herself came in, carrying a tray with Thea's breakfast set out on one of the best white napkins. Thea sat up with some embarrassment and pulled her nightgown together across her chest. Mrs. Kronborg was always busy downstairs in the morning, and Thea could not remember when her mother had come to her room before.

"I thought you'd be tired, after traveling, and might like to take it easy for once." Mrs. Kronborg put the tray on the edge of the bed. "I took some thick cream for you before the boys got at it. They raised a howl." She chuckled and sat down in the big wooden rocking chair. Her visit made Thea feel grown-up, and, somehow, important.

Mrs. Kronborg asked her about Bowers and the Harsanyis. She felt a great change in Thea, in her face and in her manner. Mr. Kronborg had noticed it, too, and had spoken of it to his wife with great satisfaction while they were undressing last night. Mrs. Kronborg sat looking at her daughter, who lay on her side, supporting herself on her elbow and lazily drinking her coffee from the tray before her. Her short-sleeved nightgown had come open at the throat again, and Mrs. Kronborg noticed how white her arms and shoulders were, as if they had been dipped in new milk. Her chest was fuller than when she went away, her breasts rounder and firmer, and though she was so white where she was uncovered, they looked rosy through the thin muslin. Her body had the elasticity that comes of being highly charged with the desire to live. Her hair, hanging in two loose braids, one by either cheek, was just enough disordered to catch the light in all its curly ends.

Thea always woke with a pink flush on her cheeks, and this morning her mother thought she had never seen her eyes so wide-open and bright; like clear green springs in the wood, when the early sunlight sparkles in them. She would make a very handsome woman, Mrs. Kronborg said to herself, if she would only get rid of that fierce look she had sometimes. Mrs. Kronborg took great pleasure in good looks, wherever she found them. She still remembered that, as a baby, Thea had been the "best-formed" of any of her children.

"I'll have to get you a longer bed," she remarked, as she set the tray on the table. "You're getting too long for that one."

Thea looked up at her mother and laughed, dropping back on her pillow with a magnificent stretch of her whole body. Mrs. Kronborg sat down again.

"I don't like to press you, Thea, but I think you'd better sing at that funeral to-morrow. I'm afraid you'll always be sorry if you don't. Sometimes a little thing like that, that seems nothing at the time, comes back on one afterward and troubles one a good deal. I don't mean the church shall run you to death this summer, like they used to. I've spoken my mind to your father about that, and he's very reasonable. But Maggie talked a good deal about you to people this winter; always asked what word we'd had, and said how she missed your singing and all. I guess you ought to do that much for her."

"All right, mother, if you think so." Thea lay looking at her mother with intensely bright eyes.

"That's right, daughter." Mrs. Kronborg rose and went over to get the tray, stopping to put her hand on Thea's chest. "You're filling out nice," she said, feeling about. "No, I wouldn't bother about the buttons. Leave 'em stay off. This is a good time to harden your chest."

Thea lay still and heard her mother's firm step receding along the bare floor of the trunk loft. There was no sham about her mother, she reflected. Her mother knew a great many things of which she never talked, and all the church people were forever chattering about things of which they knew nothing. She liked her mother.

Now for Mexican Town and the Kohlers! She meant to run in on the old woman without warning, and hug her.

X

SPANISH JOHNNY had no shop of his own, but he kept a table and an order-book in one corner of the drug store where paints and wall-paper were sold, and he was sometimes to be found there for an hour or so about noon. Thea had gone into the drug store to have a friendly chat with the proprietor, who used to lend her books from his shelves. She found Johnny there, trimming rolls of wall-paper for the parlor of Banker Smith's new house. She sat down on the top of his table and watched him.

"Johnny," she said suddenly, "I want you to write down the words of that Mexican serenade you used to sing; you know, '*Rosa de Noche.*'[1] It's an unusual song. I'm going to study it. I know enough Spanish for that."

Johnny looked up from his roller with his bright, affable smile. "*Sí*, but it is low for you, I think; *voz contralto.* It is low for me."

"Nonsense. I can do more with my low voice than I used to. I'll show you. Sit down and write it out for me, please." Thea beckoned him with the short yellow pencil tied to his order-book.

Johnny ran his fingers through his curly black hair. "If you wish. I do not know if that *serenata* all right for young ladies. Down there it is more for married ladies. They sing it for husbands—or somebody else, may-bee." Johnny's eyes twinkled and he apologized gracefully with his shoulders. He sat down at the table, and while Thea looked over his arm, began to write the song down in a long, slanting script, with highly ornamental capitals. Presently he looked up. "This-a song not exactly Mexican," he said thoughtfully. "It come from farther down; Brazil, Venezuela, may-bee. I learn it from some fellow down there, and he learn it from another fellow. It is-a most like Mexican, but not quite." Thea did not release him, but pointed to the paper. There were three verses of the song in all, and when Johnny had written them down, he sat looking at them meditatively, his head on one side. "I don' think for a high voice, *señorita,*" he objected with polite persistence. "How you accompany with piano?"

"Oh, that will be easy enough."

"For you, may-bee!" Johnny smiled and drummed on the table with the tips of his agile brown fingers. "You know something? Listen, I tell you." He rose and sat down on the table beside her, putting his foot on the chair. He loved to talk at the hour of noon. "When you was a little girl, no bigger than that, you come to my house one day 'bout noon, like this, and I was in the door, playing guitar. You was barehead, barefoot; you run away from home. You stand there and make a frown at me an' listen. By 'n by you say for me to sing. I sing some lil' ting, and then I say for you to sing with me. You don' know no words, of course, but you take the air and you sing it just-a beauti-ful! I never see a child do that, outside Mexico. You was, oh, I do' know—seven year, may-bee. By 'n by the preacher come look for you and begin for scold. I say, 'Don' scold, Meester Kronborg. She come for hear guitar. She gotta some music in her, that child. Where she get?' Then he tell me 'bout your gran'papa play oboe in the old country. I never forgetta that time." Johnny chuckled softly.

Thea nodded. "I remember that day, too. I liked your music better than the church music. When are you going to have a dance over there, Johnny?"

Johnny tilted his head. "Well, Saturday night the Spanish boys have a lil' party, some *danza*. You know Miguel Ramas? He have some young cousins, two boys, very nice-a, come from Torreon. They going to Salt Lake for some job-a, and stay off with him two-three days, and he mus' have a party. You like to come?"

That was how Thea came to go to the Mexican ball. Mexican Town had been increased by half a dozen new families during the last few years, and the Mexicans had put up an adobe dance-hall, that looked exactly like one of their own dwellings, except that it was a little longer, and was so unpretentious that nobody in Moonstone knew of its existence. The "Spanish boys" are reticent about their own affairs. Ray Kennedy used to know about all their little doings, but since his death there was no one whom the Mexicans considered *simpático.*

On Saturday evening after supper Thea told her mother that she was going over to Mrs. Tellamantez's to watch the Mexicans dance for a while, and that Johnny would bring her home.

Mrs. Kronborg smiled. She noticed that Thea had put on a white dress and had done her hair up with unusual care, and that she carried her best blue scarf. "Maybe you'll take a turn yourself, eh? I wouldn't mind watching them Mexicans. They're lovely dancers."

Thea made a feeble suggestion that her mother might go with her, but Mrs. Kronborg was too wise for that. She knew that Thea would have a better time if she went alone, and she watched her daughter go out of the gate and down the sidewalk that led to the depot.

Thea walked slowly. It was a soft, rosy evening. The sand hills were lavender. The sun had gone down a glowing copper disk, and the fleecy clouds in the east were a burning rose-color, flecked with gold. Thea passed the cottonwood grove and then the depot, where she left the sidewalk and took the sandy path toward Mexican Town. She could hear the scraping of violins being tuned, the tinkle of mandolins, and the growl of a double bass. Where had they got a double bass? She did not know there was one in Moonstone. She found later that it was the property of one of Ramas's young cousins, who was taking it to Utah with him to cheer him at his "job-a."

The Mexicans never wait until it is dark to begin to dance, and Thea had no difficulty in finding the new hall, because every other house in the town was deserted. Even the babies had gone to the ball; a neighbor was always willing to hold the baby while the mother danced. Mrs. Tellamantez came out to meet Thea and led her in. Johnny bowed to her from the platform at the end of the room, where he was playing the mandolin along with two fiddles and the bass. The hall was a long low room, with white-washed walls, a fairly tight plank floor, wooden benches along the sides, and a few bracket lamps screwed to the frame timbers. There must have been fifty people there, counting the children. The Mexican dances were very much family affairs. The fathers always danced again and again with their little daughters, as well as with their wives. One of the girls came up to greet Thea, her dark cheeks glowing with pleasure and cordiality, and introduced her brother, with whom she had just been dancing. "You better take him every time he asks you," she whispered. "He's the best dancer here, except Johnny."

Thea soon decided that the poorest dancer was herself. Even Mrs. Tellamantez, who always held her shoulders so stiffly, danced better than she did. The musicians did not remain long at their post. When one of them felt like dancing, he called some other boy to take his instrument, put on his coat, and went down on the floor. Johnny, who wore a blousy white silk shirt, did not even put on his coat.

The dances the railroad men gave in Firemen's Hall were the only dances Thea had ever been allowed to go to, and they were very different from this. The boys played rough jokes and thought it smart to be clumsy and to run into each other on the floor. For the square dances there was always the bawling voice of the caller, who was also the county auctioneer.

This Mexican dance was soft and quiet. There was no calling, the conversation was very low, the rhythm of the music was smooth and engaging, the men were graceful and courteous. Some of them Thea had never before seen out of their working clothes, smeared with grease from the round-house or clay from the brickyard. Sometimes, when the music happened to be a popular Mexican waltz song, the dancers sang it softly as they moved. There were three little girls under twelve, in their first communion dresses, and one of them had an orange marigold in her black hair, just over her ear. They danced with the men and with each other. There was an atmosphere of ease and friendly pleasure in the low, dimly lit room, and Thea could not help wondering whether the Mexicans had no jealousies or neighborly grudges as the people in Moonstone had. There was no constraint of any kind there to-night, but a kind of natural harmony about their movements, their greetings, their low conversation, their smiles.

Ramas brought up his two young cousins, Silvo and Felipe, and presented them. They were handsome, smiling youths, of eighteen and twenty, with pale-gold skins, smooth cheeks, aquiline features, and wavy black hair, like Johnny's. They were dressed alike, in black velvet jackets and soft silk shirts, with opal shirt-buttons and flowing black ties looped through gold rings. They had charming manners, and low, guitar-like voices. They knew almost no English, but a Mexican boy can pay a great many compliments with a very limited vocabulary. The Ramas boys thought Thea dazzlingly beautiful. They had never seen a Scandinavian girl before, and her hair and fair skin bewitched them. "*Blanco y oro, semejante la Pascua!*" (White and gold, like Easter!) they exclaimed to each other. Silvo, the younger, declared that he could never go on to Utah; that he and his double bass had reached their ultimate destination. The elder was more crafty; he asked Miguel Ramas whether there would be "plenty more girls like that *a* Salt Lake, may-bee?"

Silvo, overhearing, gave his brother a contemptuous glance. "Plenty more *a Paraíso*[2] may-bee!" he retorted. When they were not dancing with her, their eyes followed her, over the coiffures of their other partners. That was not difficult; one blonde head moving among so many dark ones.

Thea had not meant to dance much, but the Ramas boys danced so well and were so handsome and adoring that she yielded to their entreaties. When she sat out a dance with them, they talked to her about their family at home, and told her how their mother had once punned upon their name. *Rama,* in Spanish, meant a branch, they explained. Once when they were little lads their mother took them along when she went to help the women decorate the church for Easter. Some one asked her whether she had brought any flowers, and she replied that she had brought her "ramas." This was evidently a cherished family story.

When it was nearly midnight, Johnny announced that every one was going to his house to have "some lil' ice-cream and some lil' *música.*" He began to put out the lights and Mrs. Tellamantez led the way across the square to her *casa.* The Ramas brothers escorted Thea, and as they stepped out of the door, Silvo exclaimed, "*Hace frío!*"[3] and threw his velvet coat about her shoulders.

Most of the company followed Mrs. Tellamantez, and they sat about on the gravel in her little yard while she and Johnny and Mrs. Miguel Ramas served the ice-cream. Thea sat on Felipe's coat, since Silvo's was already about her shoulders. The youths lay down on the shining gravel beside her, one on her right and one on her left. Johnny already called them "*los acólitos,*" the altar-boys. The talk all about them was low, and indolent. One of the girls was playing on Johnny's guitar, another was picking lightly at a mandolin. The moonlight was so bright that one could see every glance and smile, and the flash of their teeth. The moonflowers over Mrs. Tellamantez's door were wide open and of an unearthly white. The moon itself looked like a great pale flower in the sky.

After all the ice-cream was gone, Johnny approached Thea, his guitar under his arm, and the elder Ramas boy politely gave up his place. Johnny sat down, took a long breath, struck a fierce chord, and then hushed it with his other hand. "Now we have some lil' *serenata,* eh? You wan' a try?"

When Thea began to sing, instant silence fell upon the company. She felt all those dark eyes fix themselves upon her intently. She could see them shine. The faces came out of the shadow like the white flowers over the door. Felipe leaned his head upon his hand. Silvo dropped on his back and lay looking at the moon, under the impression that he was still looking at Thea. When she finished the first verse, Thea whispered to Johnny, "Again, I can do it better than that."

She had sung for churches and funerals and teachers, but she had never before sung for a really musical people, and this was the first time she had ever felt the response that such a people can give. They turned themselves and all they had over to her. For the moment they cared about nothing in the world but what she was doing. Their faces confronted her, open, eager, unprotected. She felt as if all these warm-blooded people débouched into her. Mrs. Tellamantez's fateful resignation, Johnny's madness, the adoration of the boy who lay still in the sand; in an instant these things seemed to be within her instead of without, as if they had come from her in the first place.

When she finished, her listeners broke into excited murmur. The men began hunting feverishly for cigarettes. Famos Serraños the barytone bricklayer, touched Johnny's arm, gave him a questioning look, then heaved a deep sigh. Johnny dropped on his elbow, wiping his face and neck and hands with his handkerchief. "*Señorita,*" he panted, "if you sing like that once in the City of Mexico, they just-a go crazy. In the City of Mexico they ain't-a sit like stumps when they hear that, not-a much! When they like, they just-a give you the town."

Thea laughed. She, too, was excited. "Think so, Johnny? Come, sing something with me. *El Parreño;* I haven't sung that for a long time."

Johnny laughed and hugged his guitar. "You not-a forget him?" He began teasing his strings. "Come!" He threw back his head, "*Anoche-e-e—*"

"Anoche me confesse
Con un padre carmelite,
Y me dio penitencia
Que besaras tu boquita."

(Last night I made confession
With a Carmelite father,
And he gave me absolution
For the kisses you imprinted.)

Johnny had almost every fault that a tenor can have. His voice was thin, unsteady, husky in the middle tones. But it was distinctly a voice, and sometimes he managed to get something very sweet out of it. Certainly it made him happy to sing. Thea kept glancing down at him as he lay there on his elbow. His eyes seemed twice as large as usual and had lights in them like those the moonlight makes on black, running water. Thea remembered the old stories about his "spells." She had never seen him when his madness was on him, but she felt something tonight at her elbow that gave her an idea of what it might be like. For the first time she fully understood the cryptic explanation that Mrs. Tellamantez had made to Dr. Archie, long ago. There were the same shells along the walk; she believed she could pick out the very one. There was the same moon up yonder, and panting at her elbow was the same Johnny—fooled by the same old things!

When they had finished, Famos, the barytone, murmured something to Johnny; who replied, "Sure we can sing 'Trovatore.'[4] We have no alto, but all the girls can sing alto and make some noise."

The women laughed. Mexican women of the poorer class do not sing like the men. Perhaps they are too indolent. In the evening, when the men are singing their throats dry on the doorstep, or around the camp-fire beside the work-train, the women usually sit and comb their hair.

While Johnny was gesticulating and telling everybody what to sing and how to sing it, Thea put out her foot and touched the corpse of Silvo with the toe of her slipper. "Aren't you going to sing, Silvo?" she asked teasingly.

The boy turned on his side and raised himself on his elbow for a moment. "Not this night, *señorita,*" he pleaded softly, "not this night!" He dropped back again, and lay with his cheek on his right arm, the hand lying passive on the sand above his head.

"How does he flatten himself into the ground like that?" Thea asked herself. "I wish I knew. It's very effective, somehow."

Across the gulch the Kohlers' little house slept among its trees, a dark spot on the white face of the desert. The windows of their upstairs bedroom were open, and Paulina had listened to the dance music for a long while before she drowsed off. She was a light sleeper, and when she woke again, after midnight, Johnny's concert was at its height. She lay still until she could bear it no longer. Then she wakened Fritz and they went over to the window and leaned out. They could hear clearly there.

"Die Thea," whispered Mrs. Kohler; "it must be. *Ach, wunderschön!"*[5]

Fritz was not so wide awake as his wife. He grunted and scratched on the floor with his bare foot. They were listening to a Mexican part-song; the tenor, then the soprano, then both together; the barytone joins them, rages, is extinguished; the tenor expires in sobs, and the soprano finishes alone. When the soprano's last note died away, Fritz nodded to his wife. *"Ja,"* he said; *"schön."*

There was silence for a few moments. Then the guitar sounded fiercely, and several male voices began the sextette from "Lucia."[6] Johnny's reedy tenor they knew well, and the bricklayer's big, opaque barytone; the others might be anybody over there—just Mexican voices. Then at the appointed, at the acute, moment, the soprano voice, like a fountain jet, shot up into the light. *"Horch! Horch!"* the old people whispered, both at once. How it leaped from among those dusky male voices! How it played in and about and around and over them, like a goldfish darting among creek minnows, like a yellow butterfly soaring above a swarm of dark ones. "Ah," said Mrs. Kohler softly, "the dear man; if he could hear her now!"

XI

MRS. KRONBORG had said that Thea was not to be disturbed on Sunday morning, and she slept until noon. When she came downstairs the family were just sitting down to dinner, Mr. Kronborg at one end of the long table, Mrs. Kronborg at the other. Anna, stiff and ceremonious, in her summer silk, sat at her father's right, and the boys were strung along on either side of the table. There was a place left for Thea between her mother and Thor. During the silence which preceded the blessing, Thea felt something uncomfortable in the air. Anna and her older brothers had lowered their eyes when she came in. Mrs. Kronborg nodded cheerfully, and after the blessing, as she began to pour the coffee, turned to her.

"I expect you had a good time at that dance, Thea. I hope you got your sleep out."

"High society, that," remarked Charley, giving the mashed potatoes a vicious swat. Anna's mouth and eyebrows became half-moons.

Thea looked across the table at the uncompromising countenances of her older brothers. "Why, what's the matter with the Mexicans?" she asked, flushing. "They don't trouble anybody, and they are kind to their families and have good manners."

"Nice clean people; got some style about them. Do you really like that kind, Thea, or do you just pretend to? That's what I'd like to know." Gus looked at her with pained inquiry. But he at least looked at her.

"They're just as clean as white people, and they have a perfect right to their own ways. Of course I like 'em. I don't pretend things."

"Everybody according to their own taste," remarked Charley bitterly. "Quit crumbing your bread up, Thor. Ain't you learned how to eat yet?"

"Children, children!" said Mr. Kronborg nervously, looking up from the chicken he was dismembering. He glanced at his wife, whom he expected to maintain harmony in the family.

"That's all right, Charley. Drop it there," said Mrs. Kronborg. "No use spoiling your Sunday dinner with race prejudices. The Mexicans suit me and Thea very well. They are a useful people. Now you can just talk about something else."

Conversation, however, did not flourish at that dinner. Everybody ate as fast as possible. Charley and Gus said they had engagements and left the table as soon as they finished their apple pie. Anna sat primly and ate with great elegance. When she spoke at all she spoke to her father, about church matters, and always in a commiserating tone, as if he had met with some misfortune. Mr. Kronborg, quite innocent of her intentions, replied kindly and absent-mindedly. After the dessert he went to take his usual Sunday afternoon nap, and Mrs. Kronborg carried some dinner to a sick neighbor. Thea and Anna began to clear the table.

"I should think you would show more consideration for father's position, Thea," Anna began as soon as she and her sister were alone.

Thea gave her a sidelong glance. "Why, what have I done to father?"

"Everybody at Sunday-School was talking about you going over there and singing with the Mexicans all night, when you won't sing for the church. Somebody heard you, and told it all over town. Of course, we all get the blame for it."

"Anything disgraceful about singing?" Thea asked with a provoking yawn.

"I must say you choose your company! You always had that streak in you, Thea. We all hoped that going away would improve you. Of course, it reflects on father when you are scarcely polite to the nice people here and make up to the rowdies."

"Oh, it's my singing with the Mexicans you object to?" Thea put down a tray full of dishes. "Well, I like to sing over there, and I don't like to over here. I'll sing for them any time they ask me to. They know something about what I'm doing. They're a talented people."

"Talented!" Anna made the word sound like escaping steam. "I suppose you think it's smart to come home and throw that at your family!"

Thea picked up the tray. By this time she was as white as the Sunday tablecloth. "Well," she replied in a cold, even tone, "I'll have to throw it at them sooner or later. It's just a question of when, and it might as well be now as any time." She carried the tray blindly into the kitchen.

Tillie, who was always listening and looking out for her, took the dishes from her with a furtive, frightened glance at her stony face. Thea went slowly up the back stairs to her loft. Her legs seemed as heavy as lead as she climbed the stairs, and she felt as if everything inside her had solidified and grown hard.

After shutting her door and locking it, she sat down on the edge of her bed. This place had always been her refuge, but there was a hostility in the house now which this door could not shut out. This would be her last summer in that room. Its services were over; its time was done. She rose and put her hand on the low ceiling. Two tears ran down her cheeks, as if they came from ice that melted slowly. She was not ready to leave her little shell. She was being pulled out too soon. She would never be able to think anywhere else as well as here. She would never sleep so well or have such dreams in any other bed; even last night, such sweet, breathless dreams— Thea hid her face in the pillow. Wherever she went she would like to take that little bed with her. When she went away from it for good, she would leave something that she could never recover; memories of pleasant excitement, of happy adventures in her mind; of warm sleep on howling winter nights, and joyous awakenings on summer mornings. There were certain dreams that might refuse to come to her at all except in a little morning cave, facing the sun—where they came to her so powerfully, where they beat a triumph in her!

The room was hot as an oven. The sun was beating fiercely on the shingles behind the board ceiling. She undressed, and before she threw herself upon her bed in her chemise, she frowned at herself for a long while in her looking-glass. Yes, she and It must fight it out together. The thing that looked at her out of her own eyes was the only friend she could count on. Oh, she would make these people sorry enough! There would come a time when they would want to make it up with her. But, never again! She had no little vanities, only one big one, and she would never forgive.

Her mother was all right, but her mother was a part of the family, and she was not. In the nature of things, her mother had to be on both sides. Thea felt that she had been betrayed. A truce had been broken behind her back. She had never had much individual affection for any of her brothers except Thor, but she had never been disloyal, never felt scorn or held grudges. As a little girl she had always been good friends with Gunner and Axel, whenever she had time to play. Even before she got her own room, when they were all sleeping and dressing together, like little cubs, and breakfasting in the kitchen, she had led an absorbing personal life of her own. But she had a cub loyalty to the other cubs. She thought them nice boys and tried to make them get their lessons. She once fought a bully who "picked on" Axel at school. She never made fun of Anna's crimpings and curlings and beauty-rites.

Thea had always taken it for granted that her sister and brothers recognized that she had special abilities, and that they were proud of it. She had done them the honor, she told herself bitterly, to believe that though they had no particular endowments, *they were of her kind,* and not of the Moonstone kind. Now they had all grown up and become persons. They faced each other as individuals, and she saw that Anna and Gus and Charley were among the people whom she had always recognized as her natural enemies. Their ambitions and sacred proprieties were meaningless to her. She had neglected to congratulate Charley upon having been promoted from the grocery department of Commings's store to the drygoods department. Her mother had reproved her for this omission. And how was she to know, Thea asked herself, that Anna expected to be teased because Bert Rice now came and sat in the hammock with her every night? No, it was all clear enough. Nothing that she would ever do in the world would seem important to them, and nothing they would ever do would seem important to her.

Thea lay thinking intently all through the stifling afternoon. Tillie whispered something outside her door once, but she did not answer. She lay on her bed until the second church bell rang, and she saw the family go trooping up the sidewalk on the opposite side of the street, Anna and her father in the lead. Anna seemed to have taken on a very story-book attitude toward her father; patronizing and condescending, it seemed to Thea. The older boys were not in the family band. They now took their girls to church. Tillie had stayed at home to get supper. Thea got up, washed her hot face and arms, and put on the white organdie dress she had worn last night; it was getting too small for her, and she might as well wear it out.

After she was dressed she unlocked her door and went cautiously downstairs. She felt as if chilling hostilities might be awaiting her in the trunk loft, on the stairway, almost anywhere. In the dining-room she found Tillie, sitting by the open window, reading the dramatic news in a Denver Sunday paper. Tillie kept a scrapbook in which she pasted clippings about actors and actresses.

"Come look at this picture of Pauline Hall in tights, Thea," she called. "Ain't she cute? It's too bad you didn't go to the theater more when you was in Chicago; such a good chance! Didn't you even get to see Clara Morris or Modjeska?"[1]

"No; I didn't have time. Besides, it costs money, Tillie," Thea replied wearily, glancing at the paper Tillie held out to her.

Tillie looked up at her niece. "Don't you go and be upset about any of Anna's notions. She's one of these narrow kind. Your father and mother don't pay any attention to what she says. Anna's fussy; she is with me, but I don't mind her."

"Oh, I don't mind her. That's all right, Tillie. I guess I'll take a walk."

Thea knew that Tillie hoped she would stay and talk to her for a while, and she would have liked to please her. But in a house as small as that one, everything was too intimate and mixed up together. The family was the family, an integral thing. One couldn't discuss Anna there. She felt differently toward the house and everything in it, as if the battered old furniture that seemed so kindly, and the old carpets on which she had played, had been nourishing a secret grudge against her and were not to be trusted any more.

She went aimlessly out of the front gate, not knowing what to do with herself. Mexican Town, somehow, was spoiled for her just then, and she felt that she would hide if she saw Silvo or Felipe coming toward her. She walked down through the empty main street. All the stores were closed, their blinds down. On the steps of the bank some idle boys were sitting, telling disgusting stories because there was nothing else to do. Several of them had gone to school with Thea, but when she nodded to them they hung their heads and did not speak. Thea's body was often curiously expressive of what was going on in her mind, and to-night there was something in her walk and carriage that made these boys feel that she was "stuck up." If she had stopped and talked to them, they would have thawed out on the instant and would have been friendly and grateful. But Thea was hurt afresh, and walked on, holding her chin higher than ever. As she passed the Duke Block, she saw a light in Dr. Archie's office, and she went up the stairs and opened the door into his study. She found him with a pile of papers and account-books before him. He pointed her to her old chair at the end of his desk and leaned back in his own, looking at her with satisfaction. How handsome she was growing!

"I'm still chasing the elusive metal, Thea,"—he pointed to the papers before him,—"I'm up to my neck in mines, and I'm going to be a rich man some day."

"I hope you will; awfully rich. That's the only thing that counts." She looked restlessly about the consulting-room. "To do any of the things one wants to do, one has to have lots and lots of money."

Dr. Archie was direct. "What's the matter? Do you need some?"

Thea shrugged. "Oh, I can get along, in a little way." She looked intently out of the window at the arc street-lamp that was just beginning to sputter. "But it's silly to live at all for little things," she added quietly. "Living's too much trouble unless one can get something big out of it."

Dr. Archie rested his elbows on the arms of his chair, dropped his chin on his clasped hands and looked at her. "Living is no trouble for little people, believe me!" he exclaimed. "What do you want to get out of it?"

"Oh—so many things!" Thea shivered.

"But what? Money? You mentioned that. Well, you can make money, if you care about that more than anything else." He nodded prophetically above his interlacing fingers.

"But I don't. That's only one thing. Anyhow, I couldn't if I did." She pulled her dress lower at the neck as if she were suffocating. "I only want impossible things," she said roughly. "The others don't interest me."

Dr. Archie watched her contemplatively, as if she were a beaker full of chemicals working. A few years ago, when she used to sit there, the light from under his green lampshade used to fall full upon her broad face and yellow pigtails. Now her face was in the shadow and the line of light fell below her bare throat, directly across her bosom. The shrunken white organdie rose and fell as if she were struggling to be free and to break out of it altogether. He felt that her heart must be laboring heavily in there, but he was afraid to touch her; he was, indeed. He had never seen her like this before. Her hair, piled high on her head, gave her a commanding look, and her eyes, that used to be so inquisitive, were stormy.

"Thea," he said slowly, "I won't say that you can have everything you want—that means having nothing, in reality. But if you decide what it is you want most, *you can get it.*" His eye caught hers for a moment. "Not everybody can, but you can. Only, if you want a big thing, you've got to have nerve enough to cut out all that's easy, everything that's to be had cheap." Dr. Archie paused. He picked up a paper-cutter and, feeling the edge of it softly with his fingers, he added slowly, as if to himself:—

"He either fears his fate too much,
Or his deserts are small,
Who dares not put it to the touch
To win . . . or lose it all."[2]

Thea's lips parted; she looked at him from under a frown, searching his face. "Do you mean to break loose, too, and—do something?" she asked in a low voice.

"I mean to get rich, if you call that doing anything. I've found what I can do without. You make such bargains in your mind, first."

Thea sprang up and took the paper-cutter he had put down, twisting it in her hands. "A long while first, sometimes," she said with a short laugh. "But suppose one can never get out what they've got in them? Suppose they make a mess of it in the end; then what?" She threw the paper-cutter on the desk and took a step toward the doctor, until her dress touched him. She stood looking down at him. "Oh, it's easy to fail!" She was breathing through her mouth and her throat was throbbing with excitement.

As he looked up at her, Dr. Archie's hands tightened on the arms of his chair. He had thought he knew Thea Kronborg pretty well, but he did not know the girl who was standing there. She was beautiful, as his little Swede had never been, but she frightened him. Her pale cheeks, her parted lips, her flashing eyes, seemed suddenly to mean one thing—he did not know what. A light seemed to break upon her from far away—or perhaps from far within. She seemed to grow taller, like a scarf drawn out long; looked as if she were pursued and fleeing, and—yes, she looked tormented. "It's easy to fail," he heard her say again, "and if I fail, you'd better forget about me, for I'll be one of the worst women that ever lived. I'll be an awful woman!"

In the shadowy light above the lampshade he caught her glance again and held it for a moment. Wild as her eyes were, that yellow gleam at the back of them was as hard as a diamond drill-point. He rose with a nervous laugh and dropped his hand lightly on her shoulder. "No, you won't. You'll be a splendid one!"

She shook him off before he could say anything more, and went out of his door with a kind of bound. She left so quickly and so lightly that he could not even hear her footstep in the hallway outside. Archie dropped back into his chair and sat motionless for a long while.

So it went; one loved a quaint little girl, cheerful, industrious, always on the run and hustling through her tasks; and suddenly one lost her. He had thought he knew that child like the glove on his hand. But about this tall girl who threw up her head and glittered like that all over, he knew nothing. She was goaded by desires, ambitions, revulsions that were dark to him. One thing he knew: the old highroad of life, worn safe and easy, hugging the sunny slopes, would scarcely hold her again.

After that night Thea could have asked pretty much anything of him. He could have refused her nothing. Years ago a crafty little bunch of hair and smiles had shown him what she wanted, and he had promptly married her. To-night a very different sort of girl—driven wild by doubts and youth, by poverty and riches—had let him see the fierceness of her nature. She went out still distraught, not knowing or caring what she had shown him. But to Archie knowledge of that sort was obligation. Oh, he was the same old Howard Archie!

That Sunday in July was the turning-point; Thea's peace of mind did not come back. She found it hard even to practice at home. There was something in the air there that froze her throat. In the morning, she walked as far as she could walk. In the hot afternoons she lay on her bed in her nightgown, planning fiercely. She haunted the post-office. She must have worn a path in the sidewalk that led to the post-office, that summer. She was there the moment the mail-sacks came up from the depot, morning and evening, and while the letters were being sorted and distributed she paced up and down outside, under the cottonwood trees, listening to the thump, thump, thump of Mr. Thompson's stamp. She hung upon any sort of word from Chicago; a card from Bowers, a letter from Mrs. Harsanyi, from Mr. Larsen, from her landlady,—anything to reassure her that Chicago was still there. She began to feel the same restlessness that had tortured her the last spring when she was teaching in Moonstone. Suppose she never got away again, after all? Suppose one broke a leg and had to lie in bed at home for weeks, or had pneumonia and died there. The desert was so big and thirsty; if one's foot slipped, it could drink one up like a drop of water.

This time, when Thea left Moonstone to go back to Chicago, she went alone. As the train pulled out, she looked back at her mother and father and Thor. They were calm and cheerful; they did not know, they did not understand. Something pulled in her—and broke. She cried all the way to Denver, and that night, in her berth, she kept sobbing and waking herself. But when the sun rose in the morning, she was far away. It was all behind her, and she knew that she would never cry like that again. People live through such pain only once; pain comes again, but it finds a tougher surface. Thea remembered how she had gone away the first time, with what confidence in everything, and what pitiful ignorance. Such a silly! She felt resentful toward that stupid, good-natured child. How much older she was now, and how much harder! She was going away to fight, and she was going away forever.

PART III

STUPID FACES

I

SO MANY GRINNING, stupid faces! Thea was sitting by the window in Bowers's studio, waiting for him to come back from lunch. On her knee was the latest number of an illustrated musical journal in which musicians great and little stridently advertised their wares. Every afternoon she played accompaniments for people who looked and smiled like these. She was getting tired of the human countenance.

Thea had been in Chicago for two months. She had a small church position which partly paid her living expenses, and she paid for her singing lessons by playing Bowers's accompaniments every afternoon from two until six. She had been compelled to leave her old friends Mrs. Lorch and Mrs. Andersen, because the long ride from North Chicago to Bowers's studio on Michigan Avenue took too much time—an hour in the morning, and at night, when the cars were crowded, an hour and a half. For the first month she had clung to her old room, but the bad air in the cars, at the end of a long day's work, fatigued her greatly and was bad for her voice. Since she left Mrs. Lorch, she had been staying at a students' club to which she was introduced by Miss Adler, Bowers's morning accompanist, an intelligent Jewish girl from Evanston.

Thea took her lesson from Bowers every day from eleven-thirty until twelve. Then she went out to lunch with an Italian grammar under her arm, and came back to the studio to begin her work at two. In the afternoon Bowers coached professionals and taught his advanced pupils. It was his theory that Thea ought to be able to learn a great deal by keeping her ears open while she played for him.

The concert-going public of Chicago still remembers the long, sallow, discontented face of Madison Bowers. He seldom missed an evening concert, and was usually to be seen lounging somewhere at the back of the concert hall, reading a newspaper or review, and conspicuously ignoring the efforts of the performers. At the end of a number he looked up from his paper long enough to sweep the applauding audience with a contemptuous eye. His face was intelligent, with a narrow lower jaw, a thin nose, faded gray eyes, and a close-cut brown mustache. His hair was iron-gray, thin and dead-looking. He went to concerts chiefly to satisfy himself as to how badly things were done and how gullible the public was. He hated the whole race of artists; the work they did, the wages they got, and the way they spent their money. His father, old Hiram Bowers, was still alive and at work, a genial old choirmaster in Boston, full of enthusiasm at seventy. But Madison was of the colder stuff of his grandfathers, a long line of New Hampshire farmers; hard workers, close traders, with good minds, mean natures, and flinty eyes. As a boy Madison had a fine barytone voice, and his father made great sacrifices for him, sending him to Germany at an early age and keeping him abroad at his studies for years. Madison worked under the best teachers, and afterward sang in England in oratorio. His cold nature and academic methods were against him. His audiences were always aware of the contempt he felt for them. A dozen poorer singers succeeded, but Bowers did not.

Bowers had all the qualities which go to make a good teacher—except generosity and warmth. His intelligence was of a high order, his taste never at fault. He seldom worked with a voice without improving it, and in teaching the delivery of oratorio he was without a rival. Singers came from far and near to study Bach and Handel with him. Even the fashionable sopranos and contraltos of Chicago, St. Paul, and St. Louis (they were usually ladies with very rich husbands, and Bowers called them the "pampered jades of Asia") humbly endured his sardonic humor for the sake of what he could do for them. He was not at all above helping a very lame singer across, if her husband's check-book warranted it. He had a whole bag of tricks for stupid people, "life-preservers," he called them. "Cheap repairs for a cheap 'un," he used to say, but the husbands never found the repairs very cheap. Those were the days when lumbermen's daughters and brewers' wives contended in song; studied in Germany and then floated from *Sängerfest* to *Sängerfest.* Choral societies flourished in all the rich lake cities and river cities. The soloists came to Chicago to coach with Bowers, and he often took long journeys to hear and instruct a chorus. He was intensely avaricious, and from these semi-professionals he reaped a golden harvest. They fed his pockets and they fed his ever-hungry contempt, his scorn of himself and his accomplices. The more money he made, the more parsimonious he became. His wife was so shabby that she never went anywhere with him, which suited him exactly. Because his clients were luxurious and extravagant, he took a revengeful pleasure in having his shoes half-soled a second time, and in getting the last wear out of a broken collar. He had first been interested in Thea Kronborg because of her bluntness, her country roughness, and her manifest carefulness about money. The mention of Harsanyi's name always made him pull a wry face. For the first time Thea had a friend who, in his own cool and guarded way, liked her for whatever was least admirable in her.

Thea was still looking at the musical paper, her grammar unopened on the window-sill, when Bowers sauntered in a little before two o'clock. He was smoking a cheap cigarette and wore the same soft felt hat he had worn all last winter. He never carried a cane or wore gloves.

Thea followed him from the reception-room into the studio. "I may cut my lesson out to-morrow, Mr. Bowers. I have to hunt a new boarding-place."

Bowers looked up languidly from his desk where he had begun to go over a pile of letters. "What's the matter with the Studio Club? Been fighting with them again?"

"The Club's all right for people who like to live that way. I don't."

Bowers lifted his eyebrows. "Why so tempery?" he asked as he drew a check from an envelope postmarked "Minneapolis."

"I can't work with a lot of girls around. They're too familiar. I never could get along with girls of my own age. It's all too chummy. Gets on my nerves. I didn't come here to play kindergarten games." Thea began energetically to arrange the scattered music on the piano.

Bowers grimaced good-humoredly at her over the three checks he was pinning together. He liked to play at a rough game of banter with her. He flattered himself that he had made her harsher than she was when she first came to him; that he had got off a little of the sugar-coating Harsanyi always put on his pupils.

"The art of making yourself agreeable never comes amiss, Miss Kronborg. I should say you rather need a little practice along that line. When you come to marketing your wares in the world, a little smoothness goes farther than a great deal of talent sometimes. If you happen to be cursed with a real talent, then you've got to be very smooth indeed, or you'll never get your money back." Bowers snapped the elastic band around his bank-book.

Thea gave him a sharp, recognizing glance. "Well, that's the money I'll have to go without," she replied.

"Just what do you mean?"

"I mean the money people have to grin for. I used to know a railroad man who said there was money in every profession that you couldn't take. He'd tried a good many jobs," Thea added musingly; "perhaps he was too particular about the kind he could take, for he never picked up much. He was proud, but I liked him for that."

Bowers rose and closed his desk. "Mrs. Priest is late again. By the way, Miss Kronborg, remember not to frown when you are playing for Mrs. Priest. You did not remember yesterday."

"You mean when she hits a tone with her breath like that? Why do you let her? You wouldn't let me."

"I certainly would not. But that is a mannerism of Mrs. Priest's. The public like it, and they pay a great deal of money for the pleasure of hearing her do it. There she is. Remember!"

Bowers opened the door of the reception-room and a tall, imposing woman rustled in, bringing with her a glow of animation which pervaded the room as if half a dozen persons, all talking gayly, had come in instead of one. She was large, handsome, expansive, uncontrolled; one felt this the moment she crossed the threshold. She shone with care and cleanliness, mature vigor, unchallenged authority, gracious good-humor, and absolute confidence in her person, her powers, her position, and her way of life; a glowing, overwhelming self-satisfaction, only to be found where human society is young and strong and without yesterdays. Her face had a kind of heavy, thoughtless beauty, like a pink peony just at the point of beginning to fade. Her brown hair was waved in front and done up behind in a great twist, held by a tortoiseshell comb with gold filigree. She wore a beautiful little green hat with three long green feathers sticking straight up in front, a little cape made of velvet and fur with a yellow satin rose on it. Her gloves, her shoes, her veil, somehow made themselves felt. She gave the impression of wearing a cargo of splendid merchandise.

Mrs. Priest nodded graciously to Thea, coquettishly to Bowers, and asked him to untie her veil for her. She threw her splendid wrap on a chair, the yellow lining out. Thea was already at the piano. Mrs. Priest stood behind her.

" 'Rejoice Greatly'[1] first, please. And please don't hurry it in there," she put her arm over Thea's shoulder, and indicated the passage by a sweep of her white glove. She threw out her chest, clasped her hands over her abdomen, lifted her chin, worked the muscles of her cheeks back and forth for a moment, and then began with conviction, "Re-jo-oice! Re-jo-oice!"

Bowers paced the room with his catlike tread. When he checked Mrs. Priest's vehemence at all, he handled her roughly; poked and hammered her massive person with cold satisfaction, almost as if he were taking out a grudge on this splendid creation. Such treatment the imposing lady did not at all resent. She tried harder and harder, her eyes growing all the while more lustrous and her lips redder. Thea played on as she was told, ignoring the singer's struggles.

When she first heard Mrs. Priest sing in church, Thea admired her. Since she had found out how dull the good-natured soprano really was, she felt a deep contempt for her. She felt that Mrs. Priest ought to be reproved and even punished for her shortcomings; that she ought to be exposed,—at least to herself,—and not be permitted to live and shine in happy ignorance of what a poor thing it was she brought across so radiantly. Thea's cold looks of reproof were lost upon Mrs. Priest; although the lady did murmur one day when she took Bowers home in her carriage, "How handsome your afternoon girl would be if she did not have that unfortunate squint; it gives her that vacant Swede look, like an animal."[2] That amused Bowers. He liked to watch the germination and growth of antipathies.

One of the first disappointments Thea had to face when she returned to Chicago that fall, was the news that the Harsanyis were not coming back. They had spent the summer in a camp in the Adirondacks and were moving to New York. An old teacher and friend of Harsanyi's, one of the best-known piano teachers in New York, was about to retire because of failing health and had arranged to turn his pupils over to Harsanyi. Andor was to give two recitals in New York in November, to devote himself to his new students until spring, and then to go on a short concert tour. The Harsanyis had taken a furnished apartment in New York, as they would not attempt to settle a place of their own until Andor's recitals were over. The first of December, however, Thea received a note from Mrs. Harsanyi, asking her to call at the old studio, where she was packing their goods for shipment.

The morning after this invitation reached her, Thea climbed the stairs and knocked at the familiar door. Mrs. Harsanyi herself opened it, and embraced her visitor warmly. Taking Thea into the studio, which was littered with excelsior and packing-cases, she stood holding her hand and looking at her in the strong light from the big window before she allowed her to sit down. Her quick eye saw many changes. The girl was taller, her figure had become definite, her carriage positive. She had got used to living in the body of a young woman, and she no longer tried to ignore it and behave as if she were a little girl. With that increased independence of body there had come a change in her face; an indifference, something hard and skeptical. Her clothes, too, were different, like the attire of a shopgirl who tries to follow the fashions; a purple suit, a piece of cheap fur, a three-cornered purple hat with a pompon sticking up in front. The queer country clothes she used to wear suited her much better, Mrs. Harsanyi thought. But such trifles, after all, were accidental and remediable. She put her hand on the girl's strong shoulder.

"How much the summer has done for you! Yes, you are a young lady at last. Andor will be so glad to hear about you."

Thea looked about at the disorder of the familiar room. The pictures were piled in a corner, the piano and the *chaise longue* were gone. "I suppose I ought to be glad you have gone away," she said, "but I'm not. It's a fine thing for Mr. Harsanyi, I suppose."

Mrs. Harsanyi gave her a quick glance that said more than words. "If you knew how long I have wanted to get him away from here, Miss Kronborg! He is never tired, never discouraged, now."

Thea sighed. "I'm glad for that, then." Her eyes traveled over the faint discolorations on the walls where the pictures had hung. "I may run away myself. I don't know whether I can stand it here without you."

"We hope that you can come to New York to study before very long. We have thought of that. And you must tell me how you are getting on with Bowers. Andor will want to know all about it."

"I guess I get on more or less. But I don't like my work very well. It never seems serious as my work with Mr. Harsanyi did. I play Bowers's accompaniments in the afternoons, you know. I thought I would learn a good deal from the people who work with him, but I don't think I get much."

Mrs. Harsanyi looked at her inquiringly. Thea took out a carefully folded handkerchief from the bosom of her dress and began to draw the corners apart. "Singing doesn't seem to be a very brainy profession, Mrs. Harsanyi," she said slowly. "The people I see now are not a bit like the ones I used to meet here. Mr. Harsanyi's pupils, even the dumb ones, had more—well, more of everything, it seems to me. The people I have to play accompaniments for are discouraging. The professionals, like Katharine Priest and Miles Murdstone, are worst of all. If I have to play 'The Messiah' much longer for Mrs. Priest, I'll go out of my mind!" Thea brought her foot down sharply on the bare floor.

Mrs. Harsanyi looked down at the foot in perplexity. "You mustn't wear such high heels, my dear. They will spoil your walk and make you mince along. Can't you at least learn to avoid what you dislike in these singers? I was never able to care for Mrs. Priest's singing."

Thea was sitting with her chin lowered. Without moving her head she looked up at Mrs. Harsanyi and smiled; a smile much too cold and desperate to be seen on a young face, Mrs. Harsanyi felt. "Mrs. Harsanyi, it seems to me that what I learn is just *to dislike.* I dislike so much and so hard that it tires me out. I've got no heart for anything." She threw up her head suddenly and sat in defiance, her hand clenched on the arm of the chair. "Mr. Harsanyi couldn't stand these people an hour, I know he couldn't. He'd put them right out of the window there, frizzes and feathers and all. Now, take that new soprano they're all making such a fuss about, Jessie Darcey. She's going on tour with a symphony orchestra and she's working up her repertory with Bowers. She's singing some Schumann songs Mr. Harsanyi used to go over with me. Well, I don't know what he *would* do if he heard her."

"But if your own work goes well, and you know these people are wrong, why do you let them discourage you?"

Thea shook her head. "That's just what I don't understand myself. Only, after I've heard them all afternoon, I come out frozen up. Somehow it takes the shine off of everything. People want Jessie Darcey and the kind of thing she does; so what's the use?"

Mrs. Harsanyi smiled. "That stile you must simply vault over. You must not begin to fret about the successes of cheap people. After all, what have they to do with you?"

"Well, if I had somebody like Mr. Harsanyi, perhaps I wouldn't fret about them. He was the teacher for me. Please tell him so."

Thea rose and Mrs. Harsanyi took her hand again. "I am sorry you have to go through this time of discouragement. I wish Andor could talk to you, he would understand it so well. But I feel like urging you to keep clear of Mrs. Priest and Jessie Darcey and all their works."

Thea laughed discordantly. "No use urging me. I don't get on with them *at all.* My spine gets like a steel rail when they come near me. I liked them at first, you know. Their clothes and their manners were so fine, and Mrs. Priest *is* handsome. But now I keep wanting to tell them how stupid they are. Seems like they ought to be informed, don't you think so?" There was a flash of the shrewd grin that Mrs. Harsanyi remembered. Thea pressed her hand. "I must go now. I had to give my lesson hour this morning to a Duluth woman who has come on to coach, and I must go and play 'On Mighty Pens'[3] for her. Please tell Mr. Harsanyi that I think oratorio is a great chance for bluffers."

Mrs. Harsanyi detained her. "But he will want to know much more than that about you. You are free at seven? Come back this evening, then, and we will go to dinner somewhere, to some cheerful place. I think you need a party."

Thea brightened. "Oh, I do! I'll love to come; that will be like old times. You see," she lingered a moment, softening, "I wouldn't mind if there were only *one* of them I could really admire."

"How about Bowers?" Mrs. Harsanyi asked as they were approaching the stairway.

"Well, there's nothing he loves like a good fakir, and nothing he hates like a good artist. I always remember something Mr. Harsanyi said about him. He said Bowers was the cold muffin that had been left on the plate."

Mrs. Harsanyi stopped short at the head of the stairs and said decidedly: "I think Andor made a mistake. I can't believe that is the right atmosphere for you. It would hurt you more than most people. It's all wrong."

"Something's wrong," Thea called back as she clattered down the stairs in her high heels.

II

DURING THAT WINTER Thea lived in so many places that sometimes at night when she left Bowers's studio and emerged into the street she had to stop and think for a moment to remember where she was living now and what was the best way to get there.

When she moved into a new place her eyes challenged the beds, the carpets, the food, the mistress of the house. The boarding-houses were wretchedly conducted and Thea's complaints sometimes took an insulting form. She quarreled with one landlady after another and moved on. When she moved into a new room, she was almost sure to hate it on sight and to begin planning to hunt another place before she unpacked her trunk. She was moody and contemptuous toward her fellow boarders, except toward the young men, whom she treated with a careless familiarity which they usually misunderstood. They liked her, however, and when she left the house after a storm, they helped her to move her things and came to see her after she got settled in a new place. But she moved so often that they soon ceased to follow her. They could see no reason for keeping up with a girl who, under her jocularity, was cold, self-centered, and unimpressionable. They soon felt that she did not admire them.

Thea used to waken up in the night and wonder why she was so unhappy. She would have been amazed if she had known how much the people whom she met in Bowers's studio had to do with her low spirits. She had never been conscious of those instinctive standards which are called ideals, and she did not know that she was suffering for them. She often found herself sneering when she was on a street-car, or when she was brushing out her hair before her mirror, as some inane remark or too familiar mannerism flitted across her mind.

She felt no creature kindness, no tolerant good-will for Mrs.

Priest or Jessie Darcey. After one of Jessie Darcey's concerts the glowing press notices, and the admiring comments that floated about Bowers's studio, caused Thea bitter unhappiness. It was not the torment of personal jealousy. She had never thought of herself as even a possible rival of Miss Darcey. She was a poor music student, and Jessie Darcey was a popular and petted professional. Mrs. Priest, whatever one held against her, had a fine, big, showy voice and an impressive presence. She read indifferently, was inaccurate, and was always putting other people wrong, but she at least had the material out of which singers can be made. But people seemed to like Jessie Darcey exactly because she could not sing; because, as they put it, she was "so natural and unprofessional." Her singing was pronounced "artless," her voice "birdlike." Miss Darcey was thin and awkward in person, with a sharp, sallow face. Thea noticed that her plainness was accounted to her credit, and that people spoke of it affectionately. Miss Darcey was singing everywhere just then; one could not help hearing about her. She was backed by some of the packing-house people and by the Chicago Northwestern Railroad. Only one critic raised his voice against her. Thea went to several of Jessie Darcey's concerts. It was the first time she had had an opportunity to observe the whims of the public which singers live by interesting. She saw that people liked in Miss Darcey every quality a singer ought not to have, and especially the nervous complacency that stamped her as a commonplace young woman. They seemed to have a warmer feeling for Jessie than for Mrs. Priest, an affectionate and cherishing regard. Chicago was not so very different from Moonstone, after all, and Jessie Darcey was only Lily Fisher under another name.

Thea particularly hated to accompany for Miss Darcey because she sang off pitch and didn't mind it in the least. It was excruciating to sit there day after day and hear her; there was something shameless and indecent about not singing true.

One morning Miss Darcey came by appointment to go over the programme for her Peoria concert. She was such a frail-looking girl that Thea ought to have felt sorry for her. True, she had an arch, sprightly little manner, and a flash of salmon-pink on either brown cheek. But a narrow upper jaw gave her face a pinched look, and her eyelids were heavy and relaxed. By the morning light, the purplish brown circles under her eyes were pathetic enough, and foretold no long or brilliant future. A singer with a poor digestion and low vitality; she needed no seer to cast her horoscope. If Thea had ever taken the pains to study her, she would have seen that, under all her smiles and archness, poor Miss Darcey was really frightened to death. She could not understand her success any more than Thea could; she kept catching her breath and lifting her eyebrows and trying to believe that it was true. Her loquacity was not natural, she forced herself to it, and when she confided to you how many defects she could overcome by her unusual command of head resonance, she was not so much trying to persuade you as to persuade herself.

When she took a note that was high for her, Miss Darcey always put her right hand out into the air, as if she were indicating height, or giving an exact measurement. Some early teacher had told her that she could "place" a tone more surely by the help of such a gesture, and she firmly believed that it was of great assistance to her. (Even when she was singing in public, she kept her right hand down with difficulty, nervously clasping her white kid fingers together when she took a high note. Thea could always see her elbows stiffen.) She unvaryingly executed this gesture with a smile of gracious confidence, as if she were actually putting her finger on the tone: "There it is, friends!"

This morning, in Gounod's "Ave Maria," as Miss Darcey approached her B natural,—

Dans —— nos a-lár — — — mes![1]

out went the hand, with the sure airy gesture, though it was little above A she got with her voice, whatever she touched with her finger. Often Bowers let such things pass—with the right people—but this morning he snapped his jaws together and muttered, "God!" Miss Darcey tried again, with the same gesture as of putting the crowning touch, tilting her head and smiling radiantly at Bowers, as if to say, "It is for you I do all this!"

Dans —— nos a — lár — — — mes!

This time she made B flat, and went on in the happy belief that she had done well enough, when she suddenly found that her accompanist was not going on with her, and this put her out completely.

She turned to Thea, whose hands had fallen in her lap. "Oh why did you stop just there! It *is* too trying! Now we'd better go back to that other *crescendo* and try it from there."

"I beg your pardon," Thea muttered. "I thought you wanted to get that B natural." She began again, as Miss Darcey indicated.

After the singer was gone, Bowers walked up to Thea and asked languidly, "Why do you hate Jessie so? Her little variations from pitch are between her and her public; they don't hurt you. Has she ever done anything to you except be very agreeable?"

"Yes, she has done things to me," Thea retorted hotly.

Bowers looked interested. "What, for example?"

"I can't explain, but I've got it in for her."

Bowers laughed. "No doubt about that. I'll have to suggest that you conceal it a little more effectually. That is—necessary, Miss Kronborg," he added, looking back over the shoulder of the overcoat he was putting on.

He went out to lunch and Thea thought the subject closed. But late in the afternoon, when he was taking his dyspepsia tablet and a glass of water between lessons, he looked up and said in a voice ironically coaxing:—

"Miss Kronborg, I wish you would tell me why you hate Jessie."

Taken by surprise Thea put down the score she was reading and answered before she knew what she was saying, "I hate her for the sake of what I used to think a singer might be."

Bowers balanced the tablet on the end of his long forefinger and whistled softly. "And how did you form your conception of what a singer ought to be?" he asked.

"I don't know." Thea flushed and spoke under her breath; "but I suppose I got most of it from Harsanyi."

Bowers made no comment upon this reply, but opened the door for the next pupil, who was waiting in the reception-room.

It was dark when Thea left the studio that night. She knew she had offended Bowers. Somehow she had hurt herself, too. She felt unequal to the boarding-house table, the sneaking divinity student who sat next her and had tried to kiss her on the stairs last night. She went over to the waterside of Michigan Avenue and walked along beside the lake. It was a clear, frosty winter night. The great empty space over the water was restful and spoke of freedom. If she had any money at all, she would go away. The stars glittered over the wide black water. She looked up at them wearily and shook her head. She believed that what she felt was despair, but it was only one of the forms of hope. She felt, indeed, as if she were bidding the stars good-bye; but she was renewing a promise. Though their challenge is universal and eternal, the stars get no answer but that,—the brief light flashed back to them from the eyes of the young who unaccountably aspire.

The rich, noisy city, fat with food and drink, is a spent thing; its chief concern is its digestion and its little game of hide-and-seek with the undertaker. Money and office and success are the consolations of impotence. Fortune turns kind to such solid people and lets them suck their bone in peace. She flecks her whip upon flesh that is more alive, upon that stream of hungry boys and girls who tramp the streets of every city, recognizable by their pride and discontent, who are the Future, and who possess the treasure of creative power.

III

WHILE HER LIVING ARRANGEMENTS were so casual and fortuitous, Bowers's studio was the one fixed thing in Thea's life. She went out from it to uncertainties, and hastened to it from nebulous confusion. She was more influenced by Bowers than she knew. Unconsciously she began to take on something of his dry contempt, and to share his grudge without understanding exactly what it was about. His cynicism seemed to her honest, and the amiability of his pupils artificial. She admired his drastic treatment of his dull pupils. The stupid deserved all they got, and more. Bowers knew that she thought him a very clever man.

One afternoon when Bowers came in from lunch Thea handed him a card on which he read the name, "Mr. Philip Frederick Ottenburg."

"He said he would be in again to-morrow and that he wanted some time. Who is he? I like him better than the others."

Bowers nodded. "So do I. He's not a singer. He's a beer prince: son of the big brewer in St. Louis. He's been in Germany with his mother. I didn't know he was back."

"Does he take lessons?"

"Now and again. He sings rather well. He's at the head of the Chicago branch of the Ottenburg business, but he can't stick to work and is always running away. He has great ideas in beer, people tell me. He's what they call an imaginative business man; goes over to Bayreuth[1] and seems to do nothing but give parties and spend money, and brings back more good notions for the brewery than the fellows who sit tight dig out in five years. I was born too long ago to be much taken in by these chesty boys with flowered vests, but I like Fred, all the same."

"So do I," said Thea positively.

Bowers made a sound between a cough and a laugh. "Oh, he's a lady-killer, all right! The girls in here are always making eyes at him. You won't be the first." He threw some sheets of music on the piano. "Better look that over; accompaniment's a little tricky. It's for that new woman from Detroit. And Mrs. Priest will be in this afternoon."

Thea sighed. " 'I Know that my Redeemer Liveth'?"[2]

"The same. She starts on her concert tour next week, and we'll have a rest. Until then, I suppose we'll have to be going over her programme."

The next day Thea hurried through her luncheon at a German bakery and got back to the studio at ten minutes past one. She felt sure that the young brewer would come early, before it was time for Bowers to arrive. He had not said he would, but yesterday, when he opened the door to go, he had glanced about the room and at her, and something in his eye had conveyed that suggestion.

Sure enough, at twenty minutes past one the door of the reception-room opened, and a tall, robust young man with a cane and an English hat and ulster looked in expectantly. "Ah—ha!" he exclaimed, "I thought if I came early I might have good luck. And how are you to-day, Miss Kronborg?"

Thea was sitting in the window chair. At her left elbow there was a table, and upon this table the young man sat down, holding his hat and cane in his hand, loosening his long coat so that it fell back from his shoulders. He was a gleaming, florid young fellow. His hair, thick and yellow, was cut very short, and he wore a closely trimmed beard, long enough on the chin to curl a little. Even his eyebrows were thick and yellow, like fleece. He had lively blue eyes—Thea looked up at them with great interest as he sat chatting and swinging his foot rhythmically. He was easily familiar, and frankly so. Wherever people met young Ottenburg, in his office, on shipboard, in a foreign hotel or railway compartment, they always felt (and usually liked) that artless presumption which seemed to say, "In this case we may waive formalities. We really haven't time. This is to-day, but it will soon be to-morrow, and then we may be very different people, and in some other country." He had a way of floating people out of dull or awkward situations, out of their own torpor or constraint or discouragement. It was a marked personal talent, of almost incalculable value in the representative of a great business founded on social amenities. Thea had liked him yesterday for the way in which he had picked her up out of herself and her German grammar for a few exciting moments.

"By the way, will you tell me your first name, please? Thea? Oh, then you *are* a Swede, sure enough! I thought so. Let me call you Miss Thea, after the German fashion. You won't mind? Of course not!" He usually made his assumption of a special understanding seem a tribute to the other person and not to himself.

"How long have you been with Bowers here? Do you like the old grouch? So do I. I've come to tell him about a new soprano I heard at Bayreuth. He'll pretend not to care, but he does. Do you warble with him? Have you anything of a voice? Honest? You look it, you know. What are you going in for, something big? Opera?"

Thea blushed crimson. "Oh, I'm not going in for anything. I'm trying to learn to sing at funerals."

Ottenburg leaned forward. His eyes twinkled. "I'll engage you to sing at mine. You can't fool me, Miss Thea. May I hear you take your lesson this afternoon?"

"No, you may not. I took it this morning."

He picked up a roll of music that lay behind him on the table. "Is this yours? Let me see what you are doing." He snapped back the clasp and began turning over the songs. "All very fine, but tame. What's he got you at this Mozart stuff for? I shouldn't think it would suit your voice. Oh, I can make a pretty good guess at what will suit you! This from 'Gioconda'[3] is more in your line. What's this Grieg?[4] It looks interesting. *Tak for Dit Råd.* What does that mean?"

" 'Thanks for your Advice.' Don't you know it?"

"No; not at all. Let's try it." He rose, pushed open the door into the music-room, and motioned Thea to enter before him. She hung back.

"I couldn't give you much of an idea of it. It's a big song."

Ottenburg took her gently by the elbow and pushed her into the other room. He sat down carelessly at the piano and looked over the music for a moment. "I think I can get you through it. But how stupid not to have the German words. Can you really sing the Norwegian? What an infernal language to sing. Translate the text for me." He handed her the music.

Thea looked at it, then at him, and shook her head. "I can't. The truth is I don't know either English or Swedish very well, and Norwegian's still worse," she said confidentially. She not infrequently refused to do what she was asked to do, but it was not like her to explain her refusal, even when she had a good reason.

"I understand. We immigrants never speak any language well. But you know what it means, don't you?"

"Of course I do!"

"Then don't frown at me like that, but tell me."

Thea continued to frown, but she also smiled. She was confused, but not embarrassed. She was not afraid of Ottenburg. He was not one of those people who made her spine like a steel rail. On the contrary, he made one venturesome.

"Well, it goes something like this: *Thanks for your advice! But I prefer to steer my boat into the din of roaring breakers. Even if the journey is my last, I may find what I have never found before. Onward must I go, for I yearn for the wild sea. I long to fight my way through the angry waves, and to see how far, and how long I can make them carry me.*"

Ottenburg took the music and began: "Wait a moment. Is that too fast? How do you take it? That right?" He pulled up his cuffs and began the accompaniment again. He had become entirely serious, and he played with fine enthusiasm and with understanding.

Fred's talent was worth almost as much to old Otto Ottenburg as the steady industry of his older sons. When Fred sang the Prize Song[5] at an interstate meet of the *Turnverein,*[6] ten thousand *Turners* went forth pledged to Ottenburg beer.

As Thea finished the song Fred turned back to the first page, without looking up from the music. "Now, once more," he called. They began again, and did not hear Bowers when he came in and stood in the doorway. He stood still, blinking like an owl at their two heads shining in the sun. He could not see their faces, but there was something about his girl's back that he had not noticed before: a very slight and yet very free motion, from the toes up. Her whole back seemed plastic, seemed to be moulding itself to the galloping rhythm of the song. Bowers perceived such things sometimes—unwillingly. He had known to-day that there was something afoot. The river of sound which had its source in his pupil had caught him two flights down. He had stopped and listened with a kind of sneering admiration. From the door he watched her with a half-incredulous, half-malicious smile.

When he had struck the keys for the last time, Ottenburg

dropped his hands on his knees and looked up with a quick breath. "I got you through. What a stunning song! Did I play it right?"

Thea studied his excited face. There was a good deal of meaning in it, and there was a good deal in her own as she answered him. "You suited me," she said ungrudgingly.

After Ottenburg was gone, Thea noticed that Bowers was more agreeable than usual. She had heard the young brewer ask Bowers to dine with him at his club that evening, and she saw that he looked forward to the dinner with pleasure. He dropped a remark to the effect that Fred knew as much about food and wines as any man in Chicago. He said this boastfully.

"If he's such a grand business man, how does he have time to run around listening to singing-lessons?" Thea asked suspiciously.

As she went home to her boarding-house through the February slush, she wished she were going to dine with them. At nine o'clock she looked up from her grammar to wonder what Bowers and Ottenburg were having to eat. At that moment they were talking of her.

IV

THEA NOTICED that Bowers took rather more pains with her now that Fred Ottenburg often dropped in at eleven-thirty to hear her lesson. After the lesson the young man took Bowers off to lunch with him, and Bowers liked good food when another man paid for it. He encouraged Fred's visits, and Thea soon saw that Fred knew exactly why.

One morning, after her lesson, Ottenburg turned to Bowers. "If you'll lend me Miss Thea, I think I have an engagement for her. Mrs. Henry Nathanmeyer is going to give three musical evenings in April, first three Saturdays, and she has consulted me about soloists. For the first evening she has a young violinist, and she would be charmed to have Miss Kronborg. She will pay fifty dollars. Not much, but Miss Thea would meet some people there who might be useful. What do you say?"

Bowers passed the question on to Thea. "I guess you could use the fifty, couldn't you, Miss Kronborg? You can easily work up some songs."

Thea was perplexed. "I need the money awfully," she said frankly; "but I haven't got the right clothes for that sort of thing. I suppose I'd better try to get some."

Ottenburg spoke up quickly, "Oh, you'd make nothing out of it if you went to buying evening clothes. I've thought of that. Mrs. Nathanmeyer has a troop of daughters, a perfect seraglio, all ages and sizes. She'll be glad to fit you out, if you aren't sensitive about wearing kosher clothes.[1] Let me take you to see her, and you'll find that she'll arrange that easily enough. I told her she must produce something nice, blue or yellow, and properly cut. I brought half a dozen Worth gowns[2] through the customs for her two weeks ago, and she's not ungrateful. When can we go to see her?"

"I haven't any time free, except at night," Thea replied in some confusion.

"To-morrow evening, then? I shall call for you at eight. Bring all your songs along; she will want us to give her a little rehearsal, perhaps. I'll play your accompaniments, if you've no objection. That will save money for you and for Mrs. Nathanmeyer. She needs it." Ottenburg chuckled as he took down the number of Thea's boarding-house.

The Nathanmeyers were so rich and great that even Thea had heard of them, and this seemed a very remarkable opportunity. Ottenburg had brought it about by merely lifting a finger, apparently. He was a beer prince sure enough, as Bowers had said.

The next evening at a quarter to eight Thea was dressed and waiting in the boarding-house parlor. She was nervous and fidgety and found it difficult to sit still on the hard, convex upholstery of the chairs. She tried them one after another, moving about the dimly lighted, musty room, where the gas always leaked gently and sang in the burners. There was no one in the parlor but the medical student, who was playing one of Sousa's marches so vigorously that the china ornaments on the top of the piano rattled. In a few moments some of the pension-office girls would come in and begin to two-step. Thea wished that Ottenburg would come and let her escape. She glanced at herself in the long, somber mirror. She was wearing her pale-blue broadcloth church dress, which was not unbecoming but was certainly too heavy to wear to anybody's house in the evening. Her slippers were run over at the heel and she had not had time to have them mended, and her white gloves were not so clean as they should be. However, she knew that she would forget these annoying things as soon as Ottenburg came.

Mary, the Hungarian chambermaid, came to the door, stood between the plush portières, beckoned to Thea, and made an inarticulate sound in her throat. Thea jumped up and ran into the hall, where Ottenburg stood smiling, his caped cloak open, his silk hat in his white-kid hand. The Hungarian girl stood like a monument on her flat heels, staring at the pink carnation in Ottenburg's coat. Her broad, pockmarked face wore the only expression of which it was capable, a kind of animal wonder. As the young man followed Thea out, he glanced back over his shoulder through the crack of the door; the Hun clapped her hands over her stomach, opened her mouth, and made another raucous sound in her throat.

"Isn't she awful?" Thea exclaimed. "I think she's half-witted. Can you understand her?"

Ottenburg laughed as he helped her into the carriage. "Oh, yes; I can understand her!" He settled himself on the front seat opposite Thea. "Now, I want to tell you about the people we are going to see. We may have a musical public in this country some day, but as yet there are only the Germans and the Jews. All the other people go to hear Jessie Darcey sing, 'O, Promise Me!'[3] The Nathanmeyers are the finest kind of Jews. If you do anything for Mrs. Henry Nathanmeyer, you must put yourself into her hands. Whatever she says about music, about clothes, about life, will be correct. And you may feel at ease with her. She expects nothing of people; she has lived in Chicago twenty years. If you were to behave like the Magyar who was so interested in my buttonhole, she would not be surprised. If you were to sing like Jessie Darcey, she would not be surprised; but she would manage not to hear you again."

"Would she? Well, that's the kind of people I want to find." Thea felt herself growing bolder.

"You will be all right with her so long as you do not try to be anything that you are not. Her standards have nothing to do with Chicago. Her perceptions—or her grandmother's, which is the same thing—were keen when all this was an Indian village. So merely be yourself, and you will like her. She will like you because the Jews always sense talent, and," he added ironically, "they admire certain qualities of feeling that are found only in the white-skinned races."

Thea looked into the young man's face as the light of a street lamp flashed into the carriage. His somewhat academic manner amused her.

"What makes you take such an interest in singers?" she asked curiously. "You seem to have a perfect passion for hearing music-lessons. I wish I could trade jobs with you!"

"I'm not interested in singers." His tone was offended. "I am interested in talent. There are only two interesting things in the world, anyhow; and talent is one of them."

"What's the other?" The question came meekly from the figure opposite him. Another arc-light flashed in at the window.

Fred saw her face and broke into a laugh. "Why, you're guying me, you little wretch! You won't let me behave properly." He dropped his gloved hand lightly on her knee, took it away and let it hang between his own. "Do you know," he said confidentially, "I believe I'm more in earnest about all this than you are."

"About all what?"

"All you've got in your throat there."

"Oh! I'm in earnest all right; only I never was much good at talking. Jessie Darcey is the smooth talker. 'You notice the effect I get there—' If she only got 'em, she'd be a wonder, you know!"

Mr. and Mrs. Nathanmeyer were alone in their great library. Their three unmarried daughters had departed in successive carriages, one to a dinner, one to a Nietzsche club,[4] one to a ball given for the girls employed in the big department stores. When Ottenburg and Thea entered, Henry Nathanmeyer and his wife were sitting at a table at the farther end of the long room, with a reading-lamp and a tray of cigarettes and cordial-glasses between them. The overhead lights were too soft to bring out the colors of the big rugs, and none of the picture lights were on. One could merely see that there were pictures there. Fred whispered that they were Rousseaus and Corots, very fine ones which the old banker had bought long ago for next to nothing. In the hall Ottenburg had stopped Thea before a painting of a woman eating grapes out of a paper bag, and had told her gravely that there was the most beautiful Manet in the world.[5] He made her take off her hat and gloves in the hall, and looked her over a little before he took her in. But once they were in the library he seemed perfectly satisfied with her and led her down the long room to their hostess.

Mrs. Nathanmeyer was a heavy, powerful old Jewess, with a great pompadour of white hair, a swarthy complexion, an eagle nose, and sharp, glittering eyes. She wore a black velvet dress with a long train, and a diamond necklace and earrings. She took Thea to the other side of the table and presented her to Mr. Nathanmeyer, who apologized for not rising, pointing to a slippered foot on a cushion; he said that he suffered from gout. He had a very soft voice and spoke with an accent which would have been heavy if it had not been so caressing. He kept Thea standing beside him for some time. He noticed that she stood easily, looked straight down into his face, and was not embarrassed. Even when Mrs. Nathanmeyer told Ottenburg to bring a chair for Thea, the old man did not release her hand, and she did not sit down. He admired her just as she was, as she happened to be standing, and she felt it. He was much handsomer than his wife, Thea thought. His forehead was high, his hair soft and white, his skin pink, a little puffy under his clear blue eyes. She noticed how warm and delicate his hands were, pleasant to touch and beautiful to look at. Ottenburg had told her that Mr. Nathanmeyer had a very fine collection of medals and cameos, and his fingers looked as if they had never touched anything but delicately cut surfaces.

He asked Thea where Moonstone was; how many inhabitants it had; what her father's business was; from what part of Sweden her grandfather came; and whether she spoke Swedish as a child. He was interested to hear that her mother's mother was still living, and that her grandfather had played the oboe. Thea felt at home standing there beside him; she felt that he was very wise, and that he some way took one's life up and looked it over kindly, as if it were a story. She was sorry when they left him to go into the music-room.

As they reached the door of the music-room, Mrs. Nathanmeyer turned a switch that threw on many lights. The room was even larger than the library, all glittering surfaces, with two Steinway pianos.

Mrs. Nathanmeyer rang for her own maid. "Selma will take you upstairs, Miss Kronborg, and you will find some dresses on the bed. Try several of them, and take the one you like best. Selma will help you. She has a great deal of taste. When you are dressed, come down and let us go over some of your songs with Mr. Ottenburg."

After Thea went away with the maid, Ottenburg came up to Mrs. Nathanmeyer and stood beside her, resting his hand on the high back of her chair.

"Well, *gnädige Frau,*[6] do you like her?"

"I think so. I liked her when she talked to father. She will always get on better with men."

Ottenburg leaned over her chair. "Prophetess! Do you see what I meant?"

"About her beauty? She has great possibilities, but you can never tell about those Northern women. They look so strong, but they are easily battered. The face falls so early under those wide cheekbones. A single idea—hate or greed, or even love—can tear them to shreds. She is nineteen? Well, in ten years she may have quite a regal beauty, or she may have a heavy, discontented face, all dug out in channels. That will depend upon the kind of ideas she lives with."

"Or the kind of people?" Ottenburg suggested.

The old Jewess folded her arms over her massive chest, drew back her shoulders, and looked up at the young man. "With that hard glint in her eye? The people won't matter much, I fancy. They will come and go. She is very much interested in herself—as she should be."

Ottenburg frowned. "Wait until you hear her sing. Her eyes are different then. That gleam that comes in them is curious, isn't it? As you say, it's impersonal."

The object of this discussion came in, smiling. She had chosen neither the blue nor the yellow gown, but a pale rose-color, with silver butterflies. Mrs. Nathanmeyer lifted her lorgnette and studied her as she approached. She caught the characteristic things at once: the free, strong walk, the calm carriage of the head, the milky whiteness of the girl's arms and shoulders.

"Yes, that color is good for you," she said approvingly. "The yellow one probably killed your hair? Yes; this does very well indeed, so we need think no more about it."

Thea glanced questioningly at Ottenburg. He smiled and bowed, seemed perfectly satisfied. He asked her to stand in the elbow of the piano, in front of him, instead of behind him as she had been taught to do.

"Yes," said the hostess with feeling. "That other position is barbarous."

Thea sang an aria from 'Gioconda,' some songs by Schumann which she had studied with Harsanyi, and the *"Tak for Dit Råd,"* which Ottenburg liked.

"That you must do again," he declared when they finished this song. "You did it much better the other day. You accented it more, like a dance or a galop. How did you do it?"

Thea laughed, glancing sidewise at Mrs. Nathanmeyer. "You want it rough-house, do you? Bowers likes me to sing it more seriously, but it always makes me think about a story my grandmother used to tell."

Fred pointed to the chair behind her. "Won't you rest a moment and tell us about it? I thought you had some notion about it when you first sang it for me."

Thea sat down. "In Norway my grandmother knew a girl who was awfully in love with a young fellow. She went into service on a big dairy farm to make enough money for her outfit. They were married at Christmas-time, and everybody was glad, because they'd been sighing around about each other for so long. That very summer, the day before St. John's Day,[7] her husband caught her carrying on with another farm-hand. The next night all the farm people had a bonfire and a big dance up on the mountain, and everybody was dancing and singing. I guess they were all a little drunk, for they got to seeing how near they could make the girls dance to the edge of the cliff. Ole—he was the girl's husband—seemed the jolliest and the drunkest of anybody. He danced his wife nearer and nearer the edge of the rock, and his wife began to scream so that the others stopped dancing and the music stopped; but Ole went right on singing, and he danced her over the edge of the cliff and they fell hundreds of feet and were all smashed to pieces."

Ottenburg turned back to the piano. "That's the idea! Now, come Miss Thea. Let it go!"

Thea took her place. She laughed and drew herself up out of her corsets, threw her shoulders high and let them drop again. She had never sung in a low dress before, and she found it comfortable. Ottenburg jerked his head and they began the song. The accompaniment sounded more than ever like the thumping and scraping of heavy feet.

When they stopped, they heard a sympathetic tapping at the end of the room. Old Mr. Nathanmeyer had come to the door and was sitting back in the shadow, just inside the library, applauding with his cane. Thea threw him a bright smile. He continued to sit there, his slippered foot on a low chair, his cane between his fingers, and she glanced at him from time to time. The doorway made a frame for him, and he looked like a man in a picture, with the long, shadowy room behind him.

Mrs. Nathanmeyer summoned the maid again. "Selma will pack that gown in a box for you, and you can take it home in Mr. Ottenburg's carriage."

Thea turned to follow the maid, but hesitated. "Shall I wear gloves?" she asked, turning again to Mrs. Nathanmeyer.

"No, I think not. Your arms are good, and you will feel freer without. You will need light slippers, pink—or white, if you have them, will do quite as well."

Thea went upstairs with the maid and Mrs. Nathanmeyer rose, took Ottenburg's arm, and walked toward her husband. "That's the first real voice I have heard in Chicago," she said decidedly. "I don't count that stupid Priest woman. What do you say, father?"

Mr. Nathanmeyer shook his white head and smiled softly, as if he were thinking about something very agreeable. "*Svensk sommar,*" he murmured. "She is like a Swedish summer. I spent nearly a year there when I was a young man," he explained to Ottenburg.

When Ottenburg got Thea and her big box into the carriage, it occurred to him that she must be hungry, after singing so much. When he asked her, she admitted that she was very hungry, indeed.

He took out his watch. "Would you mind stopping somewhere with me? It's only eleven."

"Mind? Of course, I wouldn't mind. I wasn't brought up like that. I can take care of myself."

Ottenburg laughed. "And I can take care of myself, so we can do lots of jolly things together." He opened the carriage door and spoke to the driver. "I'm stuck on the way you sing that Grieg song," he declared.

When Thea got into bed that night she told herself that this was the happiest evening she had had in Chicago. She had enjoyed the Nathanmeyers and their grand house, her new dress, and Ottenburg, her first real carriage ride, and the good supper when she was so hungry. And Ottenburg *was* jolly! He made you want to come back at him. You weren't always being caught up and mystified. When you started in with him, you went; you cut the breeze, as Ray used to say. He had some go in him.

Philip Frederick Ottenburg was the third son of the great brewer. His mother was Katarina Fürst, the daughter and heiress of a brewing business older and richer than Otto Ottenburg's. As a young woman she had been a conspicuous figure in German-American society in New York, and not untouched by scandal. She was a handsome, headstrong girl, a rebellious and violent force in a provincial society. She was brutally sentimental and heavily romantic. Her free speech, her Continental ideas, and her proclivity for championing new causes, even when she did not know much about them, made her an object of suspicion. She was always going abroad to seek out intellectual affinities, and was one of the group of young women who followed Wagner about in his old age, keeping at a respectful distance, but receiving now and then a gracious acknowledgment that he appreciated their homage. When the composer died, Katarina, then a matron with a family, took to her bed and saw no one for a week.

After having been engaged to an American actor, a Welsh socialist agitator, and a German army officer, Fräulein Fürst at last placed herself and her great brewery interests into the trustworthy hands of Otto Ottenburg, who had been her suitor ever since he was a clerk, learning his business in her father's office.

Her first two sons were exactly like their father. Even as children they were industrious, earnest little tradesmen. As Frau Ottenburg said, "she had to wait for her Fred, but she got him at last," the first man who had altogether pleased her. Frederick entered Harvard when he was eighteen. When his mother went to Boston to visit him, she not only got him everything he wished for, but she made handsome and often embarrassing presents to all his friends. She gave dinners and supper parties for the Glee Club, made the crew break training, and was a generally disturbing influence. In his third year Fred left the university because of a serious escapade which had somewhat hampered his life ever since. He went at once into his father's business, where, in his own way, he had made himself very useful.

Fred Ottenburg was now twenty-eight, and people could only say of him that he had been less hurt by his mother's indulgence than most boys would have been. He had never wanted anything that he could not have it, and he might have had a great many things that he had never wanted. He was extravagant, but not prodigal. He turned most of the money his mother gave him into the business, and lived on his generous salary.

Fred had never been bored for a whole day in his life. When he was in Chicago or St. Louis, he went to ballgames, prize-fights, and horse-races. When he was in Germany, he went to concerts and to the opera. He belonged to a long list of sporting-clubs and hunting-clubs, and was a good boxer. He had so many natural interests that he had no affectations. At Harvard he kept away from the aesthetic circle that had already discovered Francis Thompson.[8] He liked no poetry but German poetry.[9] Physical energy was the thing he was full to the brim of, and music was one of its natural forms of expression. He had a healthy love of sport and art, of eating and drinking. When he was in Germany, he scarcely knew where the soup ended and the symphony began.

V

MARCH BEGAN BADLY FOR THEA. She had a cold during the first week, and after she got through her church duties on Sunday she had to go to bed with tonsilitis. She was still in the boarding-house at which young Ottenburg had called when he took her to see Mrs. Nathanmeyer. She had stayed on there because her room, although it was inconvenient and very small, was at the corner of the house and got the sunlight.

Since she left Mrs. Lorch, this was the first place where she had got away from a north light. Her rooms had all been as damp and mouldy as they were dark, with deep foundations of dirt under the carpets, and dirty walls. In her present room there was no running water and no clothes closet, and she had to have the dresser moved out to make room for her piano. But there were two windows, one on the south and one on the west, a light wall-paper with morning-glory vines, and on the floor a clean matting. The landlady had tried to make the room look cheerful, because it was hard to let. It was so small that Thea could keep it clean herself, after the Hun had done her worst. She hung her dresses on the door under a sheet, used the washstand for a dresser, slept on a cot, and opened both the windows when she practiced. She felt less walled in than she had in the other houses.

Wednesday was her third day in bed. The medical student who lived in the house had been in to see her, had left some tablets and a foamy gargle, and told her that she could probably go back to work on Monday. The landlady stuck her head in once a day, but Thea did not encourage her visits. The Hungarian chambermaid brought her soup and toast. She made a sloppy pretense of putting the room in order, but she was such a dirty creature that Thea would not let her touch her cot; she got up every morning and turned the mattress and made the bed herself. The exertion made her feel miserably ill, but at least she could lie still contentedly for a long while afterward. She hated the poisoned feeling in her throat, and no matter how often she gargled she felt unclean and disgusting. Still, if she had to be ill, she was almost glad that she had a contagious illness. Otherwise she would have been at the mercy of the people in the house. She knew that they disliked her, yet now that she was ill, they took it upon themselves to tap at her door, send her messages, books, even a miserable flower or two. Thea knew that their sympathy was an expression of self-righteousness, and she hated them for it. The divinity student, who was always whispering soft things to her, sent her "The Kreutzer Sonata."[1]

The medical student had been kind to her: he knew that she did not want to pay a doctor. His gargle had helped her, and he gave her things to make her sleep at night. But he had been a cheat, too. He had exceeded his rights. She had no soreness in her chest, and had told him so clearly. All this thumping of her back, and listening to her breathing, was done to satisfy personal curiosity. She had watched him with a contemptuous smile. She was too sick to care; if it amused him— She made him wash his hands before he touched her; he was never very clean. All the same, it wounded her and made her feel that the world was a pretty disgusting place. "The Kreutzer Sonata" did not make her feel any more cheerful. She threw it aside with hatred. She could not believe it was written by the same man who wrote the novel that had thrilled her.

Her cot was beside the south window, and on Wednesday afternoon she lay thinking about the Harsanyis, about old Mr. Nathanmeyer, and about how she was missing Fred Ottenburg's visits to the studio. That was much the worst thing about being sick. If she were going to the studio every day, she might be having pleasant encounters with Fred. He was always running away, Bowers said, and he might be planning to go away as soon as Mrs. Nathanmeyer's evenings were over. And here she was losing all this time!

After a while she heard the Hun's clumsy trot in the hall, and then a pound on the door. Mary came in, making her usual uncouth sounds, carrying a long box and a big basket. Thea sat up in bed and tore off the strings and paper. The basket was full of fruit, with a big Hawaiian pineapple in the middle, and in the box there were layers of pink roses with long, woody stems and dark-green leaves. They filled the room with a cool smell that made another air to breathe. Mary stood with her apron full of paper and cardboard. When she saw Thea take an envelope out from under the flowers, she uttered an exclamation, pointed to the roses, and then to the bosom of her own dress, on the left side. Thea laughed and nodded. She understood that Mary associated the color with Ottenburg's *boutonnière.* She pointed to the water pitcher,—she had nothing else big enough to hold the flowers,—and made Mary put it on the window sill beside her.

After Mary was gone Thea locked the door. When the landlady knocked, she pretended that she was asleep. She lay still all afternoon and with drowsy eyes watched the roses open. They were the first hothouse flowers she had ever had. The cool fragrance they released was soothing, and as the pink petals curled back, they were the only things between her and the gray sky. She lay on her side, putting the room and the boarding-house behind her. Fred knew where all the pleasant things in the world were, she reflected, and knew the road to them. He had keys to all the nice places in his pocket, and seemed to jingle them from time to time. And then, he was young; and her friends had always been old. Her mind went back over them. They had all been teachers; wonderfully kind, but still teachers. Ray Kennedy, she knew, had wanted to marry her, but he was the most protecting and teacher-like of them all. She moved impatiently in her cot and threw her braids away from her hot neck, over her pillow. "I don't want him for a teacher," she thought, frowning petulantly out of the window. "I've had such a string of them. I want him for a sweetheart."

VI

"THEA," SAID FRED OTTENBURG one drizzly afternoon in April, while they sat waiting for their tea at a restaurant in the Pullman Building, overlooking the lake, "what are you going to do this summer?"

"I don't know. Work, I suppose."

"With Bowers, you mean? Even Bowers goes fishing for a month. Chicago's no place to work, in the summer. Haven't you made any plans?"

Thea shrugged her shoulders. "No use having any plans when you haven't any money. They are unbecoming."

"Aren't you going home?"

She shook her head. "No. It won't be comfortable there till I've got something to show for myself. I'm not getting on at all, you know. This year has been mostly wasted."

"You're stale; that's what's the matter with you. And just now you're dead tired. You'll talk more rationally after you've had some tea. Rest your throat until it comes." They were sitting by a window. As Ottenburg looked at her in the gray light, he remembered what Mrs. Nathanmeyer had said about the Swedish face "breaking early." Thea was as gray as the weather. Her skin looked sick. Her hair, too, though on a damp day it curled charmingly about her face, looked pale.

Fred beckoned the waiter and increased his order for food. Thea did not hear him. She was staring out of the window, down at the roof of the Art Institute and the green lions, dripping in the rain. The lake was all rolling mist, with a soft shimmer of robin's-egg blue in the gray. A lumber boat, with two very tall masts, was emerging gaunt and black out of the fog. When the tea came Thea ate hungrily, and Fred watched her. He thought her eyes became a little less bleak. The kettle sang cheerfully over the spirit lamp, and she seemed to concentrate her attention upon that pleasant sound. She kept looking toward it listlessly and indulgently, in a way that gave him a realization of her loneliness. Fred lit a cigarette and smoked thoughtfully. He and Thea were alone in the quiet, dusky room full of white tables. In those days Chicago people never stopped for tea. "Come," he said at last, "what would you do this summer, if you could do whatever you wished?"

"I'd go a long way from here! West, I think. Maybe I could get some of my spring back. All this cold, cloudy weather,"—she looked out at the lake and shivered,—"I don't know, it does things to me," she ended abruptly.

Fred nodded. "I know. You've been going down ever since you had tonsilitis. I've seen it. What you need is to sit in the sun and bake for three months. You've got the right idea. I remember once when we were having dinner somewhere you kept asking me about the Cliff-Dweller ruins. Do they still interest you?"

"Of course they do. I've always wanted to go down there—long before I ever got in for this."

"I don't think I told you, but my father owns a whole canyon full of Cliff-Dweller ruins. He has a big worthless ranch down in Arizona, near a Navajo reservation, and there's a canyon on the place they call Panther Canyon, chock full of that sort of thing. I often go down there to hunt. Henry Biltmer and his wife live there and keep a tidy place. He's an old German who worked in the brewery until he lost his health. Now he runs a few cattle. Henry likes to do me a favor. I've done a few for him." Fred drowned his cigarette in his saucer and studied Thea's expression, which was wistful and intent, envious and admiring. He continued with satisfaction: "If you went down there and stayed with them for two or three months, they wouldn't let you pay anything. I might send Henry a new gun, but even I couldn't offer him money for putting up a friend of mine. I'll get you transportation. It would make a new girl of you. Let me write to Henry, and you pack your trunk. That's all that's necessary. No red tape about it. What do you say, Thea?"

She bit her lip, and sighed as if she were waking up.

Fred crumpled his napkin impatiently. "Well, isn't it easy enough?"

"That's the trouble; it's *too* easy. Doesn't sound probable. I'm not used to getting things for nothing."

Ottenburg laughed. “Oh, if that’s all, I’ll show you how to begin. You won’t get this for nothing, quite. I’ll ask you to let me stop off and see you on my way to California. Perhaps by that time you will be glad to see me. Better let me break the news to Bowers. I can manage him. He needs a little transportation himself now and then. You must get corduroy riding-things and leather leggings. There are a few snakes about. Why do you keep frowning?”

“Well, I don’t exactly see why you take the trouble. What do you get out of it? You haven’t liked me so well the last two or three weeks.”

Fred dropped his third cigarette and looked at his watch. “If you don’t see that, it’s because you need a tonic. I’ll show you what I’ll get out of it. Now I’m going to get a cab and take you home. You are too tired to walk a step. You’d better get to bed as soon as you get there. Of course, I don’t like you so well when you’re half anæsthetized all the time. What have you been doing to yourself?”

Thea rose. “I don’t know. Being bored eats the heart out of me, I guess.” She walked meekly in front of him to the elevator. Fred noticed for the hundredth time how vehemently her body proclaimed her state of feeling. He remembered how remarkably brilliant and beautiful she had been when she sang at Mrs. Nathanmeyer’s: flushed and gleaming, round and supple, something that couldn’t be dimmed or downed. And now she seemed a moving figure of discouragement. The very waiters glanced at her apprehensively. It was not that she made a fuss, but her back was most extraordinarily vocal. One never needed to see her face to know what she was full of that day. Yet she was certainly not mercurial. Her flesh seemed to take a mood and to “set,” like plaster. As he put her into the cab, Fred reflected once more that he “gave her up.” He would attack her when his lance was brighter.

PART IV

THE ANCIENT PEOPLE

I

THE SAN FRANCISCO MOUNTAIN lies in Northern Arizona, above Flagstaff, and its blue slopes and snowy summit entice the eye for a hundred miles across the desert. About its base lie the pine forests of the Navajos, where the great red-trunked trees live out their peaceful centuries in that sparkling air. The *piñons* and scrub begin only where the forest ends, where the country breaks into open, stony clearings and the surface of the earth cracks into deep canyons. The great pines stand at a considerable distance from each other. Each tree grows alone, murmurs alone, thinks alone. They do not intrude upon each other. The Navajos are not much in the habit of giving or of asking help. Their language is not a communicative one, and they never attempt an interchange of personality in speech. Over their forests there is the same inexorable reserve. Each tree has its exalted power to bear.

That was the first thing Thea Kronborg felt about the forest, as she drove through it one May morning in Henry Biltmer's democrat wagon[1]—and it was the first great forest she had ever seen. She had got off the train at Flagstaff that morning, rolled off into the high, chill air when all the pines on the mountain were fired by sunrise, so that she seemed to fall from sleep directly into the forest.

Old Biltmer followed a faint wagon trail which ran southeast, and which, as they traveled, continually dipped lower, falling away from the high plateau on the slope of which Flagstaff sits. The white peak of the mountain, the snow gorges above the timber, now disappeared from time to time as the road dropped and dropped, and the forest closed behind the wagon. More than the mountain disappeared as the forest closed thus. Thea seemed to be taking very little through the wood with her. The personality of which she was so tired seemed to let go of her. The high, sparkling air drank it up like blotting-paper. It was lost in the thrilling blue of the new sky and the song of the thin wind in the *piñons.* The old, fretted lines which marked one off, which defined her,—made her Thea Kronborg, Bowers's accompanist, a soprano with a faulty middle voice,—were all erased.

So far she had failed. Her two years in Chicago had not resulted in anything. She had failed with Harsanyi, and she had made no great progress with her voice. She had come to believe that whatever Bowers had taught her was of secondary importance, and that in the essential things she had made no advance. Her student life closed behind her, like the forest, and she doubted whether she could go back to it if she tried. Probably she would teach music in little country towns all her life. Failure was not so tragic as she would have supposed; she was tired enough not to care.

She was getting back to the earliest sources of gladness that she could remember. She had loved the sun, and the brilliant solitudes of sand and sun, long before these other things had come along to fasten themselves upon her and torment her. That night, when she clambered into her big German feather bed, she felt completely released from the enslaving desire to get on in the world. Darkness had once again the sweet wonder that it had in childhood.

II

THEA'S LIFE at the Ottenburg ranch was simple and full of light, like the days themselves. She awoke every morning when the first fierce shafts of sunlight darted through the curtainless windows of her room at the ranch house. After breakfast she took her lunch-basket and went down to the canyon. Usually she did not return until sunset.

Panther Canyon[1] was like a thousand others—one of those abrupt fissures with which the earth in the Southwest is riddled; so abrupt that you might walk over the edge of any one of them on a dark night and never know what had happened to you. This canyon headed on the Ottenburg ranch, about a mile from the ranch house, and it was accessible only at its head. The canyon walls, for the first two hundred feet below the surface, were perpendicular cliffs, striped with even-running strata of rock. From there on to the bottom the sides were less abrupt, were shelving, and lightly fringed with *piñons* and dwarf cedars. The effect was that of a gentler canyon within a wilder one. The dead city lay at the point where the perpendicular outer wall ceased and the V-shaped inner gorge began. There a stratum of rock, softer than those above, had been hollowed out by the action of time until it was like a deep groove running along the sides of the canyon. In this hollow (like a great fold in the rock) the Ancient People[2] had built their houses of yellowish stone and mortar. The over-hanging cliff above made a roof two hundred feet thick. The hard stratum below was an everlasting floor. The houses stood along in a row, like the buildings in a city block, or like a barracks.

In both walls of the canyon the same streak of soft rock had been washed out, and the long horizontal groove had been built up with houses. The dead city had thus two streets, one set in either cliff, facing each other across the ravine, with a river of blue air between them.

The canyon twisted and wound like a snake, and these two streets went on for four miles or more, interrupted by the abrupt turnings of the gorge, but beginning again within each turn. The canyon had a dozen of these false endings near its head. Beyond, the windings were larger and less perceptible, and it went on for a hundred miles, too narrow, precipitous, and terrible for man to follow it. The Cliff-Dwellers liked wide canyons, where the great cliffs caught the sun. Panther Canyon had been deserted for hundreds of years when the first Spanish missionaries came into Arizona, but the masonry of the houses was still wonderfully firm; had crumbled only where a landslide or a rolling boulder had torn it.

All the houses in the canyon were clean with the cleanness of sun-baked, wind-swept places, and they all smelled of the tough little cedars that twisted themselves into the very doorways. One of these rock-rooms Thea took for her own. Fred had told her how to make it comfortable. The day after she came old Henry brought over on one of the pack-ponies a roll of Navajo blankets that belonged to Fred, and Thea lined her cave with them. The room was not more than eight by ten feet, and she could touch the stone roof with her finger-tips. This was her old idea: a nest in a high cliff, full of sun. All morning long the sun beat upon her cliff, while the ruins on the opposite side of the canyon were in shadow. In the afternoon, when she had the shade of two hundred feet of rock wall, the ruins on the other side of the gulf stood out in the blazing sunlight. Before her door ran the narrow, winding path that had been the street of the Ancient People. The yucca and niggerhead cactus grew everywhere. From her doorstep she looked out on the ocher-colored slope that ran down several hundred feet to the stream, and this hot rock was sparsely grown with dwarf trees. Their colors were so pale that the shadows of the little trees on the rock stood out sharper than the trees themselves. When Thea first came, the chokecherry bushes were in blossom, and the scent of them was almost sickeningly sweet after a shower. At the very bottom of the canyon, along the stream, there was a thread of bright, flickering, golden-green,—cottonwood seedlings. They made a living, chattering screen behind which she took her bath every morning.

Thea went down to the stream by the Indian water trail. She had found a bathing-pool with a sand bottom, where the creek was dammed by fallen trees. The climb back was long and steep, and when she reached her little house in the cliff she always felt fresh delight in its comfort and inaccessibility. By the time she got there, the woolly red-and-gray blankets were saturated with sunlight, and she sometimes fell asleep as soon as she stretched her body on their warm surfaces. She used to wonder at her own inactivity. She could lie there hour after hour in the sun and listen to the strident whir of the big locusts, and to the light, ironical laughter of the quaking asps. All her life she had been hurrying and sputtering, as if she had been born behind time and had been trying to catch up. Now, she reflected, as she drew herself out long upon the rugs, it was as if she were waiting for something to catch up with her. She had got to a place where she was out of the stream of meaningless activity and undirected effort.

Here she could lie for half a day undistracted, holding pleasant and incomplete conceptions in her mind—almost in her hands. They were scarcely clear enough to be called ideas. They had something to do with fragrance and color and sound, but almost nothing to do with words. She was singing very little now, but a song would go through her head all morning, as a spring keeps welling up, and it was like a pleasant sensation indefinitely prolonged. It was much more like a sensation than like an idea, or an act of remembering. Music had never come to her in that sensuous form before. It had always been a thing to be struggled with, had always brought anxiety and exaltation and chagrin—never content and indolence. Thea began to wonder whether people could not utterly lose the power to work, as they can lose their voice or their memory. She had always been a little drudge, hurrying from one task to another—as if it mattered! And now her power to think seemed converted into a power of sustained sensation. She could become a mere receptacle for heat, or become a color, like the bright lizards that darted about on the hot stones outside her door; or she could become a continuous repetition of sound, like the cicadas.

III

THE FACULTY OF OBSERVATION was never highly developed in Thea Kronborg. A great deal escaped her eye as she passed through the world. But the things which were for her, she saw; she experienced them physically and remembered them as if they had once been a part of herself. The roses she used to see in the florists' shops in Chicago were merely roses. But when she thought of the moon-flowers that grew over Mrs. Tellamantez's door, it was as if she had been that vine and had opened up in white flowers every night. There were memories of light on the sand hills, of masses of prickly-pear blossoms she had found in the desert in early childhood, of the late afternoon sun pouring through the grape leaves and the mint bed in Mrs. Kohler's garden, which she would never lose. These recollections were a part of her mind and personality. In Chicago she had got almost nothing that went into her subconscious self and took root there. But here, in Panther Canyon, there were again things which seemed destined for her.

Panther Canyon was the home of innumerable swallows. They built nests in the wall far above the hollow groove in which Thea's own rock chamber lay. They seldom ventured above the rim of the canyon, to the flat, wind-swept tableland. Their world was the blue air-river between the canyon walls. In that blue gulf the arrow-shaped birds swam all day long, with only an occasional movement of the wings. The only sad thing about them was their timidity; the way in which they lived their lives between the echoing cliffs and never dared to rise out of the shadow of the canyon walls. As they swam past her door, Thea often felt how easy it would be to dream one's life out in some cleft in the world.

From the ancient dwelling there came always a dignified, unobtrusive sadness; now stronger, now fainter,—like the aromatic smell which the dwarf cedars gave out in the sun,—but always present, a part of the air one breathed. At night, when Thea dreamed about the canyon,—or in the early morning when she hurried toward it, anticipating it,—her conception of it was of yellow rocks baking in sunlight, the swallows, the cedar smell, and that peculiar sadness—a voice out of the past, not very loud, that went on saying a few simple things to the solitude eternally.

Standing up in her lodge, Thea could with her thumb nail dislodge flakes of carbon from the rock roof—the cooking-smoke of the Ancient People. They were that near! A timid, nest-building folk, like the swallows. How often Thea remembered Ray Kennedy's moralizing about the cliff cities. He used to say that he never felt the hardness of the human struggle or the sadness of history as he felt it among those ruins. He used to say, too, that it made one feel an obligation to do one's best. On the first day that Thea climbed the water trail she began to have intuitions about the women who had worn the path, and who had spent so great a part of their lives going up and down it. She found herself trying to walk as they must have walked, with a feeling in her feet and knees and loins which she had never known before,—which must have come up to her out of the accustomed dust of that rocky trail. She could feel the weight of an Indian baby hanging to her back as she climbed.

The empty houses, among which she wandered in the afternoon, the blanketed one in which she lay all morning, were haunted by certain fears and desires; feelings about warmth and cold and water and physical strength. It seemed to Thea that a certain understanding of those old people came up to her out of the rock shelf on which she lay; that certain feelings were transmitted to her, suggestions that were simple, insistent, and monotonous, like the beating of Indian drums. They were not expressible in words, but seemed rather to translate themselves into attitudes of body, into degrees of muscular tension or relaxation; the naked strength of youth, sharp as the sunshafts; the crouching timorousness of age, the sullenness of women who waited for their captors. At the first turning of the canyon there was a half-ruined tower of yellow masonry, a watch-tower upon which the young men used to entice eagles and snare them with nets. Sometimes for a whole morning Thea could see the coppery breast and shoulders of an Indian youth there against the sky; see him throw the net, and watch the struggle with the eagle.

Old Henry Biltmer, at the ranch, had been a great deal among the Pueblo Indians who are the descendants of the Cliff-Dwellers. After supper he used to sit and smoke his pipe by the kitchen stove and talk to Thea about them. He had never found any one before who was interested in his ruins. Every Sunday the old man prowled about in the canyon, and he had come to know a good deal more about it than he could account for. He had gathered up a whole chestful of Cliff-Dweller relics which he meant to take back to Germany with him some day. He taught Thea how to find things among the ruins: grinding-stones, and drills and needles made of turkey-bones. There were fragments of pottery everywhere. Old Henry explained to her that the Ancient People had developed masonry and pottery far beyond any other crafts. After they had made houses for themselves, the next thing was to house the precious water. He explained to her how all their customs and ceremonies and their religion went back to water. The men provided the food, but water was the care of the women. The stupid women carried water for most of their lives; the cleverer ones made the vessels to hold it. Their pottery was their most direct appeal to water, the envelope and sheath of the precious element itself. The strongest Indian need was expressed in those graceful jars, fashioned slowly by hand, without the aid of a wheel.

When Thea took her bath at the bottom of the canyon, in the sunny pool behind the screen of cottonwoods, she sometimes felt as if the water must have sovereign qualities, from having been the object of so much service and desire. That stream was the only living thing left of the drama that had been played out in the canyon centuries ago. In the rapid, restless heart of it, flowing swifter than the rest, there was a continuity of life that reached back into the old time. The glittering thread of current had a kind of lightly worn, loosely knit personality, graceful and laughing. Thea's bath came to have a ceremonial gravity. The atmosphere of the canyon was ritualistic.

One morning, as she was standing upright in the pool, splashing water between her shoulder-blades with a big sponge, something flashed through her mind that made her draw herself up and stand still until the water had quite dried upon her flushed skin. The stream and the broken pottery: what was any art but an effort to make a sheath, a mould in which to imprison for a moment the shining, elusive element which is life itself,—life hurrying past us and running away, too strong to stop, too sweet to lose? The Indian women had held it in their jars. In the sculpture she had seen in the Art Institute, it had been caught in a flash of arrested motion. In singing, one made a vessel of one's throat and nostrils and held it on one's breath, caught the stream in a scale of natural intervals.

IV

THEA HAD A SUPERSTITIOUS FEELING about the potsherds, and liked better to leave them in the dwellings where she found them. If she took a few bits back to her own lodge and hid them under the blankets, she did it guiltily, as if she were being watched. She was a guest in these houses, and ought to behave as such. Nearly every afternoon she went to the chambers which contained the most interesting fragments of pottery, sat and looked at them for a while. Some of them were beautifully decorated. This care, expended upon vessels that could not hold food or water any better for the additional labor put upon them, made her heart go out to those ancient potters. They had not only expressed their desire, but they had expressed it as beautifully as they could. Food, fire, water, and something else—even here, in this crack in the world, so far back in the night of the past! Down here at the beginning that painful thing was already stirring; the seed of sorrow, and of so much delight.

There were jars done in a delicate overlay, like pine cones; and there were many patterns in a low relief, like basket-work. Some of the pottery was decorated in color, red and brown, black and white, in graceful geometrical patterns. One day, on a fragment of a shallow bowl, she found a crested serpent's head, painted in red on terra-cotta. Again she found half a bowl with a broad band of white cliff-houses painted on a black ground. They were scarcely conventionalized at all; there they were in the black border, just as they stood in the rock before her. It brought her centuries nearer to these people to find that they saw their houses exactly as she saw them.

Yes, Ray Kennedy was right. All these things made one feel that one ought to do one's best, and help to fulfill some desire of the dust that slept there. A dream had been dreamed there long ago, in the night of ages, and the wind had whispered some promise to the sadness of the savage. In their own way, those people had felt the beginnings of what was to come. These potsherds were like fetters that bound one to a long chain of human endeavor.

Not only did the world seem older and richer to Thea now, but she herself seemed older. She had never been alone for so long before, or thought so much. Nothing had ever engrossed her so deeply as the daily contemplation of that line of pale-yellow houses tucked into the wrinkle of the cliff. Moonstone and Chicago had become vague. Here everything was simple and definite, as things had been in childhood. Her mind was like a ragbag into which she had been frantically thrusting whatever she could grab. And here she must throw this lumber away. The things that were really hers separated themselves from the rest. Her ideas were simplified, became sharper and clearer. She felt united and strong.

When Thea had been at the Ottenburg ranch for two months, she got a letter from Fred announcing that he "might be along at almost any time now." The letter came at night, and the next morning she took it down into the canyon with her. She was delighted that he was coming soon. She had never felt so grateful to any one, and she wanted to tell him everything that had happened to her since she had been there—more than had happened in all her life before. Certainly she liked Fred better than any one else in the world. There was Harsanyi, of course—but Harsanyi was always tired. Just now, and here, she wanted some one who had never been tired, who could catch an idea and run with it.

She was ashamed to think what an apprehensive drudge she must always have seemed to Fred, and she wondered why he had concerned himself about her at all. Perhaps she would never be so happy or so good-looking again, and she would like Fred to see her, for once, at her best. She had not been singing much, but she knew that her voice was more interesting than it had ever been before. She had begun to understand that—with her, at least—voice was, first of all, vitality; a lightness in the body and a driving power in the blood. If she had that, she could sing. When she felt so keenly alive, lying on that insensible shelf of stone, when her body bounded like a rubber ball away from its hardness, then she could sing. This, too, she could explain to Fred. He would know what she meant.

Another week passed. Thea did the same things as before, felt the same influences, went over the same ideas; but there was a livelier movement in her thoughts, and a freshening of sensation, like the brightness which came over the underbrush after a shower. A persistent affirmation—or denial—was going on in her, like the tapping of the woodpecker in the one tall pine tree across the chasm. Musical phrases drove each other rapidly through her mind, and the song of the cicada was now too long and too sharp. Everything seemed suddenly to take the form of a desire for action.

It was while she was in this abstracted state, waiting for the clock to strike, that Thea at last made up her mind what she was going to try to do in the world, and that she was going to Germany to study without further loss of time. Only by the merest chance had she ever got to Panther Canyon. There was certainly no kindly Providence that directed one's life; and one's parents did not in the least care what became of one, so long as one did not misbehave and endanger their comfort. One's life was at the mercy of blind chance. She had better take it in her own hands and lose everything than meekly draw the plough under the rod of parental guidance. She had seen it when she was at home last summer,—the hostility of comfortable, self-satisfied people toward any serious effort. Even to her father it seemed indecorous. Whenever she spoke seriously, he looked apologetic. Yet she had clung fast to whatever was left of Moonstone in her mind. No more of that! The Cliff-Dwellers had lengthened her past. She had older and higher obligations.

V

ONE SUNDAY AFTERNOON late in July old Henry Biltmer was rheumatically descending into the head of the canyon. The Sunday before had been one of those cloudy days—fortunately rare—when the life goes out of that country and it becomes a gray ghost, an empty, shivering uncertainty. Henry had spent the day in the barn; his canyon was a reality only when it was flooded with the light of its great lamp, when the yellow rocks cast purple shadows, and the resin was fairly cooking in the corkscrew cedars. The yuccas were in blossom now. Out of each clump of sharp bayonet leaves rose a tall stalk hung with greenish-white bells with thick, fleshy petals. The niggerhead cactus was thrusting its crimson blooms up out of every crevice in the rocks.

Henry had come out on the pretext of hunting a spade and pick-axe that young Ottenburg had borrowed, but he was keeping his eyes open. He was really very curious about the new occupants of the canyon, and what they found to do there all day long. He let his eye travel along the gulf for a mile or so to the first turning, where the fissure zigzagged out and then receded behind a stone promontory on which stood the yellowish, crumbling ruin of the old watch-tower.

From the base of this tower, which now threw its shadow forward, bits of rock kept flying out into the open gulf—skating upon the air until they lost their momentum, then falling like chips until they rang upon the ledges at the bottom of the gorge or splashed into the stream. Biltmer shaded his eyes with his hand. There on the promontory, against the cream-colored cliff, were two figures nimbly moving in the light, both slender and agile, entirely absorbed in their game. They looked like two boys. Both were hatless and both wore white shirts.

Henry forgot his pick-axe and followed the trail before the cliff-houses toward the tower. Behind the tower, as he well knew, were heaps of stones, large and small, piled against the face of the cliff. He had always believed that the Indian watchmen piled them there for ammunition. Thea and Fred had come upon these missiles and were throwing them for distance. As Biltmer approached he could hear them laughing, and he caught Thea's voice, high and excited, with a ring of vexation in it. Fred was teaching her to throw a heavy stone like a discus. When it was Fred's turn, he sent a triangular-shaped stone out into the air with considerable skill. Thea watched it enviously, standing in a half-defiant posture, her sleeves rolled above her elbows and her face flushed with heat and excitement. After Fred's third missile had rung upon the rocks below, she snatched up a stone and stepped impatiently out on the ledge in front of him. He caught her by the elbows and pulled her back.

"Not so close, you silly! You'll spin yourself off in a minute."

"You went that close. There's your heel-mark," she retorted.

"Well, I know how. That makes a difference." He drew a mark in the dust with his toe. "There, that's right. Don't step over that. Pivot yourself on your spine, and make a half turn. When you've swung your length, let it go."

Thea settled the flat piece of rock between her wrist and fingers, faced the cliff wall, stretched her arm in position, whirled round on her left foot to the full stretch of her body, and let the missile spin out over the gulf. She hung expectantly in the air, forgetting to draw back her arm, her eyes following the stone as if it carried her fortunes with it. Her comrade watched her; there weren't many girls who could show a line like that from the toe to the thigh, from the shoulder to the tip of the outstretched hand. The stone spent itself and began to fall. Thea drew back and struck her knee furiously with her palm.

"There it goes again! Not nearly so far as yours. What *is* the matter with me? Give me another." She faced the cliff and whirled again. The stone spun out, not quite so far as before.

Ottenburg laughed. "Why do you keep on working *after* you've thrown it? You can't help it along then."

Without replying, Thea stooped and selected another stone, took a deep breath and made another turn. Fred watched the disk, exclaiming, "Good girl! You got past the pine that time. That's a good throw."

She took out her handkerchief and wiped her glowing face and throat, pausing to feel her right shoulder with her left hand.

"Ah—ha, you've made yourself sore, haven't you? What did I tell you? You go at things too hard. I'll tell you what I'm going to do, Thea," Fred dusted his hands and began tucking in the blouse of his shirt, "I'm going to make some single-sticks and teach you to fence. You'd be all right there. You're light and quick and you've got lots of drive in you. I'd like to have you come at me with foils; you'd look so fierce," he chuckled.

She turned away from him and stubbornly sent out another stone, hanging in the air after its flight. Her fury amused Fred, who took all games lightly and played them well. She was breathing hard, and little beads of moisture had gathered on her upper lip. He slipped his arm about her. "If you will look as pretty as that—" he bent his head and kissed her. Thea was startled, gave him an angry push, drove at him with her free hand in a manner quite hostile. Fred was on his mettle in an instant. He pinned both her arms down and kissed her resolutely.

When he released her, she turned away and spoke over her shoulder. "That was mean of you, but I suppose I deserved what I got."

"I should say you did deserve it," Fred panted, "turning savage on me like that! I should say you did deserve it!"

He saw her shoulders harden. "Well, I just said I deserved it, didn't I? What more do you want?"

"I want you to tell me why you flew at me like that! You weren't playing; you looked as if you'd like to murder me."

She brushed back her hair impatiently. "I didn't mean anything, really. You interrupted me when I was watching the stone. I can't jump from one thing to another. I pushed you without thinking."

Fred thought her back expressed contrition. He went up to her, stood behind her with his chin above her shoulder, and said something in her ear. Thea laughed and turned toward him. They left the stone-pile carelessly, as if they had never been interested in it, rounded the yellow tower, and disappeared into the second turn of the canyon, where the dead city, interrupted by the jutting promontory, began again.

Old Biltmer had been somewhat embarrassed by the turn the game had taken. He had not heard their conversation, but the pantomime against the rocks was clear enough. When the two young people disappeared, their host retreated rapidly toward the head of the canyon.

"I guess that young lady can take care of herself," he chuckled. "Young Fred, though, he has quite a way with them."

VI

DAY WAS breaking over Panther Canyon. The gulf was cold and full of heavy, purplish twilight. The wood smoke which drifted from one of the cliff-houses hung in a blue scarf across the chasm, until the draft caught it and whirled it away. Thea was crouching in the doorway of her rock house, while Ottenburg looked after the crackling fire in the next cave. He was waiting for it to burn down to coals before he put the coffee on to boil.

They had left the ranch house that morning a little after three o'clock, having packed their camp equipment the day before, and had crossed the open pasture land with their lantern while the stars were still bright. During the descent into the canyon by lantern-light, they were chilled through their coats and sweaters. The lantern crept slowly along the rock trail, where the heavy air seemed to offer resistance. The voice of the stream at the bottom of the gorge was hollow and threatening, much louder and deeper than it ever was by day—another voice altogether. The sullenness of the place seemed to say that the world could get on very well without people, red or white; that under the human world there was a geological world, conducting its silent, immense operations which were indifferent to man. Thea had often seen the desert sunrise,—a light-hearted affair, where the sun springs out of bed and the world is golden in an instant. But this canyon seemed to waken like an old man, with rheum and stiffness of the joints, with heaviness, and a dull, malignant mind. She crouched against the wall while the stars faded, and thought what courage the early races must have had to endure so much for the little they got out of life.

At last a kind of hopefulness broke in the air. In a moment the pine trees up on the edge of the rim were flashing with coppery fire. The thin red clouds which hung above their pointed tops began to boil and move rapidly, weaving in and out like smoke. The swallows darted out of their rock houses as at a signal, and flew upward, toward the rim. Little brown birds began to chirp in the bushes along the watercourse down at the bottom of the ravine, where everything was still dusky and pale. At first the golden light seemed to hang like a wave upon the rim of the canyon; the trees and bushes up there, which one scarcely noticed at noon, stood out magnified by the slanting rays. Long, thin streaks of light began to reach quiveringly down into the canyon. The red sun rose rapidly above the tops of the blazing pines, and its glow burst into the gulf, about the very doorstep on which Thea sat. It bored into the wet, dark underbrush. The dripping cherry bushes, the pale aspens, and the frosty *piñons* were glittering and trembling, swimming in the liquid gold. All the pale, dusty little herbs of the bean family, never seen by any one but a botanist, became for a moment individual and important, their silky leaves quite beautiful with dew and light. The arch of sky overhead, heavy as lead a little while before, lifted, became more and more transparent, and one could look up into depths of pearly blue.

The savor of coffee and bacon mingled with the smell of wet cedars drying, and Fred called to Thea that he was ready for her. They sat down in the doorway of his kitchen, with the warmth of the live coals behind them and the sunlight on their faces, and began their breakfast, Mrs. Biltmer's thick coffee cups and the cream bottle between them, the coffee-pot and frying-pan conveniently keeping hot among the embers.

"I thought you were going back on the whole proposition, Thea, when you were crawling along with that lantern. I couldn't get a word out of you."

"I know. I was cold and hungry, and I didn't believe there was going to be any morning, anyway. Didn't you feel queer, at all?"

Fred squinted above his smoking cup. "Well, I am never strong for getting up before the sun. The world looks unfurnished. When I first lit the fire and had a square look at you, I thought I'd got the wrong girl. Pale, grim—you were a sight!"

Thea leaned back into the shadow of the rock room and warmed her hands over the coals. "It was dismal enough. How warm these walls are, all the way round; and your breakfast is so good. I'm all right now, Fred."

"Yes, you're all right now." Fred lit a cigarette and looked at her critically as her head emerged into the sun again. "You get up every morning just a little bit handsomer than you were the day before. I'd love you just as much if you were not turning into one of the loveliest women I've ever seen; but you are, and that's a fact to be reckoned with." He watched her across the thin line of smoke he blew from his lips. "What are you going to do with all that beauty and all that talent, Miss Kronborg?"

She turned away to the fire again. "I don't know what you're talking about," she muttered with an awkwardness which did not conceal her pleasure.

Ottenburg laughed softly. "Oh, yes, you do! Nobody better! You're a close one, but you give yourself away sometimes, like everybody else. Do you know, I've decided that you never do a single thing without an ulterior motive." He threw away his cigarette, took out his tobacco-pouch and began to fill his pipe. "You ride and fence and walk and climb, but I know that all the while you're getting somewhere in your mind. All these things are instruments; and I, too, am an instrument." He looked up in time to intercept a quick, startled glance from Thea. "Oh, I don't mind," he chuckled; "not a bit. Every woman, every interesting woman, has ulterior motives, many of 'em less creditable than yours. It's your constancy that amuses me. You must have been doing it ever since you were two feet high."

Thea looked slowly up at her companion's good-humored face. His eyes, sometimes too restless and sympathetic in town, had grown steadier and clearer in the open air. His short curly beard and yellow hair had reddened in the sun and wind. The pleasant vigor of his person was always delightful to her, something to signal to and laugh with in a world of negative people. With Fred she was never becalmed. There was always life in the air, always something coming and going, a rhythm of feeling and action,—stronger than the natural accord of youth. As she looked at him, leaning against the sunny wall, she felt a desire to be frank with him. She was not willfully holding anything back. But, on the other hand, she could not force things that held themselves back. "Yes, it was like that when I was little," she said at last. "I had to be close, as you call it, or go under. But I didn't know I had been like that since you came. I've had nothing to be close about. I haven't thought about anything but having a good time with you. I've just drifted."

Fred blew a trail of smoke out into the breeze and looked knowing. "Yes, you drift like a rifle ball, my dear. It's your—your direction that I like best of all. Most fellows wouldn't, you know. I'm unusual."

They both laughed, but Thea frowned questioningly. "Why wouldn't most fellows? Other fellows have liked me."

"Yes, serious fellows. You told me yourself they were all old, or solemn. But jolly fellows want to be the whole target. They would say you were all brain and muscle; that you have no feeling."

She glanced at him sidewise. "Oh, they would, would they?"

"Of course they would," Fred continued blandly. "Jolly fellows have no imagination. They want to be the animating force. When they are not around, they want a girl to be—extinct," he waved his hand. "Old fellows like Mr. Nathanmeyer understand your kind; but among the young ones, you are rather lucky to have found me. Even I wasn't always so wise. I've had my time of thinking it would not bore me to be the Apollo of a homey flat, and I've paid out a trifle to learn better. All those things get very tedious unless they are hooked up with an idea of some sort. It's because we *don't* come out here only to look at each other and drink coffee that it's so pleasant to—look at each other." Fred drew on his pipe for a while, studying Thea's abstraction. She was staring up at the far wall of the canyon with a troubled expression that drew her eyes narrow and her mouth hard. Her hands lay in her lap, one over the other, the fingers interlacing. "Suppose," Fred came out at length,—"suppose I were to offer you what most of the young men I know would offer a girl they'd been sitting up nights about: a comfortable flat in Chicago, a summer camp up in the woods, musical evenings, and a family to bring up. Would it look attractive to you?"

Thea sat up straight and stared at him in alarm, glared into his eyes. "Perfectly hideous!" she exclaimed.

Fred dropped back against the old stonework and laughed deep in his chest. "Well, don't be frightened. I won't offer them. You're not a nest-building bird. You know I always liked your song, 'Me for the jolt of the breakers!' I understand."

She rose impatiently and walked to the edge of the cliff. "It's not that so much. It's waking up every morning with the feeling that your life is your own, and your strength is your own, and your talent is your own; that you're all there, and there's no sag in you." She stood for a moment as if she were tortured by uncertainty, then turned suddenly back to him. "Don't talk about these things any more now," she entreated. "It isn't that I want to keep anything from you. The trouble is that I've got nothing to keep—except (you know as well as I) that feeling. I told you about it in Chicago once. But it always makes me unhappy to talk about it. It will spoil the day. Will you go for a climb with me?" She held out her hands with a smile so eager that it made Ottenburg feel how much she needed to get away from herself.

He sprang up and caught the hands she put out so cordially, and stood swinging them back and forth. "I won't tease you. A word's enough to me. But I love it, all the same. Understand?" He pressed her hands and dropped them. "Now, where are you going to drag me?"

"I want you to drag me. Over there, to the other houses. They are more interesting than these." She pointed across the gorge to the row of white houses in the other cliff. "The trail is broken away, but I got up there once. It's possible. You have to go to the bottom of the canyon, cross the creek, and then go up hand-over-hand."

Ottenburg, lounging against the sunny wall, his hands in the pockets of his jacket, looked across at the distant dwellings. "It's an awful climb," he sighed, "when I could be perfectly happy here with my pipe. However—" He took up his stick and hat and followed Thea down the water trail. "Do you climb this path every day? You surely earn your bath. I went down and had a look at your pool the other afternoon. Neat place, with all those little cottonwoods. Must be very becoming."

"Think so?" Thea said over her shoulder, as she swung round a turn.

"Yes, and so do you, evidently. I'm becoming expert at reading your meaning in your back. I'm behind you so much on these single-foot trails. You don't wear stays, do you?"

"Not here."

"I wouldn't, anywhere, if I were you. They will make you less elastic. The side muscles get flabby. If you go in for opera, there's a fortune in a flexible body. Most of the German singers are clumsy, even when they're well set up."

Thea switched a *piñon* branch back at him. "Oh, I'll never get fat! That I can promise you."

Fred smiled, looking after her. "Keep that promise, no matter how many others you break," he drawled.

The upward climb, after they had crossed the stream, was at first a breathless scramble through underbrush. When they reached the big boulders, Ottenburg went first because he had the longer leg-reach, and gave Thea a hand when the step was quite beyond her, swinging her up until she could get a foothold. At last they reached a little platform among the rocks, with only a hundred feet of jagged, sloping wall between them and the cliff-houses.

Ottenburg lay down under a pine tree and declared that he was going to have a pipe before he went any farther. "It's a good thing to know when to stop, Thea," he said meaningly.

"I'm not going to stop now until I get there," Thea insisted. "I'll go on alone."

Fred settled his shoulder against the tree-trunk. "Go on if you like, but I'm here to enjoy myself. If you meet a rattler on the way, have it out with him."

She hesitated, fanning herself with her felt hat. "I never have met one."

"There's reasoning for you," Fred murmured languidly.

Thea turned away resolutely and began to go up the wall, using an irregular cleft in the rock for a path. The cliff, which looked almost perpendicular from the bottom, was really made up of ledges and boulders, and behind these she soon disappeared. For a long while Fred smoked with half-closed eyes, smiling to himself now and again. Occasionally he lifted an eyebrow as he heard the rattle of small stones among the rocks above. "In a temper," he concluded; "do her good." Then he subsided into warm drowsiness and listened to the locusts in the yuccas, and the tap-tap of the old woodpecker that was never weary of assaulting the big pine.

Fred had finished his pipe and was wondering whether he wanted another, when he heard a call from the cliff far above him. Looking up, he saw Thea standing on the edge of a projecting crag. She waved to him and threw her arm over her head, as if she were snapping her fingers in the air.

As he saw her there between the sky and the gulf, with that great wash of air and the morning light about her, Fred recalled the brilliant figure at Mrs. Nathanmeyer's. Thea was one of those people who emerge, unexpectedly, larger than we are accustomed to see them. Even at this distance one got the impression of muscular energy and audacity,—a kind of brilliancy of motion,—of a personality that carried across big spaces and expanded among big things. Lying still, with his hands under his head, Ottenburg rhetorically addressed the figure in the air. "You are the sort that used to run wild in Germany, dressed in their hair and a piece of skin. Soldiers caught 'em in nets. Old Nathanmeyer," he mused, "would like a peep at her now. Knowing old fellow. Always buying those Zorn etchings of peasant girls bathing.[1] No sag in them either. Must be the cold climate." He sat up. "She'll begin to pitch rocks on me if I don't move." In response to another impatient gesture from the crag, he rose and began swinging slowly up the trail.

It was the afternoon of that long day. Thea was lying on a blanket in the door of her rock house. She and Ottenburg had come back from their climb and had lunch, and he had gone off for a nap in one of the cliff-houses farther down the path. He was sleeping peacefully, his coat under his head and his face turned toward the wall.

Thea, too, was drowsy, and lay looking through half-closed eyes up at the blazing blue arch over the rim of the canyon. She was thinking of nothing at all. Her mind, like her body, was full of warmth, lassitude, physical content. Suddenly an eagle, tawny and of great size, sailed over the cleft in which she lay, across the arch of sky. He dropped for a moment into the gulf between the walls, then wheeled, and mounted until his plumage was so steeped in light that he looked like a golden bird. He swept on, following the course of the canyon a little way and then disappearing beyond the rim. Thea sprang to her feet as if she had been thrown up from the rock by volcanic action. She stood rigid on the edge of the stone shelf, straining her eyes after that strong, tawny flight. O eagle of eagles! Endeavor, achievement, desire, glorious striving of human art! From a cleft in the heart of the world she saluted it. . . . It had come all the way; when men lived in caves, it was there. A vanished race; but along the trails, in the stream, under the spreading cactus, there still glittered in the sun the bits of their frail clay vessels, fragments of their desire.

VII

FROM THE DAY of Fred's arrival, he and Thea were unceasingly active. They took long rides into the Navajo pine forests, bought turquoises and silver bracelets from the wandering Indian herdsmen, and rode twenty miles to Flagstaff upon the slightest pretext. Thea had never felt this pleasant excitement about any man before, and she found herself trying very hard to please young Ottenburg. She was never tired, never dull. There was a zest about waking up in the morning and dressing, about walking, riding, even about sleep.

One morning when Thea came out from her room at seven o'clock, she found Henry and Fred on the porch, looking up at the sky. The day was already hot and there was no breeze. The sun was shining, but heavy brown clouds were hanging in the west, like the smoke of a forest fire. She and Fred had meant to ride to Flagstaff that morning, but Biltmer advised against it, foretelling a storm. After breakfast they lingered about the house, waiting for the weather to make up its mind. Fred had brought his guitar, and as they had the dining-room to themselves, he made Thea go over some songs with him. They got interested and kept it up until Mrs. Biltmer came to set the table for dinner. Ottenburg knew some of the Mexican things Spanish Johnny used to sing. Thea had never before happened to tell him about Spanish Johnny, and he seemed more interested in Johnny than in Dr. Archie or Wunsch.

After dinner they were too restless to endure the ranch house any longer, and ran away to the canyon to practice with singlesticks. Fred carried a slicker and a sweater, and he made Thea wear one of the rubber hats that hung in Biltmer's gun-room. As they crossed the pasture land the clumsy slicker kept catching in the lacings of his leggings.

"Why don't you drop that thing?" Thea asked. "I won't mind a shower. I've been wet before."

"No use taking chances."

From the canyon they were unable to watch the sky, since only a strip of the zenith was visible. The flat ledge about the watch-tower was the only level spot large enough for single-stick exercise, and they were still practicing there when, at about four o'clock, a tremendous roll of thunder echoed between the cliffs and the atmosphere suddenly became thick.

Fred thrust the sticks in a cleft in the rock. "We're in for it, Thea. Better make for your cave where there are blankets." He caught her elbow and hurried her along the path before the cliff-houses. They made the half-mile at a quick trot, and as they ran the rocks and the sky and the air between the cliffs turned a turbid green, like the color in a moss agate. When they reached the blanketed rock room, they looked at each other and laughed. Their faces had taken on a greenish pallor. Thea's hair, even, was green.

"Dark as pitch in here," Fred exclaimed as they hurried over the old rock doorstep. "But it's warm. The rocks hold the heat. It's going to be terribly cold outside, all right." He was interrupted by a deafening peal of thunder. "Lord, what an echo! Lucky you don't mind. It's worth watching out there. We needn't come in yet."

The green light grew murkier and murkier. The smaller vegetation was blotted out. The yuccas, the cedars, and *piñons* stood dark and rigid, like bronze. The swallows flew up with sharp, terrified twitterings. Even the quaking asps were still. While Fred and Thea watched from the doorway, the light changed to purple. Clouds of dark vapor, like chlorine gas, began to float down from the head of the canyon and hung between them and the cliff-houses in the opposite wall. Before they knew it, the wall itself had disappeared. The air was positively venomous-looking, and grew colder every minute. The thunder seemed to crash against one cliff, then against the other, and to go shrieking off into the inner canyon.

The moment the rain broke, it beat the vapors down. In the gulf before them the water fell in spouts, and dashed from the high cliffs overhead. It tore aspens and chokecherry bushes out of the ground and left the yuccas hanging by their tough roots. Only the little cedars stood black and unmoved in the torrents that fell from so far above. The rock chamber was full of fine spray from the streams of water that shot over the doorway. Thea crept to the back wall and rolled herself in a blanket, and Fred threw the heavier blankets over her. The wool of the Navajo sheep was soon kindled by the warmth of her body, and was impenetrable to dampness. Her hair, where it hung below the rubber hat, gathered the moisture like a sponge. Fred put on the slicker, tied the sweater about his neck, and settled himself cross-legged beside her. The chamber was so dark that, although he could see the outline of her head and shoulders, he could not see her face. He struck a wax match to light his pipe. As he sheltered it between his hands, it sizzled and sputtered, throwing a yellow flicker over Thea and her blankets.

"You look like a gypsy," he said as he dropped the match. "Any one you'd rather be shut up with than me? No? Sure about that?"

"I think I am. Aren't you cold?"

"Not especially." Fred smoked in silence, listening to the roar of the water outside. "We may not get away from here right away," he remarked.

"I shan't mind. Shall you?"

He laughed grimly and pulled on his pipe. "Do you know where you're at, Miss Thea Kronborg?" he said at last. "You've got me going pretty hard, I suppose you know. I've had a lot of sweethearts, but I've never been so much—engrossed before. What are you going to do about it?" He heard nothing from the blankets. "Are you going to play fair, or is it about my cue to cut away?"

"I'll play fair. I don't see why you want to go."

"What do you want me around for?—to play with?"

Thea struggled up among the blankets. "I want you for everything. I don't know whether I'm what people call in love with you or not. In Moonstone that meant sitting in a hammock with somebody. I don't want to sit in a hammock with you, but I want to do almost everything else. Oh, hundreds of things!"

"If I run away, will you go with me?"

"I don't know. I'll have to think about that. Maybe I would." She freed herself from her wrappings and stood up. "It's not raining so hard now. Hadn't we better start this minute? It will be night before we get to Biltmer's."

Fred struck another match. "It's seven. I don't know how much of the path may be washed away. I don't even know whether I ought to let you try it without a lantern."

Thea went to the doorway and looked out. "There's nothing else to do. The sweater and the slicker will keep me dry, and this will be my chance to find out whether these shoes are really water-tight.

They cost a week's salary." She retreated to the back of the cave. "It's getting blacker every minute."

Ottenburg took a brandy flask from his coat pocket. "Better have some of this before we start. Can you take it without water?"

Thea lifted it obediently to her lips. She put on the sweater and Fred helped her to get the clumsy slicker on over it. He buttoned it and fastened the high collar. She could feel that his hands were hurried and clumsy. The coat was too big, and he took off his necktie and belted it in at the waist. While she tucked her hair more securely under the rubber hat he stood in front of her, between her and the gray doorway, without moving.

"Are you ready to go?" she asked carelessly.

"If you are," he spoke quietly, without moving, except to bend his head forward a little.

Thea laughed and put her hands on his shoulders. "You know how to handle me, don't you?" she whispered. For the first time, she kissed him without constraint or embarrassment.

"Thea, Thea, Thea!" Fred whispered her name three times, shaking her a little as if to waken her. It was too dark to see, but he could feel that she was smiling.

When she kissed him she had not hidden her face on his shoulder,—she had risen a little on her toes, and stood straight and free. In that moment when he came close to her actual personality, he felt in her the same expansion that he had noticed at Mrs. Nathanmeyer's. She became freer and stronger under impulses. When she rose to meet him like that, he felt her flash into everything that she had ever suggested to him, as if she filled out her own shadow.

She pushed him away and shot past him out into the rain. "Now for it, Fred," she called back exultantly. The rain was pouring steadily down through the dying gray twilight, and muddy streams were spouting and foaming over the cliff.

Fred caught her and held her back. "Keep behind me, Thea. I don't know about the path. It may be gone altogether. Can't tell what there is under this water."

But the path was older than the white man's Arizona. The rush of water had washed away the dust and stones that lay on the surface, but the rock skeleton of the Indian trail was there, ready for the foot. Where the streams poured down through gullies, there was always a cedar or a *piñon* to cling to. By wading and slipping and climbing, they got along. As they neared the head of the canyon, where the path lifted and rose in steep loops to the surface of the plateau, the climb was more difficult. The earth above had broken away and washed down over the trail, bringing rocks and bushes and even young trees with it. The last ghost of daylight was dying and there was no time to lose. The canyon behind them was already black.

"We've got to go right through the top of this pine tree, Thea. No time to hunt a way around. Give me your hand." After they had crashed through the mass of branches, Fred stopped abruptly. "Gosh, what a hole! Can you jump it? Wait a minute."

He cleared the washout, slipped on the wet rock at the farther side, and caught himself just in time to escape a tumble. "If I could only find something to hold to, I could give you a hand. It's so cursed dark, and there are no trees here where they're needed. Here's something; it's a root. It will hold all right." He braced himself on the rock, gripped the crooked root with one hand and swung himself across toward Thea, holding out his arm. "Good jump! I must say you don't lose your nerve in a tight place. Can you keep at it a little longer? We're almost out. Have to make that next ledge. Put your foot on my knee and catch something to pull by."

Thea went up over his shoulder. "It's hard ground up here," she panted. "Did I wrench your arm when I slipped then? It was a cactus I grabbed, and it startled me."

"Now, one more pull and we're on the level."

They emerged gasping upon the black plateau. In the last five minutes the darkness had solidified and it seemed as if the skies were pouring black water. They could not see where the sky ended or the plain began. The light at the ranch house burned a steady spark through the rain. Fred drew Thea's arm through his and they struck off toward the light. They could not see each other, and the rain at their backs seemed to drive them along. They kept laughing as they stumbled over tufts of grass or stepped into slippery pools. They were delighted with each other and with the adventure which lay behind them.

"I can't even see the whites of your eyes, Thea. But I'd know who was here stepping out with me, anywhere. Part coyote you are, by the feel of you. When you make up your mind to jump, you jump! My gracious, what's the matter with your hand?"

"Cactus spines. Didn't I tell you when I grabbed the cactus? I thought it was a root. Are we going straight?"

"I don't know. Somewhere near it, I think. I'm very comfortable, aren't you? You're warm, except your cheeks. How funny they are when they're wet. Still, you always feel like you. I like this. I could walk to Flagstaff. It's fun, not being able to see anything. I feel surer of you when I can't see you. Will you run away with me?"

Thea laughed. "I won't run far to-night. I'll think about it. Look, Fred, there's somebody coming."

"Henry, with his lantern. Good enough! Halloo! Hallo—o—o!" Fred shouted.

The moving light bobbed toward them. In half an hour Thea was in her big feather bed, drinking hot lentil soup, and almost before the soup was swallowed she was asleep.

VIII

ON THE FIRST DAY OF SEPTEMBER Fred Ottenburg and Thea Kronborg left Flagstaff by the east-bound express. As the bright morning advanced, they sat alone on the rear platform of the observation car, watching the yellow miles unfold and disappear. With complete content they saw the brilliant, empty country flash by. They were tired of the desert and the dead races, of a world without change or ideas. Fred said he was glad to sit back and let the Santa Fé do the work for a while.

"And where are we going, anyhow?" he added.

"To Chicago, I suppose. Where else would we be going?" Thea hunted for a handkerchief in her handbag.

"I wasn't sure, so I had the trunks checked to Albuquerque. We can recheck there to Chicago, if you like. Why Chicago? You'll never go back to Bowers. Why wouldn't this be a good time to make a run for it? We could take the southern branch at Albuquerque, down to El Paso, and then over into Mexico. We are exceptionally free. Nobody waiting for us anywhere."

Thea sighted along the steel rails that quivered in the light behind them. "I don't see why I couldn't marry you in Chicago, as well as any place," she brought out with some embarrassment.

Fred took the handbag out of her nervous clasp and swung it about on his finger. "You've no particular love for that spot, have you? Besides, as I've told you, my family would make a row. They are an excitable lot. They discuss and argue everlastingly. The only way I can ever put anything through is to go ahead, and convince them afterward."

"Yes; I understand. I don't mind that. I don't want to marry your family. I'm sure you wouldn't want to marry mine. But I don't see why we have to go so far."

"When we get to Winslow, you look about the freight yards and you'll probably see several yellow cars with my name on them. That's why, my dear. When your visiting-card is on every beer bottle, you can't do things quietly. Things get into the papers." As he watched her troubled expression, he grew anxious. He leaned forward on his camp-chair, and kept twirling the handbag between his knees. "Here's a suggestion, Thea," he said presently. "Dismiss it if you don't like it: suppose we go down to Mexico on the chance. You've never seen anything like Mexico City; it will be a lark for you, anyhow. If you change your mind, and don't want to marry me, you can go back to Chicago, and I'll take a steamer from Vera Cruz and go up to New York. When I get to Chicago, you'll be at work, and nobody will ever be the wiser. No reason why we shouldn't both travel in Mexico, is there? You'll be traveling alone. I'll merely tell you the right places to stop, and come to take you driving. I won't put any pressure on you. Have I ever?" He swung the bag toward her and looked up under her hat.

"No, you haven't," she murmured. She was thinking that her own position might be less difficult if he had used what he called pressure. He clearly wished her to take the responsibility.

"You have your own future in the back of your mind all the time," Fred began, "and I have it in mine. I'm not going to try to carry you off, as I might another girl. If you wanted to quit me, I couldn't hold you, no matter how many times you had married me. I don't want to over-persuade you. But I'd like mighty well to get you down to that jolly old city, where everything would please you, and give myself a chance. Then, if you thought you could have a better time with me than without me, I'd try to grab you before you changed your mind. You are not a sentimental person."

Thea drew her veil down over her face. "I think I am, a little; about you," she said quietly. Fred's irony somehow hurt her.

"What's at the bottom of your mind, Thea?" he asked hurriedly. "I can't tell. Why do you consider it at all, if you're not sure? Why are you here with me now?"

Her face was half-averted. He was thinking that it looked older and more firm—almost hard—under a veil.

"Isn't it possible to do things without having any very clear reason?" she asked slowly. "I have no plan in the back of my mind. Now that I'm with you, I want to be with you; that's all. I can't settle down to being alone again. I am here to-day because I want to be with you to-day." She paused. "One thing, though; if I gave you my word, I'd keep it. And you could hold me, though you don't seem to think so. Maybe I'm not sentimental, but I'm not very light, either. If I went off with you like this, it wouldn't be to amuse myself."

Ottenburg's eyes fell. His lips worked nervously for a moment. "Do you mean that you really care for me, Thea Kronborg?" he asked unsteadily.

"I guess so. It's like anything else. It takes hold of you and you've got to go through with it, even if you're afraid. I was afraid to leave Moonstone, and afraid to leave Harsanyi. But I had to go through with it."

"And are you afraid now?" Fred asked slowly.

"Yes; more than I've ever been. But I don't think I could go back. The past closes up behind one, somehow. One would rather have a new kind of misery. The old kind seems like death or unconsciousness. You can't force your life back into that mould again. No, one can't go back." She rose and stood by the back grating of the platform, her hand on the brass rail.

Fred went to her side. She pushed up her veil and turned her most glowing face to him. Her eyes were wet and there were tears on her lashes, but she was smiling the rare, whole-hearted smile he had seen once or twice before. He looked at her shining eyes, her parted lips, her chin a little lifted. It was as if they were colored by a sunrise he could not see. He put his hand over hers and clasped it with a strength she felt. Her eyelashes trembled, her mouth softened, but her eyes were still brilliant.

"Will you always be like you were down there, if I go with you?" she asked under her breath.

His fingers tightened on hers. "By God, I will!" he muttered.

"That's the only promise I'll ask you for. Now go away for a while and let me think about it. Come back at lunch-time and I'll tell you. Will that do?"

"Anything will do, Thea, if you'll only let me keep an eye on you. The rest of the world doesn't interest me much. You've got me in deep."

Fred dropped her hand and turned away. As he glanced back from the front end of the observation car, he saw that she was still standing there, and any one would have known that she was brooding over something. The earnestness of her head and shoulders had a certain nobility. He stood looking at her for a moment.

When he reached the forward smoking-car, Fred took a seat at the end, where he could shut the other passengers from his sight. He put on his traveling-cap and sat down wearily, keeping his head near the window. "In any case, I shall help her more than I shall hurt her," he kept saying to himself. He admitted that this was not the only motive which impelled him, but it was one of them. "I'll make it my business in life to get her on. There's nothing else I care about so much as seeing her have her chance. She hasn't touched her real force yet. She isn't even aware of it. Lord, don't I know something about them? There isn't one of them that has such a depth to draw from. She'll be one of the great artists of our time. Playing accompaniments for that cheese-faced sneak! I'll get her off to Germany this winter, or take her. She hasn't got any time to waste now. I'll make it up to her, all right."

Ottenburg certainly meant to make it up to her, in so far as he could. His feeling was as generous as strong human feelings are likely to be. The only trouble was, that he was married already, and had been since he was twenty.

His older friends in Chicago, people who had been friends of his family, knew of the unfortunate state of his personal affairs; but they were people whom in the natural course of things Thea Kronborg would scarcely meet. Mrs. Frederick Ottenburg lived in California, at Santa Barbara, where her health was supposed to be better than elsewhere, and her husband lived in Chicago. He visited his wife every winter to reinforce her position, and his devoted mother, although her hatred for her daughter-in-law was scarcely approachable in words, went to Santa Barbara every year to make things look better and to relieve her son.

When Frederick Ottenburg was beginning his junior year at Harvard, he got a letter from Dick Brisbane, a Kansas City boy he knew, telling him that his *fiancée,* Miss Edith Beers, was going to New York to buy her trousseau. She would be at the Holland House, with her aunt and a girl from Kansas City who was to be a bridesmaid, for two weeks or more. If Ottenburg happened to be going down to New York, would he call upon Miss Beers and "show her a good time"?

Fred did happen to be going to New York. He was going down from New Haven, after the Thanksgiving game. He called on Miss

Beers and found her, as he that night telegraphed Brisbane, a "ripping beauty, no mistake." He took her and her aunt and her uninteresting friend to the theater and to the opera, and he asked them to lunch with him at the Waldorf. He took no little pains in arranging the luncheon with the head waiter. Miss Beers was the sort of girl with whom a young man liked to seem experienced. She was dark and slender and fiery. She was witty and slangy; said daring things and carried them off with *nonchalance.* Her childish extravagance and contempt for all the serious facts of life could be charged to her father's generosity and his long packing-house purse. Freaks that would have been vulgar and ostentatious in a more simple-minded girl, in Miss Beers seemed whimsical and picturesque. She darted about in magnificent furs and pumps and close-clinging gowns, though that was the day of full skirts. Her hats were large and floppy. When she wriggled out of her moleskin coat at luncheon, she looked like a slim black weasel. Her satin dress was a mere sheath, so conspicuous by its severity and scantness that every one in the dining-room stared. She ate nothing but alligator-pear salad and hothouse grapes, drank a little champagne, and took cognac in her coffee. She ridiculed, in the raciest slang, the singers they had heard at the opera the night before, and when her aunt pretended to reprove her, she murmured indifferently, "What's the matter with you, old sport?" She rattled on with a subdued loquaciousness, always keeping her voice low and monotonous, always looking out of the corner of her eye and speaking, as it were, in asides, out of the corner of her mouth. She was scornful of everything,—which became her eyebrows. Her face was mobile and discontented, her eyes quick and black. There was a sort of smouldering fire about her, young Ottenburg thought. She entertained him prodigiously.

After luncheon Miss Beers said she was going uptown to be fitted, and that she would go alone because her aunt made her nervous. When Fred held her coat for her, she murmured, "Thank you, Alphonse," as if she were addressing the waiter. As she stepped into a hansom, with a long stretch of thin silk stocking, she said negligently, over her fur collar, "Better let me take you along and drop you somewhere." He sprang in after her, and she told the driver to go to the Park.

It was a bright winter day, and bitterly cold. Miss Beers asked Fred to tell her about the game at New Haven, and when he did so paid no attention to what he said. She sank back into the hansom and held her muff before her face, lowering it occasionally to utter laconic remarks about the people in the carriages they passed, interrupting Fred's narrative in a disconcerting manner. As they entered the Park he happened to glance under her wide black hat at her black eyes and hair—the muff hid everything else—and discovered that she was crying. To his solicitous inquiry she replied that it "was enough to make you damp, to go and try on dresses to marry a man you weren't keen about."

Further explanations followed. She had thought she was "perfectly cracked" about Brisbane, until she met Fred at the Holland House three days ago. Then she knew she would scratch Brisbane's eyes out if she married him. What was she going to do?

Fred told the driver to keep going. What did she want to do? Well, she didn't know. One had to marry somebody, after all the machinery had been put in motion. Perhaps she might as well scratch Brisbane as anybody else; for scratch she would, if she didn't get what she wanted.

Of course, Fred agreed, one had to marry somebody. And certainly this girl beat anything he had ever been up against before. Again he told the driver to go ahead. Did she mean that she would think of marrying him, by any chance? Of course she did, Alphonse. Hadn't he seen that all over her face three days ago? If he hadn't, he was a snowball.

By this time Fred was beginning to feel sorry for the driver. Miss Beers, however, was compassionless. After a few more turns, Fred suggested tea at the Casino. He was very cold himself, and remembering the shining silk hose and pumps, he wondered that the girl was not frozen. As they got out of the hansom, he slipped the driver a bill and told him to have something hot while he waited.

At the tea-table, in a snug glass enclosure, with the steam sputtering in the pipes beside them and a brilliant winter sunset without, they developed their plan. Miss Beers had with her plenty of money, destined for tradesmen, which she was quite willing to divert into other channels—the first excitement of buying a trousseau had worn off, anyway. It was very much like any other shopping. Fred had his allowance and a few hundred he had won on the game. She would meet him to-morrow morning at the Jersey ferry. They could take one of the west-bound Pennsylvania trains and go—anywhere, some place where the laws weren't too fussy.—Fred had not even thought about the laws!—It would be all right with her father; he knew Fred's family.

Now that they were engaged, she thought she would like to drive a little more. They were jerked about in the cab for another hour through the deserted Park. Miss Beers, having removed her hat, reclined upon Fred's shoulder.

The next morning they left Jersey City by the latest fast train out. They had some misadventures, crossed several States before they found a justice obliging enough to marry two persons whose names automatically instigated inquiry. The bride's family were rather pleased with her originality; besides, any one of the Ottenburg boys was clearly a better match than young Brisbane. With Otto Ottenburg, however, the affair went down hard, and to his wife, the once proud Katarina Fürst, such a disappointment was almost unbearable. Her sons had always been clay in her hands, and now the *geliebter Sohn*[1] had escaped her.

Beers, the packer, gave his daughter a house in St. Louis, and Fred went into his father's business. At the end of a year, he was mutely appealing to his mother for sympathy. At the end of two, he was drinking and in open rebellion. He had learned to detest his wife. Her wastefulness and cruelty revolted him. The ignorance and the fatuous conceit which lay behind her grimacing mask of slang and ridicule humiliated him so deeply that he became absolutely reckless. Her grace was only an uneasy wriggle, her audacity was the result of insolence and envy, and her wit was restless spite. As her personal mannerisms grew more and more odious to him, he began to dull his perceptions with champagne. He had it for tea, he drank it with dinner, and during the evening he took enough to insure that he would be well insulated when he got home. This behavior spread alarm among his friends. It was scandalous, and it did not occur among brewers. He was violating the *noblesse oblige* of his guild. His father and his father's partners looked alarmed.

When Fred's mother went to him and with clasped hands entreated an explanation, he told her that the only trouble was that he couldn't hold enough wine to make life endurable, so he was going to get out from under and enlist in the navy. He didn't want anything but the shirt on his back and clean salt air. His mother could look out; he was going to make a scandal.

Mrs. Otto Ottenburg went to Kansas City to see Mr. Beers, and had the satisfaction of telling him that he had brought up his daughter like a savage, *eine Ungebildete*.[2] All the Ottenburgs and all the Beers, and many of their friends, were drawn into the quarrel. It was to public opinion, however, and not to his mother's activities, that Fred owed his partial escape from bondage. The cosmopolitan brewing world of St. Louis had conservative standards. The Ottenburgs' friends were not predisposed in favor of the plunging Kansas City set, and they disliked young Fred's wife from the day that she was brought among them. They found her ignorant and ill-bred and insufferably impertinent. When they became aware of how matters were going between her and Fred, they omitted no opportunity to snub her. Young Fred had always been popular, and St. Louis people took up his cause with warmth. Even the younger men, among whom Mrs. Fred tried to draft a following, at first avoided and then ignored her. Her defeat was so conspicuous, her life became such a desert, that she at last consented to accept the house in Santa Barbara which Mrs. Otto Ottenburg had long owned and cherished. This villa, with its luxuriant gardens, was the price of Fred's furlough. His mother was only too glad to offer it in his behalf. As soon as his wife was established in California, Fred was transferred from St. Louis to Chicago.

A divorce was the one thing Edith would never, never, give him. She told him so, and she told his family so, and her father stood behind her. She would enter into no arrangement that might eventually lead to divorce. She had insulted her husband before guests and servants, had scratched his face, thrown hand-mirrors and hairbrushes and nail-scissors at him often enough, but she knew that Fred was hardly the fellow who would go into court and offer that sort of evidence. In her behavior with other men she was discreet.

After Fred went to Chicago, his mother visited him often, and dropped a word to her old friends there, who were already kindly disposed toward the young man. They gossiped as little as was compatible with the interest they felt, undertook to make life agreeable for Fred, and told his story only where they felt it would do good: to girls who seemed to find the young brewer attractive. So far, he had behaved well, and had kept out of entanglements.

Since he was transferred to Chicago, Fred had been abroad several times, and had fallen more and more into the way of going about among young artists,—people with whom personal relations were incidental. With women, and even girls, who had careers to follow, a young man might have pleasant friendships without being regarded as a prospective suitor or lover. Among artists his position was not irregular, because with them his marriageableness was not an issue. His tastes, his enthusiasm, and his agreeable personality made him welcome.

With Thea Kronborg he had allowed himself more liberty than he usually did in his friendships or gallantries with young artists, because she seemed to him distinctly not the marrying kind. She impressed him as equipped to be an artist, and to be nothing else; already directed, concentrated, formed as to mental habit. He was generous and sympathetic, and she was lonely and needed friendship; needed cheerfulness. She had not much power of reaching out toward useful people or useful experiences, did not see opportunities. She had no tact about going after good positions or enlisting the interest of influential persons. She antagonized people rather than conciliated them. He discovered at once that she had a merry side, a robust humor that was deep and hearty, like her laugh, but it slept most of the time under her own doubts and the dullness of her life. She had not what is called a "sense of humor." That is, she had no intellectual humor; no power to enjoy the absurdities of people, no relish of their pretentiousness and inconsistencies—which only depressed her. But her joviality, Fred felt, was an asset, and ought to be developed. He discovered that she was more receptive and more effective under a pleasant stimulus than she was under the gray grind which she considered her salvation. She was still Methodist enough to believe that if a thing were hard and irksome, it must be good for her. And yet, whatever she did well was spontaneous. Under the least glow of excitement, as at Mrs. Nathanmeyer's, he had seen the apprehensive, frowning drudge of Bowers's studio flash into a resourceful and consciously beautiful woman.

His interest in Thea was serious, almost from the first, and so sincere that he felt no distrust of himself. He believed that he knew a great deal more about her possibilities than Bowers knew, and he liked to think that he had given her a stronger hold on life. She had never seen herself or known herself as she did at Mrs. Nathanmeyer's musical evenings. She had been a different girl ever since. He had not anticipated that she would grow more fond of him than his immediate usefulness warranted. He thought he knew the ways of artists, and, as he said, she must have been "at it from her cradle."

He had imagined, perhaps, but never really believed, that he would find her waiting for him sometime as he found her waiting on the day he reached the Biltmer ranch. Once he found her so—well, he did not pretend to be anything more or less than a reasonably well-intentioned young man. A lovesick girl or a flirtatious woman he could have handled easily enough. But a personality like that, unconsciously revealing itself for the first time under the exaltation of a personal feeling,—what could one do but watch it? As he used to say to himself, in reckless moments back there in the canyon, "You can't put out a sunrise." He had to watch it, and then he had to share it.

Besides, was he really going to do her any harm? The Lord knew he would marry her if he could! Marriage would be an incident, not an end with her; he was sure of that. If it were not he, it would be some one else; some one who would be a weight about her neck, probably; who would hold her back and beat her down and divert her from the first plunge for which he felt she was gathering all her energies. He meant to help her, and he could not think of another man who would. He went over his unmarried friends, East and West, and he could not think of one who would know what she was driving at—or care. The clever ones were selfish, the kindly ones were stupid.

"Damn it, if she's going to fall in love with somebody, it had better be me than any of the others—of the sort she'd find. Get her tied up with some conceited ass who'd try to make her over, train her like a puppy! Give one of 'em a big nature like that, and he'd be horrified. He wouldn't show his face in the clubs until he'd gone after her and combed her down to conform to some fool idea in his own head—put there by some other woman, too, his first sweetheart or his grandmother or a maiden aunt. At least, I understand her. I know what she needs and where she's bound, and I mean to see that she has a fighting chance."

His own conduct looked crooked, he admitted; but he asked himself whether, between men and women, all ways were not more or less crooked. He believed those which are called straight were the most dangerous of all. They seemed to him, for the most part, to lie between windowless stone walls, and their rectitude had been achieved at the expense of light and air. In their unquestioned regularity lurked every sort of human cruelty and meanness, and every kind of humiliation and suffering. He would rather have any woman he cared for wounded than crushed. He would deceive her not once, he told himself fiercely, but a hundred times, to keep her free.

When Fred went back to the observation car at one o'clock, after the luncheon call, it was empty, and he found Thea alone on the platform. She put out her hand, and met his eyes.

"It's as I said. Things have closed behind me. I can't go back, so I am going on—to Mexico?" She lifted her face with an eager, questioning smile.

Fred met it with a sinking heart. Had he really hoped she would give him another answer? He would have given pretty much anything— But there, that did no good. He could give only what he had. Things were never complete in this world; you had to snatch at them as they came or go without. Nobody could look into her face and draw back, nobody who had any courage. She had courage enough for anything—look at her mouth and chin and eyes! Where did it come from, that light? How could a face, a familiar face, become so the picture of hope, be painted with the very colors of youth's exaltation? She was right; she was not one of those who draw back. Some people get on by avoiding dangers, others by riding through them.

They stood by the railing looking back at the sand levels, both feeling that the train was steaming ahead very fast. Fred's mind was a confusion of images and ideas. Only two things were clear to him: the force of her determination, and the belief that, handicapped as he was, he could do better by her than another man would do. He knew he would always remember her, standing there with that expectant, forward-looking smile, enough to turn the future into summer.

PART V

DOCTOR ARCHIE'S VENTURE

I

DR. HOWARD ARCHIE had come down to Denver for a meeting of the stockholders in the San Felipe silver mine.[1] It was not absolutely necessary for him to come, but he had no very pressing cases at home. Winter was closing down in Moonstone, and he dreaded the dullness of it. On the 10th day of January, therefore, he was registered at the Brown Palace Hotel. On the morning of the 11th he came down to breakfast to find the streets white and the air thick with snow. A wild northwester was blowing down from the mountains, one of those beautiful storms that wrap Denver in dry, furry snow, and make the city a loadstone to thousands of men in the mountains and on the plains. The brakemen out on their box-cars, the miners up in their diggings, the lonely homesteaders in the sand hills of Yucca and Kit Carson Counties, begin to think of Denver, muffled in snow, full of food and drink and good cheer, and to yearn for her with that admiration which makes her, more than other American cities, an object of sentiment.

Howard Archie was glad he had got in before the storm came. He felt as cheerful as if he had received a legacy that morning, and he greeted the clerk with even greater friendliness than usual when he stopped at the desk for his mail. In the dining-room he found several old friends seated here and there before substantial breakfasts: cattlemen and mining engineers from odd corners of the State, all looking fresh and well pleased with themselves. He had a word with one and another before he sat down at the little table by a window, where the Austrian head waiter stood attentively behind a chair. After his breakfast was put before him, the doctor began to run over his letters. There was one directed in Thea Kronborg's handwriting, forwarded from Moonstone. He saw with astonishment, as he put another lump of sugar into his cup, that this letter bore a New York postmark. He had known that Thea was in Mexico, traveling with some Chicago people, but New York, to a Denver man, seems much farther away than Mexico City. He put the letter behind his plate, upright against the stem of his water goblet, and looked at it thoughtfully while he drank his second cup of coffee. He had been a little anxious about Thea; she had not written to him for a long while.

As he never got good coffee at home, the doctor always drank three cups for breakfast when he was in Denver. Oscar knew just when to bring him a second pot, fresh and smoking. "And more cream, Oscar, please. You know I like lots of cream," the doctor murmured, as he opened the square envelope, marked in the upper right-hand corner, "Everett House, Union Square." The text of the letter was as follows:—

DEAR DOCTOR ARCHIE:—

I have not written to you for a long time, but it has not been unintentional. I could not write you frankly, and so I would not write at all. I can be frank with you now, but not by letter. It is a great deal to ask, but I wonder if you could come to New York to help me out? I have got into difficulties, and I need your advice. I need your friendship. I am afraid I must even ask you to lend me money, if you can without serious inconvenience. I have to go to Germany to study, and it can't be put off any longer. My voice is ready. Needless to say, I don't want any word of this to reach my family. They are the last people I would turn to, though I love my mother dearly. If you can come, please telegraph me at this hotel. Don't despair of me. I'll make it up to you yet.

Your old friend,
THEA KRONBORG.

This in a bold, jagged handwriting with a Gothic turn to the letters,—something between a highly sophisticated hand and a very unsophisticated one,—not in the least smooth or flowing.

The doctor bit off the end of a cigar nervously and read the letter through again, fumbling distractedly in his pockets for matches, while the waiter kept trying to call his attention to the box he had just placed before him. At last Oscar came out, as if the idea had just struck him, "Matches, sir?"

"Yes, thank you." The doctor slipped a coin into his palm and rose, crumpling Thea's letter in his hand and thrusting the others into his pocket unopened. He went back to the desk in the lobby and beckoned the clerk, upon whose kindness he threw himself apologetically.

"Harry, I've got to pull out unexpectedly. Call up the Burlington, will you, and ask them to route me to New York the quickest way, and to let us know. Ask for the hour I'll get in. I have to wire."

"Certainly, Dr. Archie. Have it for you in a minute." The young man's pallid, clean-scraped face was all sympathetic interest as he reached for the telephone. Dr. Archie put out his hand and stopped him.

"Wait a minute. Tell me, first, is Captain Harris down yet?"

"No, sir. The Captain hasn't come down yet this morning."

"I'll wait here for him. If I don't happen to catch him, nail him and get me. Thank you, Harry."

The doctor spoke gratefully and turned away. He began to pace the lobby, his hands behind him, watching the bronze elevator doors like a hawk. At last Captain Harris issued from one of them, tall and imposing, wearing a Stetson and fierce mustaches, a fur coat on his arm, a solitaire glittering upon his little finger and another in his black satin ascot. He was one of the grand old bluffers of those good old days. As gullible as a schoolboy, he had managed, with his sharp eye and knowing air and twisted blond mustaches, to pass himself off for an astute financier, and the Denver papers respectfully referred to him as the Rothschild of Cripple Creek.[2]

Dr. Archie stopped the Captain on his way to breakfast. "Must see you a minute, Captain. Can't wait. Want to sell you some shares in the San Felipe. Got to raise money."

The Captain grandly bestowed his hat upon an eager porter who had already lifted his fur coat tenderly from his arm and stood nursing it. In removing his hat, the Captain exposed a bald, flushed dome, thatched about the ears with yellowish gray hair. "Bad time to sell, doctor. You want to hold on to San Felipe, and buy more. What have you got to raise?"

"Oh, not a great sum. Five or six thousand. I've been buying up close and have run short."

"I see, I see. Well, doctor, you'll have to let me get through that door. I was out last night, and I'm going to get my bacon, if you lose your mine." He clapped Archie on the shoulder and pushed him along in front of him. "Come ahead with me, and we'll talk business."

Dr. Archie attended the Captain and waited while he gave his order, taking the seat the old promoter indicated.

"Now, sir," the Captain turned to him, "you don't want to sell anything. You must be under the impression that I'm one of these damned New England sharks that get their pound of flesh off the widow and orphan. If you're a little short, sign a note and I'll write a check. That's the way gentlemen do business. If you want to put up some San Felipe as collateral, let her go, but I shan't touch a share of it. Pens and ink, please, Oscar,"—he lifted a large forefinger to the Austrian.

The Captain took out his checkbook and a book of blank notes, and adjusted his nose-nippers.[3] He wrote a few words in one book and Archie wrote a few in the other. Then they each tore across perforations and exchanged slips of paper.

"That's the way. Saves office rent," the Captain commented with satisfaction, returning the books to his pocket. "And now, Archie, where are you off to?"

"Got to go East to-night. A deal waiting for me in New York." Dr. Archie rose.

The Captain's face brightened as he saw Oscar approaching with a tray, and he began tucking the corner of his napkin inside his collar, over his ascot. "Don't let them unload anything on you back there, doctor," he said genially, "and don't let them relieve you of anything, either. Don't let them get any Cripple stuff off you. We can manage our own silver out here, and we're going to take it out by the ton, sir!"

The doctor left the dining-room, and after another consultation with the clerk, he wrote his first telegram to Thea:—

MISS THEA KRONBORG,
Everett House, New York.
Will call at your hotel eleven o'clock Friday morning. Glad to come. Thank you.

ARCHIE

He stood and heard the message actually clicked off on the wire, with the feeling that she was hearing the click at the other end. Then he sat down in the lobby and wrote a note to his wife and one to the other doctor in Moonstone. When he at last issued out into the storm, it was with a feeling of elation rather than of anxiety. Whatever was wrong, he could make it right. Her letter had practically said so.

He tramped about the snowy streets, from the bank to the Union Station, where he shoved his money under the grating of the ticket window as if he could not get rid of it fast enough. He had never been in New York, never been farther east than Buffalo. "That's rather a shame," he reflected boyishly as he put the long tickets in his pocket, "for a man nearly forty years old." However, he thought as he walked up toward the club, he was on the whole glad that his first trip had a human interest, that he was going for something, and because he was wanted. He loved holidays. He felt as if he were going to Germany himself. "Queer,"—he went over it with the snow blowing in his face,—"but that sort of thing is more interesting than mines and making your daily bread. It's worth paying out to be in on it,—for a fellow like me. And when it's Thea— Oh, I back her!" he laughed aloud as he burst in at the door of the Athletic Club, powdered with snow.

Archie sat down before the New York papers and ran over the advertisements of hotels, but he was too restless to read. Probably he had better get a new overcoat, and he was not sure about the shape of his collars. "I don't want to look different to her from everybody else there," he mused. "I guess I'll go down and have Van look me over. He'll put me right."

So he plunged out into the snow again and started for his tailor's. When he passed a florist's shop he stopped and looked in at the window, smiling; how naturally pleasant things recalled one another. At the tailor's he kept whistling, "Flow gently, Sweet Afton,"[4] while Van Dusen advised him, until that resourceful tailor and haberdasher exclaimed, "You must have a date back there, doctor; you behave like a bridegroom," and made him remember that he wasn't one.

Before he let him go, Van put his finger on the Masonic pin in his client's lapel. "Mustn't wear that, doctor. Very bad form back there."[5]

II

FRED OTTENBURG, smartly dressed for the afternoon with a long black coat and gaiters was sitting in the dusty parlor of the Everett House. His manner was not in accord with his personal freshness, the good lines of his clothes, and the shining smoothness of his hair. His attitude was one of deep dejection, and his face, though it had the cool, unimpeachable fairness possible only to a very blond young man, was by no means happy. A page shuffled into the room and looked about. When he made out the dark figure in a shadowy corner, tracing over the carpet pattern with a cane, he droned, "The lady says you can come up, sir."

Fred picked up his hat and gloves and followed the creature, who seemed an aged boy in uniform, through dark corridors that smelled of old carpets. The page knocked at the door of Thea's sitting-room, and then wandered away. Thea came to the door with a telegram in her hand. She asked Ottenburg to come in and pointed to one of the clumsy, sullen-looking chairs that were as thick as they were high. The room was brown with time, dark in spite of two windows that opened on Union Square, with dull curtains and carpet, and heavy, respectable-looking furniture in somber colors. The place was saved from utter dismalness by a coal fire under the black marble mantelpiece,—brilliantly reflected in a long mirror that hung between the two windows. This was the first time Fred had seen the room, and he took it in quickly, as he put down his hat and gloves.

Thea seated herself at the walnut writing-desk, still holding the slip of yellow paper. "Dr. Archie is coming," she said. "He will be here Friday morning."

"Well, that's good, at any rate," her visitor replied with a determined effort at cheerfulness. Then, turning to the fire, he added blankly, "If you want him."

"Of course I want him. I would never have asked such a thing of him if I hadn't wanted him a great deal. It's a very expensive trip." Thea spoke severely. Then she went on, in a milder tone. "He doesn't say anything about the money, but I think his coming means that he can let me have it."

Fred was standing before the mantel, rubbing his hands together nervously. "Probably. You are still determined to call on him?" He sat down tentatively in the chair Thea had indicated. "I don't see why you won't borrow from me, and let him sign with you, for instance. That would constitute a perfectly regular business transaction. I could bring suit against either of you for my money."

Thea turned toward him from the desk. "We won't take that up again, Fred. I should have a different feeling about it if I went on your money. In a way I shall feel freer on Dr. Archie's, and in another way I shall feel more bound. I shall try even harder." She paused. "He is almost like my father," she added irrelevantly.

"Still, he isn't, you know," Fred persisted. "It wouldn't be anything new. I've loaned money to students before, and got it back, too."

"Yes; I know you're generous," Thea hurried over it, "but this will be the best way. He will be here on Friday, did I tell you?"

"I think you mentioned it. That's rather soon. May I smoke?" he took out a small cigarette case. "I suppose you'll be off next week?" he asked as he struck a match.

"Just as soon as I can," she replied with a restless movement of her arms, as if her dark-blue dress were too tight for her. "It seems as if I'd been here forever."

"And yet," the young man mused, "we got in only four days ago. Facts really don't count for much, do they? It's all in the way people feel: even in little things."

Thea winced, but she did not answer him. She put the telegram back in its envelope and placed it carefully in one of the pigeonholes of the desk.

"I suppose," Fred brought out with effort, "that your friend is in your confidence?"

"He always has been. I shall have to tell him about myself. I wish I could without dragging you in."

Fred shook himself. "Don't bother about where you drag me, please," he put in, flushing. "I don't give—" he subsided suddenly.

"I'm afraid," Thea went on gravely, "that he won't understand. He'll be hard on you."

Fred studied the white ash of his cigarette before he flicked it off. "You mean he'll see me as even worse than I am. Yes, I suppose I shall look very low to him: a fifth-rate scoundrel. But that only matters in so far as it hurts his feelings."

Thea sighed. "We'll both look pretty low. And after all, we must really be just about as we shall look to him."

Ottenburg started up and threw his cigarette into the grate. "That I deny. Have you ever been really frank with this preceptor of your childhood, even when you *were* a child? Think a minute, have you? Of course not! From your cradle, as I once told you, you've been 'doing it' on the side, living your own life, admitting to yourself things that would horrify him. You've always deceived him to the extent of letting him think you different from what you are. He couldn't understand then, he can't understand now. So why not spare yourself and him?"

She shook her head. "Of course, I've had my own thoughts. Maybe he has had his, too. But I've never done anything before that he would much mind. I must put myself right with him,—as right as I can,—to begin over. He'll make allowances for me. He always has. But I'm afraid he won't for you."

"Leave that to him and me. I take it you want me to see him?" Fred sat down again and began absently to trace the carpet pattern with his cane. "At the worst," he spoke wanderingly, "I thought you'd perhaps let me go in on the business end of it and invest along with you. You'd put in your talent and ambition and hard work, and I'd put in the money and—well, nobody's good wishes are to be scorned, not even mine. Then, when the thing panned out big, we could share together. Your doctor friend hasn't cared half so much about your future as I have."

"He's cared a good deal. He doesn't know as much about such things as you do. Of course you've been a great deal more help to me than any one else ever has," Thea said quietly. The black clock on the mantel began to strike. She listened to the five strokes and then said, "I'd have liked your helping me eight months ago. But now, you'd simply be keeping me."

"You weren't ready for it eight months ago." Fred leaned back at last in his chair. "You simply weren't ready for it. You were too tired. You were too timid. Your whole tone was too low. You couldn't rise from a chair like that,"—she had started up apprehensively and gone toward the window.—"You were fumbling and awkward. Since then you've come into your personality. You were always locking horns with it before. You were a sullen little drudge eight months ago, afraid of being caught at either looking or moving like yourself. Nobody could tell anything about you. A voice is not an instrument that's found ready-made. A voice is personality. It can be as big as a circus and as common as dirt.—There's good money in that kind, too, but I don't happen to be interested in them.—Nobody could tell much about what you might be able to do, last winter. I divined more than anybody else."

"Yes, I know you did." Thea walked over to the old-fashioned mantel and held her hands down to the glow of the fire. "I owe so much to you, and that's what makes things hard. That's why I have to get away from you altogether. I depend on you for so many things. Oh, I did even last winter, in Chicago!" She knelt down by the grate and held her hands closer to the coals. "And one thing leads to another."

Ottenburg watched her as she bent toward the fire. His glance brightened a little. "Anyhow, you couldn't look as you do now, before you knew me. You *were* clumsy. And whatever you do now, you do splendidly. And you can't cry enough to spoil your face for more than ten minutes. It comes right back, in spite of you. It's only since you've known me that you've let yourself be beautiful."

Without rising she turned her face away. Fred went on impetuously. "Oh, you can turn it away from me, Thea; you can take it away from me! All the same—" his spurt died and he fell back. "How can you turn on me so, after all!" he sighed.

"I haven't. But when you arranged with yourself to take me in like that, you couldn't have been thinking very kindly of me. I can't understand how you carried it through, when I was so easy, and all the circumstances were so easy."

Her crouching position by the fire became threatening. Fred got up, and Thea also rose.

"No," he said, "I can't make you see that now. Some time later, perhaps, you will understand better. For one thing, I honestly could not imagine that words, names, meant so much to you." Fred was talking with the desperation of a man who has put himself in the wrong and who yet feels that there was an idea of truth in his conduct. "Suppose that you had married your brakeman and lived with him year after year, caring for him even less than you do for your doctor, or for Harsanyi. I suppose you would have felt quite all right about it, because that relation has a name in good standing. To me, that seems—sickening!" He took a rapid turn about the room and then as Thea remained standing, he rolled one of the elephantine chairs up to the hearth for her.

"Sit down and listen to me for a moment, Thea." He began pacing from the hearthrug to the window and back again, while she sat down compliantly. "Don't you know most of the people in the world are not individuals at all? They never have an individual idea or experience. A lot of girls go to boarding-school together, come out the same season, dance at the same parties, are married off in groups, have their babies at about the same time, send their children to school together, and so the human crop renews itself. Such women know as much about the reality of the forms they go through as they know about the wars they learn the dates of. They get their most personal experiences out of novels and plays. Everything is second-hand with them. Why, you *couldn't* live like that."

Thea sat looking toward the mantel, her eyes half closed, her chin level, her head set as if she were enduring something. Her hands, very white, lay passive on her dark gown. From the window corner Fred looked at them and at her. He shook his head and flashed an angry, tormented look out into the blue twilight over the Square, through which muffled cries and calls and the clang of car bells came up from the street. He turned again and began to pace the floor, his hands in his pockets.

"Say what you will, Thea Kronborg, you are not that sort of person. You will never sit alone with a pacifier and a novel. You won't subsist on what the old ladies have put into the bottle for you. You will always break through into the realities. That was the first thing Harsanyi found out about you; that you couldn't be kept on the outside. If you'd lived in Moonstone all your life and got on with the discreet brakeman, you'd have had just the same nature. Your children would have been the realities then, probably. If they'd been commonplace, you'd have killed them with driving. You'd have managed some way to live twenty times as much as the people around you."

Fred paused. He sought along the shadowy ceiling and heavy mouldings for words. When he began again, his voice was lower, and at first he spoke with less conviction, though again it grew on him. "Now I knew all this—oh, knew it better than I can ever make you understand! You've been running a handicap. You had no time to lose. I wanted you to have what you need and to get on fast—get through with me, if need be; I counted on that. You've no time to sit round and analyze your conduct or your feelings. Other women give their whole lives to it. They've nothing else to do. Helping a man to get his divorce is a career for them; just the sort of intellectual exercise they like."

Fred dived fiercely into his pockets as if he would rip them out and scatter their contents to the winds. Stopping before her, he took a deep breath and went on again, this time slowly. "All that sort of thing is foreign to you. You'd be nowhere at it. You haven't that kind of mind. The grammatical niceties of conduct are dark to you. You're simple—and poetic." Fred's voice seemed to be wandering about in the thickening dusk. "You won't play much. You won't, perhaps, love many times." He paused. "And you did love me, you know. Your railroad friend would have understood me. I *could* have thrown you back. The reverse was there,—it stared me in the face,—but I couldn't pull it. I let you drive ahead." He threw out his hands. What Thea noticed, oddly enough, was the flash of the firelight on his cuff link. He turned again. "And you'll always drive ahead," he muttered. "It's your way."

There was a long silence. Fred had dropped into a chair. He seemed, after such an explosion, not to have a word left in him. Thea put her hand to the back of her neck and pressed it, as if the muscles there were aching.

"Well," she said at last, "I at least overlook more in you than I do in myself. I am always excusing you to myself. I don't do much else."

"Then why, in Heaven's name, won't you let me be your friend? You make a scoundrel of me, borrowing money from another man to get out of my clutches."

"If I borrow from him, it's to study. Anything I took from you would be different. As I said before, you'd be keeping me."

"Keeping! I like your language. It's pure Moonstone, Thea,—like your point of view. I wonder how long you'll be a Methodist." He turned away bitterly.

"Well, I've never said I wasn't Moonstone, have I? I am, and that's why I want Dr. Archie. I can't see anything so funny about Moonstone, you know." She pushed her chair back a little from the hearth and clasped her hands over her knee, still looking thoughtfully into the red coals. "We always come back to the same thing, Fred. The name, as you call it, makes a difference to me how I feel about myself. You would have acted very differently with a girl of your own kind, and that's why I can't take anything from you now. You've made everything impossible. Being married is one thing and not being married is the other thing, and that's all there is to it. I can't see how you reasoned with yourself, if you took the trouble to reason. You say I was too much alone, and yet what you did was to cut me off more than I ever had been. Now I'm going to try to make good to my friends out there. That's all there is left for me."

"Make good to your friends!" Fred burst out. "What one of them cares as I care, or believes as I believe? I've told you I'll never ask a gracious word from you until I can ask it with all the churches in Christendom at my back."

Thea looked up, and when she saw Fred's face, she thought sadly that he, too, looked as if things were spoiled for him. "If you know me as well as you say you do, Fred," she said slowly, "then you are not being honest with yourself. You know that I can't do things halfway. If you kept me at all—you'd keep me." She dropped her head wearily on her hand and sat with her forehead resting on her fingers.

Fred leaned over her and said just above his breath, "Then, when I get that divorce, you'll take it up with me again? You'll at least let me know, warn me, before there is a serious question of anybody else?"

Without lifting her head, Thea answered him. "Oh, I don't think there will ever be a question of anybody else. Not if I can help it. I suppose I've given you every reason to think there will be,—at once, on shipboard, any time."

Ottenburg drew himself up like a shot. "Stop it, Thea!" he said sharply. "That's one thing you've never done. That's like any common woman." He saw her shoulders lift a little and grow calm. Then he went to the other side of the room and took up his hat and gloves from the sofa. He came back cheerfully. "I didn't drop in to bully you this afternoon. I came to coax you to go out for tea with me somewhere." He waited, but she did not look up or lift her head, still sunk on her hand.

Her handkerchief had fallen. Fred picked it up and put it on her knee, pressing her fingers over it. "Good-night, dear and wonderful," he whispered,—"wonderful and dear! How can you ever get away from me when I will always follow you, through every wall, through every door, wherever you go." He looked down at her bent head, and the curve of her neck that was so sad. He stooped, and with his lips just touched her hair where the firelight made it ruddiest. "I didn't know I had it in me, Thea. I thought it was all a fairy tale. I don't know myself any more." He closed his eyes and breathed deeply. "The salt's all gone out of your hair. It's full of sun and wind again. I believe it has memories." Again she heard him take a deep breath. "I could do without you for a lifetime, if that would give you to yourself. A woman like you doesn't find herself, alone."

She thrust her free hand up to him. He kissed it softly, as if she were asleep and he were afraid of waking her.

From the door he turned back irrelevantly. "As to your old friend, Thea, if he's to be here on Friday, why,"—he snatched out his watch and held it down to catch the light from the grate,—"he's on the train now! That ought to cheer you. Good-night." She heard the door close.

III

ON FRIDAY AFTERNOON Thea Kronborg was walking excitedly up and down her sitting-room, which at that hour was flooded by thin, clear sunshine. Both windows were open, and the fire in the grate was low, for the day was one of those false springs that sometimes blow into New York from the sea in the middle of winter, soft, warm, with a persuasive salty moisture in the air and a relaxing thaw under foot. Thea was flushed and animated, and she seemed as restless as the sooty sparrows that chirped and cheeped distractingly about the windows. She kept looking at the black clock, and then down into the Square. The room was full of flowers, and she stopped now and then to arrange them or to move them into the sunlight. After the bellboy came to announce a visitor, she took some Roman hyacinths from a glass and stuck them in the front of her dark-blue dress.

When at last Fred Ottenburg appeared in the doorway, she met him with an exclamation of pleasure. "I am glad you've come, Fred. I was afraid you might not get my note, and I wanted to see you before you see Dr. Archie. He's so nice!" She brought her hands together to emphasize her statement.

"Is he? I'm glad. You see I'm quite out of breath. I didn't wait for the elevator, but ran upstairs. I was so pleased at being sent for." He dropped his hat and overcoat. "Yes, I should say he is nice! I don't seem to recognize all of these," waving his handkerchief about at the flowers.

"Yes, he brought them himself, in a big box. He brought lots with him besides flowers. Oh, lots of things! The old Moonstone feeling,"—Thea moved her hand back and forth in the air, fluttering her fingers,—"the feeling of starting out, early in the morning, to take my lesson."

"And you've had everything out with him?"

"No, I haven't."

"Haven't?" He looked up in consternation.

"No, I haven't!" Thea spoke excitedly, moving about over the sunny patches on the grimy carpet. "I've lied to him, just as you said I had always lied to him, and that's why I'm so happy. I've let him think what he likes to think. Oh, I couldn't do anything else, Fred,"—she shook her head emphatically. "If you'd seen him when he came in, so pleased and excited! You see this is a great adventure for him. From the moment I began to talk to him, he entreated me not to say too much, not to spoil his notion of me. Not in so many words, of course. But if you'd seen his eyes, his face, his kind hands! Oh, no! I couldn't." She took a deep breath, as if with a renewed sense of her narrow escape.

"Then, what did you tell him?" Fred demanded.

Thea sat down on the edge of the sofa and began shutting and opening her hands nervously. "Well, I told him enough, and not too much. I told him all about how good you were to me last winter, getting me engagements and things, and how you had helped me with my work more than anybody. Then I told him about how you sent me down to the ranch when I had no money or anything." She paused and wrinkled her forehead. "And I told him that I wanted to marry you and ran away to Mexico with you, and that I was awfully happy until you told me that you couldn't marry me because—well, I told him why." Thea dropped her eyes and moved the toe of her shoe about restlessly on the carpet.

"And he took it from you, like that?" Fred asked, almost with awe.

"Yes, just like that, and asked no questions. He was hurt; he had some wretched moments. I could see him squirming and squirming and trying to get past it. He kept shutting his eyes and rubbing his forehead. But when I told him that I absolutely knew you wanted to marry me, that you would whenever you could, that seemed to help him a good deal."

"And that satisfied him?" Fred asked wonderingly. He could not quite imagine what kind of person Dr. Archie might be.

"He took me by the shoulders once and asked, oh, in such a frightened way, 'Thea, was he *good* to you, this young man?' When I told him you were, he looked at me again: 'And you care for him a great deal, you believe in him?' Then he seemed satisfied." Thea paused. "You see, he's just tremendously good, and tremendously afraid of things—of some things. Otherwise he would have got rid of Mrs. Archie." She looked up suddenly: "You were right, though; one can't tell people about things they don't know already."

Fred stood in the window, his back to the sunlight, fingering the jonquils. "Yes, you can, my dear. But you must tell it in such a way that they don't know you're telling it, and that they don't know they're hearing it."

Thea smiled past him, out into the air. "I see. It's a secret. Like the sound in the shell."

"What's that?" Fred was watching her and thinking how moving that faraway expression, in her, happened to be. "What did you say?"

She came back. "Oh, something old and Moonstony! I have almost forgotten it myself. But I feel better than I thought I ever could again. I can't wait to be off. Oh, Fred," she sprang up, "I want to get at it!"

As she broke out with this, she threw up her head and lifted herself a little on her toes. Fred colored and looked at her fearfully, hesitatingly. Her eyes, which looked out through the window, were bright—they had no memories. No, she did not remember. That momentary elevation had no associations for her. It was unconscious.

He looked her up and down and laughed and shook his head. "You are just all I want you to be—and that is,—not for me! Don't worry, you'll get at it. You are at it. My God! have you ever, for one moment, been at anything else?"

Thea did not answer him, and clearly she had not heard him. She was watching something out in the thin light of the false spring and its treacherously soft air.

Fred waited a moment. "Are you going to dine with your friend to-night?"

"Yes. He has never been in New York before. He wants to go about. Where shall I tell him to go?"

"Wouldn't it be a better plan, since you wish me to meet him, for you both to dine with me? It would seem only natural and friendly. You'll have to live up a little to his notion of us." Thea seemed to consider the suggestion favorably. "If you wish him to be easy in his mind," Fred went on, "that would help. I think, myself, that we are rather nice together. Put on one of the new dresses you got down there, and let him see how lovely you can be. You owe him some pleasure, after all the trouble he has taken."

Thea laughed, and seemed to find the idea exciting and pleasant. "Oh, very well! I'll do my best. Only don't wear a dress coat, please. He hasn't one, and he's nervous about it."

Fred looked at his watch. "Your monument up there is fast. I'll be here with a cab at eight. I'm anxious to meet him. You've given me the strangest idea of his callow innocence and aged indifference."

She shook her head. "No, he's none of that. He's very good, and he won't admit things. I love him for it. Now, as I look back on it, I see that I've always, even when I was little, shielded him."

As she laughed, Fred caught the bright spark in her eye that he knew so well, and held it for a happy instant. Then he blew her a kiss with his finger-tips and fled.

IV

AT NINE O'CLOCK THAT EVENING our three friends were seated in the balcony of a French restaurant, much gayer and more intimate than any that exists in New York to-day. This old restaurant was built by a lover of pleasure, who knew that to dine gayly human beings must have the reassurance of certain limitations of space and of a certain definite style; that the walls must be near enough to suggest shelter, the ceiling high enough to give the chandeliers a setting. The place was crowded with the kind of people who dine late and well, and Dr. Archie, as he watched the animated groups in the long room below the balcony, found this much the most festive scene he had ever looked out upon. He said to himself, in a jovial mood somewhat sustained by the cheer of the board, that this evening alone was worth his long journey. He followed attentively the orchestra, ensconced at the farther end of the balcony, and told Thea it made him feel "quite musical" to recognize "The Invitation to the Dance" or "The Blue Danube," and that he could remember just what kind of day it was when he heard her practicing them at home, and lingered at the gate to listen.

For the first few moments, when he was introduced to young Ottenburg in the parlor of the Everett House, the doctor had been awkward and unbending. But Fred, as his father had often observed, "was not a good mixer for nothing." He had brought Dr. Archie around during the short cab ride, and in an hour they had become old friends.

From the moment when the doctor lifted his glass and, looking consciously at Thea, said, "To your success," Fred liked him. He felt his quality; understood his courage in some directions and what Thea called his timidity in others, his unspent and miraculously preserved youthfulness. Men could never impose upon the doctor, he guessed, but women always could. Fred liked, too, the doctor's manner with Thea, his bashful admiration and the little hesitancy by which he betrayed his consciousness of the change in her. It was just this change that, at present, interested Fred more than anything else. That, he felt, was his "created value," and it was his best chance for any peace of mind. If that were not real, obvious to an old friend like Archie, then he cut a very poor figure, indeed.

Fred got a good deal, too, out of their talk about Moonstone. From her questions and the doctor's answers he was able to form some conception of the little world that was almost the measure of Thea's experience, the one bit of the human drama that she had followed with sympathy and understanding. As the two ran over the list of their friends, the mere sound of a name seemed to recall volumes to each of them, to indicate mines of knowledge and observation they had in common. At some names they laughed delightedly, at some indulgently and even tenderly.

"You two young people must come out to Moonstone when Thea gets back," the doctor said hospitably.

"Oh, we shall!" Fred caught it up. "I'm keen to know all these people. It is very tantalizing to hear only their names."

"Would they interest an outsider very much, do you think, Dr. Archie?" Thea leaned toward him. "Isn't it only because we've known them since I was little?"

The doctor glanced at her deferentially. Fred had noticed that he seemed a little afraid to look at her squarely—perhaps a trifle embarrassed by a mode of dress to which he was unaccustomed. "Well, you are practically an outsider yourself, Thea, now," he observed smiling. "Oh, I know," he went on quickly in response to her gesture of protest,—"I know you don't change toward your old friends, but you can see us all from a distance now. It's all to your advantage that you can still take your old interest, isn't it, Mr. Ottenburg?"

"That's exactly one of her advantages, Dr. Archie. Nobody can ever take that away from her, and none of us who came later can ever hope to rival Moonstone in the impression we make. Her scale of values will always be the Moonstone scale. And, with an artist, that *is* an advantage." Fred nodded.

Dr. Archie looked at him seriously. "You mean it keeps them from getting affected?"

"Yes; keeps them from getting off the track generally."

While the waiter filled the glasses, Fred pointed out to Thea a big black French barytone who was eating anchovies by their tails at one of the tables below, and the doctor looked about and studied his fellow diners.

"Do you know, Mr. Ottenburg," he said deeply, "these people all look happier to me than our Western people do. Is it simply good manners on their part, or do they get more out of life?"

Fred laughed to Thea above the glass he had just lifted. "Some of them are getting a good deal out of it now, doctor. This is the hour when bench-joy brightens."[1]

Thea chuckled and darted him a quick glance. "Bench-joy! Where did you get that slang?"

"That happens to be very old slang, my dear. Older than Moonstone or the sovereign State of Colorado. Our old friend Mr. Nathanmeyer could tell us why it happens to hit you." He leaned forward and touched Thea's wrist, "See that fur coat just coming in, Thea. It's D'Albert.[2] He's just back from his Western tour. Fine head, hasn't he?"

"To go back," said Dr. Archie; "I insist that people do look happier here. I've noticed it even on the street, and especially in the hotels."

Fred turned to him cheerfully. "New York people live a good deal in the fourth dimension, Dr. Archie. It's that you notice in their faces."

The doctor was interested. "The fourth dimension," he repeated slowly; "and is that slang, too?"

"No,"—Fred shook his head,—"that's merely a figure. I mean that life is not quite so personal here as it is in your part of the world. People are more taken up by hobbies, interests that are less subject to reverses than their personal affairs. If you're interested in Thea's voice, for instance, or in voices in general, that interest is just the same, even if your mining stocks go down."

The doctor looked at him narrowly. "You think that's about the principal difference between country people and city people, don't you?"

Fred was a little disconcerted at being followed up so resolutely, and he attempted to dismiss it with a pleasantry. "I've never thought much about it, doctor. But I should say, on the spur of the moment, that that is one of the principal differences between people anywhere. It's the consolation of fellows like me who don't accomplish much. The fourth dimension is not good for business, but we think we have a better time."

Dr. Archie leaned back in his chair. His heavy shoulders were contemplative. "And she," he said slowly; "should you say that she is one of the kind you refer to?" He inclined his head toward the shimmer of the pale-green dress beside him. Thea was leaning, just then, over the balcony rail, her head in the light from the chandeliers below.

"Never, never!" Fred protested. "She's as hard-headed as the worst of you—with a difference."

The doctor sighed. "Yes, with a difference; something that makes a good many revolutions to the second. When she was little I used to feel her head to try to locate it."

Fred laughed. "Did you, though? So you were on the track of it? Oh, it's there! We can't get round it, miss," as Thea looked back inquiringly. "Dr. Archie, there's a fellow townsman of yours I feel a real kinship for." He pressed a cigar upon Dr. Archie and struck a match for him. "Tell me about Spanish Johnny."

The doctor smiled benignantly through the first waves of smoke. "Well, Johnny's an old patient of mine, and he's an old admirer of Thea's. She was born a cosmopolitan, and I expect she learned a good deal from Johnny when she used to run away and go to Mexican Town. We thought it a queer freak then."

The doctor launched into a long story, in which he was often eagerly interrupted or joyously confirmed by Thea, who was drinking her coffee and forcing open the petals of the roses with an ardent and rather rude hand. Fred settled down into enjoying his comprehension of his guests. Thea, watching Dr. Archie and interested in his presentation, was unconsciously impersonating her suave, gold-tinted friend. It was delightful to see her so radiant and responsive again. She had kept her promise about looking her best; when one could so easily get together the colors of an apple branch in early spring, that was not hard to do. Even Dr. Archie felt, each time he looked at her, a fresh consciousness. He recognized the fine texture of her mother's skin, with the difference that, when she reached across the table to give him a bunch of grapes, her arm was not only white, but somehow a little dazzling. She seemed to him taller, and freer in all her movements. She had now a way of taking a deep breath when she was interested, that made her seem very strong, somehow, and brought her at one quite overpoweringly. If he seemed shy, it was not that he was intimidated by her worldly clothes, but that her greater positiveness, her whole augmented self, made him feel that his accustomed manner toward her was inadequate.

Fred, on his part, was reflecting that the awkward position in which he had placed her would not confine or chafe her long. She looked about at other people, at other women, curiously. She was not quite sure of herself, but she was not in the least afraid or apologetic. She seemed to sit there on the edge, emerging from one world into another, taking her bearings, getting an idea of the concerted movement about her, but with absolute self-confidence. So far from shrinking, she expanded. The mere kindly effort to please Dr. Archie was enough to bring her out.

There was much talk of auræ at that time, and Fred mused that every beautiful, every compellingly beautiful woman, had an aura, whether other people did or no. There was, certainly, about the woman he had brought up from Mexico, such an emanation. She existed in more space than she occupied by measurement. The enveloping air about her head and shoulders was subsidized—was more moving than she herself, for in it lived the awakenings, all the first sweetness that life kills in people. One felt in her such a wealth of *Jugendzeit,*[3] all those flowers of the mind and the blood that bloom and perish by the myriad in the few exhaustless years when the imagination first kindles. It was in watching her as she emerged like this, in being near and not too near, that one got, for a moment, so much that one had lost; among other legendary things the legendary theme of the absolutely magical power of a beautiful woman.

After they had left Thea at her hotel, Dr. Archie admitted to Fred, as they walked up Broadway through the rapidly chilling air, that once before he had seen their young friend flash up into a more potent self, but in a darker mood. It was in his office one night, when she was at home the summer before last. "And then I got the idea," he added simply, "that she would not live like other people: that, for better or worse, she had uncommon gifts."

"Oh, we'll see that it's for better, you and I," Fred reassured him. "Won't you come up to my hotel with me? I think we ought to have a long talk."

"Yes, indeed," said Dr. Archie gratefully; "I think we ought."

V

THEA WAS TO SAIL on Tuesday, at noon, and on Saturday Fred Ottenburg arranged for her passage, while she and Dr. Archie went shopping. With rugs and sea-clothes she was already provided; Fred had got everything of that sort she needed for the voyage up from Vera Cruz. On Sunday afternoon Thea went to see the Harsanyis. When she returned to her hotel, she found a note from Ottenburg, saying that he had called and would come again to-morrow.

On Monday morning, while she was at breakfast, Fred came in. She knew by his hurried, distracted air as he entered the dining-room that something had gone wrong. He had just got a telegram from home. His mother had been thrown from her carriage and hurt; a concussion of some sort, and she was unconscious. He was leaving for St. Louis that night on the eleven o'clock train. He had a great deal to attend to during the day. He would come that evening, if he might, and stay with her until train time, while she was doing her packing. Scarcely waiting for her consent, he hurried away.

All day Thea was somewhat cast down. She was sorry for Fred, and she missed the feeling that she was the one person in his mind. He had scarcely looked at her when they exchanged words at the breakfast-table. She felt as if she were set aside, and she did not seem so important even to herself as she had yesterday. Certainly, she reflected, it was high time that she began to take care of herself again. Dr. Archie came for dinner, but she sent him away early, telling him that she would be ready to go to the boat with him at half-past ten the next morning. When she went upstairs, she looked gloomily at the open trunk in her sitting-room, and at the trays piled on the sofa. She stood at the window and watched a quiet snowstorm spending itself over the city. More than anything else, falling snow always made her think of Moonstone; of the Kohlers' garden, of Thor's sled, of dressing by lamplight and starting off to school before the paths were broken.

When Fred came, he looked tired, and he took her hand almost without seeing her.

"I'm so sorry, Fred. Have you had any more word?"

"She was still unconscious at four this afternoon. It doesn't look very encouraging." He approached the fire and warmed his hands. He seemed to have contracted, and he had not at all his habitual ease of manner. "Poor mother!" he exclaimed; "nothing like this should have happened to her. She has so much pride of person. She's not at all an old woman, you know. She's never got beyond vigorous and rather dashing middle age." He turned abruptly to Thea and for the first time really looked at her. "How badly things come out! She'd have liked you for a daughter-in-law. Oh, you'd have fought like the devil, but you'd have respected each other." He sank into a chair and thrust his feet out to the fire. "Still," he went on thoughtfully, seeming to address the ceiling, "it might have been bad for you. Our big German houses, our good German cooking—you might have got lost in the upholstery. That substantial comfort might take the temper out of you, dull your edge. Yes," he sighed, "I guess you were meant for the jolt of the breakers."

"I guess I'll get plenty of jolt," Thea murmured, turning to her trunk.

"I'm rather glad I'm not staying over until to-morrow," Fred reflected. "I think it's easier for me to glide out like this. I feel now as if everything were rather casual, anyhow. A thing like that dulls one's feelings."

Thea, standing by her trunk, made no reply. Presently he shook himself and rose. "Want me to put those trays in for you?"

"No, thank you. I'm not ready for them yet."

Fred strolled over to the sofa, lifted a scarf from one of the trays and stood abstractedly drawing it through his fingers. "You've been so kind these last few days, Thea, that I began to hope you might soften a little; that you might ask me to come over and see you this summer."

"If you thought that, you were mistaken," she said slowly. "I've hardened, if anything. But I shan't carry any grudge away with me, if you mean that."

He dropped the scarf. "And there's nothing—nothing at all you'll let me do?"

"Yes, there is one thing, and it's a good deal to ask. If I get knocked out, or never get on, I'd like you to see that Dr. Archie gets his money back. I'm taking three thousand dollars of his."

"Why, of course I shall. You may dismiss that from your mind. How fussy you are about money, Thea. You make such a point of it." He turned sharply and walked to the windows.

Thea sat down in the chair he had quitted. "It's only poor people who feel that way about money, and who are really honest," she said gravely. "Sometimes I think that to be really honest, you must have been so poor that you've been tempted to steal."

"To what?"

"To steal. I used to be, when I first went to Chicago and saw all the things in the big stores there. Never anything big, but little things, the kind I'd never seen before and could never afford. I did take something once, before I knew it."

Fred came toward her. For the first time she had his whole attention, in the degree to which she was accustomed to having it. "Did you? What was it?" he asked with interest.

"A sachet. A little blue silk bag of orris-root powder. There was a whole counterful of them, marked down to fifty cents. I'd never seen any before, and they seemed irresistible. I took one up and wandered about the store with it. Nobody seemed to notice, so I carried it off."

Fred laughed. "Crazy child! Why, your things always smell of orris; is it a penance?"

"No, I love it. But I saw that the firm didn't lose anything by me. I went back and bought it there whenever I had a quarter to spend. I got a lot to take to Arizona. I made it up to them."

"I'll bet you did!" Fred took her hand. "Why didn't I find you that first winter? I'd have loved you just as you came!"

Thea shook her head. "No, you wouldn't, but you might have found me amusing. The Harsanyis said yesterday afternoon that I wore such a funny cape and that my shoes always squeaked. They think I've improved. I told them it was your doing if I had, and then they looked scared."

"Did you sing for Harsanyi?"

"Yes. He thinks I've improved there, too. He said nice things to me. Oh, he was very nice! He agrees with you about my going to Lehmann,[1] if she'll take me. He came out to the elevator with me, after we had said good-bye. He said something nice out there, too, but he seemed sad."

"What was it that he said?"

"He said, 'When people, serious people, believe in you, they give you some of their best, so—take care of it, Miss Kronborg.' Then he waved his hands and went back."

"If you sang, I wish you had taken me along. Did you sing well?" Fred turned from her and went back to the window. "I wonder when I shall hear you sing again." He picked up a bunch of violets and smelled them. "You know, your leaving me like this—well, it's almost inhuman to be able to do it so kindly and unconditionally."

"I suppose it is. It was almost inhuman to be able to leave home, too,—the last time, when I knew it was for good. But all the same, I cared a great deal more than anybody else did. I lived through it. I have no choice now. No matter how much it breaks me up, I have to go. Do I seem to enjoy it?"

Fred bent over her trunk and picked up something which proved to be a score, clumsily bound. "What's this? Did you ever try to sing this?" He opened it and on the engraved title-page read Wunsch's inscription, "*Einst, O Wunder!*" He looked up sharply at Thea.

"Wunsch gave me that when he went away. I've told you about him, my old teacher in Moonstone. He loved that opera."

Fred went toward the fireplace, the book under his arm, singing softly:—

> *"Einst, O Wunder, entblüht auf meinem Grabe,*
> *Eine Blume der Asche meines Herzens,*[2]

"You have no idea at all where he is, Thea?" He leaned against the mantel and looked down at her.

"No, I wish I had. He may be dead by this time. That was five years ago, and he used himself hard. Mrs. Kohler was always afraid he would die off alone somewhere and be stuck under the prairie. When we last heard of him, he was in Kansas."

"If he were to be found, I'd like to do something for him. I seem to get a good deal of him from this." He opened the book again, where he kept the place with his finger, and scrutinized the purple ink. "How like a German! Had he ever sung the song for you?"

"No. I didn't know where the words were from until once, when Harsanyi sang it for me, I recognized them."

Fred closed the book. "Let me see, what was your noble brakeman's name?"

Thea looked up with surprise. "Ray, Ray Kennedy."

"Ray Kennedy!" he laughed. "It couldn't well have been better! Wunsch and Dr. Archie, and Ray, and I,"—he told them off on his fingers,—"your whistling-posts! You haven't done so badly. We've backed you as we could, some in our weakness and some in our might. In your dark hours—and you'll have them—you may like to remember us." He smiled whimsically and dropped the score into the trunk. "You are taking that with you?"

"Surely I am. I haven't so many keepsakes that I can afford to leave that. I haven't got many that I value so highly."

"That you value so highly?" Fred echoed her gravity playfully. "You are delicious when you fall into your vernacular." He laughed half to himself.

"What's the matter with that? Isn't it perfectly good English?"

"Perfectly good Moonstone, my dear. Like the ready-made clothes that hang in the windows, made to fit everybody and fit nobody, a phrase that can be used on all occasions. Oh,"—he started across the room again,—"that's one of the fine things about your going! You'll be with the right sort of people and you'll learn a good, live, warm German, that will be like yourself. You'll get a new speech full of shades and color like your voice; alive, like your mind. It will be almost like being born again, Thea."

She was not offended. Fred had said such things to her before, and she wanted to learn. In the natural course of things she would never have loved a man from whom she could not learn a great deal.

"Harsanyi said once," she remarked thoughtfully, "that if one became an artist one had to be born again, and that one owed nothing to anybody."

"Exactly. And when I see you again I shall not see you, but your daughter. May I?" He held up his cigarette case questioningly and then began to smoke, taking up again the song which ran in his head:—

"Deutlich schimmert auf jedem Purpurblättchen, Adelaide!"[3]

"I have half an hour with you yet, and then, exit Fred." He walked about the room, smoking and singing the words under his breath.

"You'll like the voyage," he said abruptly. "That first approach to a foreign shore, stealing up on it and finding it—there's nothing like it. It wakes up everything that's asleep in you. You won't mind my writing to some people in Berlin? They'll be nice to you."

"I wish you would." Thea gave a deep sigh. "I wish one could look ahead and see what is coming to one."

"Oh, no!" Fred was smoking nervously; "that would never do. It's the uncertainty that makes one try. You've never had any sort of chance, and now I fancy you'll make it up to yourself. You'll find the way to let yourself out in one long flight."

Thea put her hand on her heart. "And then drop like the rocks we used to throw—anywhere." She left the chair and went over to the sofa, hunting for something in the trunk trays. When she came back she found Fred sitting in her place. "Here are some handkerchiefs of yours. I've kept one or two. They're larger than mine and useful if one has a headache."

"Thank you. How nicely they smell of your things!" He looked at the white squares for a moment and then put them in his pocket. He kept the low chair, and as she stood beside him he took her hands and sat looking intently at them, as if he were examining them for some special purpose, tracing the long round fingers with the tips of his own. "Ordinarily, you know, there are reefs that a man catches to and keeps his nose above water. But this is a case by itself. There seems to be no limit as to how much I can be in love with you. I keep going." He did not lift his eyes from her fingers, which he continued to study with the same fervor. "Every kind of stringed instrument there is plays in your hands, Thea," he whispered, pressing them to his face.

She dropped beside him and slipped into his arms, shutting her eyes and lifting her cheek to his. "Tell me one thing," Fred whispered. "You said that night on the boat, when I first told you, that if you could you would crush it all up in your hands and throw it into the sea. Would you, all those weeks?"

She shook her head.

"Answer me, would you?"

"No, I was angry then. I'm not now. I'd never give them up. Don't make me pay too much." In that embrace they lived over again all the others. When Thea drew away from him, she dropped her face in her hands. "You are good to me," she breathed, "you are!"

Rising to his feet, he put his hands under her elbows and lifted her gently. He drew her toward the door with him. "Get all you can. Be generous with yourself. Don't stop short of splendid things. I want them for you more than I want anything else, more than I want one splendid thing for myself. I can't help feeling that you'll gain, somehow, by my losing so much. That you'll gain the very thing I lose. Take care of her, as Harsanyi said. She's wonderful!" He kissed her and went out of the door without looking back, just as if he were coming again to-morrow.

Thea went quickly into her bedroom. She brought out an armful of muslin things, knelt down, and began to lay them in the trays. Suddenly she stopped, dropped forward and leaned against the open trunk, her head on her arms. The tears fell down on the dark old carpet. It came over her how many people must have said good-bye and been unhappy in that room. Other people, before her time, had hired this room to cry in. Strange rooms and strange streets and faces, how sick at heart they made one! Why was she going so far, when what she wanted was some familiar place to hide in?—the rock house, her little room in Moonstone, her own bed. Oh, how good it would be to lie down in that little bed, to cut the nerve that kept one struggling, that pulled one on and on, to sink into peace there, with all the family safe and happy downstairs. After all, she was a Moonstone girl, one of the preacher's children. Everything else was in Fred's imagination. Why was she called upon to take such chances? Any safe, humdrum work that did not compromise her would be better. But if she failed now, she would lose her soul. There was nowhere to fall, after one took that step, except into abysses of wretchedness. She knew what abysses, for she could still hear the old man playing in the snowstorm, "*Ach, ich habe sie verloren!*" That melody was released in her like a passion of longing. Every nerve in her body thrilled to it. It brought her to her feet, carried her somehow to bed and into troubled sleep.

That night she taught in Moonstone again: she beat her pupils in hideous rages, she kept on beating them. She sang at funerals, and struggled at the piano with Harsanyi. In one dream she was looking into a hand-glass and thinking that she was getting better-looking, when the glass began to grow smaller and smaller and her own reflection to shrink, until she realized that she was looking into Ray Kennedy's eyes, seeing her face in that look of his which she could never forget. All at once the eyes were Fred Ottenburg's, and not

Ray's. All night she heard the shrieking of trains, whistling in and out of Moonstone, as she used to hear them in her sleep when they blew shrill in the winter air. But to-night they were terrifying,—the spectral, fated trains that "raced with death," about which the old woman from the depot used to pray.

In the morning she wakened breathless after a struggle with Mrs. Livery Johnson's daughter. She started up with a bound, threw the blankets back and sat on the edge of the bed, her night-dress open, her long braids hanging over her bosom, blinking at the daylight. After all, it was not too late. She was only twenty years old, and the boat sailed at noon. There was still time!

PART VI

KRONBORG

Ten Years Later

I

IT IS A GLORIOUS WINTER DAY. Denver, standing on her high plateau under a thrilling green-blue sky, is masked in snow and glittering with sunlight. The Capitol building is actually in armor, and throws off the shafts of the sun until the beholder is dazzled and the outlines of the building are lost in a blaze of reflected light. The stone terrace is a white field over which fiery reflections dance, and the trees and bushes are faithfully repeated in snow—on every black twig a soft, blurred line of white. From the terrace one looks directly over to where the mountains break in their sharp, familiar lines against the sky. Snow fills the gorges, hangs in scarfs on the great slopes, and on the peaks the fiery sunshine is gathered up as by a burning-glass.

Howard Archie is standing at the window of his private room in the offices of the San Felipe Mining Company, on the sixth floor of the Raton Building, looking off at the mountain glories of his State while he gives dictation to his secretary. He is ten years older than when we saw him last, and emphatically ten years more prosperous. A decade of coming into things has not so much aged him as it has fortified, smoothed, and assured him. His sandy hair and imperial conceal whatever gray they harbor. He has not grown heavier, but more flexible, and his massive shoulders carry fifty years and the control of his great mining interests more lightly than they carried forty years and a country practice. In short, he is one of the friends to whom we feel grateful for having got on in the world, for helping to keep up the general temperature and our own confidence in life. He is an acquaintance that one would hurry to overtake and greet among a hundred. In his warm handshake and generous smile there is the stimulating cordiality of good fellows come into good fortune and eager to pass it on; something that makes one think better of the lottery of life and resolve to try again.

When Archie had finished his morning mail, he turned away from the window and faced his secretary. "Did anything come up yesterday afternoon while I was away, T. B.?"

Thomas Burk turned over the leaf of his calendar. "Governor Alden[1] sent down to say that he wanted to see you before he sends his letter to the Board of Pardons. Asked if you could go over to the State House this morning."

Archie shrugged his shoulders. "I'll think about it."

The young man grinned.

"Anything else?" his chief continued.

T. B. swung round in his chair with a look of interest on his shrewd, clean-shaven face. "Old Jasper Flight was in, Dr. Archie. I never expected to see him alive again. Seems he's tucked away for the winter with a sister who's a housekeeper at the Oxford. He's all crippled up with rheumatism, but as fierce after it as ever. Wants to know if you or the company won't grub-stake him again. Says he's sure of it this time; had located something when the snow shut down on him in December. He wants to crawl out at the first break in the weather, with that same old burro with the split ear. He got somebody to winter the beast for him. He's superstitious about that burro, too; thinks it's divinely guided. You ought to hear the line of talk he put up here yesterday; said when he rode in his carriage, that burro was a-going to ride along with him."

Archie laughed. "Did he leave you his address?"

"He didn't neglect anything," replied the clerk cynically.

"Well, send him a line and tell him to come in again. I like to hear him. Of all the crazy prospectors I've ever known, he's the most interesting, because he's really crazy. It's a religious conviction with him, and with most of 'em it's a gambling fever or pure vagrancy. But Jasper Flight believes that the Almighty keeps the secret of the silver deposits in these hills, and gives it away to the deserving. He's a downright noble figure. Of course I'll stake him! As long as he can crawl out in the spring. He and that burro are a sight together. The beast is nearly as white as Jasper; must be twenty years old."

"If you stake him this time, you won't have to again," said T. B. knowingly. "He'll croak up there, mark my word. Says he never ties the burro at night now, for fear he might be called sudden, and the beast would starve. I guess that animal could eat a lariat rope, all right, and enjoy it."

"I guess if we knew the things those two have eaten, and haven't eaten, in their time, T. B., it would make us vegetarians." The doctor sat down and looked thoughtful. "That's the way for the old man to go. It would be pretty hard luck if he had to die in a hospital. I wish he could turn up something before he cashes in. But his kind seldom do; they're bewitched. Still, there was Stratton. I've been meeting Jasper Flight, and his side meat and tin pans, up in the mountains for years, and I'd miss him. I always halfway believe the fairy tales he spins me. Old Jasper Flight," Archie murmured, as if he liked the name or the picture it called up.

A clerk came in from the outer office and handed Archie a card. He sprang up and exclaimed, "Mr. Ottenburg? Bring him in."

Fred Ottenburg entered, clad in a long, fur-lined coat, holding a checked-cloth hat in his hand, his cheeks and eyes bright with the outdoor cold. The two men met before Archie's desk and their handclasp was longer than friendship prompts except in regions where the blood warms and quickens to meet the dry cold. Under the general keying-up of the altitude, manners take on a heartiness, a vivacity, that is one expression of the half-unconscious excitement which Colorado people miss when they drop into lower strata of air. The heart, we are told, wears out early in that high atmosphere, but while it pumps it sends out no sluggish stream. Our two friends stood gripping each other by the hand and smiling.

"When did you get in, Fred? And what have you come for?" Archie gave him a quizzical glance.

"I've come to find out what you think you're doing out here," the younger man declared emphatically. "I want to get next, I do. When can you see me?"

"Anything on to-night? Then suppose you dine with me. Where can I pick you up at five-thirty?"

"Bixby's office, general freight agent of the Burlington." Ottenburg began to button his overcoat and drew on his gloves. "I've got to have one shot at you before I go, Archie. Didn't I tell you Pinky Alden was a cheap squirt?"

Alden's backer laughed and shook his head. "Oh, he's worse than that, Fred. It isn't polite to mention what he is, outside of the Arabian Nights. I guessed you'd come to rub it into me."

Ottenburg paused, his hand on the doorknob, his high color challenging the doctor's calm. "I'm disgusted with you, Archie, for training with such a pup. A man of your experience!"

"Well, he's been an experience," Archie muttered. "I'm not coy about admitting it, am I?"

Ottenburg flung open the door. "Small credit to you. Even the women are out for capital and corruption, I hear. Your Governor's done more for the United Breweries in six months than I've been able to do in six years. He's the lily-livered sort we're looking for. Good-morning."

That afternoon at five o'clock Dr. Archie emerged from the State House after his talk with Governor Alden, and crossed the terrace under a saffron sky. The snow, beaten hard, was blue in the dusk; a day of blinding sunlight had not even started a thaw. The lights of the city twinkled pale below him in the quivering violet air, and the dome of the State House behind him was still red with the light from the west. Before he got into his car, the doctor paused to look about him at the scene of which he never tired. Archie lived in his own house on Colfax Avenue, where he had roomy grounds and a rose garden and a conservatory. His housekeeping was done by three Japanese boys, devoted and resourceful, who were able to manage Archie's dinner parties, to see that he kept his engagements, and to make visitors who stayed at the house so comfortable that they were always loath to go away.

Archie had never known what comfort was until he became a widower, though with characteristic delicacy, or dishonesty, he insisted upon accrediting his peace of mind to the San Felipe, to Time, to anything but his release from Mrs. Archie.

Mrs. Archie died just before her husband left Moonstone and came to Denver to live, six years ago. The poor woman's fight against dust was her undoing at last. One summer day when she was rubbing the parlor upholstery with gasoline,—the doctor had often forbidden her to use it on any account, so that was one of the pleasures she seized upon in his absence,—an explosion occurred. Nobody ever knew exactly how it happened, for Mrs. Archie was dead when the neighbors rushed in to save her from the burning house. She must have inhaled the burning gas and died instantly.

Moonstone severity relented toward her somewhat after her death. But even while her old cronies at Mrs. Smiley's millinery store said that it was a terrible thing, they added that nothing but a powerful explosive *could* have killed Mrs. Archie, and that it was only right the doctor should have a chance.

Archie's past was literally destroyed when his wife died. The house burned to the ground, and all those material reminders which have such power over people disappeared in an hour. His mining interests now took him to Denver so often that it seemed better to make his headquarters there. He gave up his practice and left Moonstone for good. Six months afterward, while Dr. Archie was living at the Brown Palace Hotel, the San Felipe mine began to give up that silver hoard which old Captain Harris had always accused it of concealing, and San Felipe headed the list of mining quotations in every daily paper, East and West. In a few years Dr. Archie was a very rich man. His mine was such an important item in the mineral output of the State, and Archie had a hand in so many of the new industries of Colorado and New Mexico, that his political influence was considerable. He had thrown it all, two years ago, to the new reform party, and had brought about the election of a governor of whose conduct he was now heartily ashamed. His friends believed that Archie himself had ambitious political plans.

II

When Ottenburg and his host reached the house on Colfax Avenue, they went directly to the library, a long double room on the second floor which Archie had arranged exactly to his own taste. It was full of books and mounted specimens of wild game, with a big writing-table at either end, stiff, old-fashioned engravings, heavy hangings and deep upholstery.

When one of the Japanese boys brought the cocktails, Fred turned from the fine specimen of peccoray he had been examining and said, "A man is an owl to live in such a place alone, Archie. Why don't you marry? As for me, just because I can't marry, I find the world full of charming, unattached women, any one of whom I could fit up a house for with alacrity."

"You're more knowing than I." Archie spoke politely. "I'm not very wide awake about women. I'd be likely to pick out one of the uncomfortable ones—and there are a few of them, you know." He drank his cocktail and rubbed his hands together in a friendly way. "My friends here have charming wives, and they don't give me a chance to get lonely. They are very kind to me, and I have a great many pleasant friendships."

Fred put down his glass. "Yes, I've always noticed that women have confidence in you. You have the doctor's way of getting next. And you enjoy that kind of thing?"

"The friendship of attractive women? Oh, dear, yes! I depend upon it a great deal."

The butler announced dinner, and the two men went downstairs to the dining-room. Dr. Archie's dinners were always good and well served, and his wines were excellent.

"I saw the Fuel and Iron people to-day," Ottenburg said, looking up from his soup. "Their heart is in the right place. I can't see why in the mischief you ever got mixed up with that reform gang,

Archie. You've got nothing to reform out here. The situation has always been as simple as two and two in Colorado; mostly a matter of a friendly understanding."

"Well,"—Archie spoke tolerantly,—"some of the young fellows seemed to have red-hot convictions, and I thought it was better to let them try their ideas out."

Ottenburg shrugged his shoulders. "A few dull young men who haven't ability enough to play the old game the old way, so they want to put on a new game which doesn't take so much brains and gives away more advertising; that's what your anti-saloon league and vice commission amounts to. They provide notoriety for the fellows who can't distinguish themselves at running a business or practicing law or developing an industry. Here you have a mediocre lawyer with no brains and no practice, trying to get a look-in on something. He comes up with the novel proposition that the prostitute has a hard time of it, puts his picture in the paper, and the first thing you know, he's a celebrity. He gets the rake-off and she's just where she was before. How could you fall for a mouse-trap like Pink Alden, Archie?"

Dr. Archie laughed as he began to carve. "Pink seems to get under your skin. He's not worth talking about. He's gone his limit. People won't read about his blameless life any more. I knew those interviews he gave out would cook him. They were a last resort. I could have stopped him, but by that time I'd come to the conclusion that I'd let the reformers down. I'm not against a general shaking-up, but the trouble with Pinky's crowd is they never get beyond a general writing-up. We gave them a chance to do something, and they just kept on writing about each other and what temptations they had overcome."

While Archie and his friend were busy with Colorado politics, the impeccable Japanese attended swiftly and intelligently to his duties, and the dinner, as Ottenburg at last remarked, was worthy of more profitable conversation.

"So it is," the doctor admitted. "Well, we'll go upstairs for our coffee and cut this out. Bring up some cognac and arak, Tai," he added as he rose from the table.

They stopped to examine a moose's head on the stairway, and when they reached the library the pine logs in the fireplace had been lighted, and the coffee was bubbling before the hearth. Tai placed two chairs before the fire and brought a tray of cigarettes.

"Bring the cigars in my lower desk drawer, boy," the doctor directed. "Too much light in here, isn't there, Fred? Light the lamp there on my desk, Tai." He turned off the electric glare and settled himself deep into the chair opposite Ottenburg's.

"To go back to our conversation, doctor," Fred began while he waited for the first steam to blow off his coffee; "why don't you make up your mind to go to Washington? There'd be no fight made against you. I needn't say the United Breweries would back you. There'd be some *kudos* coming to us, too; backing a reform candidate."

Dr. Archie measured his length in his chair and thrust his large boots toward the crackling pitch-pine. He drank his coffee and lit a big black cigar while his guest looked over the assortment of cigarettes on the tray. "You say why don't I," the doctor spoke with the deliberation of a man in the position of having several courses to choose from, "but, on the other hand, why should I?" He puffed away and seemed, through his half-closed eyes, to look down several long roads with the intention of luxuriously rejecting all of them and remaining where he was. "I'm sick of politics. I'm disillusioned about serving my crowd, and I don't particularly want to serve yours. Nothing in it that I particularly want; and a man's not effective in politics unless he wants something for himself, and wants it hard. I can reach my ends by straighter roads. There are plenty of things to keep me busy. We haven't begun to develop our resources in this State; we haven't had a look in on them yet. That's the only thing that isn't fake—making men and machines go, and actually turning out a product."

The doctor poured himself some white cordial and looked over the little glass into the fire with an expression which led Ottenburg to believe that he was getting at something in his own mind. Fred lit a cigarette and let his friend grope for his idea.

"My boys, here," Archie went on, "have got me rather interested in Japan. Think I'll go out there in the spring, and come back the other way, through Siberia. I've always wanted to go to Russia." His eyes still hunted for something in his big fireplace. With a slow turn of his head he brought them back to his guest and fixed them upon him. "Just now, I'm thinking of running on to New York for a few weeks," he ended abruptly.

Ottenburg lifted his chin. "Ah!" he exclaimed, as if he began to see Archie's drift. "Shall you see Thea?"

"Yes." The doctor replenished his cordial glass. "In fact, I suspect I am going exactly *to* see her. I'm getting stale on things here, Fred. Best people in the world and always doing things for me. I'm fond of them, too, but I've been with them too much. I'm getting ill-tempered, and the first thing I know I'll be hurting people's feelings. I snapped Mrs. Dandridge up over the telephone this afternoon when she asked me to go out to Colorado Springs on Sunday to meet some English people who are staying at the Antlers. Very nice of her to want me, and I was as sour as if she'd been trying to work me for something. I've got to get out for a while, to save my reputation."

To this explanation Ottenburg had not paid much attention. He seemed to be looking at a fixed point: the yellow glass eyes of a fine wildcat over one of the bookcases. "You've never heard her at all, have you?" he asked reflectively. "Curious, when this is her second season in New York."

"I was going on last March. Had everything arranged. And then old Cap Harris thought he could drive his car and me through a lamp-post and I was laid up with a compound fracture for two months. So I didn't get to see Thea."

Ottenburg studied the red end of his cigarette attentively. "She might have come out to see you. I remember you covered the distance like a streak when she wanted you."

Archie moved uneasily. "Oh, she couldn't do that. She had to get back to Vienna to work on some new parts for this year. She sailed two days after the New York season closed."

"Well, then she couldn't, of course." Fred smoked his cigarette close and tossed the end into the fire. "I'm tremendously glad you're going now. If you're stale, she'll jack you up. That's one of her specialties. She got a rise out of me last December that lasted me all winter."

"Of course," the doctor apologized, "you know so much more about such things. I'm afraid it will be rather wasted on me. I'm no judge of music."

"Never mind that." The younger man pulled himself up in his chair. "She gets it across to people who aren't judges. That's just what she does." He relapsed into his former lassitude. "If you were stone deaf, it wouldn't all be wasted. It's a great deal to watch her. Incidentally, you know, she is very beautiful. Photographs give you no idea."

Dr. Archie clasped his large hands under his chin. "Oh, I'm counting on that. I don't suppose her voice will sound natural to me. Probably I wouldn't know it."

Ottenburg smiled. "You'll know it, if you ever knew it. It's the same voice, only more so. You'll know it."

"Did you, in Germany that time, when you wrote me? Seven years ago, now. That must have been at the very beginning."

"Yes, somewhere near the beginning. She sang one of the Rhine daughters." Fred paused and drew himself up again. "Sure, I knew it from the first note. I'd heard a good many young voices come up out of the Rhine, but, by gracious, I hadn't heard one like that!" He fumbled for another cigarette. "Mahler[1] was conducting that night. I met him as he was leaving the house and had a word with him. 'Interesting voice you tried out this evening,' I said. He stopped and smiled. 'Miss Kronborg, you mean? Yes, very. She seems to sing for the idea. Unusual in a young singer.' I'd never heard him admit before that a singer could have an idea. She not only had it, but she got it across. The Rhine music, that I'd known since I was a boy, was fresh to me, vocalized for the first time. You realized that she was beginning that long story, adequately, with the end in view. Every phrase she sang was basic. She simply *was* the idea of the Rhine music." Ottenburg rose and stood with his back to the fire. "And at the end, where you don't see the maidens at all, the same thing again: two pretty voices *and* the Rhine voice." Fred snapped his fingers and dropped his hand.

The doctor looked up at him enviously. "You see, all that would be lost on me," he said modestly. "I don't know the dream nor the interpretation thereof. I'm out of it. It's too bad that so few of her old friends can appreciate her."

"Take a try at it," Fred encouraged him. "You'll get in deeper than you can explain to yourself. People with no personal interest do that."

"I suppose," said Archie diffidently, "that college German, gone to seed, wouldn't help me out much. I used to be able to make my German patients understand me."

"Sure it would!" cried Ottenburg heartily. "Don't be above knowing your libretto. That's all very well for musicians, but common mortals like you and me have got to know what she's singing about. Get out your dictionary and go at it as you would at any other proposition. Her diction is beautiful, and if you know the text you'll get a great deal. So long as you're going to hear her, get all that's coming to you. You bet in Germany people know their librettos by heart! You Americans are so afraid of stooping to learn anything."

"I *am* a little ashamed," Archie admitted. "I guess that's the way we mask our general ignorance. However, I'll stoop this time; I'm more ashamed not to be able to follow her. The papers always say she's such a fine actress." He took up the tongs and began to rearrange the logs that had burned through and fallen apart. "I suppose she has changed a great deal?" he asked absently.

"We've all changed, my dear Archie,—she more than most of us. Yes, and no. She's all there, only there's a great deal more of her. I've had only a few words with her in several years. It's better not, when I'm tied up this way. The laws are barbarous, Archie."

"Your wife is—still the same?" the doctor asked sympathetically.

"Absolutely. Hasn't been out of a sanitarium for seven years now. No prospect of her ever being out, and as long as she's there I'm tied hand and foot. What does society get out of such a state of things, I'd like to know, except a tangle of irregularities? If you want to reform, there's an opening for you!"

"It's bad, oh, very bad; I agree with you!" Dr. Archie shook his head. "But there would be complications under another system, too. The whole question of a young man's marrying has looked pretty grave to me for a long while. How have they the courage to keep on doing it? It depresses me now to buy wedding presents." For some time the doctor watched his guest, who was sunk in bitter reflections. "Such things used to go better than they do now, I believe. Seems to me all the married people I knew when I was a boy were happy enough." He paused again and bit the end off a fresh cigar. "You never saw Thea's mother, did you, Ottenburg? That's a pity. Mrs. Kronborg was a fine woman. I've always been afraid Thea made a mistake, not coming home when Mrs. Kronborg was ill, no matter what it cost her."

Ottenburg moved about restlessly. "She couldn't, Archie, she positively couldn't. I felt you never understood that, but I was in Dresden at the time, and though I wasn't seeing much of her, I could size up the situation for myself. It was by just a lucky chance that she got to sing *Elisabeth*[2] that time at the Dresden Opera, a complication of circumstances. If she'd run away, for any reason, she might have waited years for such a chance to come again. She gave a wonderful performance and made a great impression. They offered her certain terms; she had to take them and follow it up then and there. In that game you can't lose a single trick. She was ill herself, but she sang. Her mother was ill, and she sang. No, you mustn't hold that against her, Archie. She did the right thing there." Ottenburg drew out his watch. "Hello! I must be traveling. You hear from her regularly?"

"More or less regularly. She was never much of a letter-writer. She tells me about her engagements and contracts, but I know so little about that business that it doesn't mean much to me beyond the figures, which seem very impressive. We've had a good deal of business correspondence, about putting up a stone to her father and mother, and, lately, about her youngest brother, Thor. He is with me now; he drives my car. To-day he's up at the mine."

Ottenburg, who had picked up his overcoat, dropped it. "Drives your car?" he asked incredulously.

"Yes. Thea and I have had a good deal of bother about Thor. We tried a business college, and an engineering school, but it was no good. Thor was born a chauffeur before there were cars to drive. He was never good for anything else; lay around home and collected postage stamps and took bicycles to pieces, waiting for the automobile to be invented. He's just as much a part of a car as the steering-gear. I can't find out whether he likes his job with me or not, or whether he feels any curiosity about his sister. You can't find anything out from a Kronborg nowadays. The mother was different."

Fred plunged into his coat. "Well, it's a queer world, Archie. But you'll think better of it, if you go to New York. Wish I were going with you. I'll drop in on you in the morning at about eleven. I want a word with you about this Interstate Commerce Bill. Good-night."

Dr. Archie saw his guest to the motor which was waiting below, and then went back to his library, where he replenished the fire and sat down for a long smoke. A man of Archie's modest and rather credulous nature develops late, and makes his largest gain between forty and fifty. At thirty, indeed, as we have seen, Archie was a soft-hearted boy under a manly exterior, still whistling to keep up his courage. Prosperity and large responsibilities—above all, getting free of poor Mrs. Archie—had brought out a good deal more than he knew was in him. He was thinking tonight as he sat before the fire, in the comfort he liked so well, that but for lucky chances, and lucky holes in the ground, he would still be a country practitioner, reading his old books by his office lamp. And yet, he was not so fresh and energetic as he ought to be. He was tired of business and of politics. Worse than that, he was tired of the men with whom he had to do and of the women who, as he said, had been kind to him. He felt as if he were still hunting for something, like old Jasper Flight. He knew that this was an unbecoming and ungrateful state of mind, and he reproached himself for it. But he could not help wondering why it was that life, even when it gave so much, after all gave so little. What was it that he had expected and missed? Why was he, more than he was anything else, disappointed?

He fell to looking back over his life and asking himself which years of it he would like to live over again,—just as they had been,—and they were not many. His college years he would live again, gladly. After them there was nothing he would care to repeat until he came to Thea Kronborg. There had been something stirring about those years in Moonstone, when he was a restless young man on the verge of breaking into larger enterprises, and when she was a restless child on the verge of growing up into something unknown. He realized now that she had counted for a great deal more to him than he knew at the time. It was a continuous sort of relationship. He was always on the lookout for her as he went about the town, always vaguely expecting her as he sat in his office at night. He had never asked himself then if it was strange that he should find a child of twelve the most interesting and companionable person in Moonstone. It had seemed a pleasant, natural kind of solicitude. He explained it then by the fact that he had no children of his own. But now, as he looked back at those years, the other interests were faded and inanimate. The thought of them was heavy. But wherever his life had touched Thea Kronborg's, there was still a little warmth left, a little sparkle. Their friendship seemed to run over those discontented years like a leafy pattern, still bright and fresh when the other patterns had faded into the dull background. Their walks and drives and confidences, the night they watched the rabbit in the moonlight,—why were these things stirring to remember? Whenever he thought of them, they were distinctly different from the other memories of his life; always seemed humorous, gay, with a little thrill of anticipation and mystery about them. They came nearer to being tender secrets than any others he possessed. Nearer than anything else they corresponded to what he had hoped to find in the world, and had not found. It came over him now that the unexpected favors of fortune, no matter how dazzling, do not mean very much to us. They may excite or divert us for a time, but when we look back, the only things we cherish are those which in some way met our original want; the desire which formed in us in early youth, undirected, and of its own accord.

III

For the first four years after Thea went to Germany things went on as usual with the Kronborg family. Mrs. Kronborg's land in Nebraska increased in value and brought her in a good rental. The family drifted into an easier way of living, half without realizing it, as families will. Then Mr. Kronborg, who had never been ill, died suddenly of cancer of the liver, and after his death Mrs. Kronborg went, as her neighbors said, into a decline. Hearing discouraging reports of her from the physician who had taken over his practice, Dr. Archie went up from Denver to see her. He found her in bed, in the room where he had more than once attended her, a handsome woman of sixty with a body still firm and white, her hair, faded now to a very pale primrose, in two thick braids down her back, her eyes clear and calm. When the doctor arrived, she was sitting up in her bed, knitting. He felt at once how glad she was to see him, but he soon gathered that she had made no determination to get well. She told him, indeed, that she could not very well get along without Mr. Kronborg. The doctor looked at her with astonishment. Was it possible that she could miss the foolish old man so much? He reminded her of her children.

"Yes," she replied; "the children are all very well, but they are not father. We were married young."

The doctor watched her wonderingly as she went on knitting, thinking how much she looked like Thea. The difference was one of degree rather than of kind. The daughter had a compelling enthusiasm, the mother had none. But their framework, their foundation, was very much the same.

In a moment Mrs. Kronborg spoke again. "Have you heard anything from Thea lately?"

During his talk with her, the doctor gathered that what Mrs. Kronborg really wanted was to see her daughter Thea. Lying there day after day, she wanted it calmly and continuously. He told her that, since she felt so, he thought they might ask Thea to come home.

"I've thought a good deal about it," said Mrs. Kronborg slowly. "I hate to interrupt her, now that she's begun to get advancement. I expect she's seen some pretty hard times, though she was never one to complain. Perhaps she'd feel that she would like to come. It would be hard, losing both of us while she's off there."

When Dr. Archie got back to Denver he wrote a long letter to Thea, explaining her mother's condition and how much she wished to see her, and asking Thea to come, if only for a few weeks. Thea had repaid the money she had borrowed from him, and he assured her that if she happened to be short of funds for the journey, she had only to cable him.

A month later he got a frantic sort of reply from Thea. Complications in the opera at Dresden had given her an unhoped-for opportunity to go on in a big part. Before this letter reached the doctor, she would have made her début as *Elisabeth,* in "Tannhäuser." She wanted to go to her mother more than she wanted anything else in the world, but, unless she failed,—which she would not,—she absolutely could not leave Dresden for six months. It was not that she chose to stay; she had to stay—or lose everything. The next few months would put her five years ahead, or would put her back so far that it would be of no use to struggle further. As soon as she was free, she would go to Moonstone and take her mother back to Germany with her. Her mother, she was sure, could live for years yet, and she would like German people and German ways, and could be hearing music all the time. Thea said she was writing her mother and begging her to help her one last time; to get strength and to wait for her six months, and then she (Thea) would do everything. Her mother would never have to make an effort again.

Dr. Archie went up to Moonstone at once. He had great confidence in Mrs. Kronborg's power of will, and if Thea's appeal took hold of her enough, he believed she might get better. But when he was shown into the familiar room off the parlor, his heart sank. Mrs. Kronborg was lying serene and fateful on her pillows. On the dresser at the foot of her bed there was a large photograph of Thea in the character in which she was to make her début. Mrs. Kronborg pointed to it.

"Isn't she lovely, doctor? It's nice that she hasn't changed much. I've seen her look like that many a time."

They talked for a while about Thea's good fortune. Mrs. Kronborg had had a cablegram saying, "First performance well received. Great relief." In her letter Thea said: "If you'll only get better, dear mother, there's nothing I can't do. I will make a really great success, if you'll try with me. You shall have everything you want, and we will always be together. I have a little house all picked out where we are to live."

"Bringing up a family is not all it's cracked up to be," said Mrs. Kronborg with a flicker of irony, as she tucked the letter back under her pillow. "The children you don't especially need, you have always with you, like the poor. But the bright ones get away from you. They have their own way to make in the world. Seems like the brighter they are, the farther they go. I used to feel sorry that you had no family, doctor, but maybe you're as well off."

"Thea's plan seems sound to me, Mrs. Kronborg. There's no reason I can see why you shouldn't pull up and live for years yet, under proper care. You'd have the best doctors in the world over there, and it would be wonderful to live with anybody who looks like that." He nodded at the photograph of the young woman who must have been singing "*Dich, teure Halle, grüss' ich wieder,*"[1] her eyes looking up, her beautiful hands outspread with pleasure.

Mrs. Kronborg laughed quite cheerfully. "Yes, wouldn't it? If father were here, I might rouse myself. But sometimes it's hard to come back. Or if she were in trouble, maybe I could rouse myself."

"But, dear Mrs. Kronborg, she is in trouble," her old friend expostulated. "As she says, she's never needed you as she needs you now. I make my guess that she's never begged anybody to help her before."

Mrs. Kronborg smiled. "Yes, it's pretty of her. But that will pass. When these things happen far away they don't make such a mark; especially if your hands are full and you've duties of your own to think about. My own father died in Nebraska when Gunner was born,—we were living in Iowa then,—and I was sorry, but the baby made it up to me. I was father's favorite, too. That's the way it goes, you see."

The doctor took out Thea's letter to him, and read it over to Mrs. Kronborg. She seemed to listen, and not to listen.

When he finished, she said thoughtfully: "I'd counted on hearing her sing again. But I always took my pleasures as they come. I always enjoyed her singing when she was here about the house. While she was practicing I often used to leave my work and sit down in a rocker and give myself up to it, the same as if I'd been at an entertainment. I was never one of these housekeepers that let their work drive them to death. And when she had the Mexicans over here, I always took it in. First and last,"—she glanced judicially at the photograph,—"I guess I got about as much out of Thea's voice as anybody will ever get."

"I guess you did!" the doctor assented heartily; "and I got a good deal myself. You remember how she used to sing those Scotch songs for me, and lead us with her head, her hair bobbing?"

" 'Flow Gently, Sweet Afton,'—I can hear it now," said Mrs. Kronborg; "and poor father never knew when he sang sharp! He used to say, 'Mother, how do you always know when they make mistakes practicing?' " Mrs. Kronborg chuckled.

Dr. Archie took her hand, still firm like the hand of a young woman. "It was lucky for her that you did know. I always thought she got more from you than from any of her teachers."

"Except Wunsch; he was a real musician," said Mrs. Kronborg respectfully. "I gave her what chance I could, in a crowded house. I kept the other children out of the parlor for her. That was about all I could do. If she wasn't disturbed, she needed no watching. She went after it like a terrier after rats from the first, poor child. She was downright afraid of it. That's why I always encouraged her taking Thor off to outlandish places. When she was out of the house, then she was rid of it."

After they had recalled many pleasant memories together, Mrs. Kronborg said suddenly: "I always understood about her going off without coming to see us that time. Oh, I know! You had to keep your own counsel. You were a good friend to her. I've never forgot that." She patted the doctor's sleeve and went on absently. "There was something she didn't want to tell me, and that's why she didn't come. Something happened when she was with those people in Mexico. I worried for a good while, but I guess she's come out of it all right. She'd had a pretty hard time, scratching along alone like that when she was so young, and my farms in Nebraska were down so low that I couldn't help her none. That's no way to send a girl out. But I guess, whatever there was, she wouldn't be afraid to tell me now." Mrs. Kronborg looked up at the photograph with a smile.

"She doesn't look like she was beholding to anybody, does she?"

"She isn't, Mrs. Kronborg. She never has been. That was why she borrowed the money from me."

"Oh, I knew she'd never have sent for you if she'd done anything to shame us. She was always proud." Mrs. Kronborg paused and turned a little on her side. "It's been quite a satisfaction to you and me, doctor, having her voice turn out so fine. The things you hope for don't always turn out like that, by a long sight. As long as old Mrs. Kohler lived, she used always to translate what it said about Thea in the German papers she sent. I could make some of it out myself,—it's not very different from Swedish,—but it pleased the old lady. She left Thea her piece-picture of the burning of Moscow. I've got it put away in moth-balls for her, along with the oboe her grandfather brought from Sweden. I want her to take father's oboe back there some day." Mrs. Kronborg paused a moment and compressed her lips. "But I guess she'll take a finer instrument than that with her, back to Sweden!" she added.

Her tone fairly startled the doctor, it was so vibrating with a fierce, defiant kind of pride he had heard often in Thea's voice. He looked down wonderingly at his old friend and patient. After all, one never knew people to the core. Did she, within her, hide some of that still passion of which her daughter was all-compact?

"That last summer at home wasn't very nice for her," Mrs. Kronborg began as placidly as if the fire had never leaped up in her. "The other children were acting-up because they thought I might make a fuss over her and give her the big-head. We gave her the dare, somehow, the lot of us, because we couldn't understand her changing teachers and all that. That's the trouble about giving the dare to them quiet, unboastful children; you never know how far it'll take 'em. Well, we ought not to complain, doctor; she's given us a good deal to think about."

The next time Dr. Archie came to Moonstone, he came to be a pallbearer at Mrs. Kronborg's funeral. When he last looked at her, she was so serene and queenly that he went back to Denver feeling almost as if he had helped to bury Thea Kronborg herself. The handsome head in the coffin seemed to him much more really Thea than did the radiant young woman in the picture, looking about at the Gothic vaultings and greeting the Hall of Song.

IV

ONE BRIGHT MORNING late in February Dr. Archie was breakfasting comfortably at the Waldorf. He had got into Jersey City on an early train, and a red, windy sunrise over the North River had given him a good appetite. He consulted the morning paper while he drank his coffee and saw that "Lohengrin"[1] was to be sung at the opera that evening. In the list of the artists who would appear was the name "Kronborg." Such abruptness rather startled him. "Kronborg": it was impressive and yet, somehow, disrespectful; somewhat rude and brazen, on the back page of the morning paper. After breakfast he went to the hotel ticket office and asked the girl if she could give him something for "Lohengrin," "near the front." His manner was a trifle awkward and he wondered whether the girl noticed it. Even if she did, of course, she could scarcely suspect. Before the ticket stand he saw a bunch of blue posters announcing the opera casts for the week. There was "Lohengrin," and under it he saw:—

Elsa von Brabant . . . . Thea Kronborg.

That looked better. The girl gave him a ticket for a seat which she said was excellent. He paid for it and went out to the cabstand. He mentioned to the driver a number on Riverside Drive and got into a taxi. It would not, of course, be the right thing to call upon Thea when she was going to sing in the evening. He knew that much, thank goodness! Fred Ottenburg had hinted to him that, more than almost anything else, that would put one in wrong.

When he reached the number to which he directed his letters, he dismissed the cab and got out for a walk. The house in which Thea lived was as impersonal as the Waldorf, and quite as large. It was above 116th Street, where the Drive narrows, and in front of it the shelving bank dropped to the North River. As Archie strolled about the paths which traversed this slope, below the street level, the fourteen stories of the apartment hotel rose above him like a perpendicular cliff. He had no idea on which floor Thea lived, but he reflected, as his eye ran over the many windows, that the outlook would be fine from any floor. The forbidding hugeness of the house made him feel as if he had expected to meet Thea in a crowd and had missed her. He did not really believe that she was hidden away behind any of those glittering windows, or that he was to hear her this evening. His walk was curiously uninspiring and unsuggestive. Presently remembering that Ottenburg had encouraged him to study his lesson, he went down to the opera house and bought a libretto. He had even brought his old "Adler's German and English" in his trunk, and after luncheon he settled down in his gilded suite at the Waldorf with a big cigar and the text of "Lohengrin."

The opera was announced for seven-forty-five, but at half-past seven Archie took his seat in the right front of the orchestra circle. He had never been inside the Metropolitan Opera House[2] before, and the height of the audience room, the rich color, and the sweep of the balconies were not without their effect upon him. He watched the house fill with a growing feeling of expectation. When the steel curtain rose and the men of the orchestra took their places, he felt distinctly nervous. The burst of applause which greeted the conductor keyed him still higher. He found that he had taken off his gloves and twisted them to a string. When the lights went down and the violins began the overture, the place looked larger than ever; a great pit, shadowy and solemn. The whole atmosphere, he reflected, was somehow more serious than he had anticipated.

After the curtains were drawn back upon the scene beside the Scheldt,[3] he got readily into the swing of the story. He was so much interested in the bass who sang *King Henry* that he had almost forgotten for what he was waiting so nervously, when the *Herald* began in stentorian tones to summon *Elsa Von Brabant.* Then he began to realize that he was rather frightened. There was a flutter of white at the back of the stage, and women began to come in: two, four, six, eight, but not the right one. It flashed across him that this was something like buck-fever, the paralyzing moment that comes upon a man when his first elk looks at him through the bushes, under its great antlers; the moment when a man's mind is so full of shooting that he forgets the gun in his hand until the buck nods adieu to him from a distant hill.

All at once, before the buck had left him, she was there. Yes, unquestionably it was she. Her eyes were downcast, but the head, the cheeks, the chin—there could be no mistake; she advanced slowly, as if she were walking in her sleep. Some one spoke to her; she only inclined her head. He spoke again, and she bowed her head still lower. Archie had forgotten his libretto, and he had not counted upon these long pauses. He had expected her to appear and sing and reassure him. They seemed to be waiting for her. Did she ever forget? Why in thunder didn't she— She made a sound, a faint one. The people on the stage whispered together and seemed confounded. His nervousness was absurd. She must have done this often before; she knew her bearings. She made another sound, but he could make nothing of it. Then the King sang to her, and Archie began to remember where they were in the story. She came to the front of the stage, lifted her eyes for the first time, clasped her hands and began, "*Einsam in trüben Tagen.*"[4]

Yes, it was exactly like buck-fever. Her face was there, toward the house now, before his eyes, and he positively could not see it. She was singing, at last, and he positively could not hear her. He was conscious of nothing but an uncomfortable dread and a sense of crushing disappointment. He had, after all, missed her. Whatever was there, she was not there—for him.

The King interrupted her. She began again, "*In lichter Waffen Scheine.*" Archie did not know when his buck-fever passed, but presently he found that he was sitting quietly in a darkened house, not listening to but dreaming upon a river of silver sound. He felt apart from the others, drifting alone on the melody, as if he had been alone with it for a long while and had known it all before. His power of attention was not great just then, but in so far as it went he seemed to be looking through an exalted calmness at a beautiful woman from far away, from another sort of life and feeling and understanding than his own, who had in her face something he had known long ago, much brightened and beautified. As a lad he used to believe that the faces of people who died were like that in the next world; the same faces, but shining with the light of a new understanding. No, Ottenburg had not prepared him!

What he felt was admiration and estrangement. The homely reunion, that he had somehow expected, now seemed foolish. Instead of feeling proud that he knew her better than all these people about him, he felt chagrined at his own ingenuousness. For he did not know her better. This woman he had never known; she had somehow devoured his little friend, as the wolf ate up Red Ridinghood. Beautiful, radiant, tender as she was, she chilled his old affection; that sort of feeling was not appropriate. She seemed much, much farther away from him than she had seemed all those years when she was in Germany. The ocean he could cross, but there was something here he could not cross. There was a moment, when she turned to the King and smiled that rare, sunrise smile of her childhood, when he thought she was coming back to him. After the *Herald*'s second call for her champion, when she knelt in her impassioned prayer, there was again something familiar, a kind of wild wonder that she had had the power to call up long ago. But she merely reminded him of Thea; this was not the girl herself.

After the tenor came on, the doctor ceased trying to make the woman before him fit into any of his cherished recollections. He took her, in so far as he could, for what she was then and there. When the knight raised the kneeling girl and put his mailed hand on her hair, when she lifted to him a face full of worship and passionate humility, Archie gave up his last reservation. He knew no more about her than did the hundreds around him, who sat in the shadow and looked on, as he looked, some with more understanding, some with less. He knew as much about *Ortrud* or *Lohengrin* as he knew about *Elsa*—more, because she went further than they, she sustained the legendary beauty of her conception more consistently. Even he could see that. Attitudes, movements, her face, her white arms and fingers, everything was suffused with a rosy tenderness, a warm humility, a gracious and yet—to him—wholly estranging beauty.

During the balcony singing in the second act the doctor's thoughts were as far away from Moonstone as the singer's doubtless were. He had begun, indeed, to feel the exhilaration of getting free from personalities, of being released from his own past as well as from Thea Kronborg's. It was very much, he told himself, like a military funeral, exalting and impersonal. Something old died in one, and out of it something new was born. During the duet with *Ortrud*, and the splendors of the wedding processional, this new feeling grew and grew. At the end of the act there were many curtain calls and *Elsa* acknowledged them, brilliant, gracious, spirited, with her far-breaking smile; but on the whole she was harder and more self-contained before the curtain than she was in the scene behind it. Archie did his part in the applause that greeted her, but it was the new and wonderful he applauded, not the old and dear. His personal, proprietary pride in her was frozen out.

He walked about the house during the *entr'acte*, and here and there among the people in the foyer he caught the name "Kronborg." On the staircase, in front of the coffee-room, a long-haired youth with a fat face was discoursing to a group of old women about "die Kronborg." Dr. Archie gathered that he had crossed on the boat with her.

After the performance was over, Archie took a taxi and started for Riverside Drive. He meant to see it through to-night. When he entered the reception hall of the hotel before which he had strolled that morning, the hall porter challenged him. He said he was waiting for Miss Kronborg. The porter looked at him suspiciously and asked whether he had an appointment. He answered brazenly that he had. He was not used to being questioned by hall boys. Archie sat first in one tapestry chair and then in another, keeping a sharp eye on the people who came in and went up in the elevators. He walked about and looked at his watch. An hour dragged by. No one had come in from the street now for about twenty minutes, when two women entered, carrying a great many flowers and followed by a tall young man in chauffeur's uniform. Archie advanced toward the taller of the two women, who was veiled and carried her head very firmly. He confronted her just as she reached the elevator. Although he did not stand directly in her way, something in his attitude compelled her to stop. She gave him a piercing, defiant glance through the white scarf that covered her face. Then she lifted her hand and brushed the scarf back from her head. There was still black on her brows and lashes. She was very pale and her face was drawn and deeply lined. She looked, the doctor told himself with a sinking heart, forty years old. Her suspicious, mystified stare cleared slowly.

"Pardon me," the doctor murmured, not knowing just how to address her here before the porters, "I came up from the opera. I merely wanted to say good-night to you."

Without speaking, still looking incredulous, she pushed him into the elevator. She kept her hand on his arm while the cage shot up, and she looked away from him, frowning, as if she were trying to remember or realize something. When the cage stopped, she pushed him out of the elevator through another door, which a maid opened, into a square hall. There she sank down on a chair and looked up at him.

"Why didn't you let me know?" she asked in a hoarse voice.

Archie heard himself laughing the old, embarrassed laugh that seldom happened to him now. "Oh, I wanted to take my chance with you, like anybody else. It's been so long, now!"

She took his hand through her thick glove and her head dropped forward. "Yes, it has been long," she said in the same husky voice, "and so much has happened."

"And you are so tired, and I am a clumsy old fellow to break in on you to-night," the doctor added sympathetically. "Forgive me, this time." He bent over and put his hand soothingly on her shoulder. He felt a strong shudder run through her from head to foot.

Still bundled in her fur coat as she was, she threw both arms about him and hugged him. "Oh, Dr. Archie, *Dr. Archie,*"—she shook him,—"don't let me go. Hold on, now you're here," she laughed, breaking away from him at the same moment and sliding out of her fur coat. She left it for the maid to pick up and pushed the doctor into the sitting-room, where she turned on the lights. "Let me *look* at you. Yes; hands, feet, head, shoulders—just the same. You've grown no older. You can't say as much for me, can you?"

She was standing in the middle of the room, in a white silk shirtwaist and a short black velvet skirt, which somehow suggested that they had 'cut off her petticoats all round about.'[5] She looked distinctly clipped and plucked. Her hair was parted in the middle and done very close to her head, as she had worn it under the wig. She looked like a fugitive, who had escaped from something in clothes caught up at hazard. It flashed across Dr. Archie that she was running away from the other woman down at the opera house, who had used her hardly.

He took a step toward her. "I can't tell a thing in the world about you, Thea—if I may still call you that."

She took hold of the collar of his overcoat. "Yes, call me that. Do: I like to hear it. You frighten me a little, but I expect I frighten you more. I'm always a scarecrow after I sing a long part like that—so high, too." She absently pulled out the handkerchief that protruded from his breast pocket and began to wipe the black paint off her eyebrows and lashes. "I can't take you in much to-night, but I must see you for a little while." She pushed him to a chair. "I shall be more recognizable to-morrow. You mustn't think of me as you see me to-night. Come at four to-morrow afternoon and have tea with me. Can you? That's good."

She sat down in a low chair beside him and leaned forward, drawing her shoulders together. She seemed to him inappropriately young and inappropriately old, shorn of her long tresses at one end and of her long robes at the other.

"How do you happen to be here?" she asked abruptly. "How can you leave a silver mine? I couldn't! Sure nobody'll cheat you? But you can explain everything to-morrow." She paused. "You remember how you sewed me up in a poultice, once? I wish you could to-night. I need a poultice, from top to toe. Something very disagreeable happened down there. You said you were out front? Oh, don't say anything about it. I always know exactly how it goes, unfortunately. I was rotten in the balcony. I never get that. You didn't notice it? Probably not, but I did."

Here the maid appeared at the door and her mistress rose. "My supper? Very well, I'll come. I'd ask you to stay, doctor, but there wouldn't be enough for two. They seldom send up enough for one,"—she spoke bitterly. "I haven't got a sense of you yet,"—turning directly to Archie again. "You haven't been here. You've only announced yourself, and told me you are coming to-morrow. You haven't seen me, either. This is not I. But I'll be here waiting for you to-morrow, my whole works! Good-night, till then." She patted him absently on the sleeve and gave him a little shove toward the door.

V

WHEN ARCHIE GOT BACK to his hotel at two o'clock in the morning, he found Fred Ottenburg's card under his door, with a message scribbled across the top: "When you come in, please call up room 811, this hotel." A moment later Fred's voice reached him over the telephone.

"That you, Archie? Won't you come up? I'm having some supper and I'd like company. Late? What does that matter? I won't keep you long."

Archie dropped his overcoat and set out for room 811. He found Ottenburg in the act of touching a match to a chafing-dish, at a table laid for two in his sitting-room. "I'm catering here," he announced cheerfully. "I let the waiter off at midnight, after he'd set me up. You'll have to account for yourself, Archie."

The doctor laughed, pointing to three wine-coolers under the table. "Are you expecting guests?"

"Yes, two." Ottenburg held up two fingers,—"you, and my higher self. He's a thirsty boy, and I don't invite him often. He has been known to give me a headache. Now, where have you been, Archie, until this shocking hour?"

"Bah, you've been banting!" the doctor exclaimed, pulling out his white gloves as he searched for his handkerchief and throwing them into a chair. Ottenburg was in evening clothes and very pointed dress shoes. His white waistcoat, upon which the doctor had fixed a challenging eye, went down straight from the top button, and he wore a camelia. He was conspicuously brushed and trimmed and polished. His smoothly controlled excitement was wholly different from his usual easy cordiality, though he had his face, as well as his figure, well in hand. On the serving-table there was an empty champagne pint and a glass. He had been having a little starter, the doctor told himself, and would probably be running on high gear before he got through. There was even now an air of speed about him.

"Been, Freddy?"—the doctor at last took up his question. "I expect I've been exactly where you have. Why didn't you tell me you were coming on?"

"I wasn't, Archie." Fred lifted the cover of the chafing-dish and stirred the contents. He stood behind the table, holding the lid with his handkerchief. "I had never thought of such a thing. But Landry, a young chap who plays her accompaniments and who keeps an eye out for me, telegraphed me that Madame Rheinecker had gone to Atlantic City with a bad throat, and Thea might have a chance to sing *Elsa*. She has sung it only twice here before, and I missed it in Dresden. So I came on. I got in at four this afternoon and saw you registered, but I thought I wouldn't butt in. How lucky you got here just when she was coming on for this. You couldn't have hit a better time." Ottenburg stirred the contents of the dish faster and put in more sherry. "And where have you been since twelve o'clock, may I ask?"

Archie looked rather self-conscious, as he sat down on a fragile gilt chair that rocked under him, and stretched out his long legs. "Well, if you'll believe me, I had the brutality to go to see her. I wanted to identify her. Couldn't wait."

Ottenburg placed the cover quickly on the chafing-dish and took a step backward. "You did, old sport? My word! None but the brave deserve the fair. Well,"—he stooped to turn the wine,—"and how was she?"

"She seemed rather dazed, and pretty well used up. She seemed disappointed in herself, and said she hadn't done herself justice in the balcony scene."

"Well, if she didn't, she's not the first. Beastly stuff to sing right in there; lies just on the 'break' in the voice." Fred pulled a bottle out of the ice and drew the cork. Lifting his glass he looked meaningly at Archie. "You know who, doctor. Here goes!" He drank off his glass with a sigh of satisfaction. After he had turned the lamp low under the chafing-dish, he remained standing, looking pensively down at the food on the table. "Well, she rather pulled it off! As a backer, you're a winner, Archie. I congratulate you." Fred poured himself another glass. "Now you must eat something, and so must I. Here, get off that bird cage and find a steady chair. This stuff ought to be rather good; head waiter's suggestion. Smells all right." He bent over the chafing-dish and began to serve the contents. "Perfectly innocuous: mushrooms and truffles and a little crab-meat. And now, on the level, Archie, how did it hit you?"

Archie turned a frank smile to his friend and shook his head. "It was all miles beyond me, of course, but it gave me a pulse. The general excitement got hold of me, I suppose. I like your wine, Freddy." He put down his glass. "It goes to the spot to-night. She *was* all right, then? You weren't disappointed?"

"Disappointed? My dear Archie, that's the high voice we dream of; so pure and yet so virile and human. That combination hardly ever happens with sopranos." Ottenburg sat down and turned to the doctor, speaking calmly and trying to dispel his friend's manifest bewilderment. "You see, Archie, there's the voice itself, so beautiful and individual, and then there's something else; the thing in it which responds to every shade of thought and feeling, spontaneously, almost unconsciously. That color has to be born in a singer, it can't be acquired; lots of beautiful voices haven't a vestige of it. It's almost like another gift—the rarest of all. The voice simply is the mind and is the heart. It can't go wrong in interpretation, because it has in it the thing that makes all interpretation. That's why you feel so sure of her. After you've listened to her for an hour or so, you aren't afraid of anything. All the little dreads you have with other artists vanish. You lean back and you say to yourself, 'No, *that* voice will never betray.' *Treulich geführt, treulich bewacht.*"[1]

Archie looked envyingly at Fred's excited, triumphant face. How satisfactory it must be, he thought, to really know what she was doing and not to have to take it on hearsay. He took up his glass with a sigh. "I seem to need a good deal of cooling off to-night. I'd just as lief forget the Reform Party for once.

"Yes, Fred," he went on seriously; "I thought it sounded very beautiful, and I thought she was very beautiful, too. I never imagined she could be as beautiful as that."

"Wasn't she? Every attitude a picture, and always the right kind of picture, full of that legendary, supernatural thing she gets into it. I never heard the prayer sung like that before. That look that came in her eyes; it went right out through the back of the roof. Of course, you get an *Elsa* who can look through walls like that, and visions and Grail-knights happen naturally. She becomes an abbess, that girl, after *Lohengrin* leaves her. She's made to live with ideas and enthusiasms, not with a husband." Fred folded his arms, leaned back in his chair, and began to sing softly:—

> "*In lichter Waffen Scheine,*
> *Ein Ritter nahte da.*"[2]

"Doesn't she die, then, at the end?" the doctor asked guardedly.

Fred smiled, reaching under the table. "Some *Elsas* do; she didn't. She left me with the distinct impression that she was just beginning.[3] Now, doctor, here's a cold one." He twirled a napkin smoothly about the green glass, the cork gave and slipped out with a soft explosion. "And now we must have another toast. It's up to you, this time."

The doctor watched the agitation in his glass. "The same," he said without lifting his eyes. "That's good enough. I can't raise you."

Fred leaned forward, and looked sharply into his face. "That's the point; how *could* you raise me? Once again!"

"Once again, and always the same!" The doctor put down his glass. "This doesn't seem to produce any symptoms in me tonight." He lit a cigar. "Seriously, Freddy, I wish I knew more about what she's driving at. It makes me jealous, when you are so in it and I'm not."

"In it?" Fred started up. "My God, haven't you seen her this blessed night?—when she'd have kicked any other man down the elevator shaft, if I know her. Leave me something; at least what I can pay my five bucks for."

"Seems to me you get a good deal for your five bucks," said Archie ruefully. "And that, after all, is what she cares about,—what people get."

Fred lit a cigarette, took a puff or two, and then threw it away. He was lounging back in his chair, and his face was pale and drawn hard by that mood of intense concentration which lurks under the sunny shallows of the vineyard. In his voice there was a longer perspective than usual, a slight remoteness. "You see, Archie, it's all very simple, a natural development. It's exactly what Mahler said back there in the beginning, when she sang *Woglinde*.[4] It's the idea, the basic idea, pulsing behind every bar she sings. She simplifies a character down to the musical idea it's built on, and makes everything conform to that. The people who chatter about her being a great actress don't seem to get the notion of where *she* gets the notion. It all goes back to her original endowment, her tremendous musical talent. Instead of inventing a lot of business and expedients to suggest character, she knows the thing at the root, and lets the musical pattern take care of her. The score pours her into all those lovely postures, makes the light and shadow go over her face, lifts her and drops her. She lies on it, the way she used to lie on the Rhine music. Talk about rhythm!"

The doctor frowned dubiously as a third bottle made its appearance above the cloth. "Aren't you going in rather strong?"

Fred laughed. "No, I'm becoming too sober. You see this is breakfast now; kind of wedding breakfast. I feel rather weddingish. I don't mind. You know," he went on as the wine gurgled out, "I was thinking to-night when they sprung the wedding music, how any fool can have that stuff played over him when he walks up the aisle with some dough-faced little hussy who's hooked him. But it isn't every fellow who can see—well, what we saw to-night. There are compensations in life, Dr. Howard Archie, though they come in disguise. Did you notice her when she came down the stairs? Wonder where she gets that bright-and-morning star look? Carries to the last row of the family circle. I moved about all over the house. I'll tell you a secret, Archie: that carrying power was one of the first things that put me wise. Noticed it down there in Arizona, in the open. That, I said, belongs only to the big ones." Fred got up and began to move rhythmically about the room, his hands in his pockets. The doctor was astonished at his ease and steadiness, for there were slight lapses in his speech. "You see, Archie, *Elsa* isn't a part that's particularly suited to Thea's voice at all, as I see her voice. It's over-lyrical for her. She makes it, but there's nothing in it that fits her like a glove, except, maybe, that long duet in the third act. There, of course,"—he held out his hands as if he were measuring something,—"we know exactly where we are. But wait until they give her a chance at something that lies properly in her voice, and you'll see me rosier than I am to-night."

Archie smoothed the tablecloth with his hand. "I am sure I don't want to see you any rosier, Fred."

Ottenburg threw back his head and laughed. "It's enthusiasm, doctor. It's not the wine. I've got as much inflated as this for a dozen trashy things: brewers' dinners and political orgies. You, too, have your extravagances, Archie. And what I like best in you is this particular enthusiasm, which is not at all practical or sensible, which is downright Quixotic. You are not altogether what you seem, and you have your reservations. Living among the wolves, you have not become one. *Lupibus vivendi non lupus sum.*"

The doctor seemed embarrassed. "I was just thinking how tired she looked, plucked of all her fine feathers, while we get all the fun. Instead of sitting here carousing, we ought to go solemnly to bed."

"I get your idea." Ottenburg crossed to the window and threw it open. "Fine night outside; a hag of a moon just setting. It begins to smell like morning. After all, Archie, think of the lonely and rather solemn hours we've spent waiting for all this, while she's been—reveling."

Archie lifted his brows. "I somehow didn't get the idea to-night that she revels much."

"I don't mean this sort of thing." Fred turned toward the light and stood with his back to the window. "That," with a nod toward the wine-cooler, "is only a cheap imitation, that any poor stiff-fingered fool can buy and feel his shell grow thinner. But take it from me, no matter what she pays, or how much she may see fit to lie about it, the real, the master revel is hers." He leaned back against the window sill and crossed his arms. "Anybody with all that voice and all that talent and all that beauty, has her hour. Her hour," he went on deliberately, "when she can say, 'there it is, at last, *wie im Traum ich*—[5]

" 'As in my dream I dreamed it,
As in my will it was.' "

He stood silent a moment, twisting the flower from his coat by the stem and staring at the blank wall with haggard abstraction. "Even I can say to-night, Archie," he brought out slowly,

" 'As in my dream I dreamed it,
As in my will it was.'

Now, doctor, you may leave me. I'm beautifully drunk, but not with anything that ever grew in France."

The doctor rose. Fred tossed his flower out of the window behind him and came toward the door. "I say," he called, "have you a date with anybody?"

The doctor paused, his hand on the knob. "With Thea, you

mean? Yes. I'm to go to her at four this afternoon—if you haven't paralyzed me."

"Well, you won't eat me, will you, if I break in and send up my card? She'll probably turn me down cold, but that won't hurt my feelings. If she ducks me, you tell her for me, that to spite me now she'd have to cut off more than she can spare. Good-night, Archie."

VI

IT WAS LATE on the morning after the night she sang *Elsa*, when Thea Kronborg stirred uneasily in her bed. The room was darkened by two sets of window shades, and the day outside was thick and cloudy. She turned and tried to recapture unconsciousness, knowing that she would not be able to do so. She dreaded waking stale and disappointed after a great effort. The first thing that came was always the sense of the futility of such endeavor, and of the absurdity of trying too hard. Up to a certain point, say eighty degrees, artistic endeavor could be fat and comfortable, methodical and prudent. But if you went further than that, if you drew yourself up toward ninety degrees, you parted with your defenses and left yourself exposed to mischance. The legend was that in those upper reaches you might be divine; but you were much likelier to be ridiculous. Your public wanted just about eighty degrees; if you gave it more it blew its nose and put a crimp in you. In the morning, especially, it seemed to her very probable that whatever struggled above the good average was not quite sound. Certainly very little of that superfluous ardor, which cost so dear, ever got across the footlights. These misgivings waited to pounce upon her when she wakened. They hovered about her bed like vultures.

She reached under her pillow for her handkerchief, without opening her eyes. She had a shadowy memory that there was to be something unusual, that this day held more disquieting possibilities than days commonly held. There was something she dreaded; what was it? Oh, yes, Dr. Archie was to come at four.

A reality like Dr. Archie, poking up out of the past, reminded one of disappointments and losses, of a freedom that was no more: reminded her of blue, golden mornings long ago, when she used to waken with a burst of joy at recovering her precious self and her precious world; when she never lay on her pillows at eleven o'clock like something the waves had washed up. After all, why had he come? It had been so long, and so much had happened. The things she had lost, he would miss readily enough. What she had gained, he would scarcely perceive. He, and all that he recalled, lived for her as memories. In sleep, and in hours of illness or exhaustion, she went back to them and held them to her heart. But they were better as memories. They had nothing to do with the struggle that made up her actual life. She felt drearily that she was not flexible enough to be the person her old friend expected her to be, the person she herself wished to be with him.

Thea reached for the bell and rang twice,—a signal to her maid to order her breakfast. She rose and ran up the window shades and turned on the water in her bathroom, glancing into the mirror apprehensively as she passed it. Her bath usually cheered her, even on low mornings like this. Her white bathroom, almost as large as her sleeping-room, she regarded as a refuge. When she turned the key behind her, she left care and vexation on the other side of the door. Neither her maid nor the management nor her letters nor her accompanist could get at her now.

When she pinned her braids about her head, dropped her nightgown and stepped out to begin her Swedish movements, she was a natural creature again, and it was so that she liked herself best. She slid into the tub with anticipation and splashed and tumbled about a good deal. Whatever else she hurried, she never hurried her bath. She used her brushes and sponges and soaps like toys, fairly playing in the water. Her own body was always a cheering sight to her. When she was careworn, when her mind felt old and tired, the freshness of her physical self, her long, firm lines, the smoothness of her skin, reassured her. This morning, because of awakened memories, she looked at herself more carefully than usual, and was not discouraged. While she was in the tub she began to whistle softly the tenor aria, "*Ah! Fuyez, douce image,*"[1] somehow appropriate to the bath. After a noisy moment under the cold shower, she stepped out on the rug flushed and glowing, threw her arms above her head, and rose on her toes, keeping the elevation as long as she could. When she dropped back on her heels and began to rub herself with the towels, she took up the aria again, and felt quite in the humor for seeing Dr. Archie. After she had returned to her bed, the maid brought her letters and the morning papers with her breakfast.

"Telephone Mr. Landry and ask him if he can come at half-past three, Theresa, and order tea to be brought up at five."

When Howard Archie was admitted to Thea's apartment that afternoon, he was shown into the music-room back of the little reception room. Thea was sitting in a davenport behind the piano, talking to a young man whom she later introduced as her friend Mr. Landry. As she rose, and came to meet him, Archie felt a deep relief, a sudden thankfulness. She no longer looked clipped and plucked, or dazed and fleeing.

Dr. Archie neglected to take account of the young man to whom he was presented. He kept Thea's hands and held her where he met her, taking in the light, lively sweep of her hair, her clear green eyes and her throat that came up strong and dazzlingly white from her green velvet gown. The chin was as lovely as ever, the cheeks as smooth. All the lines of last night had disappeared. Only at the outer corners of her eyes, between the eye and the temple, were the faintest indications of a future attack—mere kitten scratches that playfully hinted where one day the cat would claw her. He studied her without any embarrassment. Last night everything had been awkward; but now, as he held her hands, a kind of harmony came between them, a reëstablishment of confidence.

"After all, Thea,—in spite of all, I still know you," he murmured.

She took his arm and led him up to the young man who was standing beside the piano. "Mr. Landry knows all about you, Dr. Archie. He has known about you for many years." While the two men shook hands she stood between them, drawing them together by her presence and her glances. "When I first went to Germany, Landry was studying there. He used to be good enough to work with me when I could not afford to have an accompanist for more than two hours a day. We got into the way of working together. He is a singer, too, and has his own career to look after, but he still manages to give me some time. I want you to be friends." She smiled from one to the other.

The rooms, Archie noticed, full of last night's flowers, were furnished in light colors, the hotel bleakness of them a little softened by a magnificent Steinway piano, white bookshelves full of books and scores, some drawings of ballet dancers, and the very deep sofa behind the piano.

"Of course," Archie asked apologetically, "you have seen the papers?"

"Very cordial, aren't they? They evidently did not expect as much as I did. *Elsa* is not really in my voice. I can sing the music, but I have to go after it."

"That is exactly," the doctor came out boldly, "what Fred Ottenburg said this morning."

They had remained standing, the three of them, by the piano, where the gray afternoon light was strongest. Thea turned to the doctor with interest. "Is Fred in town? They were from him, then—some flowers that came last night without a card." She indicated the white lilacs on the window sill. "Yes, he would know, certainly," she said thoughtfully. "Why don't we sit down? There will be some tea for you in a minute, Landry. He's very dependent upon it," disapprovingly to Archie. "Now tell me, Doctor, did you really have a good time last night, or were you uncomfortable? Did you feel as if I were trying to hold my hat on by my eyebrows?"

He smiled. "I had all kinds of a time. But I had no feeling of that sort. I couldn't be quite sure that it was you at all. That was why I came up here last night. I felt as if I'd lost you."

She leaned toward him and brushed his sleeve reassuringly. "Then I didn't give you an impression of painful struggle? Landry was singing at Weber and Fields'[2] last night. He didn't get in until the performance was half over. But I see the *Tribune* man[3] felt that I was working pretty hard. Did you see that notice, Oliver?"

Dr. Archie looked closely at the red-headed young man for the first time, and met his lively brown eyes, full of a droll, confiding sort of humor. Mr. Landry was not prepossessing. He was undersized and clumsily made, with a red, shiny face and a sharp little nose that looked as if it had been whittled out of wood and was always in the air, on the scent of something. Yet it was this queer little beak, with his eyes, that made his countenance anything a face at all. From a distance he looked like the grocery-man's delivery boy in a small town. His dress seemed an acknowledgment of his grotesqueness: a short coat, like a little boy's roundabout, and a vest fantastically sprigged and dotted, over a lavender shirt.

At the sound of a muffled buzz, Mr. Landry sprang up. "May I answer the telephone for you?" He went to the writing-table and took up the receiver. "Mr. Ottenburg is downstairs," he said, turning to Thea and holding the mouthpiece against his coat.

"Tell him to come up," she replied without hesitation. "How long are you going to be in town, Dr. Archie?"

"Oh, several weeks, if you'll let me stay. I won't hang around and be a burden to you, but I want to try to get educated up to you, though I expect it's late to begin."

Thea rose and touched him lightly on the shoulder. "Well, you'll never be any younger, will you?"

"I'm not so sure about that," the doctor replied gallantly.

The maid appeared at the door and announced Mr. Frederick Ottenburg. Fred came in, very much got up, the doctor reflected, as he watched him bending over Thea's hand. He was still pale and looked somewhat chastened, and the lock of hair that hung down over his forehead was distinctly moist. But his black afternoon coat, his gray tie and gaiters were of a correctness that Dr. Archie could never attain for all the efforts of his faithful slave, Van Deusen, the Denver haberdasher. To be properly up to those tricks, the doctor supposed, you had to learn them young. If he were to buy a silk hat that was the twin of Ottenburg's, it would be shaggy in a week, and he could never carry it as Fred held his.

Ottenburg had greeted Thea in German, and as she replied in the same language, Archie joined Mr. Landry at the window. "You know Mr. Ottenburg, he tells me?"

Mr. Landry's eyes twinkled. "Yes, I regularly follow him about, when he's in town. I would, even if he didn't send me such wonderful Christmas presents: Russian vodka by the half-dozen!"

Thea called to them, "Come, Mr. Ottenburg is calling on all of us. Here's the tea."

The maid opened the door and two waiters from downstairs appeared with covered trays. The tea-table was in the parlor. Thea drew Ottenburg with her and went to inspect it. "Where's the rum? Oh, yes, in that thing! Everything seems to be here, but send up some currant preserves and cream cheese for Mr. Ottenburg. And in about fifteen minutes, bring some fresh toast. That's all, thank you."

For the next few minutes there was a clatter of teacups and responses about sugar. "Landry always takes rum. I'm glad the rest of you don't. I'm sure it's bad." Thea poured the tea standing and got through with it as quickly as possible, as if it were a refreshment snatched between trains. The tea-table and the little room in which it stood seemed to be out of scale with her long step, her long reach, and the energy of her movements. Dr. Archie, standing near her, was pleasantly aware of the animation of her figure. Under the clinging velvet, her body seemed independent and unsubdued.

They drifted, with their plates and cups, back to the music-room. When Thea followed them, Ottenburg put down his tea suddenly. "Aren't you taking anything? Please let me." He started back to the table.

"No, thank you, nothing. I'm going to run over that aria for you presently, to convince you that I can do it. How did the duet go, with Schlag?"

She was standing in the doorway and Fred came up to her: "That you'll never do any better. You've worked your voice into it perfectly. Every *nuance*—wonderful!"

"Think so?" She gave him a sidelong glance and spoke with a certain gruff shyness which did not deceive anybody, and was not meant to deceive. The tone was equivalent to "Keep it up. I like it, but I'm awkward with it."

Fred held her by the door and did keep it up, furiously, for full five minutes. She took it with some confusion, seeming all the while to be hesitating, to be arrested in her course and trying to pass him. But she did not really try to pass, and her color deepened. Fred spoke in German, and Archie caught from her an occasional *Ja? So?* muttered rather than spoken.

When they rejoined Landry and Dr. Archie, Fred took up his tea again. "I see you're singing *Venus*[4] Saturday night. Will they never let you have a chance at *Elisabeth*?"

She shrugged her shoulders. "Not here. There are so many singers here, and they try us out in such a stingy way. Think of it, last year I came over in October, and it was the first of December before I went on at all! I'm often sorry I left Dresden."

"Still," Fred argued, "Dresden is limited."

"Just so, and I've begun to sigh for those very limitations. In New York everything is impersonal. Your audience never knows its own mind, and its mind is never twice the same. I'd rather sing where the people are pig-headed and throw carrots at you if you don't do it the way they like it. The house here is splendid, and the night audiences are exciting. I hate the matinées; like singing at a *Kaffeeklatsch.*" She rose and turned on the lights.

"Ah!" Fred exclaimed, "why do you do that? That is a signal that tea is over." He got up and drew out his gloves.

"Not at all. Shall you be here Saturday night?" She sat down on the piano bench and leaned her elbow back on the keyboard. "Necker[5] sings *Elisabeth*. Make Dr. Archie go. Everything she sings is worth hearing."

"But she's failing so. The last time I heard her she had no voice at all. She *is* a poor vocalist!"

Thea cut him off. "She's a great artist, whether she's in voice or not, and she's the only one here. If you want a big voice, you can take my *Ortrud* of last night; that's big enough, and vulgar enough."

Fred laughed and turned away, this time with decision. "I don't want her!" he protested energetically. "I only wanted to get a rise out of you. I like Necker's *Elisabeth* well enough. I like your *Venus* well enough, too."

"It's a beautiful part, and it's often dreadfully sung. It's very hard to sing, of course."

Ottenburg bent over the hand she held out to him. "For an uninvited guest, I've fared very well. You were nice to let me come up. I'd have been terribly cut up if you'd sent me away. May I?" He kissed her hand lightly and backed toward the door, still smiling, and promising to keep an eye on Archie. "He can't be trusted at all, Thea. One of the waiters at Martin's worked a Tourainian hare off on him at luncheon yesterday, for seven twenty-five."

Thea broke into a laugh, the deep one he recognized. "Did he have a ribbon on, this hare? Did they bring him in a gilt cage?"

"No,"—Archie spoke up for himself,—"they brought him in a brown sauce, which was very good. He didn't taste very different from any rabbit."

"Probably came from a push-cart on the East Side." Thea looked at her old friend commiseratingly. "Yes, *do* keep an eye on him, Fred. I had no idea," shaking her head. "Yes, I'll be obliged to you."

"Count on me!" Their eyes met in a gay smile, and Fred bowed himself out.

VII

ON SATURDAY NIGHT Dr. Archie went with Fred Ottenburg to hear "Tannhäuser." Thea had a rehearsal on Sunday afternoon, but as she was not on the bill again until Wednesday, she promised to dine with Archie and Ottenburg on Monday, if they could make the dinner early.

At a little after eight on Monday evening, the three friends returned to Thea's apartment and seated themselves for an hour of quiet talk.

"I'm sorry we couldn't have had Landry with us tonight," Thea said, "but he's on at Weber and Fields' every night now. You ought to hear him, Dr. Archie. He often sings the old Scotch airs you used to love."

"Why not go down this evening?" Fred suggested hopefully, glancing at his watch. "That is, if you'd like to go. I can telephone and find what time he comes on."

Thea hesitated. "No, I think not. I took a long walk this afternoon and I'm rather tired. I think I can get to sleep early and be so much ahead. I don't mean at once, however," seeing Dr. Archie's disappointed look. "I always like to hear Landry," she added. "He never had much voice, and it's worn, but there's a sweetness about it, and he sings with such taste."

"Yes, doesn't he? May I?" Fred took out his cigarette case. "It really doesn't bother your throat?"

"A little doesn't. But cigar smoke does. Poor Dr. Archie! Can you do with one of those?"

"I'm learning to like them," the doctor declared, taking one from the case Fred proffered him.

"Landry's the only fellow I know in this country who can do that sort of thing," Fred went on. "Like the best English ballad singers. He can sing even popular stuff by higher lights, as it were."

Thea nodded. "Yes; sometimes I make him sing his most foolish things for me. It's restful, as he does it. That's when I'm homesick, Dr. Archie."

"You knew him in Germany, Thea?" Dr. Archie had quietly abandoned his cigarette as a comfortless article. "When you first went over?"

"Yes. He was a good friend to a green girl. He helped me with my German and my music and my general discouragement. Seemed to care more about my getting on than about himself. He had no money, either. An old aunt had loaned him a little to study on.— Will you answer that, Fred?"

Fred caught up the telephone and stopped the buzz while Thea went on talking to Dr. Archie about Landry. Telling some one to hold the wire, he presently put down the instrument and approached Thea with a startled expression on his face.

"It's the management," he said quietly. "Gloeckler has broken down: fainting fits. Madame Rheinecker is in Atlantic City and Schramm is singing in Philadelphia tonight. They want to know whether you can come down and finish *Sieglinde*."[1]

"What time is it?"

"Eight fifty-five. The first act is just over. They can hold the curtain twenty-five minutes."

Thea did not move. "Twenty-five and thirty-five makes sixty," she muttered. "Tell them I'll come if they hold the curtain till I am in the dressing-room. Say I'll have to wear her costumes, and the dresser must have everything ready. Then call a taxi, please."

Thea had not changed her position since he first interrupted her, but she had grown pale and was opening and shutting her hands rapidly. She looked, Fred thought, terrified. He half turned toward the telephone, but hung on one foot.

"Have you ever sung the part?" he asked.

"No, but I've rehearsed it. That's all right. Get the cab." Still she made no move. She merely turned perfectly blank eyes to Dr. Archie and said absently, "It's curious, but just at this minute I can't remember a bar of 'Walküre' after the first act. And I let my maid go out." She sprang up and beckoned Archie without so much, he felt sure, as knowing who he was. "Come with me." She went quickly into her sleeping-chamber and threw open a door into a trunk-room. "See that white trunk? It's not locked. It's full of wigs, in boxes. Look until you find one marked 'Ring 2.' Bring it quick!"

While she directed him, she threw open a square trunk and began tossing out shoes of every shape and color.

Ottenburg appeared at the door. "Can I help you?"

She threw him some white sandals with long laces and silk stockings pinned to them. "Put those in something, and then go to the piano and give me a few measures in there—you know." She was behaving somewhat like a cyclone now, and while she wrenched open drawers and closet doors, Ottenburg got to the piano as quickly as possible and began to herald the reappearance of the Volsung pair, trusting to memory.

In a few moments Thea came out enveloped in her long fur coat with a scarf over her head and knitted woolen gloves on her hands. Her glassy eye took in the fact that Fred was playing from memory, and even in her distracted state, a faint smile flickered over her colorless lips. She stretched out a woolly hand, "The score, please. Behind you, there."

Dr. Archie followed with a canvas box and a satchel. As they went through the hall, the men caught up their hats and coats. They left the music-room, Fred noticed, just seven minutes after he got the telephone message. In the elevator Thea said in that husky whisper which had so perplexed Dr. Archie when he first heard it, "Tell the driver he must do it in twenty minutes, less if he can. He must leave the light on in the cab. I can do a good deal in twenty minutes. If only you hadn't made me eat—Damn that duck!" she broke out bitterly; "why did you?"

"Wish I had it back! But it won't bother you, to-night. You need strength," he pleaded consolingly.

But she only muttered angrily under her breath, "Idiot, idiot!"

Ottenburg shot ahead and instructed the driver, while the doctor put Thea into the cab and shut the door. She did not speak to either of them again. As the driver scrambled into his seat she opened the score and fixed her eyes upon it. Her face, in the white light, looked as bleak as a stone quarry.

As her cab slid away, Ottenburg shoved Archie into a second taxi that waited by the curb. "We'd better trail her," he explained. "There might be a hold-up of some kind." As the cab whizzed off he broke into an eruption of profanity.

"What's the matter, Fred?" the doctor asked. He was a good deal dazed by the rapid evolutions of the last ten minutes.

"Matter enough!" Fred growled, buttoning his overcoat with a shiver. "What a way to sing a part for the first time! That duck really is on my conscience. It will be a wonder if she can do anything but quack! Scrambling on in the middle of a performance like this, with no rehearsal! The stuff she has to sing in there is a fright—rhythm, pitch,—and terribly difficult intervals."

"She looked frightened," Dr. Archie said thoughtfully, "but I thought she looked—determined."

Fred sniffed. "Oh, determined! That's the kind of rough deal that makes savages of singers. Here's a part she's worked on and got ready for for years, and now they give her a chance to go on and butcher it. Goodness knows when she's looked at the score last, or whether she can use the business she's studied with this cast. Necker's singing *Brünnhilde*;[2] she may help her, if it's not one of her sore nights."

"Is she sore at Thea?" Dr. Archie asked wonderingly.

"My dear man, Necker's sore at everything. She's breaking up; too early; just when she ought to be at her best. There's one story that she is struggling under some serious malady, another that she learned a bad method at the Prague Conservatory and has ruined her organ. She's the sorest thing in the world. If she weathers this winter through, it'll be her last. She's paying for it with the last rags of her voice. And then—" Fred whistled softly.

"Well, what then?"

"Then our girl may come in for some of it. It's dog eat dog, in this game as in every other."

The cab stopped and Fred and Dr. Archie hurried to the box office. The Monday-night house was sold out. They bought standing room and entered the auditorium just as the press representative of the house was thanking the audience for their patience and telling them that although Madame Gloeckler was too ill to sing, Miss Kronborg had kindly consented to finish her part. This announcement was met with vehement applause from the upper circles of the house.

"She has her—constituents," Dr. Archie murmured.

"Yes, up there, where they're young and hungry. These people down here have dined too well. They won't mind, however. They like fires and accidents and *divertissements.* Two *Sieglindes* are more unusual than one, so they'll be satisfied."

***

After the final disappearance of the mother of Siegfried, Ottenburg and the doctor slipped out through the crowd and left the house. Near the stage entrance Fred found the driver who had brought Thea down. He dismissed him and got a larger car. He and Archie waited on the sidewalk, and when Kronborg came out alone they gathered her into the cab and sprang in after her.

Thea sank back into a corner of the back seat and yawned. "Well, I got through, eh?" Her tone was reassuring. "On the whole, I think I've given you gentlemen a pretty lively evening, for one who has no social accomplishments."

"Rather! There was something like a popular uprising at the end of the second act. Archie and I couldn't keep it up as long as the rest of them did. A howl like that ought to show the management which way the wind is blowing. You probably know you were magnificent."

"I thought it went pretty well," she spoke impartially. "I was rather smart to catch his tempo there, at the beginning of the first recitative, when he came in too soon, don't you think? It's tricky in there, without a rehearsal. Oh, I was all right! He took that syncopation too fast in the beginning. Some singers take it fast there—think it sounds more impassioned. That's one way!" She sniffed, and Fred shot a mirthful glance at Archie. Her boastfulness would have been childish in a schoolboy. In the light of what she had done, of the strain they had lived through during the last two hours, it made one laugh,—almost cry. She went on, robustly: "And I didn't feel my dinner, really, Fred. I am hungry again, I'm ashamed to say,—and I forgot to order anything at my hotel."

Fred put his hand on the door. "Where to? You must have food."

"Do you know any quiet place, where I won't be stared at? I've still got make-up on."

"I do. Nice English chop-house on Forty-fourth Street. Nobody there at night but theater people after the show, and a few bachelors." He opened the door and spoke to the driver.

As the car turned, Thea reached across to the front seat and drew Dr. Archie's handkerchief out of his breast pocket. "This comes to me naturally," she said, rubbing her cheeks and eyebrows. "When I was little I always loved your handkerchiefs because they were silk and smelled of Cologne water. I think they must have been the only really clean handkerchiefs in Moonstone. You were always wiping my face with them, when you met me out in the dust, I remember. Did I never have any?"

"I think you'd nearly always used yours up on your baby brother."

Thea sighed. "Yes, Thor had such a way of getting messy. You say he's a good chauffeur?" She closed her eyes for a moment as if they were tired. Suddenly she looked up. "Isn't it funny, how we travel in circles? Here you are, still getting me clean, and Fred is still feeding me. I would have died of starvation at that boarding-house on Indiana Avenue if he hadn't taken me out to the Buckingham and filled me up once in a while. What a cavern I was to fill, too. The waiters used to look astonished. I'm still singing on that food."

Fred alighted and gave Thea his arm as they crossed the icy sidewalk. They were taken upstairs in an antiquated lift and found the cheerful chop-room half full of supper parties. An English company playing at the Empire had just come in. The waiters, in red waistcoats, were hurrying about. Fred got a table at the back of the room, in a corner, and urged his waiter to get the oysters on at once.

"Takes a few minutes to open them, sir," the man expostulated.

"Yes, but make it as few as possible, and bring the lady's first. Then grilled chops with kidneys, and salad."

Thea began eating celery stalks at once, from the base to the foliage. "Necker said something nice to me tonight. You might have thought the management would say something, but not they." She looked at Fred from under her blackened lashes. "It *was* a stunt, to jump in and sing that second act without rehearsal. It doesn't sing itself."

Ottenburg was watching her brilliant eyes and her face. She was much handsomer than she had been early in the evening. Excitement of this sort enriched her. It was only under such excitement, he reflected, that she was entirely illuminated, or wholly present. At other times there was something a little cold and empty, like a big room with no people in it. Even in her most genial moods there was a shadow of restlessness, as if she were waiting for something and were exercising the virtue of patience. During dinner she had been as kind as she knew how to be, to him and to Archie, and had given them as much of herself as she could. But, clearly, she knew only one way of being really kind, from the core of her heart out; and there was but one way in which she could give herself to people largely and gladly, spontaneously. Even as a girl she had been at her best in vigorous effort, he remembered; physical effort, when there was no other kind at hand. She could be expansive only in explosions. Old Nathanmeyer had seen it. In the very first song Fred had ever heard her sing, she had unconsciously declared it.

Thea Kronborg turned suddenly from her talk with Archie and peered suspiciously into the corner where Ottenburg sat with folded arms, observing her. "What's the matter with you, Fred? I'm afraid of you when you're quiet,—fortunately you almost never are. What are you thinking about?"

"I was wondering how you got right with the orchestra so quickly, there at first. I had a flash of terror," he replied easily.

She bolted her last oyster and ducked her head. "So had I! I don't know how I did catch it. Desperation, I suppose; same way the Indian babies swim when they're thrown into the river. I *had* to. Now it's over, I'm glad I had to. I learned a whole lot to-night."

Archie, who usually felt that it behooved him to be silent during such discussions, was encouraged by her geniality to venture, "I don't see how you can learn anything in such a turmoil; or how you can keep your mind on it, for that matter."

Thea glanced about the room and suddenly put her hand up to her hair. "Mercy, I've no hat on! Why didn't you tell me? And I seem to be wearing a rumpled dinner dress, with all this paint on my face! I must look like something you picked up on Second Avenue. I hope there are no Colorado reformers about, Dr. Archie. What a dreadful old pair these people must be thinking you! Well, I had to eat." She sniffed the savor of the grill as the waiter uncovered it. "Yes, draught beer, please. No, thank you, Fred, *no* champagne.—To go back to your question, Dr. Archie, you can believe I keep my mind on it. That's the whole trick, in so far as stage experience goes; keeping right there every second. If I think of anything else for a flash, I'm gone, done for. But at the same time, one can take things in—with another part of your brain, maybe. It's different from what you get in study, more practical and conclusive. There are some things you learn best in calm, and some in storm. You learn the delivery of a part only before an audience."

"Heaven help us," gasped Ottenburg. "Weren't you hungry, though! It's beautiful to see you eat."

"Glad you like it. Of course I'm hungry. Are you staying over for 'Rheingold' Friday afternoon?"

"My dear Thea,"—Fred lit a cigarette,—"I'm a serious business man now. I have to sell beer. I'm due in Chicago on Wednesday. I'd come back to hear you, but *Fricka*[3] is not an alluring part."

"Then you've never heard it well done." She spoke up hotly. "Fat German woman scolding her husband, eh? That's not my idea. Wait till you hear my *Fricka*. It's a beautiful part." Thea leaned forward on the table and touched Archie's arm. "You remember, Dr. Archie, how my mother always wore her hair, parted in the middle and done low on her neck behind, so you got the shape of her head and such a calm, white forehead? I wear mine like that for *Fricka*. A little more coronet effect, built up a little higher at the sides, but the idea's the same. I think you'll notice it." She turned to Ottenburg reproachfully: "It's noble music, Fred, from the first measure. There's nothing lovelier than the *wonniger Hausrath*.[4] It's all such comprehensive sort of music—fateful. Of course, *Fricka knows*," Thea ended quietly.

Fred sighed. "There, you've spoiled my itinerary. Now I'll have to come back, of course. Archie, you'd better get busy about seats to-morrow."

"I can get you box seats, somewhere. I know nobody here, and I never ask for any." Thea began hunting among her wraps. "Oh, how funny! I've only these short woolen gloves, and no sleeves. Put on my coat first. Those English people can't make out where you got your lady, she's so made up of contradictions." She rose laughing and plunged her arms into the coat Dr. Archie held for her. As she settled herself into it and buttoned it under her chin, she gave him an old signal with her eyelid. "I'd like to sing another part to-night. This is the sort of evening I fancy, when there's something to do. Let me see: I have to sing in 'Trovatore' Wednesday night, and there are rehearsals for the 'Ring' every day this week. Consider me dead until Saturday, Dr. Archie. I invite you both to dine with me on Saturday night, the day after 'Rheingold.' And Fred must leave early, for I want to talk to you alone. You've been here nearly a week, and I haven't had a serious word with you. *Tak for mad*,[5] Fred, as the Norwegians say."

VIII

THE "RING OF THE NIEBELUNGS" was to be given at the Metropolitan on four successive Friday afternoons. After the first of these performances, Fred Ottenburg went home with Landry for tea. Landry was one of the few public entertainers who own real estate in New York. He lived in a little three-story brick house on Jane Street, in Greenwich Village, which had been left to him by the same aunt who paid for his musical education.

Landry was born, and spent the first fifteen years of his life, on a rocky Connecticut farm not far from Cos Cob. His father was an ignorant, violent man, a bungling farmer and a brutal husband. The farmhouse, dilapidated and damp, stood in a hollow beside a marshy pond. Oliver had worked hard while he lived at home, although he was never clean or warm in winter and had wretched food all the year round. His spare, dry figure, his prominent larynx, and the peculiar red of his face and hands belonged to the choreboy he had never outgrown. It was as if the farm, knowing he would escape from it as early as he could, had ground its mark on him deep. When he was fifteen Oliver ran away and went to live with his Catholic aunt, on Jane Street, whom his mother was never allowed to visit. The priest of St. Joseph's Parish discovered that he had a voice.

Landry had an affection for the house on Jane Street, where he had first learned what cleanliness and order and courtesy were. When his aunt died he had the place done over, got an Irish housekeeper, and lived there with a great many beautiful things he had collected. His living expenses were never large, but he could not restrain himself from buying graceful and useless objects. He was a collector for much the same reason that he was a Catholic, and he was a Catholic chiefly because his father used to sit in the kitchen and read aloud to his hired men disgusting "exposures" of the Roman Church, enjoying equally the hideous stories and the outrage to his wife's feelings.

At first Landry bought books; then rugs, drawings, china. He had a beautiful collection of old French and Spanish fans. He kept them in an escritoire he had brought from Spain, but there were always a few of them lying about in his sitting-room.

While Landry and his guest were waiting for the tea to be brought, Ottenburg took up one of these fans from the low marble mantel-shelf and opened it in the firelight. One side was painted with a pearly sky and floating clouds. On the other was a formal garden where an elegant shepherdess with a mask and crook was fleeing on high heels from a satin-coated shepherd.

"You ought not to keep these things about, like this, Oliver. The dust from your grate must get at them."

"It does, but I get them to enjoy them, not to have them. They're pleasant to glance at and to play with at odd times like this, when one is waiting for tea or something."

Fred smiled. The idea of Landry stretched out before his fire playing with his fans, amused him. Mrs. McGinnis brought the tea and put it before the hearth: old teacups that were velvety to the touch and a pot-bellied silver cream pitcher of an Early Georgian pattern, which was always brought, though Landry took rum.

Fred drank his tea walking about, examining Landry's sumptuous writing-table in the alcove and the Boucher drawing[1] in red chalk over the mantel. "I don't see how you can stand this place without a heroine. It would give me a raging thirst for gallantries."

Landry was helping himself to a second cup of tea. "Works quite the other way with me. It consoles me for the lack of her. It's just feminine enough to be pleasant to return to. Not any more tea? Then sit down and play for me. I'm always playing for other people, and I never have a chance to sit here quietly and listen."

Ottenburg opened the piano and began softly to boom forth the shadowy introduction to the opera they had just heard. "Will that do?" he asked jokingly. "I can't seem to get it out of my head."

"Oh, excellently! Thea told me it was quite wonderful, the way you can do Wagner scores on the piano. So few people can give one any idea of the music. Go ahead, as long as you like. I can smoke, too." Landry flattened himself out on his cushions and abandoned himself to ease with the circumstance of one who has never grown quite accustomed to ease.

Ottenburg played on, as he happened to remember. He understood now why Thea wished him to hear her in "Rheingold." It had been clear to him as soon as *Fricka* rose from sleep and looked out over the young world, stretching one white arm toward the new Götterburg shining on the heights. "*Wotan! Gemahl! erwache!*"[2] She was pure Scandinavian, this *Fricka:* "Swedish summer"! he remembered old Mr. Nathanmeyer's phrase. She had wished him to see her because she had a distinct kind of loveliness for this part, a shining beauty like the light of sunset on distant sails. She seemed to take on the look of immortal loveliness, the youth of the golden apples, the shining body and the shining mind. *Fricka* had been a jealous spouse to him for so long that he had forgot she meant wisdom before she meant domestic order, and that, in any event, she was always a goddess. The *Fricka* of that afternoon was so clear and sunny, so nobly conceived, that she made a whole atmosphere about herself and quite redeemed from shabbiness the helplessness and unscrupulousness of the gods. Her reproaches to *Wotan* were the pleadings of a tempered mind, a consistent sense of beauty. In the long silences of her part, her shining presence was a visible complement to the discussion of the orchestra. As the themes which were to help in weaving the drama to its end first came vaguely upon the ear, one saw their import and tendency in the face of this clearest-visioned of the gods.

In the scene between *Fricka* and *Wotan,* Ottenburg stopped. "I can't seem to get the voices, in there."

Landry chuckled. "Don't try. I know it well enough. I expect I've been over that with her a thousand times. I was playing for her almost every day when she was first working on it. When she begins with a part she's hard to work with: so slow you'd think she was stupid if you didn't know her. Of course she blames it all on her accompanist. It goes on like that for weeks sometimes. This did. She kept shaking her head and staring and looking gloomy. All at once, she got her line—it usually comes suddenly, after stretches of not getting anywhere at all—and after that it kept changing and clearing. As she worked her voice into it, it got more and more of that 'gold' quality that makes her *Fricka* so different."

Fred began *Fricka*'s first aria again. "It's certainly different. Curious how she does it. Such a beautiful idea, out of a part that's always been so ungrateful. She's a lovely thing, but she was never so beautiful as that, really. Nobody is." He repeated the loveliest phrase. "How does she manage it, Landry? You've worked with her."

Landry drew cherishingly on the last cigarette he meant to permit himself before singing. "Oh, it's a question of a big personality—and all that goes with it. Brains, of course. Imagination, of course. But the important thing is that she was born full of color, with a rich personality. That's a gift of the gods, like a fine nose. You have it, or you haven't. Against it, intelligence and musicianship and habits of industry don't count at all. Singers are a conventional race. When Thea was studying in Berlin the other girls were mortally afraid of her. She has a pretty rough hand with women, dull ones, and she could be rude, too! The girls used to call her *die Wölfin.*"[3]

Fred thrust his hands into his pockets and leaned back against the piano. "Of course, even a stupid woman could get effects with such machinery: such a voice and body and face. But they couldn't possibly belong to a stupid woman, could they?"

Landry shook his head. "It's personality; that's as near as you can come to it. That's what constitutes real equipment. What she does is interesting because she does it. Even the things she discards are suggestive. I regret some of them. Her conceptions are colored in so many different ways. You've heard her *Elisabeth?* Wonderful, isn't it? She was working on that part years ago when her mother was ill. I could see her anxiety and grief getting more and more into the part. The last act is heart-breaking. It's as homely as a country prayer meeting: might be any lonely woman getting ready to die.[4] It's full of the thing every plain creature finds out for himself, but that never gets written down. It's unconscious memory, maybe; inherited memory, like folk-music. I call it personality."

Fred laughed, and turning to the piano began coaxing the *Fricka* music again. "Call it anything you like, my boy. I have a name for it myself, but I shan't tell you." He looked over his shoulder at Landry, stretched out by the fire. "You have a great time watching her, don't you?"

"Oh, yes!" replied Landry simply. "I'm not interested in much that goes on in New York. Now, if you'll excuse me, I'll have to dress." He rose with a reluctant sigh. "Can I get you anything? Some whiskey?"

"Thank you, no. I'll amuse myself here. I don't often get a chance at a good piano when I'm away from home. You haven't had this one long, have you? Action's a bit stiff. I say," he stopped Landry in the doorway, "has Thea ever been down here?"

Landry turned back. "Yes. She came several times when I had erysipelas. I was a nice mess, with two nurses. She brought down some inside window-boxes, planted with crocuses and things. Very cheering, only I couldn't see them or her."

"Didn't she like your place?"

"She thought she did, but I fancy it was a good deal cluttered up for her taste. I could hear her pacing about like something in a cage. She pushed the piano back against the wall and the chairs into corners, and she broke my amber elephant." Landry took a yellow object some four inches high from one of his low bookcases. "You can see where his leg is glued on,—a souvenir. Yes, he's lemon amber, very fine."

Landry disappeared behind the curtains and in a moment Fred heard the wheeze of an atomizer. He put the amber elephant on the piano beside him and seemed to get a great deal of amusement out of the beast.

IX

When Archie and Ottenburg dined with Thea on Saturday evening, they were served downstairs in the hotel dining-room, but they were to have their coffee in her own apartment. As they were going up in the elevator after dinner, Fred turned suddenly to Thea. "And why, please, did you break Landry's amber elephant?"

She looked guilty and began to laugh. "Hasn't he got over that yet? I didn't really mean to break it. I was perhaps careless. His things are so over-petted that I was tempted to be careless with a lot of them."

"How can you be so heartless, when they're all he has in the world?"

"He has me. I'm a great deal of diversion for him; all he needs. There," she said as she opened the door into her own hall, "I shouldn't have said that before the elevator boy."

"Even an elevator boy couldn't make a scandal about Oliver. He's such a catnip man."

Dr. Archie laughed, but Thea, who seemed suddenly to have thought of something annoying, repeated blankly, "Catnip man?"

"Yes, he lives on catnip, and rum tea. But he's not the only one. You are like an eccentric old woman I know in Boston, who goes about in the spring feeding catnip to street cats. You dispense it to a lot of fellows. Your pull seems to be more with men than with women, you know; with seasoned men, about my age, or older. Even on Friday afternoon I kept running into them, old boys I hadn't seen for years, thin at the part and thick at the girth, until I stood still in the draft and held my hair on. They're always there; I hear them talking about you in the smoking-room. Probably we don't get to the point of apprehending anything good until we're about forty. Then, in the light of what is going, and of what, God help us! is coming, we arrive at understanding."

"I don't see why people go to the opera, anyway,—serious people." She spoke discontentedly. "I suppose they get something, or think they do. Here's the coffee. There, please," she directed the waiter. Going to the table she began to pour the coffee, standing. She wore a white dress trimmed with crystals which had rattled a good deal during dinner, as all her movements had been impatient and nervous, and she had twisted the dark velvet rose at her girdle until it looked rumpled and weary. She poured the coffee as if it were a ceremony in which she did not believe. "Can you make anything of Fred's nonsense, Dr. Archie?" she asked, as he came to take his cup.

Fred approached her. "My nonsense is all right. The same brand has gone with you before. It's you who won't be jollied. What's the matter? You have something on your mind."

"I've a good deal. Too much to be an agreeable hostess." She turned quickly away from the coffee and sat down on the piano bench, facing the two men. "For one thing, there's a change in the cast for Friday afternoon. They're going to let me sing *Sieglinde*." Her frown did not conceal the pleasure with which she made this announcement.

"Are you going to keep us dangling about here forever, Thea? Archie and I are supposed to have other things to do." Fred looked at her with an excitement quite as apparent as her own.

"Here I've been ready to sing *Sieglinde* for two years, kept in torment, and now it comes off within two weeks, just when I want to be seeing something of Dr. Archie. I don't know what their plans are down there. After Friday they may let me cool for several weeks, and they may rush me. I suppose it depends somewhat on how things go Friday afternoon."

"Oh, they'll go fast enough! That's better suited to your voice than anything you've sung here. That gives you every opportunity I've waited for." Ottenburg crossed the room and standing beside her began to play "*Du bist der Lenz.*"[1]

With a violent movement Thea caught his wrists and pushed his hands away from the keys.

"Fred, can't you be serious? A thousand things may happen between this and Friday to put me out. Something will happen. If that part were sung well, as well as it ought to be, it would be one of the most beautiful things in the world. That's why it never is sung right, and never will be." She clenched her hands and opened them despairingly, looking out of the open window. "It's inaccessibly beautiful!" she brought out sharply.

Fred and Dr. Archie watched her. In a moment she turned back to them. "It's impossible to sing a part like that well for the first time, except for the sort who will never sing it any better. Everything hangs on that first night, and that's bound to be bad. There you are," she shrugged impatiently. "For one thing, they change the cast at the eleventh hour and then rehearse the life out of me."

Ottenburg put down his cup with exaggerated care. "Still, you really want to do it, you know."

"Want to?" she repeated indignantly; "of course I want to! If this were only next Thursday night—But between now and Friday I'll do nothing but fret away my strength. Oh, I'm not saying I don't need the rehearsals! But I don't need them strung out through a week. That system's well enough for phlegmatic singers; it only drains me. Every single feature of operatic routine is detrimental to me. I usually go on like a horse that's been fixed to lose a race. I have to work hard to do my worst, let alone my best. I wish you could hear me sing well, once," she turned to Fred defiantly; "I have, a few times in my life, when there was nothing to gain by it."

Fred approached her again and held out his hand. "I recall my instructions, and now I'll leave you to fight it out with Archie. He can't possibly represent managerial stupidity to you as I seem to have a gift for doing."

As he smiled down at her, his good humor, his good wishes, his understanding, embarrassed her and recalled her to herself. She kept her seat, still holding his hand. "All the same, Fred, isn't it too bad, that there are so many things—" She broke off with a shake of the head.

"My dear girl, if I could bridge over the agony between now and Friday for you— But you know the rules of the game; why torment yourself? You saw the other night that you had the part under your thumb. Now walk, sleep, play with Archie, keep your tiger hungry, and she'll spring all right on Friday. I'll be there to see her, and there'll be more than I, I suspect. Harsanyi's on the Wilhelm der Grosse; gets in on Thursday."

"Harsanyi?" Thea's eye lighted. "I haven't seen him for years. We always miss each other." She paused, hesitating. "Yes, I should like that. But he'll be busy, maybe?"

"He gives his first concert at Carnegie Hall, week after next. Better send him a box if you can."

"Yes, I'll manage it." Thea took his hand again. "Oh, I should like that, Fred!" she added impulsively. "Even if I were put out, he'd get the idea,"—she threw back her head,—"for there is an idea!"

"Which won't penetrate here," he tapped his brow and began to laugh. "You are an ungrateful huzzy, *comme les autres!*"[2]

Thea detained him as he turned away. She pulled a flower out of a bouquet on the piano and absently drew the stem through the lapel of his coat. "I shall be walking in the Park to-morrow afternoon, on the reservoir path, between four and five, if you care to join me. You know that after Harsanyi I'd rather please you than anyone else. You know a lot, but he knows even more than you."

"Thank you. Don't try to analyze it. *Schlafen Sie wohl!*"[3] he kissed her fingers and waved from the door, closing it behind him.

"He's the right sort, Thea." Dr. Archie looked warmly after his disappearing friend. "I've always hoped you'd make it up with Fred."

"Well, haven't I? Oh, marry him, you mean! Perhaps it may come about, some day. Just at present he's not in the marriage market any more than I am, is he?"

"No, I suppose not. It's a damned shame that a man like Ottenburg should be tied up as he is, wasting all the best years of his life. A woman with general paresis ought to be legally dead."

"Don't let us talk about Fred's wife, please. He had no business to get into such a mess, and he had no business to stay in it. He's always been a softy where women were concerned."

"Most of us are, I'm afraid," Dr. Archie admitted meekly.

"Too much light in here, isn't there? Tires one's eyes. The stage lights are hard on mine." Thea began turning them out. "We'll leave the little one, over the piano." She sank down by Archie on the deep sofa. "We two have so much to talk about that we keep away from it altogether; have you noticed? We don't even nibble the edges. I wish we had Landry here to-night to play for us. He's very comforting."

"I'm afraid you don't have enough personal life, outside your work, Thea." The doctor looked at her anxiously.

She smiled at him with her eyes half closed. "My dear doctor, I don't have any. Your work becomes your personal life. You are not much good until it does. It's like being woven into a big web. You can't pull away, because all your little tendrils are woven into the picture. It takes you up, and uses you, and spins you out; and that is your life. Not much else can happen to you."

"Didn't you think of marrying, several years ago?"

"You mean Nordquist? Yes; but I changed my mind. We had been singing a good deal together. He's a splendid creature."

"Were you much in love with him, Thea?" the doctor asked hopefully.

She smiled again. "I don't think I know just what that expression means. I've never been able to find out. I think I was in love with you when I was little, but not with any one since then. There are a great many ways of caring for people. It's not, after all, a simple state, like measles or tonsilitis. Nordquist is a taking sort of man. He and I were out in a rowboat once in a terrible storm. The lake was fed by glaciers,—ice water,—and we couldn't have swum a stroke if the boat had filled. If we hadn't both been strong and kept our heads, we'd have gone down. We pulled for every ounce there was in us, and we just got off with our lives. We were always being thrown together like that, under some kind of pressure. Yes, for a while I thought he would make everything right." She paused and sank back, resting her head on a cushion, pressing her eyelids down with her fingers. "You see," she went on abruptly, "he had a wife and two children. He hadn't lived with her for several years, but when she heard that he wanted to marry again, she began to make trouble. He earned a good deal of money, but he was careless and always wretchedly in debt. He came to me one day and told me he thought his wife would settle for a hundred thousand marks and consent to a divorce. I got very angry and sent him away. Next day he came back and said he thought she'd take fifty thousand."

Dr. Archie drew away from her, to the end of the sofa. "Good God, Thea,"—He ran his handkerchief over his forehead. "What sort of people—" He stopped and shook his head.

Thea rose and stood beside him, her hand on his shoulder. "That's exactly how it struck me," she said quietly. "Oh, we have things in common, things that go away back, under everything. You understand, of course. Nordquist didn't. He thought I wasn't willing to part with the money. I couldn't let myself buy him from Fru Nordquist, and he couldn't see why. He had always thought I was close about money, so he attributed it to that. I am careful,"—she ran her arm through Archie's and when he rose began to walk about the room with him. "I can't be careless with money. I began the world on six hundred dollars, and it was the price of a man's life. Ray Kennedy had worked hard and been sober and denied himself, and when he died he had six hundred dollars to show for it. I always measure things by that six hundred dollars, just as I measure high buildings by the Moonstone standpipe. There are standards we can't get away from."

Dr. Archie took her hand. "I don't believe we should be any happier if we did get away from them. I think it gives you some of your poise, having that anchor. You look," glancing down at her head and shoulders, "sometimes so like your mother."

"Thank you. You couldn't say anything nicer to me than that. On Friday afternoon, didn't you think?"

"Yes, but at other times, too. I love to see it. Do you know what I thought about that first night when I heard you sing? I kept remembering the night I took care of you when you had pneumonia, when you were ten years old. You were a terribly sick child, and I was a country doctor without much experience. There were no oxygen tanks about then. You pretty nearly slipped away from me. If you had—"

Thea dropped her head on his shoulder. "I'd have saved myself and you a lot of trouble, wouldn't I? Dear Dr. Archie!" she murmured.

"As for me, life would have been a pretty bleak stretch, with you left out." The doctor took one of the crystal pendants that hung from her shoulder and looked into it thoughtfully. "I guess I'm a romantic old fellow, underneath. And you've always been my romance. Those years when you were growing up were my happiest. When I dream about you, I always see you as a little girl."

They paused by the open window. "Do you? Nearly all my dreams, except those about breaking down on the stage or missing trains, are about Moonstone. You tell me the old house has been pulled down, but it stands in my mind, every stick and timber. In my sleep I go all about it, and look in the right drawers and cupboards for everything. I often dream that I'm hunting for my rubbers in that pile of overshoes that was always under the hatrack in the hall. I pick up every overshoe and know whose it is, but I can't find my own. Then the school bell begins to ring and I begin to cry.

That's the house I rest in when I'm tired. All the old furniture and the worn spots in the carpet—it rests my mind to go over them."

They were looking out of the window. Thea kept his arm. Down on the river four battleships were anchored in line, brilliantly lighted, and launches were coming and going, bringing the men ashore. A searchlight from one of the ironclads was playing on the great headland up the river, where it makes its first resolute turn. Overhead the night-blue sky was intense and clear.

"There's so much that I want to tell you," she said at last, "and it's hard to explain. My life is full of jealousies and disappointments, you know. You get to hating people who do contemptible work and who get on just as well as you do. There are many disappointments in my profession, and bitter, bitter contempts!" Her face hardened, and looked much older. "If you love the good thing vitally, enough to give up for it all that one must give up for it, then you must hate the cheap thing just as hard. I tell you, there is such a thing as creative hate! A contempt that drives you through fire, makes you risk everything and lose everything, makes you a long sight better than you ever knew you could be." As she glanced at Dr. Archie's face, Thea stopped short and turned her own face away. Her eyes followed the path of the searchlight up the river and rested upon the illumined headland.

"You see," she went on more calmly, "voices are accidental things. You find plenty of good voices in common women, with common minds and common hearts. Look at that woman who sang *Ortrud* with me last week. She's new here and the people are wild about her. 'Such a beautiful volume of tone!' they say. I give you my word she's as stupid as an owl and as coarse as a pig, and any one who knows anything about singing would see that in an instant. Yet she's quite as popular as Necker, who's a great artist. How can I get much satisfaction out of the enthusiasm of a house that likes her atrociously bad performance at the same time that it pretends to like mine? If they like her, then they ought to hiss me off the stage. We stand for things that are irreconcilable, absolutely. You can't try to do things right and not despise the people who do them wrong. How can I be indifferent? If that doesn't matter, then nothing matters. Well, sometimes I've come home as I did the other night when you first saw me, so full of bitterness that it was as if my mind were full of daggers. And I've gone to sleep and wakened up in the Kohlers' garden, with the pigeons and the white rabbits, so happy! And that saves me." She sat down on the piano bench. Archie thought she had forgotten all about him, until she called his name. Her voice was soft now, and wonderfully sweet. It seemed to come from somewhere deep within her, there were such strong vibrations in it. "You see, Dr. Archie, what one really strives for in art is not the sort of thing you are likely to find when you drop in for a performance at the opera. What one strives for is so far away, so deep, so beautiful"—she lifted her shoulders with a long breath, folded her hands in her lap and sat looking at him with a resignation that made her face noble,—"that there's nothing one can say about it, Dr. Archie."

Without knowing very well what it was all about, Archie was passionately stirred for her. "I've always believed in you, Thea; always believed," he muttered.

She smiled and closed her eyes. "They save me: the old things, things like the Kohlers' garden. They are in everything I do."

"In what you sing, you mean?"

"Yes. Not in any direct way,"—she spoke hurriedly,—"the light, the color, the feeling. Most of all the feeling. It comes in when I'm working on a part, like the smell of a garden coming in at the window. I try all the new things, and then go back to the old. Perhaps my feelings were stronger then. A child's attitude toward everything is an artist's attitude. I am more or less of an artist now, but then I was nothing else. When I went with you to Chicago that first time, I carried with me the essentials, the foundation of all I do now. The point to which I could go was scratched in me then. I haven't reached it yet, by a long way."

Archie had a swift flash of memory. Pictures passed before him. "You mean," he asked wonderingly, "that you knew then that you were so gifted?"

Thea looked up at him and smiled. "Oh, I didn't know anything! Not enough to ask you for my trunk when I needed it. But you see, when I set out from Moonstone with you, I had had a rich, romantic past. I had lived a long, eventful life, and an artist's life, every hour of it. Wagner says, in his most beautiful opera, that art is only a way of remembering youth.[4] And the older we grow the more precious it seems to us, and the more richly we can present that memory. When we've got it all out,—the last, the finest thrill of it, the brightest hope of it,"—she lifted her hand above her head and dropped it,—"then we stop. We do nothing but repeat after that.

The stream has reached the level of its source. That's our measure."

There was a long, warm silence. Thea was looking hard at the floor, as if she were seeing down through years and years, and her old friend stood watching her bent head. His look was one with which he used to watch her long ago, and which, even in thinking about her, had become a habit of his face. It was full of solicitude, and a kind of secret gratitude, as if to thank her for some inexpressible pleasure of the heart. Thea turned presently toward the piano and began softly to waken an old air:—

"Ca' the yowes to the knowes,
Ca' them where the heather grows,
Ca' them where the burnie rowes,
My bonnie dear-ie."[5]

Archie sat down and shaded his eyes with his hand. She turned her head and spoke to him over her shoulder. "Come on, you know the words better than I. That's right."

"We'll gae down by Clouden's side,
Through the hazels spreading wide,
O'er the waves that sweetly glide,
To the moon sae clearly.
Ghaist nor bogle shalt thou fear,
Thou'rt to love and Heav'n sae dear,
Nocht of ill may come thee near,
My bonnie dear-ie!"

"We can get on without Landry. Let's try it again, I have all the words now. Then we'll have 'Sweet Afton.' Come: '*Ca' the yowes to the knowes*'—"

X

Ottenburg dismissed his taxicab at the 91st Street entrance of the Park and floundered across the drive through a wild spring snowstorm. When he reached the reservoir path he saw Thea ahead of him, walking rapidly against the wind. Except for that one figure, the path was deserted. A flock of gulls were hovering over the reservoir, seeming bewildered by the driving currents of snow that whirled above the black water and then disappeared within it. When he had almost overtaken Thea, Fred called to her, and she turned and waited for him with her back to the wind. Her hair and furs were powdered with snowflakes, and she looked like some rich-pelted animal, with warm blood, that had run in out of the woods. Fred laughed as he took her hand.

"No use asking how you do. You surely needn't feel much anxiety about Friday, when you can look like this."

She moved close to the iron fence to make room for him beside her, and faced the wind again. "Oh, I'm *well* enough, in so far as that goes. But I'm not lucky about stage appearances. I'm easily upset, and the most perverse things happen."

"What's the matter? Do you still get nervous?"

"Of course I do. I don't mind nerves so much as getting numbed," Thea muttered, sheltering her face for a moment with her muff. "I'm under a spell, you know, hoo-dooed. It's the thing I *want* to do that I can never do. Any other effects I can get easily enough."

"Yes, you get effects, and not only with your voice. That's where you have it over all the rest of them; you're as much at home on the stage as you were down in Panther Canyon—as if you'd just been let out of a cage. Didn't you get some of your ideas down there?"

Thea nodded. "Oh, yes! For heroic parts, at least. Out of the rocks, out of the dead people. You mean the idea of standing up under things, don't you, meeting catastrophe? No fussiness. Seems to me they must have been a reserved, somber people, with only a muscular language, all their movements for a purpose; simple, strong, as if they were dealing with fate bare-handed." She put her gloved fingers on Fred's arm. "I don't know how I can ever thank you enough. I don't know if I'd ever have got anywhere without Panther Canyon. How did you know that was the one thing to do for me? It's the sort of thing nobody ever helps one to, in this world. One can learn how to sing, but no singing teacher can give anybody what I got down there. How did you know?"

"I didn't know. Anything else would have done as well. It was your creative hour. I knew you were getting a lot, but I didn't realize how much."

Thea walked on in silence. She seemed to be thinking.

"Do you know what they really taught me?" she came out suddenly. "They taught me the inevitable hardness of human life. No artist gets far who doesn't know that. And you can't know it with your mind. You have to realize it in your body, somehow; deep. It's an animal sort of feeling. I sometimes think it's the strongest of all. Do you know what I'm driving at?"

"I think so. Even your audiences feel it, vaguely: that you've sometime or other faced things that make you different."

Thea turned her back to the wind, wiping away the snow that clung to her brows and lashes. "Ugh!" she exclaimed; "no matter how long a breath you have, the storm has a longer. I haven't signed for next season, yet, Fred. I'm holding out for a big contract: forty performances. Necker won't be able to do much next winter. It's going to be one of those between seasons; the old singers are too old, and the new ones are too new. They might as well risk me as anybody. So I want good terms. The next five or six years are going to be my best."

"You'll get what you demand, if you are uncompromising. I'm safe in congratulating you now."

Thea laughed. "It's a little early. I may not get it at all. They don't seem to be breaking their necks to meet me. I can go back to Dresden."

As they turned the curve and walked westward they got the wind from the side, and talking was easier.

Fred lowered his collar and shook the snow from his shoulders. "Oh, I don't mean on the contract particularly. I congratulate you on what you can do, Thea, and on all that lies behind what you do. On the life that's led up to it, and on being able to care so much. That, after all, is the unusual thing."

She looked at him sharply, with a certain apprehension. "Care? Why shouldn't I care? If I didn't, I'd be in a bad way. What else have I got?" She stopped with a challenging interrogation, but Ottenburg did not reply. "You mean," she persisted, "that you don't care as much as you used to?"

"I care about your success, of course." Fred fell into a slower pace. Thea felt at once that he was talking seriously and had dropped the tone of half-ironical exaggeration he had used with her of late years. "And I'm grateful to you for what you demand from yourself, when you might get off so easily. You demand more and more all the time, and you'll do more and more. One is grateful to anybody for that; it makes life in general a little less sordid. But as a matter of fact, I'm not much interested in how anybody sings anything."

"That's too bad of you, when I'm just beginning to see what is worth doing, and how I want to do it!" Thea spoke in an injured tone.

"That's what I congratulate you on. That's the great difference between your kind and the rest of us. It's how long you're able to keep it up that tells the story. When you needed enthusiasm from the outside, I was able to give it to you. Now you must let me withdraw."

"I'm not tying you, am I?" she flashed out. "But withdraw to what? What do you want?"

Fred shrugged. "I might ask you, What have I got? I want things that wouldn't interest you; that you probably wouldn't understand. For one thing, I want a son to bring up."

"I can understand that. It seems to me reasonable. Have you also found somebody you want to marry?"

"Not particularly." They turned another curve, which brought the wind to their backs, and they walked on in comparative calm, with the snow blowing past them. "It's not your fault, Thea, but I've had you too much in my mind. I've not given myself a fair chance in other directions. I was in Rome when you and Nordquist were there. If that had kept up, it might have cured me."

"It might have cured a good many things," remarked Thea grimly.

Fred nodded sympathetically and went on. "In my library in St. Louis, over the fireplace, I have a property spear I had copied from one in Venice,—oh, years ago, after you first went abroad, while you were studying. You'll probably be singing *Brünnhilde* pretty soon now,[1] and I'll send it on to you, if I may. You can take it and its history for what they're worth. But I'm nearly forty years old, and I've served my turn. You've done what I hoped for you, what I was honestly willing to lose you for—then. I'm older now, and I think I was an ass. I wouldn't do it again if I had the chance, not much! But I'm not sorry. It takes a great many people to make one—*Brünnhilde.*"

Thea stopped by the fence and looked over into the black choppiness on which the snowflakes fell and disappeared with magical rapidity. Her face was both angry and troubled. "So you really feel I've been ungrateful. I thought you sent me out to get something. I didn't know you wanted me to bring in something easy. I thought you wanted something—" She took a deep breath and shrugged her shoulders. "But there! nobody on God's earth wants it, *really!* If one other person wanted it,"—she thrust her hand out before him and clenched it,—"my God, what I could do!"

Fred laughed dismally. "Even in my ashes I feel myself pushing you! How can anybody help it? My dear girl, can't you see that anybody else who wanted it as you do would be your rival, your deadliest danger? Can't you see that it's your great good fortune that other people can't care about it so much?"

But Thea seemed not to take in his protest at all. She went on vindicating herself. "It's taken me a long while to do anything, of course, and I've only begun to see daylight. But anything good is—expensive. It hasn't seemed long. I've always felt responsible to you."

Fred looked at her face intently, through the veil of snowflakes, and shook his head. "To me? You are a truthful woman, and you don't mean to lie to me. But after the one responsibility you do feel, I doubt if you've enough left to feel responsible to God! Still, if you've ever in an idle hour fooled yourself with thinking I had anything to do with it, Heaven knows I'm grateful."

"Even if I'd married Nordquist," Thea went on, turning down the path again, "there would have been something left out. There always is. In a way, I've always been married to you. I'm not very flexible; never was and never shall be. You caught me young. I could never have that over again. One can't, after one begins to know anything. But I look back on it. My life hasn't been a gay one, any more than yours. If I shut things out from you, you shut them out from me. We've been a help and a hindrance to each other. I guess it's always that way, the good and the bad all mixed up. There's only one thing that's all beautiful—and always beautiful! That's why my interest keeps up."

"Yes, I know." Fred looked sidewise at the outline of her head against the thickening atmosphere. "And you give one the impression that that is enough. I've gradually, gradually given you up."

"See, the lights are coming out." Thea pointed to where they flickered, flashes of violet through the gray tree-tops. Lower down the globes along the drives were becoming a pale lemon color. "Yes, I don't see why anybody wants to marry an artist, anyhow. I remember Ray Kennedy used to say he didn't see how any woman could marry a gambler, for she would only be marrying what the game left." She shook her shoulders impatiently. "Who marries who is a small matter, after all. But I hope I can bring back your interest in my work. You've cared longer and more than anybody else, and I'd like to have somebody human to make a report to once in a while. You can send me your spear. I'll do my best. If you're not interested, I'll do my best anyhow. I've only a few friends, but I can lose every one of them, if it has to be. I learned how to lose when my mother died.—We must hurry now. My taxi must be waiting."

The blue light about them was growing deeper and darker, and the falling snow and the faint trees had become violet. To the south, over Broadway, there was an orange reflection in the clouds. Motors and carriage lights flashed by on the drive below the reservoir path, and the air was strident with horns and shrieks from the whistles of the mounted policemen.

Fred gave Thea his arm as they descended from the embankment. "I guess you'll never manage to lose me or Archie, Thea. You do pick up queer ones. But loving you is a heroic discipline. It wears a man out. Tell me one thing: could I have kept you, once, if I'd put on every screw?"

Thea hurried him along, talking rapidly, as if to get it over. "You might have kept me in misery for a while, perhaps. I don't know. I have to think well of myself, to work. You could have made it hard. I'm not ungrateful. I was a difficult proposition to deal with. I understand now, of course. Since you didn't tell me the truth in the beginning, you couldn't very well turn back after I'd set my head. At least, if you'd been the sort who could, you wouldn't have had to,—for I'd not have cared a button for that sort, even then." She stopped beside a car that waited at the curb and gave him her hand. "There. We part friends?"

Fred looked at her. "You know. Ten years."

"I'm not ungrateful," Thea repeated as she got into her cab.

"Yes," she reflected, as the taxi cut into the Park carriage road, "we don't get fairy tales in this world, and he has, after all, cared more and longer than anybody else." It was dark outside now, and the light from the lamps along the drive flashed into the cab. The snowflakes hovered like swarms of white bees about the globes.

Thea sat motionless in one corner staring out of the window at the cab lights that wove in and out among the trees, all seeming to be bent upon joyous courses. Taxicabs were still new in New York, and the theme of popular minstrelsy. Landry had sung her a ditty he heard in some theater on Third Avenue, about

"But there passed him a bright-eyed taxi
With the girl of his heart inside."

Almost inaudibly Thea began to hum the air, though she was thinking of something serious, something that had touched her deeply. At the beginning of the season, when she was not singing often, she had gone one afternoon to hear Paderewski's recital.[2] In front of her sat an old German couple, evidently poor people who had made sacrifices to pay for their excellent seats. Their intelligent enjoyment of the music, and their friendliness with each other, had interested her more than anything on the programme. When the pianist began a lovely melody in the first movement of the Beethoven D minor sonata, the old lady put out her plump hand and touched her husband's sleeve and they looked at each other in recognition. They both wore glasses, but such a look! Like forget-me-nots, and so full of happy recollections. Thea wanted to put her arms around them and ask them how they had been able to keep a feeling like that, like a nosegay in a glass of water.

XI

DR. ARCHIE SAW NOTHING of Thea during the following week. After several fruitless efforts, he succeeded in getting a word with her over the telephone, but she sounded so distracted and driven that he was glad to say good-night and hang up the instrument. There were, she told him, rehearsals not only for "Walküre," but also for "Götterdämmerung," in which she was to sing *Waltraute*[1] two weeks later.

On Thursday afternoon Thea got home late, after an exhausting rehearsal. She was in no happy frame of mind. Madame Necker, who had been very gracious to her that night when she went on to complete Gloeckler's performance of *Sieglinde,* had, since Thea was cast to sing the part instead of Gloeckler in the production of the "Ring," been chilly and disapproving, distinctly hostile. Thea had always felt that she and Necker stood for the same sort of endeavor, and that Necker recognized it and had a cordial feeling for her. In Germany she had several times sung *Brangäne* to Necker's *Isolde,*[2] and the older artist had let her know that she thought she sang it beautifully. It was a bitter disappointment to find that the approval of so honest an artist as Necker could not stand the test of any significant recognition by the management. Madame Necker was forty, and her voice was failing just when her powers were at their height. Every fresh young voice was an enemy, and this one was accompanied by gifts which she could not fail to recognize.

Thea had her dinner sent up to her apartment, and it was a very poor one. She tasted the soup and then indignantly put on her wraps to go out and hunt a dinner. As she was going to the elevator, she had to admit that she was behaving foolishly. She took off her hat and coat and ordered another dinner. When it arrived, it was no better than the first. There was even a burnt match under the milk toast. She had a sore throat, which made swallowing painful and boded ill for the morrow. Although she had been speaking in whispers all day to save her throat, she now perversely summoned the housekeeper and demanded an account of some laundry that had been lost. The housekeeper was indifferent and impertinent, and Thea got angry and scolded violently. She knew it was very bad for her to get into a rage just before bedtime, and after the housekeeper left she realized that for ten dollars' worth of underclothing she had been unfitting herself for a performance which might eventually mean many thousands. The best thing now was to stop reproaching herself for her lack of sense, but she was too tired to control her thoughts.

While she was undressing—Thérèse was brushing out her *Sieglinde* wig in the trunk-room—she went on chiding herself bitterly. "And how am I ever going to get to sleep in this state?" she kept asking herself. "If I don't sleep, I'll be perfectly worthless to-morrow. I'll go down there to-morrow and make a fool of myself. If I'd let that laundry alone with whatever nigger has stolen it—*Why* did I undertake to reform the management of this hotel to-night? After to-morrow I could pack up and leave the place. There's the Phillamon—I liked the rooms there better, anyhow—and the Umberto—" She began going over the advantages and disadvantages of different apartment hotels. Suddenly she checked herself. "What *am* I doing this for? I can't move into another hotel to-night. I'll keep this up till morning. I shan't sleep a wink."

Should she take a hot bath, or shouldn't she? Sometimes it relaxed her, and sometimes it roused her and fairly put her beside herself. Between the conviction that she must sleep and the fear that she couldn't, she hung paralyzed. When she looked at her bed, she shrank from it in every nerve. She was much more afraid of it than she had ever been of the stage of any opera house. It yawned before her like the sunken road at Waterloo.

She rushed into her bathroom and locked the door. She would risk the bath, and defer the encounter with the bed a little longer. She lay in the bath half an hour. The warmth of the water penetrated to her bones, induced pleasant reflections and a feeling of well-being. It was very nice to have Dr. Archie in New York, after all, and to see him get so much satisfaction out of the little companionship she was able to give him. She liked people who got on, and who became more interesting as they grew older. There was Fred; he was much more interesting now than he had been at thirty. He was intelligent about music, and he must be very intelligent in his business, or he would not be at the head of the Brewers' Trust. She respected that kind of intelligence and success. Any success was good. She herself had made a good start, at any rate, and now, if she could get to sleep— Yes, they were all more interesting than they used to be. Look at Harsanyi, who had been so long retarded; what a place he had made for himself in Vienna. If she could get to sleep, she would show him something to-morrow that he would understand.

She got quickly into bed and moved about freely between the sheets. Yes, she was warm all over. A cold, dry breeze was coming in from the river, thank goodness! She tried to think about her little rock house and the Arizona sun and the blue sky. But that led to memories which were still too disturbing. She turned on her side, closed her eyes, and tried an old device.

She entered her father's front door, hung her hat and coat on the rack, and stopped in the parlor to warm her hands at the stove. Then she went out through the dining-room, where the boys were getting their lessons at the long table; through the sitting room, where Thor was asleep in his cot bed, his dress and stockings hanging on a chair. In the kitchen she stopped for her lantern and her hot brick. She hurried up the back stairs and through the windy loft to her own glacial room. The illusion was marred only by the consciousness that she ought to brush her teeth before she went to bed, and that she never used to do it. Why—? The water was frozen solid in the pitcher, so she got over that. Once between the red blankets there was a short, fierce battle with the cold; then, warmer—warmer. She could hear her father shaking down the hard-coal burner for the night, and the wind rushing and banging down the village street. The boughs of the cottonwood, hard as bone, rattled against her gable. The bed grew softer and warmer. Everybody was warm and well downstairs. The sprawling old house had gathered them all in, like a hen, and had settled down over its brood. They were all warm in her father's house. Softer and softer. She was asleep. She slept ten hours without turning over. From sleep like that, one awakes in shining armor.

On Friday afternoon there was an inspiring audience; there was not an empty chair in the house. Ottenburg and Dr. Archie had seats in the orchestra circle, got from a ticket broker. Landry had not been able to get a seat, so he roamed about in the back of the house, where he usually stood when he dropped in after his own turn in vaudeville was over. He was there so often and at such irregular hours that the ushers thought he was a singer's husband, or had something to do with the electrical plant.

Harsanyi and his wife were in a box, near the stage, in the second circle. Mrs. Harsanyi's hair was noticeably gray, but her face was fuller and handsomer than in those early years of struggle, and she was beautifully dressed. Harsanyi himself had changed very little. He had put on his best afternoon coat in honor of his pupil, and wore a pearl in his black ascot. His hair was longer and more bushy than he used to wear it, and there was now one gray lock on the right side. He had always been an elegant figure, even when he went about in shabby clothes and was crushed with work. Before the curtain rose he was restless and nervous, and kept looking at his watch and wishing he had got a few more letters off before he left his hotel. He had not been in New York since the advent of the taxicab, and had allowed himself too much time. His wife knew that he was afraid of being disappointed this afternoon. He did not often go to the opera because the stupid things that singers did vexed him so, and it always put him in a rage if the conductor held the tempo or in any way accommodated the score to the singer.

When the lights went out and the violins began to quaver their long D against the rude figure of the basses, Mrs. Harsanyi saw her husband's fingers fluttering on his knee in a rapid tattoo. At the moment when *Sieglinde* entered from the side door, she leaned toward him and whispered in his ear, "Oh, the lovely creature!" But he made no response, either by voice or gesture. Throughout the first scene he sat sunk in his chair, his head forward and his one yellow eye rolling restlessly and shining like a tiger's in the dark. His eye followed *Sieglinde* about the stage like a satellite, and as she sat at the table listening to *Siegmund*'s long narrative, it never left her. When she prepared the sleeping draught and disappeared after *Hunding*,[3] Harsanyi bowed his head still lower and put his hand over his eye to rest it. The tenor,—a young man who sang with great vigor, went on:—

> *"Wälse! Wälse!*
> *Wo ist dein Schwert?"*[4]

Harsanyi smiled, but he did not look forth again until *Sieglinde* reappeared. She went through the story of her shameful bridal feast and into the Walhall' music, which she always sang so nobly, and the entrance of the one-eyed stranger:—

> "*Mir allein*
> *Weckte das Auge.*"[5]

Mrs. Harsanyi glanced at her husband, wondering whether the singer on the stage could not feel his commanding glance. On came the *crescendo:*—

> "*Was je ich verlor,*
> *Was je ich beweint*
> *Wär' mir gewonnen.*"
>
> (All that I have lost,
> All that I have mourned,
> Would I then have won.)

Harsanyi touched his wife's arm softly.

Seated in the moonlight, the *Volsung* pair began their loving inspection of each other's beauties, and the music born of murmuring sound passed into her face, as the old poet said,—and into her body as well. Into one lovely attitude after another the music swept her, love impelled her. And the voice gave out all that was best in it. Like the spring, indeed, it blossomed into memories and prophecies, it recounted and it foretold, as she sang the story of her friendless life, and of how the thing which was truly herself, "bright as the day, rose to the surface" when in the hostile world she for the first time beheld her Friend. Fervently she rose into the hardier feeling of action and daring, the pride in hero-strength and hero-blood, until in a splendid burst, tall and shining like a Victory, she christened him:—

> "*Siegmund—*
> *So nenn ich dich!*"[6]

Her impatience for the sword swelled with her anticipation of his act, and throwing her arms above her head, she fairly tore a sword out of the empty air for him, before *Nothung* had left the tree. *In höchster Trunkenheit,*[7] indeed, she burst out with the flaming cry of their kinship: "If you are *Siegmund,* I am *Sieglinde!*"

Laughing, singing, bounding, exulting,—with their passion and their sword,—the *Volsungs* ran out into the spring night.

As the curtain fell, Harsanyi turned to his wife. "At last," he sighed, "somebody with *enough!* Enough voice and talent and beauty, enough physical power. And such a noble, noble style!"

"I can scarcely believe it, Andor. I can see her now, that clumsy girl, hunched up over your piano. I can see her shoulders. She always seemed to labor so with her back. And I shall never forget that night when you found her voice."

The audience kept up its clamor until, after many reappearances with the tenor, Kronborg came before the curtain alone. The house met her with a roar, a greeting that was almost savage in its fierceness. The singer's eyes, sweeping the house, rested for a moment on Harsanyi, and she waved her long sleeve toward his box.

"She *ought* to be pleased that you are here," said Mrs. Harsanyi. "I wonder if she knows how much she owes to you."

"She owes me nothing," replied her husband quickly. "She paid her way. She always gave something back, even then."

"I remember you said once that she would do nothing common," said Mrs. Harsanyi thoughtfully.

"Just so. She might fail, die, get lost in the pack. But if she achieved, it would be nothing common. There are people whom one can trust for that. There is one way in which they will never fail." Harsanyi retired into his own reflections.

After the second act Fred Ottenburg brought Archie to the Harsanyis' box and introduced him as an old friend of Miss Kronborg. The head of a musical publishing house joined them, bringing with him a journalist and the president of a German singing society. The conversation was chiefly about the new *Sieglinde.* Mrs. Harsanyi was gracious and enthusiastic, her husband nervous and uncommunicative. He smiled mechanically, and politely answered questions addressed to him. "Yes, quite so." "Oh, certainly." Every one, of course, said very usual things with great conviction. Mrs. Harsanyi was used to hearing and uttering the commonplaces which such occasions demanded. When her husband withdrew into the shadow, she covered his retreat by her sympathy and cordiality. In reply to a direct question from Ottenburg, Harsanyi said, flinching, "*Isolde?* Yes, why not? She will sing all the great rôles, I should think."

The chorus director said something about "dramatic temperament." The journalist insisted that it was "explosive force," "projecting power."

Ottenburg turned to Harsanyi. "What is it, Mr. Harsanyi? Miss Kronborg says if there is anything in her, you are the man who can say what it is."

The journalist scented copy and was eager. "Yes, Harsanyi. You know all about her. What's her secret?"

Harsanyi rumpled his hair irritably and shrugged his shoulders. "Her secret? It is every artist's secret,"—he waved his hand,—"passion. That is all. It is an open secret, and perfectly safe. Like heroism, it is inimitable in cheap materials."

The lights went out. Fred and Archie left the box as the second act came on.

Artistic growth is, more than it is anything else, a refining of the sense of truthfulness. The stupid believe that to be truthful is easy; only the artist, the great artist, knows how difficult it is. That afternoon nothing new came to Thea Kronborg, no enlightenment, no inspiration. She merely came into full possession of things she had been refining and perfecting for so long. Her inhibitions chanced to be fewer than usual, and, within herself, she entered into the inheritance that she herself had laid up, into the fullness of the faith she had kept before she knew its name or its meaning.

Often when she sang, the best she had was unavailable; she could not break through to it, and every sort of distraction and mischance came between it and her. But this afternoon the closed roads opened, the gates dropped. What she had so often tried to reach, lay under her hand. She had only to touch an idea to make it live.

While she was on the stage she was conscious that every movement was the right movement, that her body was absolutely the instrument of her idea. Not for nothing had she kept it so severely, kept it filled with such energy and fire. All that deep-rooted vitality flowered in her voice, her face, in her very finger-tips. She felt like a tree bursting into bloom. And her voice was as flexible as her body; equal to any demand, capable of every *nuance.* With the sense of its perfect companionship, its entire trustworthiness, she had been able to throw herself into the dramatic exigencies of the part, everything in her at its best and everything working together.

The third act came on, and the afternoon slipped by. Thea Kronborg's friends, old and new, seated about the house on different floors and levels, enjoyed her triumph according to their natures.

There was one there, whom nobody knew, who perhaps got greater pleasure out of that afternoon than Harsanyi himself. Up in the top gallery a gray-haired little Mexican, withered and bright as a string of peppers beside a 'dobe door, kept praying and cursing under his breath, beating on the brass railing and shouting "Bravó! Bravó!" until he was repressed by his neighbors.

He happened to be there because a Mexican band was to be a feature of Barnum and Bailey's circus that year. One of the managers of the show had traveled about the Southwest, signing up a lot of Mexican musicians at low wages, and had brought them to New York. Among them was Spanish Johnny. After Mrs. Tellamantez died, Johnny abandoned his trade and went out with his mandolin to pick up a living for one. His irregularities had become his regular mode of life.

When Thea Kronborg came out of the stage entrance on Fortieth Street, the sky was still flaming with the last rays of the sun that was sinking off behind the North River. A little crowd of people was lingering about the door—musicians from the orchestra who were waiting for their comrades, curious young men, and some poorly dressed girls who were hoping to get a glimpse of the singer. She bowed graciously to the group, through her veil, but she did not look to the right or left as she crossed the sidewalk to her cab. Had she lifted her eyes an instant and glanced out through her white scarf, she must have seen the only man in the crowd who had removed his hat when she emerged, and who stood with it crushed up in his hand. And she would have known him, changed as he was. His lustrous black hair was full of gray, and his face was a good deal worn by the *éxtasis*[8] so that it seemed to have shrunk away from his shining eyes and teeth and left them too prominent. But she would have known him. She passed so near that he could have touched her, and he did not put on his hat until her taxi had snorted away. Then he walked down Broadway with his hands in his overcoat pockets, wearing a smile which embraced all the stream of life that passed him and the lighted towers that rose into the limpid blue of the evening sky. If the singer, going home exhausted in her cab, was wondering what was the good of it all, that smile, could she have seen it, would have answered her. It is the only commensurate answer.

***

Here we must leave Thea Kronborg. From this time on the story of her life is the story of her achievement. The growth of an artist is an intellectual and spiritual development which can scarcely be followed in a personal narrative. This story attempts to deal only with the simple and concrete beginnings which color and accent an artist's work, and to give some account of how a Moonstone girl found her way out of a vague, easy-going world into a life of disciplined endeavor. Any account of the loyalty of young hearts to some exalted ideal, and the passion with which they strive, will always, in some of us, rekindle generous emotions.

EPILOGUE

MOONSTONE AGAIN, in the year 1909. The Methodists are giving an ice-cream sociable in the grove about the new court-house. It is a warm summer night of full moon. The paper lanterns which hang among the trees are foolish toys, only dimming, in little lurid circles, the great softness of the lunar light that floods the blue heavens and the high plateau. To the east the sand hills shine white as of old, but the empire of the sand is gradually diminishing. The grass grows thicker over the dunes than it used to, and the streets of the town are harder and firmer than they were twenty-five years ago. The old inhabitants will tell you that sandstorms are infrequent now, that the wind blows less persistently in the spring and plays a milder tune. Cultivation has modified the soil and the climate, as it modifies human life.

The people seated about under the cottonwoods are much smarter than the Methodists we used to know. The interior of the new Methodist Church looks like a theater, with a sloping floor, and as the congregation proudly say, "opera chairs." The matrons who attend to serving the refreshments to-night look younger for their years than did the women of Mrs. Kronborg's time, and the children all look like city children. The little boys wear "Buster Browns"[1] and the little girls Russian blouses. The country child, in made-overs and cut-downs, seems to have vanished from the face of the earth.

At one of the tables, with her Dutch-cut twin boys, sits a fair-haired, dimpled matron who was once Lily Fisher. Her husband is president of the new bank, and she "goes East for her summers," a practice which causes envy and discontent among her neighbors. The twins are well-behaved children, biddable, meek, neat about their clothes, and always mindful of the proprieties they have learned at summer hotels. While they are eating their ice-cream and trying not to twist the spoon in their mouths, a little shriek of laughter breaks from an adjacent table. The twins look up. There sits a spry little old spinster whom they know well. She has a long chin, a long nose, and she is dressed like a young girl, with a pink sash and a lace garden hat with pink rosebuds. She is surrounded by a crowd of boys,—loose and lanky, short and thick,—who are joking with her roughly, but not unkindly.

"Mamma," one of the twins comes out in a shrill treble, "why is Tillie Kronborg always talking about a thousand dollars?"

The boys, hearing this question, break into a roar of laughter, the women titter behind their paper napkins, and even from Tillie there is a little shriek of appreciation. The observing child's remark had made every one suddenly realize that Tillie never stopped talking about that particular sum of money. In the spring, when she went to buy early strawberries, and was told that they were thirty cents a box, she was sure to remind the grocer that though her name was Kronborg she didn't get a thousand dollars a night. In the autumn, when she went to buy her coal for the winter, she expressed amazement at the price quoted her, and told the dealer he must have got her mixed up with her niece to think she could pay such a sum. When she was making her Christmas presents, she never failed to ask the women who came into her shop what you *could* make for anybody who got a thousand dollars a night. When the Denver papers announced that Thea Kronborg had married Frederick Ottenburg, the head of the Brewers' Trust, Moonstone people expected that Tillie's vain-gloriousness would take another form. But Tillie had hoped that Thea would marry a title, and she did not boast much about Ottenburg,—at least not until after her memorable trip to Kansas City to hear Thea sing.

Tillie is the last Kronborg left in Moonstone. She lives alone in a little house with a green yard, and keeps a fancy-work and millinery store. Her business methods are informal, and she would never come out even at the end of the year, if she did not receive a draft for a good round sum from her niece at Christmas time. The arrival of this draft always renews the discussion as to what Thea would do for her aunt if she really did the right thing. Most of the Moonstone people think Thea ought to take Tillie to New York and keep her as a companion. While they are feeling sorry for Tillie because she does not live at the Plaza, Tillie is trying not to hurt their feelings by showing too plainly how much she realizes the superiority of her position. She tries to be modest when she complains to the postmaster that her New York paper is more than three days late. It means enough, surely, on the face of it, that she is the only person in Moonstone who takes a New York paper or who has any reason for taking one. A foolish young girl, Tillie lived in the splendid sorrows of "Wanda" and "Strathmore";[2] a foolish old girl, she lives in her niece's triumphs. As she often says, she just missed going on the stage herself.

That night after the sociable, as Tillie tripped home with a crowd of noisy boys and girls, she was perhaps a shade troubled. The twin's question rather lingered in her ears. Did she, perhaps, insist too much on that thousand dollars? Surely, people didn't for a minute think it was the money she cared about? As for that, Tillie tossed her head, she didn't care a rap. They must understand that this money was different.

When the laughing little group that brought her home had gone weaving down the sidewalk through the leafy shadows and had disappeared, Tillie brought out a rocking chair and sat down on her porch. On glorious, soft summer nights like this, when the moon is opulent and full, the day submerged and forgotten, she loves to sit there behind her rose-vine and let her fancy wander where it will. If you chanced to be passing down that Moonstone street and saw that alert white figure rocking there behind the screen of roses and lingering late into the night, you might feel sorry for her, and how mistaken you would be! Tillie lives in a little magic world, full of secret satisfactions. Thea Kronborg has given much noble pleasure to a world that needs all it can get, but to no individual has she given more than to her queer old aunt in Moonstone. The legend of Kronborg, the artist, fills Tillie's life; she feels rich and exalted in it. What delightful things happen in her mind as she sits there rocking! She goes back to those early days of sand and sun, when Thea was a child and Tillie was herself, so it seems to her, "young." When she used to hurry to church to hear Mr. Kronborg's wonderful sermons, and when Thea used to stand up by the organ of a bright Sunday morning and sing "Come, Ye Disconsolate." Or she thinks about that wonderful time when the Metropolitan Opera Company sang a week's engagement in Kansas City, and Thea sent for her and had her stay with her at the Coates House and go to every performance at Convention Hall. Thea let Tillie go through her costume trunks and try on her wigs and jewels. And the kindness of Mr. Ottenburg! When Thea dined in her own room, he went down to dinner with Tillie, and never looked bored or absent-minded when she chattered. He took her to the hall the first time Thea sang there, and sat in the box with her and helped her through "Lohengrin." After the first act, when Tillie turned tearful eyes to him and burst out, "I don't care, she always seemed grand like that, even when she was a girl. I expect I'm crazy, but she just seems to me full of all them old times!"—Ottenburg was so sympathetic and patted her hand and said, "But that's just what she is, full of the old times, and you are a wise woman to see it." Yes, he said that to her. Tillie often wondered how she had been able to bear it when Thea came down the stairs in the wedding robe embroidered in silver, with a train so long it took six women to carry it.

Tillie had lived fifty-odd years for that week, but she got it, and no miracle was ever more miraculous than that. When she used to be working in the fields on her father's Minnesota farm, she couldn't help believing that she would some day have to do with the "wonderful," though her chances for it had then looked so slender.

The morning after the sociable, Tillie, curled up in bed, was roused by the rattle of the milk cart down the street. Then a neighbor boy came down the sidewalk outside her window, singing "Casey Jones"[3] as if he hadn't a care in the world. By this time Tillie was wide awake. The twin's question, and the subsequent laughter, came back with a faint twinge. Tillie knew she was short-sighted about facts, but this time— Why, there were her scrapbooks, full of newspaper and magazine articles about Thea, and half-tone cuts, snap-shots of her on land and sea, and photographs of her in all her parts. There, in her parlor, was the phonograph that had come from Mr. Ottenburg last June, on Thea's birthday; she had only to go in there and turn it on, and let Thea speak for herself. Tillie finished brushing her white hair and laughed as she gave it a smart turn and brought it into her usual French twist. If Moonstone doubted, she had evidence enough: in black and white, in figures and photographs, evidence in hair lines on metal disks. For one who had so often seen two and two as making six, who had so often stretched a point, added a touch, in the good game of trying to make the world brighter than it is, there was positive bliss in having such deep foundations of support. She need never tremble in secret lest she might sometime stretch a point in Thea's favor.—Oh, the comfort, to a soul too zealous, of having at last a rose so red it could not be further painted, a lily so truly auriferous that no amount of gilding could exceed the fact!

Tillie hurried from her bedroom, threw open the doors and windows, and let the morning breeze blow through her little house.

In two minutes a cob fire was roaring in her kitchen stove, in five she had set the table. At her household work Tillie was always bursting out with shrill snatches of song, and as suddenly stopping, right in the middle of a phrase, as if she had been struck dumb. She emerged upon the back porch with one of these bursts, and bent down to get her butter and cream out of the ice-box. The cat was purring on the bench and the morning-glories were thrusting their purple trumpets in through the lattice-work in a friendly way. They reminded Tillie that while she was waiting for the coffee to boil she could get some flowers for her breakfast table. She looked out uncertainly at a bush of sweet-briar that grew at the edge of her yard, off across the long grass and the tomato vines. The front porch, to be sure, was dripping with crimson ramblers that ought to be cut for the good of the vines; but never the rose in the hand for Tillie! She caught up the kitchen shears and off she dashed through grass and drenching dew. Snip, snip; the short-stemmed sweet-briars, salmon-pink and golden-hearted, with their unique and inimitable woody perfume, fell into her apron.

After she put the eggs and toast on the table, Tillie took last Sunday's New York paper from the rack beside the cupboard and sat down, with it for company. In the Sunday paper there was always a page about singers, even in summer, and that week the musical page began with a sympathetic account of Madame Kronborg's first performance of *Isolde* in London. At the end of the notice, there was a short paragraph about her having sung for the King at Buckingham Palace and having been presented with a jewel by His Majesty.

Singing for the King; but Goodness! she was always doing things like that! Tillie tossed her head. All through breakfast she kept sticking her sharp nose down into the glass of sweet-briar, with the old incredible lightness of heart, like a child's balloon tugging at its string. She had always insisted, against all evidence, that life was full of fairy tales, and it was! She had been feeling a little down, perhaps, and Thea had answered her, from so far. From a common person, now, if you were troubled, you might get a letter. But Thea almost never wrote letters. She answered every one, friends and foes alike, in one way, her own way, her only way. Once more Tillie has to remind herself that it is all true, and is not something she has "made up." Like all romancers, she is a little terrified at seeing one of her wildest conceits admitted by the hard-headed world. If our dream comes true, we are almost afraid to believe it; for that is the best of all good fortune, and nothing better can happen to any of us.

When the people on Sylvester Street tire of Tillie's stories, she goes over to the east part of town, where her legends are always welcome. The humbler people of Moonstone still live there. The same little houses sit under the cottonwoods; the men smoke their pipes in the front doorways, and the women do their washing in the back yard. The older women remember Thea, and how she used to come kicking her express wagon along the sidewalk, steering by the tongue and holding Thor in her lap. Not much happens in that part of town, and the people have long memories. A boy grew up on one of those streets who went to Omaha and built up a great business, and is now very rich. Moonstone people always speak of him and Thea together, as examples of Moonstone enterprise. They do, however, talk oftener of Thea. A voice has even a wider appeal than a fortune. It is the one gift that all creatures would possess if they could. Dreary Maggie Evans, dead nearly twenty years, is still remembered because Thea sang at her funeral "after she had studied in Chicago."

However much they may smile at her, the old inhabitants would miss Tillie. Her stories give them something to talk about and to conjecture about, cut off as they are from the restless currents of the world. The many naked little sandbars which lie between Venice and the mainland, in the seemingly stagnant water of the lagoons, are made habitable and wholesome only because, every night, a foot and a half of tide creeps in from the sea and winds its fresh brine up through all that network of shining water-ways. So, into all the little settlements of quiet people, tidings of what their boys and girls are doing in the world bring real refreshment; bring to the old, memories, and to the young, dreams.

THE END

EXPLANATORY NOTES

For more detailed, often eccentric information about Cather's textual allusions, readers are referred to John March, *A Reader's Companion to the Fiction of Willa Cather*, ed. Marilyn Arnold (Westport, Conn.: Greenwood Press, 1993).

PART I, FRIENDS OF CHILDHOOD

CHAPTER I

1. *Moonstone . . . Duke Block over the drug store:* Moonstone, Colorado, was modeled closely on Cather's hometown of Red Cloud, Nebraska. The name Moonstone has a Red Cloud association: Cather's childhood friend Dr. G. E. McKeeby, the prototype for Dr. Archie, had his office above the drug store in the prestigious "Moon Block" in Red Cloud.

2. *twenty-five years ago:* The novel begins in the mid-1880s, when Cather herself was Thea's age.

3. *Fay Templeton:* Generously endowed American vaudeville actress and singer (1865–1939), whose vivacity and throaty delivery endeared her to a generation.

4. *Czerny's "Daily Studies":* Karl Czerny (1791–1857), Viennese composer, pianist, and pedagogue, whose 1854 *Vierzig Tägliche Studien auf dem Piano-Forte*, Op. 337, was issued in America in 1881 as *40 Daily Studies*.

5. *flaxseed jacket:* Thea's poultice is designed to relieve chest congestion. Ground flaxseed was mixed to a paste with hot water, and then applied to back and chest. A jacket was commonly sewn tightly over it to keep it in place.

CHAPTER II

1. *tower of Babel:* A reference to the story in Genesis 11:1–9. At a time when all the world spoke a single language, the Babylonians tried to distinguish themselves by building a tower that would reach to heaven. God punished their arrogance by confusing their speech (Babel = babble) and dispersing them across the face of the earth.

2. *opal:* The pearly-white stone also known as moonstone.

3. *"My native land, good-night":* From canto 1, stanza 13, of *Childe Harold's Pilgrimage* (1812–18), enormously popular work by the Romantic poet George Gordon, Lord Byron (1788–1824). Childe Harold is the earliest of the extravagant Byronic heroes—a melancholy, defiant outcast who wanders about Europe reflecting on nature, his own states of mind, and historical and contemporary people and events.

4. *"Maid of Athens":* Byron's love poem "Song" (1810) became known by its first line, "Maid of Athens, ere we part."

5. *"There was a sound of revelry":* From canto 3, stanza 21, of *Childe Harold.*

CHAPTER III

1. *"Standard Recitations": Standard Recitations by Best Authors,* compiled by Frances Pauline Sullivan (New York: M. J. Ivers, 1883–98).

CHAPTER IV

1. *Thea's favorite fairy tale:* "The Snow Queen" by the Danish author Hans Christian Andersen (1805–75).

2. *Jenny Lind:* Nicknamed "the Swedish Nightingale," Jenny Lind (1820–87) was one of the most admired singers of the Victorian era. Her two-year concert tour of the United States (1850–51) began under the aegis of the showman P. T. Barnum, who turned her tour into a media event.

3. *Clementi:* Muzio Clementi (1752–1832), Italian-born, English-educated, child-prodigy pianist, composer, teacher, and piano manufacturer, regarded as the first genuine composer for the piano.

4. *"Invitation to the Dance":* Op. 65 in D flat, composed in 1819 by Carl Maria von Weber.

5. *Lente currite . . . noctis equi:* From Ovid, *Amores,* I.xiii.40.

6. *Napoleon's retreat from Moscow:* Napoleon I (1769–1821), emperor of the French, sought to incorporate Russia into his empire, but his disastrous 1812 military campaign there marked the turning point in his career. In face of his Grande Armée of 500,000, the Russian troops fell back, sys-

tematically devastating the land and burning the city of Moscow, with all its supplies. Napoleon's troops were left starving and exposed to the brutal Russian winter; only a fifth survived the journey home.

Wunsch's "piece-picture" seems to have been modeled after one that hung in the office of an Austrian tailor in Pittsburgh, where Cather lived between 1896 and 1906. Cather wrote admiringly of it in a letter, admitting that she felt deprived at never having had it to look at when she was a child. Including it in her novel, she explained, was her revenge on Fortune (John March, *A Reader's Companion to the Fiction of Willa Cather*, pp. 520–21).

7. *Murat, in Oriental dress:* Joachim Murat (1767–1815), French cavalry commander, marshal, aide to Napoleon, and king of Naples, who was fond of wearing elaborate clothing even in battle.

8. *"Come, Ye Disconsolate":* Hymn by Thomas Moore set to a German air, published in 1816.

CHAPTER VI

1. *A Distinguished Provincial in Paris:* Part 2 of *Illusions Perdues* (*Lost Illusions*) by Honoré de Balzac (1799–1850), from *La Comédie humaine*, his series of novels about the different social classes in early-nineteenth-century France. Balzac, who is regarded as the originator of realism in the novel, had enormous influence on later writers, including Henry James; Cather deeply admired both Balzac and James.

2. "*La Golandrina*": "The Swallow," a popular song, written in 1862 by Narcisco Serradell (1843–1910).

3. *mi testa . . . La fiebre:* My head . . . The fever.

4. *Muchacha:* The little girl.

CHAPTER VII

1. *sand hills:* The sand hills of north central Nebraska are rich in minerals and fossils, but are now mostly grass-covered. Some of these details (petrified wood, desert vegetation, agates, colors) bear closer resemblance to Arizona's Painted Desert, which Cather had recently visited.

2. *stubborn as a Finn:* A mild ethnic slur, not against inhabitants of modern Finland, but against the nomadic Lapps (Sami), a group that differs racially, culturally, and linguistically from other Scandinavians. The Sami, who traverse the northern regions of Finland, Sweden, and Norway (once known as Finnmark) with their reindeer herds, have suffered a history of discrimination much like that of the Native American tribes in North America.

3. *drawn-work:* Ornamentation of linen formed by drawing parallel threads and uniting the cross-threads to form a pattern.

4. *Prescott's histories:* William Hickling Prescott (1796–1859), American historian who wrote epic-style, romanticized histories of Spain and Spanish America, including a three-volume *History of the Conquest of Mexico* (1843).

5 *Washington Irving:* American author (1783–1859) best known for his satirical *History of New York* (1809) and *The Sketch Book* (1819–20), a compilation of essays and folktales. In the 1830s he traveled to the western frontier, describing his adventures in *A Tour on the Prairies* (1834).

6. *Robert Ingersoll's speeches:* Robert Green Ingersoll (1833–99), American politician known as "the great agnostic" because of his long campaign on behalf of freethinking. His lectures, which criticized the Bible from a rationalist point of view, had an important influence on American thought in the late nineteenth century.

7. *"The Age of Reason":* Deistic work (1794–95) by the American political theorist Thomas Paine (1737–1809), which affirmed the existence of God on the basis of the arguments of design and first cause. Paine urged respect for God's creation, but not for the Bible or for institutionalized churches, which he accused of calumny and manipulation for profit.

8. *Lake Valley:* While some of the places named in this novel are fictitious, others, including the "Bridal Chamber" at Lake Valley, in Sierra County, New Mexico, are not. After a discouraging start, during the 1880s Lake Valley became one of the most spectacular silver camps in the West.

9. *Forty-niners and the Mormons:* The Forty-niners were the thousands of prospectors lured west by the California gold rush of 1849. Two years before, in 1847, sixteen thousand Mormons (members of the Church of Latter-day Saints) had made the largest single migration in the history of the nation. Following their leader, Joseph Smith, they traveled from New England west (across Nebraska) to Utah. Their wagon-wheel scars are still visible today on what is known as the Mormon Trail.

10. "*Westward the course of Empire takes its way*": This phrase, part of a longer message that was first telegraphed across the Missouri River in 1860, is the opening line of the poem "On the Prospect of Planting Arts and Learning in America" by the Anglican bishop George Berkeley (1685–1753), famous as an idealistic philosopher. The poem was written in 1723 to promote Berkeley's (never realized) scheme to found a college in Bermuda that would train American Indians to serve as Christian missionaries to their own tribes.

11. *Greaser:* Term for Mexican or Spanish-American, usually derogatory.

CHAPTER VIII

1. *"Polar Explorations" . . . Greely's party:* Adolphus Washington Greely (1844–1935) took a party of twenty-five to Ellesmere Island in 1881. Planned relief parties failed to reach them, and only seven survived until they were rescued in 1884. Greely's 1885 account, *Three Years of Arctic Service: An Account of the Lady Franklin Bay Expedition of 1881–1884 . . .*, told of buffalo-skin sleeping bags that thawed while occupied, but froze solid to the ground when living bodies left them.

2. *Outside the family, everyone . . . called her "Thee-a":* In the Scandinavian languages Thea's name is pronounced "Taya" (as her baby brother Thor's name is pronounced "Tor").

3. *W.C.T.U.:* The Women's Christian Temperance Union, which was established in 1874 "for the protection of the home, the abolition of the liquor traffic, and the triumph of the golden rule in custom and law."

4. *Ancient Mariner's:* In "The Rime of the Ancient Mariner," 1798 poem by Samuel Taylor Coleridge (1772–1834), the Mariner accosts wayfarers to hear his tale. "By thy long grey beard and glittering eye," asks the narrator, "Now wherefore stopp'st thou me?"

5. *"Ballade" by Reinecke:* Carl Heinrich Reinecke (1824–1910), German composer, teacher, and pianist, composed many exercises for young pianists.

6. *"Erminie":* A popular comic opera by Edward Jakobowski, first produced in 1885.

7. *"The Polish Boy":* An 1834 poem by Anna Sophia Winterbotham Stephens, which tells in gruesome detail about a Polish boy who stabs himself and dies on his mother's breast rather than allowing himself to be taken by the Russians. It was widely reprinted in nineteenth-century anthologies and was a favorite for school contests.

8. *Progressive Euchre Club:* A card club (named for the card game euchre), organized in Red Cloud, Nebraska, in 1886, to which Willa Cather's parents and other socially prominent townspeople belonged. Members took turns hosting meetings (held in a refinished hall over the drugstore) and serving refreshments to as many as forty tables. The club is also mentioned in two of Cather's other semiautobiographical works, *My Ántonia* and "Old Mrs. Harris."

9. *"Beloved, it is Night":* Probably "Beloved, It Is Morn" (1896), by Florence Aylward (1862–1950), whose last verse begins, "Beloved, it is night, it is night!"

10. *"Thy Sentinel Am I":* Song with music by William Michael Watson, composer of drawing-room ballads, 1884.

11. *"Rock of Ages . . . carelessly the maiden sang":* The popular hymn "Rock of Ages" (1776), by Augustus M. Toplady (1740–78), was incorporated into a recitation combining the speaking or singing voice with piano, published in *The Family Library of Poetry and Song*, vol. 2, edited by William Cullen Bryant (New York: Fords, Howard and Hulbert, 1880), p. 367. Cather quotes the text incorrectly, substituting "carelessly" for "thoughtlessly."

12. *"She sang . . . that touched my heart":* "The Song That Reached My Heart: Home Sweet Home," a popular song written by Jules Jordan, published in 1887. "Home Sweet Home" is from Henry Bishop's 1823 opera, *Clari, or The Maid of Milan.*

13. *"When [sic] Shepherds Watched":* Nahum Tate's 1700 hymn "While Shepherds Watched Their Flocks by Night" is based on the St. Luke version of the Christmas story; the most popular melody used for it in America is "Sherburne," a fugue by Daniel Read.

14. *"Musical Memories" of the Reverend H. R. Haweis:* Hugh Reginald Haweis (1838–1901), British Anglican clergyman and amateur musician, author of several works, including *My Musical Memories* (1884) and *Music and Morals* (1872). Haweis was an early enthusiast of the composer Richard Wagner (see Introduction).

Chapter IX

1. *"Among the Breakers" . . . "The Veteran of 1812": Among the Breakers* was a play by George Melville-Baker (published 1873), performed during Cather's childhood by the Red Cloud Histrionic Association (cf. Moonstone Drama Club) in 1881 and 1885. *The Veteran of 1812:* or, *Kesiah and the Scout* was a military drama by T. Trask Woodward, published in 1883.

2. *"Drummer Boy of Shiloh":* A military allegory by F. B. Wigle (1887), based on incidents from the Civil War.

3. *"Just Before the Battle, Mother":* A popular Civil War song by George Frederick Root.

Chapter X

1. *Leipsic:* Earlier spelling of the German city Leipzig.

2. *Gluck's "Orpheus": Orfeo ed Euridice*, opera by the German composer Christoph Willibald Gluck (1714–87) based on the Greek myth, first performed in Vienna in 1762. Gluck was the first effective reformer of opera before Wagner. Both composers were determined to rid the genre of the artifice and excess associated with the Italian tradition.

3. *vom blatt, mit mir:* Follow the score, accompany me.

4. *"Ach, ich habe sie verloren . . .":* "Ah, I have lost you, / Gone now is all my joy."

5. *"Euridice . . . weh dass ich auf Erden bin!":* "Euridice, why must I remain on earth above!"

6. *immer:* Always.

7. *It is written for alto . . . Only one!:* Wunsch is referring to Pauline Viardot-Garcia (1821–1910), Spanish mezzo-soprano who had tremendous success in the role of Orfeo. The role was originally written for a castrato, later for tenor; in the nineteenth century it was generally performed by female contraltos.

8. *künstlerisch:* Artistic.

9. *Hasenpfeffer:* Highly seasoned hare braised in red wine, a popular German dish.

CHAPTER XI

1. *Heine . . . Im leuchtenden Sommermorgen:* Heinrich Heine (1767–1856), Romantic German lyric poet frequently cited by Cather. The poem, accurately titled "Am leuchtenden Sommermorgen," is No. 45 from his *Lyrisches Intermezzo*. His *Buch der Lieder* (*Book of Songs*, 1827) was one of the most widely read and influential books of poetry in Germany; many of his songs were set to music.

2. *und so weiter:* And so forth.

3. *Weiter, nun:* Now continue.

4. *der Geist, die Phantasie . . . der Rhythmus:* Spirit, imagination, rhythm.

5. *Aber nicht die Americanischen Fräulein:* But not the American young ladies.

6. *Märchen:* Fairy tale.

7. *". . . giebt es keine Kunst!":* Wunsch's long tirade is against American female singers whose musicianship is technically correct but without heart, and therefore can never be "art."

CHAPTER XII

1. *Flint's Physiology . . . Oliver Wendell Holmes: A Text-Book of Human Physiology*, by Austin Flint (1812–86), was published in 1876. The American poet and essayist Oliver Wendell Holmes (1809–94) was also a physician, and many of his poems concern doctors and medicine.

2. *Waverley Novels:* Enormously influential novels by Sir Walter Scott

(1771–1832), who invented the historical novel. Scott was avidly read and imitated throughout the nineteenth century.

3. *Constance de Beverley, minstrel girl:* Constance de Beverley, heroine of Scott's *Marmion*, is a nun who breaks her vows for the sake of love, and is punished by being buried alive. In *The Fair Maid of Perth*, the minstrel girl throws herself from a cliff when her secret lover, the prince, is murdered.

4. *Duchesse de Langeais:* Eponymous heroine of a novel by Balzac; a Parisian socialite who ends her days in a convent.

5. *Robert Burns:* Scottish national poet and songwriter (1759–96), much admired by his contemporaries. He wrote equally well in English and his native Gaelic.

Chapter XIII

1. *das Mistbeet:* Hot bed, manure bed.

2. *die Bäume . . . Der Taubenschlag! Gerechter Himmel:* The trees . . . the dove house, merciful heaven.

3. *Fuszreise:* Walking tour, hike.

4. *"Leben sie wohl, mein Kind!":* "Live well, my child!"

Chapter XIV

1. *"Thanatopsis," Hamlet's soliloquy, Cato on "Immortality":* "Thanatopsis" is a poem about death by the New England poet William Cullen Bryant (1794–1878). Hamlet's most famous soliloquy, the meditation that begins "To be, or not to be," occurs in act 3, scene 2, of Shakespeare's play. Cato on "Immortality" refers to the 1713 tragedy *Cato* by the English author Joseph Addison (1672–1719) about the Roman statesman and moralist Marcus Porcius Cato the Elder (234–149 B.C.). The meditation on immortality occurs in act 5.

Chapter XV

1. *Tabor Grand Opera House:* Denver's lavishly decorated opera house, which opened in 1881 and hosted many of the world's great performers up to the turn of the century.

2. *Maggie Mitchell in Little Barefoot:* Margaret Julia Mitchell (1832–1918), a sprightly actress of great popularity. In an 1894 review, Cather complained that Mitchell, who continued to play her most popular ingenue role into her seventies, was deluded and wearisome. *Little Barefoot*, based

on a domestic drama by Augustus Waldauer, opened with Mitchell as the lead in 1863.

CHAPTER XVI

1. *"The Odalisque":* A female slave or concubine in the harem of a Turkish sultan, a subject popular among Romantic and Victorian painters. The artist most associated with odalisques is Jean-Auguste-Dominique Ingres (1780–1867), though they were also painted by Delacroix, Couture, Bouguereau, and many others. Giddy's picture may have been an engraved reproduction, a copy, or (most likely) a pornographic photograph posed in imitation of one of these paintings. A prototype can be found in Richard N. Ellis and Duane A. Smith, *Colorado: A History in Photographs* (Nimwot: University of Colorado Press, 1991), p. 45.

2. *Mrs. Langtry:* Lillie Langtry (1853–1929), native of the Isle of Jersey and known as the "Jersey Lily," was one of the first Englishwomen of elevated social rank to go on the stage. She made her theatrical debut in 1881 after her diplomat husband failed financially. Never considered a great actress, she was known for her great beauty and her affair with Edward VII. For the first English edition of *The Song of the Lark* in 1916, publisher John Murry requested that this reference be cut from the text, and it remained excised in the English 1938 revised edition.

3. *Canyon de Chelly:* One of the most spectacular cliff-dwelling remains in the American Southwest. In northeastern Arizona, the site was inhabited by Anasazi from 100 B.C. to A.D. 1600, followed by the Navajo, who were forced out by the U.S. government in 1863. In *Death Comes for the Archbishop* (1927), Cather touches on the topic of the brutal Navajo relocation, led by the Indian scout Kit Carson.

4. *prescribed for gapes:* Gapes is a disease of young fowl caused by the presence of a nematode worm in the trachea. Cure at this time was turpentine applied with a feather inside the windpipe.

5. *"the youth who bore":* From "Excelsior" by the popular American poet Henry Wadsworth Longfellow (1807–82).

6. *Harvey House manager:* Harvey Houses were a chain of restaurants established at western railroad depots, dedicated to providing high-quality food, service, and cleanliness to railroad passengers. Employees ("Harvey girls") like Katie Casey were expected to adhere to traditional values and high moral standards in the rough-hewn West.

Chapter XVII

1. *Christian Endeavor . . . Band of Hope:* The Young People's Society of Christian Endeavor, organized in Maine in 1881, was the first interdenominational youth ministry. The Band of Hope was a youth temperance movement (later an auxiliary of the W.C.T.U.) which established a chapter in Cather's hometown of Red Cloud in 1876.

2. *Doxology:* A hymn or psalm of praise to God sung before the offering at a church service or prayer meeting—specifically here, "The Old Hundredth," beginning, "Praise God from whom all blessings flow."

3. *Jules Verne:* French author (1828–1905) of a series of books combining adventure and popular science; titles include *Journey to the Center of the Earth* (1864) and *20,000 Leagues Under the Sea* (1870).

4. *Anna Karenina:* 1873–76 novel by Russian author Leo Tolstoy (1828–1910).

Chapter XVIII

1. *Naaman the leper . . . house of Rimmon:* Biblical story from 2 Kings 5:1–27, in which the servant Naaman asks God's pardon for joining his master in ceremonies of heathen worship.

2. *The Blue Danube:* Popular waltz (1867) by Viennese composer Johann Strauss II (1825–99).

3. *"Marching through Georgia":* A song written in 1865 by Henry Clay Work (1832–84), commemorating General Sherman's famous march to the sea during the Civil War.

4. *same word . . . surrender:* When the English demanded his surrender at Waterloo, Napoleon's general Pierre-Jacques-Etienne Cambronne (1770–1842) was heard to reply "*merde*" (excrement).

5. *"The Soul Awakened":* Two possibilities have been suggested for this picture. John March suggests "The Soul's Awakening" by James Sant (1820–1916), a sentimental image of a radiant-faced young girl pressing a book against her breast and marking a passage with her finger; engraved versions were widely sold in the Victorian period. Polly Duryea proposes "The Awakening Conscience" (1851–53) by the Pre-Raphaelite painter William Holman Hunt (1827–1910), a suggestive scene of a woman rising from the lap of a man seated at a piano; her expression is more conflicted than radiant. See Polly Patricia Duryea, "Paintings and Drawings in Willa Cather's Prose: A Catalogue Raisonné" (Diss. University of Nebraska, 1993), p. 185.

6. *Mary Anderson:* Popular American classical actress (1859–1940), who appealed to all social classes.

7. *old Dumas:* Alexandre Dumas *père* (1802–70), one of the pioneers of Romantic theater in France, author of *The Count of Monte Cristo*. Cather quotes his "one passion, four walls" credo again in her 1922 essay "The Novel Démeublé."

CHAPTER XX

1. *telescope:* A telescope bag, collapsible luggage.

PART II, THE SONG OF THE LARK

CHAPTER I

1. *unkept wastes of North Chicago:* The Great Chicago fire of 1871, which laid waste to much of the city, was followed by a feverish rebuilding campaign. The architectural talent at the city's disposal was unparalleled, and by the 1880s Chicago vied with Philadelphia for "second city" status. Nonetheless, it was many years before all the bare, windswept spaces were filled or landscaped.

2. *Jephthah's Daughter* (Judges 11), *Rizpah* (2 Samuel 3:7–10), *David's Lament for Absalom* (2 Samuel 18–19): All Old Testament stories of family tragedy, betrayal, and strife.

3. *Homestead Act:* One of the most important congressional acts in American history, the 1862 Homestead Act provided for public land to be distributed without charge to settlers who contracted to cultivate it. Together with the banishment of Indian tribes to reservations and the extended arm of the railroad, the act was responsible for an influx of about 150,000 German, Bohemian, and Scandinavian settlers to the prairie region where Willa Cather spent her childhood.

4. *temporarily indisposed:* Euphemism for "pregnant."

5. *Columbia wheel:* The Columbia wheel was the first American high-wheeled bicycle, manufactured in Hartford, Connecticut, beginning in 1878.

CHAPTER II

1. *Kaffeeklatsch:* Coffee party.

2. *Sängerfest:* Singing festival.

Chapter III

1. *Kinderszenen:* "Scenes from Childhood" (1838), thirteen pieces for solo piano by Robert Schumann (1810–56).

2. Of the musicians mentioned in this paragraph, Beethoven (1770–1827), Chopin (1810–49), and Robert Schumann (1810–56) and his wife, Clara, the pianist (1819–96), were part of the Romantic movement, which was modern and daring in Wunsch's youth. The others are from the earlier, classical generation.

3. *Stuttgart method:* A method developed at the Stuttgart Conservatory, in which beginning piano players used a pole, attached to the piano so that wrists and arms could rest on it, to develop finger muscles independently. It fell out of favor because it also caused rigidity in the wrist and arm.

4. *Christian fleeing from the City of Destruction:* A reference to the hero of John Bunyan's *Pilgrim's Progress*, one of Cather's favorite childhood books. An allegory in the form of a dream, the story takes Christian on a pilgrimage from the City of Destruction to the Celestial City; along the way his obstacles include the Slough of Despond, the Valley of Humiliation, and the Giant Despair.

5. *Emma Juch:* Austrian-American singer (1863–1939) who organized her own opera touring company in 1889.

6. *"The Ninety and Nine":* An 1868 hymn by Elizabeth Cecilia Douglass Clephane, also titled "The Lost Sheep."

Chapter IV

1. *"Die Lorelei":* An 1823 song by Heinrich Heine, No. 2 in the cycle *Die Heimkehr* (The Homecoming), about a siren who lures sailors on the Rhine to shipwreck. It has been set to music by many composers; the version by Friedrich Silcher has become a folk song indelibly associated with German heritage and culture.

2. *"Ich weiss nicht . . . so traurig bin":* "I do not know why this confronts me, / that I am so sorrowful."

3. *"Und das . . . Lorelei getan":* "And with this, her poignant singing, The Lorelei has done."

Chapter V

1. *Corots . . . Barbizon School:* The Barbizon School was an informal group of landscape painters who worked at the village of Barbizon in France ca. 1830–70. Though not a member, Jean-Baptiste-Camille Corot

(1796–1875) is often associated with the school, which was strongly influenced by seventeenth-century Dutch landscape painting and was very popular in America.

2. *Inter-Ocean:* A Chicago daily newspaper published 1872–1914, after which it merged with the *Record-Herald.*

3. *the casts . . . Dying Gladiator . . . Venus di Milo . . . Apollo Belvedere . . . great equestrian statue:* Nineteenth-century art schools and museums typically exhibited plaster casts of famous classical and Renaissance sculptures for educational purposes. The *Dying Gladiator* (accurately the *Dying Gaul*), at the Capitoline Museum in Rome, is a marble copy of an original Greek bronze by Epigonus, made ca. A.D. 240 to commemorate the victory over the invading Gauls in Asia Minor. Byron describes it in canto 4, stanzas 140–41, of *Childe Harold's Pilgrimage*. The *Venus di Milo* is a Greek marble dating from the second to first century B.C., found on the island of Melos and now in the Louvre. The *Apollo Belvedere* (also celebrated in *Childe Harold*, canto 4, stanzas 161–63)—named for its location in the Belvedere gallery at the Vatican—is a Roman marble copy of an original Greek bronze. The "great equestrian statue" was a Renaissance piece, a bronze portrait of the condottiere Bartolomeo Colleoni by the Florentine sculptor Andrea del Verrocchio (1435–88), designed in 1480 and erected in the piazza of the Scuola Grande of St. Mark's, Venice.

4. *Gérôme . . . "The Pasha's Grief": The Grief of the Pasha* by the French Romantic history and genre painter Jean-Léon Gérôme (1824–1904), today at the Jocelyn Art Museum, Omaha, Nebraska.

5. *newborn calf on a litter: Bringing Home the Newborn Calf* (1864) by Jean-François Millet (1814–75), realist painter of French peasant life who was sometimes linked with the Barbizon School.

6. *"The Song of the Lark":* 1885 painting by Jules Bréton (1827–1906). In a 1901 article about the Chicago Art Institute, Cather wrote that "hundreds of merchants and farmer boys all over Nebraska and Kansas and Iowa" remembered this painting, and "perhaps the ugly little peasant girl standing barefooted among the wheat fields in the early morning have taught some of these people to hear the lark sing for themselves" (*The World and the Parish*, ed. William S. Curtin [Lincoln: University of Nebraska Press, 1970], vol. 2, p. 843). Because so many readers confused Thea Kronborg with the "Lark" of the title, however, Cather removed an illustration of the painting from the book's jacket, and included a disclaimer in her 1932 Preface.

7. *Auditorium:* Dankmar Adler and Louis Sullivan's Auditorium Building, the crown jewel of the early Chicago Architectural School, was dedicated in 1889 by President Benjamin Harrison. Its lavish concert hall,

featuring Sullivan's intricate organic ornament and Adler's perfect acoustics, is one of the world's finest public spaces.

8. *Dvořák's Symphony... "From the New World":* Antonín Dvořák (1841–1904), Czech composer who spent two years (1892–94) as director of the National Conservatory in New York. In 1893, while summering in the Czech community at Spillville, Iowa, he wrote his exuberant symphony "From the New World," drawing on American scenes and folk music.

9. *entry of the gods into Walhalla... "Rhinegold"... strife between gods and men: Das Rheingold* is the first of the four sequential music dramas that constitute Wagner's *Der Ring des Nibelungen* (The *Ring of the Nibelungs*). The others are *Die Walküre* (*The Valkyries*), *Siegfried*, and *Götterdämmerung* (*Twilight of the Gods*). The stories, drawn from Scandinavian and Germanic legend, weave together myths of the gods with human heroic tragedy, and portray a struggle between greed (represented by the magic Ring) and human love (or "the feminine principle"), a battle that ends in the destruction of the world. The cycle was first produced in its entirety at Bayreuth, the German festival center Wagner designed for productions of his own work, in 1876. Its first American production was at the Metropolitan Opera in 1889.

CHAPTER VI

1. *Theodore Thomas:* German-born violinist and cunductor (1835–1905), director of the Chicago Symphony Orchestra between 1891 and 1905. Thomas was one of the most influntial advocates of Wagner's music in the United States.

2. *Festival Association:* Theodore Thomas led Cincinnati's Biennial Musical Festivals from 1873 to 1904. The "great chorus" from the 1873 season included over a thousand voices singing Haydn, Mozart, Handel, Wagner, Mendelssohn, and Gluck.

3. *Henrietta Sontag:* German operatic soprano (1806–54) whose popularity rivaled that of Jenny Lind. Theodore Thomas was first violinist in the orchestra that accompanied both Lind and Sontag on their tours.

CHAPTER VII

1. *as if its singleness gave it privileges:* Harsanyi's one-eyed acuity is a reference to the one-eyed Norse god Odin (Wotan), who guides the fates of characters Thea will later play in Wagner's operas.

Chapter X

1. *"Rosa de Noche":* On Cather's trip to Arizona in 1912 she copied the words of a serenade, learned from her Mexican friend Julio, into a letter to Elizabeth Sergeant (see Introduction). It was written for *voz contralto* and fits this description.

2. *a Paraíso:* In Paradise.

3. *Hace frío:* It's cold.

4. *"Trovatore": Il Trovatore,* 1853 opera by Giuseppe Verdi (1813–1901).

5. *wunderschön:* Exceedingly beautiful.

6. *sextette from "Lucia": Lucia di Lammermoor,* 1835 opera by Gaetano Donizetti (1797–1848), based on a novel by Sir Walter Scott.

Chapter XI

1. *Pauline Hall . . . Clara Morris . . . Modjeska:* Pauline Hall (1860–1919) was an American actress, dancer, singer, and popular pinup girl who organized her own company in 1892 and toured widely across the country. Clara Morris (1848–1925), also American, was a profoundly popular actress with limited gifts and an overwrought, melodramatic style. Helena Modjeska (1840–1909) was in an altogether different category. Born in Poland, she made her reputation in Europe as a Shakespearean actress, and immigrated to the United States in 1876. Here she introduced Ibsen to the American stage (1883), and despite faulty English, consolidated her reputation as one of the great tragic actresses of her time. Cather had lunch with Modjeska in 1898, and portrays the actress in her 1926 novel, *My Mortal Enemy.*

2. *"He either fears his fate too much . . .":* From the second stanza of "My Dear and Only Love" by James Graham, marquis of Montrose (1612–50).

PART III, STUPID FACES

Chapter I

1. *"Rejoice Greatly":* From the 1742 oratorio *Messiah* by George Frideric Handel (1685–1759).

2. *vacant Swede look:* Swedish immigrants were frequently subjected to slurs about their intelligence (e.g., "dumb Swede").

3. *"On Mighty Pens":* Aria from *Die Schöpfung* (The Creation), oratorio by Franz Josef Haydn (1732–1809), based on Milton's poem *Paradise Lost.*

Chapter II

1. *Gounod's "Ave Maria" . . . Dans——nos a-lár———mes:* In this phrase from the 1853 work by French composer Charles-François Gounod (1818–93), the syllable "lár" is written as a high A.

Chapter III

1. *Bayreuth:* German festival center designed to Richard Wagner's own specifications for the production of his music dramas.
2. *"I Know that my Redeemer Liveth":* Air from Handel's *Messiah*.
3. *"Gioconda": La Gioconda* (The Smiling One), 1876 opera about a ballad singer by Amilcare Ponchielli (1834–86), based on a novel by Victor Hugo.
4. *Grieg:* Edvard Grieg (1843–1907), Norwegian composer.
5. *Prize Song:* The hero of Wagner's *Die Meistersinger von Nürnberg* sings the "Prize Song" to gain admission to the Masters' Guild.
6. *Turnverein:* Athletic club.

Chapter IV

1. *kosher clothes:* Here, as elsewhere in the novel, derogatory references to ethnic groups are interspersed with admiring portraits of their individual members. While the slurs were a common part of public discourse in Cather's era, the admiration was not.
2. *Worth gowns:* Charles Frederick Worth (1825–95), designer for European royalty, was virtual dictator of high fashion for the last thirty years of the nineteenth century.
3. *"O[h], Promise Me":* A popular song from the comic operetta *Robin Hood* by Reginald de Koven, first produced in 1891.
4. *Nietzsche club:* Friedrich Wilhelm Nietzsche (1844–1900), German philosopher, early proponent of Wagner's music, and vehement critic of bourgeois tastes and values. Nietzsche advocated a "heroic morality," i.e., choosing to live at the highest level of passion and creativity, rather than following the "herd mentality" of the crowd.
5. *most beautiful Manet in the world: The Street Singer* (1862) by Edouard Manet (1832–83), who is associated with both the Realist and the Impressionist schools. The painting is today in the Boston Museum of Fine Arts.
6. *gnädige Frau:* Madam.
7. *St. John's Day:* The feast day of St. John the Baptist is June 24, close enough to the solstice that the two festivals are combined in Scandinavia.

There, the midsummer revelries—which include bonfires, drinking, and dancing throughout the long sunlit night—compare in intensity and ribaldry only with the winter solstice festivities at Christmas.

8. *Francis Thompson:* British poet (1859–1907) whose work was published in the 1890s; his best-known poems are "The Hound of Heaven" and "The Kingdom of God."

9. *liked no poetry but German poetry:* All aspects of German culture were fashionable in American intellectual life in this era—a fad stemming largely from Wagner's undisputed triumph over the American musical world in the 1890s.

Chapter V

1. *"The Kreutzer Sonata":* Leo Tolstoy's morbid late novel (1889) about sexual morality, named after the Beethoven Sonata played frequently by two of its characters.

PART IV, THE ANCIENT PEOPLE

Chapter I

1. *democrat wagon:* A light wagon, usually with two seats and a top, widely used in the West.

Chapter II

1. *Panther Canyon:* Cather modeled Panther Canyon on Walnut Canyon, between Flagstaff and Winslow, Arizona, which she visited in 1912.

2. *Ancient People:* The Sinagua, a member of the Hakataya regional group of prehistoric inhabitants of the American Southwest. During their peak development period (1125–1215), the Sinagua occupied Walnut Canyon and built its tiny cliff houses and storage rooms. Until it came under the management of the National Park Service in 1933, Walnut Canyon (like Canyon de Chelly, Mesa Verde, and many other prehistoric sites) was vandalized and looted by souvenir hunters (like Ray Kennedy and Henry Biltmer in this novel). Cather's 1925 novel, *The Professor's House*, treats the loss of archaeological treasures at Mesa Verde in considerable detail.

Chapter VI

1. *Zorn etchings of peasant girls bathing:* Anders Zorn (1860–1920), Swedish painter and etcher, famous for his scenes of folk life and vigorous, full-blown bathing nudes.

Chapter VIII

1. *geliebter Sohn:* Beloved son.
2. *eine Ungebildete:* Uneducated person.

PART V, DOCTOR ARCHIE'S VENTURE

Chapter I

1. *Denver . . . San Felipe silver mine:* Denver was founded in 1859 at Cherry Creek, the site of the first Colorado gold find a year before. Later, silver-bearing lead carbonite was found at Leadville and Creede, attracting more miners and financiers, and Colorado became known as "the silver state." The San Felipe mine is fictional.

2. *Rothschild of Cripple Creek:* Colorado's richest gold field was discovered at Cripple Creek in 1891. The Rothschilds, a famous family of German financiers, controlled much of the wealth of Western Europe for almost a century.

3. *nose-nippers:* Glasses that clip to the bridge of the nose.

4. *"Flow Gently, Sweet Afton":* A song by Robert Burns to an old Scottish air; a tribute to Burns's early patroness, who lived on the Afton.

5. *Masonic pin . . . bad form back there:* The international Secret Fraternal Order of Freemasons had its origins in medieval English and Scottish fraternities of stonemasons and cathedral builders. The order is founded on principles that are traditionally liberal and democratic—religious toleration, loyalty to local government, and political compromise—and the American Founding Fathers and a great many presidents and congressmen have been Masons. After Masons were charged with an 1826 murder in New York, however, the Anti-Masonic Party was organized to prevent members of the order from being elected to public office, and its influence spread to other eastern states. Thus while both Dr. Archie's Scottish ancestry and his political ambitions made it advantageous to be a Mason in Colorado, it is doubtful that would have been true in New York.

Chapter IV

1. *bench-joy:* Probably related to "bencher," an idler who spends time on ale-house benches.

2. *D'Albert:* Eugen Francis Charles D'Albert (1864–1932), German pianist and composer, whose individual physiognomy was often remarked upon.

3. *Jugendzeit:* Youth.

Chapter V

1. *Lehmann:* Lilli Lehmann (1848–1929), renowned German Wagnerian singer, teacher of Thea's prototype, Olive Fremstad (see Introduction).

2. *"Einst, O Wunder . . .":* "Soon, O wonder, upon my tomb will blossom / A flower from the ashes of my heart. . . ." The lines are from "Adelaide," a poem by Friedrich von Matthisson (1761–1831), set to music by Beethoven.

3. *"Deutlich schimmert auf jedem . . .":* The rest of the stanza of "Adelaide" continues: *"der Asche meines Herzens deutlich schimmert, . . . auf jedem Purpurblättchen, Adelaide, Adelaide":* "The ashes of my heart will shimmer brightly on every little purple petal, Adelaide."

PART VI, KRONBORG

Chapter I

1. *Governor Alden:* This character is fictitious, but the political disturbances discussed in this chapter are not.

Chapter II

1. *Mahler:* Gustav Mahler (1860–1911), Austrian composer and conductor who led orchestras at Budapest, Hamburg, and Vienna before coming to the Metropolitan Opera (1907–9). His legendary high standards of performance and refusal to compromise artistic integrity won him enemies as well as admirers. In New York in 1909 Mahler conducted *Tristan und Isolde* with Olive Fremstad as Isolde, and afterward Mahler confessed that he had "never known a performance of *Tristan* to equal this." See John Dizikes, *Opera in America: A Cultural History* (New Haven: Yale University Press, 1993), p. 350.

2. *Elisabeth:* The heroine of Wagner's early (1845, revised 1860) opera *Tannhäuser*, about the struggle between sacred and profane love.

Chapter III

1. *"Dich, teure Halle . . .":* "Dear Hall, I greet thee again!" Sung by Elisabeth as she makes her first stage appearance in the Hall of Song.

Chapter IV

1. *Lohengrine:* 1850 opera by Wagner.

2. *the Metropolitan Opera House:* The "Old Met," the opulent 3,000-

seat house at Thirty-ninth Street and Broadway in New York, which opened in 1883 and closed in 1966.

3. *scene beside the Scheldt:* The opera is set near the Scheldt River in Brabant (Belgium) in the tenth-century reign of King Henry of Saxony.

4. *"Einsam in trüben Tagen. . . . In lichter Waffen Scheine":* "Lonely, in troubled days. . . . In splendid, shining armor." Elsa, falsely accused of murdering her brother, dreams of Lohengrin, the knight who will rescue her from her enemies. The next few pages of Cather's text include references to the *Herald*, who summons Elsa to *King Henry*'s tribunal; *Ortrud*, the pagan sorceress who has plotted to rob Elsa of her crown; *Lohengrin*, her rescuing knight; and "*the wedding processional*" (or "wedding march"), music that has become a familiar standard in American church weddings.

5. *"cut off . . . round about":* Line from a nursery rhyme, "The Little Woman and the Pedlar."

Chapter V

1. *Treulich geführt, treulich bewacht:* "Faithfully guided, faithfully guarded," phrases excerpted from the wedding processional, act 3.

2. *"In lichter Waffen Scheine, / Ein Ritter nahte da":* "In splendid shining armor, a knight approached."

3. *Some Elsas do; she didn't:* In Wagner's opera, Elsa "sinks lifeless to the ground" after Lohengrin leaves her in the last scene. Most of the source material for the legend provides a similar ending, but some versions leave her fate undetermined. For a discussion of Wagner's sources for *Lohengrin* see Henry Krehbiel, *A Book of Operas* (Garden City, N.Y.: Garden City Publishing, 1909), or Ernest Newman, *Wagner Nights* (London: Putnam & Co., 1949).

4. *Woglinde:* One of the Rhine-maidens in *Das Rheingold*.

5. *wie im Traum ich—:* As in a dream—. This exact phrase does not appear in the libretto, but the sense of it corresponds to words from "Elsa's Dream" in act 1 (scene 2), "Lass mich ihn sah, wie ich ihn sah, Wie ich ihn sah, sei er mir nah" ("Let me see him as I first saw him, As I saw him may he come to me").

Chapter VI

1. *"Ah! Fuyez, douce image":* Line from *Manon*, 1884 opera by Jules Massenet (1842–1912), in which the hero, preparing to enter the priesthood, seeks to wash away memories of the woman who betrayed him.

2. *Weber and Fields':* Broadway music hall, opened in 1895 by the burlesque team Joseph Weber (1867–1942) and Lew Fields (1867–1941).

3. *the Tribune man:* Henry Edward Krehbiel (1854–1923), dean of the New York music critics in this period, was reviewer for the New York *Tribune* from 1880 to 1923, and author of many books on opera. He was also one of the era's most discerning critics of Wagner and a deep admirer of Olive Fremstad. In his glowing review of Fremstad's first portrayal of Elsa (December 18, 1909) he described her voice as "thrice admirable in its beauty, suavity and continence of tone and its tender expressiveness" (John March, *A Reader's Companion to the Fiction of Willa Cather*, pp. 773–74).

4. *Venus:* Goddess of profane love, Elisabeth's rival for Tannhäuser's devotion.

5. *Necker:* Thea's operatic colleagues referred to in these chapters (Mmes. Necker, Rheinecker, Schramm, and Gloeckler) are all fictional.

Chapter VII

1. *Sieglinde:* Sieglinde and Siegmund are the Wälsung (Volsung) twins in Wagner's *Ring* cycle, mortal children of Wotan (Odin), chief of the gods. *Die Walküre* (The Valkyries), the second opera in the cycle, tells the story of the twins, who are separated in childhood, meet again as adults, fall in love, and produce Siegfried, the hero of the *Ring*.

2. *Brünnhilde:* The Valkyrie who gives this opera its name. In Norse mythology, Valkyries are the fierce and beautiful handmaidens of Odin who live in rocky crags above the human world; their task is to mark those to be slain in battle and escort the worthy dead to Valhalla. In this opera Brünnhilde disobeys Wotan's orders and protects both Siegmund and Sieglinde, for which she is put to sleep in a ring of fire. See also notes 1–5 for Chapter XI, below.

3 *Fricka:* Wife of Wotan, Norse goddess of wedlock and fidelity. In *Das Rheingold* (and in *Die Walküre*) she upbraids Wotan for his infidelities, his untrustworthiness, and his rash promises.

4. *wonniger Hausrath:* "glorious homestead," part of Fricka's admiring tribute to Walhalla (Valhalla), the castle built for her and Wotan by the giants.

5. *Tak for mad:* Thank you for the food.

Chapter VIII

1. *Boucher drawing:* François Boucher (1703–70), French rococo painter and favorite of Madame de Pompadour, best known for his voluptuous boudoir decorations.

2. *"Wotan, Gemahl! erwache!":* "Wotan, my lord, awake!" Fricka wakens Wotan to behold the just-built Walhalla, which she hopes will entice him to spend more time at home.

3. *die Wölfin:* She-wolf.

4. *last act is heart-breaking:* In the last act Elisabeth dies brokenhearted when Tannhäuser fails to return with the other pilgrims from Rome, where he had sought absolution for his dissolute life. He has not received the Pope's blessing, but Elisabeth's steadfast faith in him—and her death on the altar of love—saves him from damnation.

Chapter IX

1. *"Du bist der Lenz":* "You are the springtime," from the love duet between Siegfried and Sieglinde.

2. *comme les autres:* Like the rest.

3. *Schlafen Sie wohl!:* Sleep well.

4. *remembering youth:* In act 3, scene 2 of *Die Meistersinger von Nürnberg* (1868), Hans Sachs coaches his pupil Walther on the rules of the mastersingers' guild: that true art embodies the memory of youth and spring.

5. *"Ca' the yowes to the knowes . . .":* 1790 poem by Robert Burns.

Chapter X

1. *You'll probably be singing Brünnhilde:* Brünnhilde is the most demanding female role in the *Ring* cycle, reserved for experienced singers with extraordinary vocal power and stamina.

2. *Paderewski's recital:* Ignacy Jan Paderewski (1860–1941), popular Polish pianist and composer.

Chapter XI

1. *Waltraute:* One of the Valkyries, a sister of Brünnhilde.

2. *Brangäne . . . Isolde:* Brangäne is the servant who gives Tristan and Isolde the potion that causes them to fall famously in love. In Wagner's opera *Tristan und Isolde* (1865) Brangäne is a contralto part, not soprano. Olive Fremstad, like Thea, had begun her career singing contralto roles.

3. *Hunding:* Sieglinde's husband. In order to learn more about the attractive stranger at her door, Sieglinde drugs her husband's evening drink. Later Hunding will challenge Siegmund to the fatal duel in which Brünnhilde tries to intervene. See also notes 1 and 2 for Chapter VII, above.

4. *"Wälse! Wälse! / Wo ist dein Schwert?":* "Volsung! Volsung! Where is

your sword?" Wotan, father of the Volsung twins, has promised Siegmund in childhood that a sword (*Nothung*, or Needful) would be made available at a time of dire need. Now, when he is a wounded and weaponless fugitive after battle with Hunding, Siegmund arrives unwittingly at his enemy's home. These lines are part of his lament that he is "needful" of the promised sword now.

5. *"Mir allein / Weckte das Auge":* "To me alone the eye suggested [sweet, longing sadness, tears, and comfort all at once]." Sieglinde is singing about the "one-eyed stranger," Wotan himself, who at her forced wedding to Hunding had thrust the sword *Nothung* into an ash tree, promising that the one who could pull it out would restore all that she had lost ("*Was je ich verlor . . .*").

6. *"Siegmund— / So nenn ich dich!":* "Siegmund—so I name you!" While Sieglinde nurses the fugitive's wounds they tell each other their unhappy histories; the twins have fallen in love by the time they discover their relation. She then calls him by name, and pointing to the sword in the tree, tells Siegmund it belongs to him.

7. *"In höchster Trunkenheit":* "In extreme intoxication."

8. *éxtasis:* Seizure, rapture.

EPILOGUE

1. *"Buster Browns":* Broad, rounded starched collars worn with a Windsor tie, a style identified with the comic-strip character Buster Brown, dating from about 1900.

2. *"Wanda" . . . "Strathmore":* Title characters of popular romances by Ouida (pseudonym for Mary Louise de la Ramée) (1839–1908), an English novelist often disparaged by Willa Cather.

3. *"Casey Jones":* Folk ballad about a railroadman, noted for his daring and resourcefulness, who was killed in a wreck of the Cannon Ball Express in 1900.

APPENDIX:
WILLA CATHER'S 1932 PREFACE TO THE 1915 EDITION

The Song of the Lark was written in the years 1914 and 1915. The title of the book is unfortunate; many readers take it for granted that the "lark song" refers to the vocal accomplishments of the heroine, which is altogether a mistake. Her song was not of the skylark order. The book was named for a rather second-rate French painting in the Chicago Art Museum; a picture in which a little peasant girl, on her way to work in the fields at early morning, stops and looks up to listen to a lark. The title was meant to suggest a young girl's awakening to something beautiful. I wanted to call the story *Artist's Youth*, but my publisher discouraged me, wisely enough.

The chief fault of the book is that it describes a descending curve; the life of a successful artist in the full tide of achievement is not so interesting as the life of a talented young girl 'fighting her way,' as we say. Success is never so interesting as struggle—not even to the successful, not even to the most mercenary forms of ambition.

The life of nearly every artist who succeeds in the true sense (succeeds in delivering himself completely to his art) is more or less like Wilde's story, *The Portrait of Dorian Gray*. As Thea Kronborg is more and more released into the dramatic and musical possibilities of her profession, as her artistic life grows fuller and richer, it becomes more interesting to her than her own life. As the gallery of her musical impersonations grows in number and beauty, and that perplexing thing called 'style' (which is a singer's very self) becomes more direct and simple and noble, the Thea Kronborg who is behind the imperishable daughters of music becomes somewhat dry and preoccupied. Her human life is made up of exacting engagements and dull business details, of shifts to evade an idle, gaping world which is determined that no artist shall ever do his best. Her artistic life is the only one in which she is happy, or free, or even very real. It is the reverse of Wilde's story; the harassed, susceptible human creature comes and goes, subject to colds, brokers, dressmakers, managers. But the free creature, who retains her youth and beauty and warm imagination, is kept shut up in the closet, along with the scores and wigs.

The interesting and important fact that, in an artist of the type I chose, personal life becomes paler as the imaginative life becomes richer, does not, however, excuse my story for becoming paler. The story set out to tell of an artist's awakening and struggle; her floundering escape from a smug, domestic, self-satisfied provincial world of utter ignorance. It should have been content to do that. I should have disregarded conventional design and stopped where my first conception stopped, telling the latter part of the story by suggestion merely. What I cared about, and still care about, was the girl's escape; the play of blind chance, the way in which commonplace occurrences fell together to liberate her from commonness. She seemed wholly at the mercy of accident; but to persons of her vitality and honesty, fortunate accidents always happen.

WILLA CATHER
NEW BRUNSWICK, CANADA
July 16, 1932